Penaeid shrimps – their biology and management

Penaeid shrimps – their biology and management

Selected papers presented at the workshop on the scientific basis for the management of penaeid shrimp held at Key West, Florida, U S A, November 1981
The workshop was sponsored by:

U S Department of Commerce, NOAA/NMFS
Southeast Fisheries Center, Miami, Florida
Gulf States Marine Fisheries Commission, Ocean Springs, Mississippi

in collaboration with the

Food and Agriculture Organization of the U N
Fishery Resources and Environmental Division, Rome, Italy

Edited by:

Dr John A Gulland
Fishery Resources and Environmental Division
Food and Agriculture Organization, Rome, Italy

and

Prof Brian J Rothschild
Chesapeake Biological Laboratory
University of Maryland, Maryland, U S A

Published by:
Fishing News Books Limited
Farnham · Surrey · England

© Fishing News Books Ltd 1984.

British Library CIP Data

Penaeid shrimps: their biology and management.
  1. Shrimp fisheries
  I. Gulland, J. A.   II. Rothschild, Brian J.
  III. United States, *Department of Commerce*
  338.3′7253843   SH380.6

  ISBN 0-85238-131-X

Published by Fishing News Books Ltd
1 Long Garden Walk, Farnham, Surrey

Printed in Great Britain by
Page Bros (Norwich) Ltd, Norwich, Norfolk.

Contents

Preface

Penaeid shrimp are distributed widely in the shallower tropical and sub-tropical waters. They are highly prized on most markets throughout the world. Thus they support many important fisheries. The United States fishery in the Gulf of Mexico lands over 100,000 tons annually, worth nearly $1 billion. Penaeid shrimp are often a major export item, the exports from India alone being worth some $250 million annually.

The high value of shrimp and its consequent attractiveness to fishermen has brought its own problems. Like all open-access fisheries, shrimp fisheries tend to suffer from excess fishing capacity, leading to a reduction in net returns from the fishery, though there have been few clear instances of a significant decline in the gross yield in weight. Shrimp fisheries also have their special problems. When fishing for shrimp, trawlers also catch large quantities of fin-fish, perhaps ten times or more the weight of shrimp. In several fisheries some or all of these fish are dumped over the side. This seems like a waste of potential valuable food, though it may be argued that the fish compete with or predate on the shrimp. In any case the shrimp fishery must affect any actual or potential fisheries directed on fin-fish.

Shrimp fisheries are complex both with regard to the life history of the shrimp and the nature of the fisheries. The typical life cycle of a penaeid shrimp species is that the younger stages are found close inshore, often in brackish waters. As they grow they move offshore, and the older and larger animals are found in relatively deep water, where the principal fishery by the larger industrial vessels usually occurs. In addition the younger inshore stages are often the target of small-scale fishermen, often using traditional methods, and fishing for subsistence, or for sport. Optimum use of the shrimp resource requires the correct balance between the different fisheries, which can raise difficult social and economic questions, as well as requiring good knowledge of the dynamics of shrimp stocks.

The assessment of shrimp resources necessary for management requires more than the simple application of the techniques already developed for the fin-fish of temperate waters. Shrimp are short-lived, and when heavily fished can be treated essentially as an annual crop. Shrimp, unlike many species of fish, do not carry, in the form of otoliths or scales, convenient birth-certificates so that the age of an individual can be determined facilitating estimation of the essential population dynamics parameters of growth and mortality. Seemingly vulnerable young stages of shrimp occur in the coastal zone where they may be particularly sensitive to natural environmental changes (*eg*, unusually heavy rainfall), and man-made changes (land reclamation, destruction of mangroves, pollution, *etc*).

With the importance of shrimp fisheries, and the problems of managing them in mind, the Southeast Center of the U S National Marine Fisheries Service, and the Gulf States Marine Fisheries Commission, in collaboration with the Food and Agriculture Organization of the United Nations organized a workshop on the scientific basis for the management of penaeid shrimp. This was held in Key West, Florida in November 1981. It was attended by 45 participants from 15 countries.

This volume contains the report of that workshop, together with a number of the papers presented at the workshop. These papers have been arranged in two groups. First are overview papers describing the shrimp fisheries of each country, and their major problems; these cover the fisheries of Australia, China, the Guianas – Brazil, the Gulfs region, India, Indonesia, Senegal and the United States. Second are the papers dealing with one or more specific aspects of shrimp biology and management. These have been arranged according to the major headings of the agenda of the meeting – biology of shrimp (Penn);

methods of analysis (Brunenmeister, Kirkwood and Pauly); interactions with other species (Sheridan *et al*, and Cushing); the effect of environmental factors (Staples *et al*, and Garcia); and management (Bowen and Hancock, Gulland, and Poffenburger).

Because of space and time limitations, it has not been possible to include all the papers presented at the meeting whilst other papers have been abridged from their original version. The editors apologize to those authors whose contributions have been omitted or shortened. Some of the papers omitted have been published in the original or modified form elsewhere. These include:

S Garcia and L LeReste. Life cycles, dynamics, exploitation and management of coastal penaeid shrimp stocks. FAO Fish. Tech. Paper 203

G R Morgan and S Garcia. The relationship between stock and recruitment in the shrimp stocks of Kuwait and Saudi Arabia. Oceanogr. trop. 17(2): 133–137

W L Griffin and W E Grant. A bioeconomic analysis of a CECAF shrimp fishery. Dakar. FAO/UNDP CECAF project. Doc. CECAF/TECH/82/41, 77 pp.

Report of the workshop on the scientific bases for the management of penaeid shrimp

1 Introduction

The annual global catch of tropical or 'penaeid' shrimp amounts to 700 000 tons. Some of the largest fisheries for shrimp appear to be in waters off Indonesia, Thailand, India and in the Gulf of Mexico. Shrimp are extremely valuable, often of importance for domestic use, but also as a valuable export item. The great value of the shrimp emphasizes the importance of shrimp management, especially since substantial increases in global shrimp production are not expected, hence making it important to improve management and make most efficient use of stocks in existing fisheries.

However, shrimp management is somewhat different in concept than the management of other fisheries. This is because of the unique life history, population dynamics, and the character of the shrimp fisheries. In terms of life history, shrimp generally spawn offshore; the young shrimp then move into estuaries which serve as a nursery area; the various species spend a variable amount of time in the estuarine areas before moving offshore and spawning. The dependence of shrimp upon estuaries raises considerable concern for the estuarine habitat. Yet, curiously, the quantitative extent to which man's activity, except for complete estuarine destruction, affects the actual production of shrimp has not been made clear. In terms of population dynamics, penaeid shrimp are fast growing and very generally live only about one year. They thus have unusually high mortality rates and because of this, determination of the best sizes at which to capture shrimp are critically sensitive to determinations of mortality and growth rates. Despite apparently intense fisheries for shrimp, it is not clear how recruitment is affected by stock size or the environment and thus there is concern as to whether high levels of fishing effort generate population instabilities or whether high levels of fishing effort push shrimp populations precipitously close to being in danger of collapse. In terms of fisheries, the best biological level of fishing effort and the best economic levels of fishing effort are difficult to determine. In addition to determining the best levels of fishing effort, it is also critical for many fisheries to determine how effort should be allocated between inshore 'artisanal' or 'small-scale' fisheries and offshore 'industrial' or 'large-scale fisheries'. Further, most shrimp fisheries in the world take substantial quantities of small fish which are sometimes discarded, and there is considerable concern as to developing feasible means for utilizing the discards.

The importance of the fisheries and the various problems associated with shrimp management suggested bringing together experts on the management of shrimp to consider the problems and bring them closer to solution.

Accordingly, the Workshop on the Scientific Basis for the Management of Penaeid Shrimp was held at Key West, Florida from 18 to 24 November. It was attended by 45 participants from 15 countries. A list of those attending is given in *Appendix 1*. The discussions were based on a set of papers reviewing the current situation in the major shrimp fishing countries, and other papers examining particular situations. A list of papers available at the meeting (most of which had been circulated to participants in advance of the meeting) is given in *Appendix 2*.

The participants were welcomed by Mr Larry Simpson on behalf of the Gulf States Marine Fisheries Commission, by Dr William W Fox on behalf of the U S National Marine Fisheries Service, and by Dr John A Gulland on behalf of the Food and Agriculture Organization of the United Nations.

The following served as Rapporteurs: Dr Donald A Hancock, Australia; Dr Scott Nichols, U S A; Dr Joan Browder, U S A; Dr Serge Garcia, FAO; Mr Terrance Leary U S A; and Mr Bernard Bowen, Australia.

The Workshop was arranged to consider and identify the problems associated with shrimp management; the biology of shrimp and rate measurements; the data base; methods of analysis; multispecies problems; environmental aspects; management and future work.

2 The problems

The Workshop began with reviews from each country of their shrimp fisheries, and of the problems that these fisheries (and those responsible for these fisheries) are facing. These can be summarized as follows:

Australia
Management in most Australian prawn fisheries, except those off eastern Australia, includes license limitation. Full exploitation in virtually all areas was achieved by 1975, but only in Western Australia did limited entry controls precede full exploitation. However, recently increased effort levels in the major fisheries resulting from larger vessels built under a federal ship-building program are causing concern. Also, two years of low catches of the brown tiger prawn in Western Australia are being examined for stock/recruitment implications. Western Australian management objectives, which are effectively those of other Australian limited entry fisheries, have been defined as 'the prime objective must be the maintenance of the resources at a level approaching the maximum sustainable yield, while giving proper attention to the economic viability of the fishing units with a view to maintaining a profitable industry' (Bowen and Hancock). Concern has been expressed about the extent of habitat modification.

Indonesia
In the 1970s the trawl fisheries expanded rapidly, resulting in increased shrimp and fish production, greater earnings of foreign exhange (up to US$200 million in 1979), and higher employment. However, there have also been negative aspects, through the over-exploitation of limited stocks. These have been particularly serious in the overcrowded areas of Java and Sumatra. Here there have been serious conflicts between traditional inshore fishermen and the trawlers.

As a result, a ban on trawling in the waters around Sumatra, Java, and Bali was introduced in 1980. The number of boats fell from 3 500 to 1 000, mostly so-called 'baby trawlers'. This has been followed by a big drop in catch, but stocks seem to have recovered. Crowding and conflict is less serious in Kalimantan, and in West Irian the only fishing is by large trawlers whose numbers are controlled.

Though the recent actions have reduced some of the immediate problems of social distress and conflict, much more study is needed to determine the best methods of management to deal with the social problems of the over-crowded areas of Java. There is also concern over the effects of the destruction of mangroves, and other changes in the coastal area, on the long-term well-being of the shrimp stocks.

China
At present the main problems in the fishery are heavy fishing effort, poor economic benefit and extravagant power consumption, which will eventually bring about population fluctuation. The goals of management are in conflict with each other, and their corresponding optimum efforts also differ greatly from each other. Attainment of optimum economic results from the fishery will require a reduction in fishing effort. Increased employment, however, will require a sacrifice of economic benefit and increased power consumption. We cannot have both at the same time. Young prawns need to be protected from illegal netting, but more so from destruction by saltworks, which in some years exceeds the numbers caught. Spawning stock is well below the numbers needed for maximum recruitment.

Thailand
Total landings of all species of shrimp have been increasing, but these include a large and probably increasing quantity of small non-penaeid shrimp. Most penaeid shrimp are taken, together with many species of demersal fish, in the mixed-species trawl fishery. This has been suffering for several years from over-exploitation, and a far too great fleet capacity, which has been exacerbated by the loss of free access to distant water fishery grounds under the new ocean regime.

India
Landings of shrimp increased rapidly until 1973. Recent catches are now below the peak years of 1973 and 1975, although fishing effort has probably continued to increase. There is therefore serious concern about depletion of the resource, as well as the severe conflict between the different groups of fishermen, especially between the traditional fishermen harvesting the small shrimp in the estuaries and lagoons, and the mechanized fleet of trawlers harvesting the larger shrimp in the offshore waters.

Gulf area
Recorded catches in the industrial fisheries along the eastern coast of Arabia have fallen considerably in recent years. The downward trend in total catch is less clear due to increased artisanal

catches, which are not well known. Recruitment appears to have decreased, though it is not known whether this is due to reduced spawning, damage to nursery grounds by land reclamation, or purely natural causes. A long closed season has been introduced.

Senegal
The stocks are fully exploited. Economic factors have caused a reduction in effort by the industrial fleet, but formal controls of the effort, as well as of mesh size, are under consideration.

United States (Gulf of Mexico)
The goal of managing the shrimp fishery is to attain the greatest overall benefit to the nation with particular reference to food production and recreational opportunities on the basis of maximum sustainable yield as modified by relevant economic, social and ecological factors (Center for Wetland Resources in Griffin *et al.*). On the one hand, a reduction in effort would almost certainly lead to economic benefits. On the other hand, an increase in effort would be of limited economic value to the fishermen and could result in increased risk of population collapse or a sustained reduction in the production of the population (Gulf of Mexico Fishery Management Council *in* Rothschild and Parrack). Concern has been expressed about the nature of the stock-and-recruitment relationship.

Mexico (Pacific Coast)
Fishing effort has doubled in ten years with no increase in catch. Management objectives are maximum catch and maximum employment. Management and conservation measures refer to closed seasons and mesh regulation so as to maximize yield per given recruitment. Revenue from the fishery is still sufficiently high to create a potential for further effort increase.

Guianas/Brazil
Fishing effort is considered to be excessive for the past ten years, possibly causing local over-exploitation. There has been a temporary reduction in the number of vessels permitted under international arrangements, but the expected licensing policy could result in an oversized total fleet which could aggravate even more the decline in the relative abundance detected for some species and in the economic yield from the fishery. Effects of human activities on nursery areas, and plans for artisanal fisheries could both be of concern for juvenile stocks. Also introduction of fin-fish trawlers will add to fishing pressure. Scientific advice for management is rather limited.

In other parts of Brazil the exploitation of shrimp resources has also reached a high level, the larger part coming from the artisanal fishery, possibly causing over-exploitation. Also, human activities may have been affecting the juvenile stocks.

Nicaragua
The present goal of management is to obtain the best economic and social benefits, and more productive catch per unit of effort. A dramatic reduction in fishing effort caused no reduction in catch, but still did not improve the economics of the individual boats. This is believed to be because fishing effort over the past ten years has been greatly in excess of the optimum required for maximum catch. This has resulted from priority having been given to economic pressures rather than biological advice, which had suggested that MSY had been reached about thirteen years ago.

General comments on problems
Most shrimp fisheries throughout the world face similar problems. The stocks are fully exploited, with little opportunity of increasing total catches. Fishing effort continues to increase, giving rise to serious economic or social problems even when the stocks themselves may be in no danger.

The meeting was therefore believed to be more than timely. Despite the growing problems being faced by the managers of shrimp fisheries, the scientists were, in many countries, not well prepared to provide the managers with the advice required. One reason for this has been lack of definition of the ultimate management objectives. Even for a single fishery these may be incompatible, contradictory, and sometimes amazingly vague. Unless the scientist has clear guidance on what the fisheries are being managed for, it is difficult for him to plan his research and frame his advice in an appropriate manner. Usually they will be biased on objectives that can be expressed in simple biological terms, *eg*, attaining MSY, and may not be helpful in determining the management actions needed to achieve other types of objectives, *eg*, economic efficiency, or the resolution of social conflicts. The meeting then discussed the problems that scientists met in advising the manager to tackle his problems. One general problem is that of anticipation and timeliness.

Without careful thought about the need for data gathering and analysis on a real-time basis, the point at which the problem becomes serious may be reached, and passed, so quickly that remedial action becomes difficult or virtually impossible. Once full exploitation is approached – not reached – a conservative view must be taken on the potential for biological danger and provisions made accordingly. In a common property resource too much emphasis may be given to particular objectives, such as maximizing employment in the harvesting section alone or maximizing throughput for processing facilities, *etc*, while missing the telltale signs that all is not well in other sections, *eg*, with the state of the stocks, or the economics of individual operating units, because of tardy or inadequate availability of data. If no action is considered until a target is achieved (directly or indirectly) by the time action is actually taken, the target will certainly have been overshot and another fishery will have been added to the long list of documentation of failure – failure to recognize and failure to act. Three examples can be offered:

(*1*) Maximum Sustainable Yield (MSY) can be a useful concept for general guidance, but as a specific target it can easily be exceeded. This has been the experience of both the rock lobster and prawn fisheries of Western Australia, even where fishing effort has been allowed to expand gradually under tight controls (Bowen and Hancock).

(*2*) Failure to identify the potential for changes in effective fishing effort can allow dramatic escalation beyond the calculated optimum in a very short time. For example, in the prawn fisheries of Australia (Walker, Bowen, and Hancock; Penn) a shipbuilding bounty scheme has led to the building of larger boats which dramatically increased the pressure on the stocks. In Western Australia this caused an increase of effective effort from a comfortable level to excess in a very few years.

(*3*) Failure to identify economic signals is, if anything, even more reprehensible. While future stock levels are usually not predictable except in specific instances (Lhomme and Garcia; Staples, Dall, and Vance) several economic factors may well be known with some reliability in advance of the season, *eg*, likely fuel prices, labor costs, market prices – which should put the emphasis on early economic advice – not years behind as often seems to be the case. The paper by Poffenberger, amongst the economic papers, provides some very useful information, but the assumptions he uses will probably need to be revised in the light of some of the practical experiences recorded in other papers.

A particularly difficult but important problem is that of stock and recruitment. For years fishery biologists rested in the assumption that recruitment was independent of the size of the adult stock and hence also of the effect of fishing. This belief has now been effectively challenged, notably at an ICES symposium held in Aarhus, Denmark in 1970, but in the only contribution to that meeting dealing with crustaceans Hancock (1973) noted that no established relations between stock and recruitment had been identified. In contrast to this, possible stock-recruit relations have been examined for several shrimp stocks in papers submitted to this meeting (Penn; Rothschild and Parrack; Morgan and Garcia; Ye; Brunenmeister; Ehrhardt). Also, the Ricker model has been shown to give a good representation of the stock and recruit relation in the Western Australia rock lobster (Morgan, Phillips, and Joll, in press). If, as now seems likely, a sufficient reduction in the adult stock of shrimp can cause a fall in recruitment, the implications for management, especially laissez-faire management, are very serious. It cannot be assumed that, whatever may happen to the economics of the fishery, the biological production will always be maintained.

In addition to the problems of stock and recruitment, the following scientific problems were identified as being important to the way in which scientific advice is given to managers.

(*1*) The variation in stocks may mask trends. In order to detect trends, observations must be made over extended periods.

(*2*) Nursery areas are separate from adult stocks, and may be especially vulnerable to effects of the environment, and man's activities, other than fishing.

(*3*) Most fisheries are carried out by several groups of fishermen, using a variety of gears. In addition to the major problems for the fishery manager in terms of conflict between the groups, the presence of distinct fisheries on different sizes of shrimp causes a number of scientific problems, particularly in calculating fishing effort.

(*4*) Many fisheries are based on several species of shrimp and techniques need to be developed for estimating population dynamic parameters and management in a multiple species setting.

(*5*) Age of individual shrimp cannot be determined directly, and, therefore, techniques which depend upon knowing the age of shrimp must be used cautiously.

(*6*) Entry into the fishery, either by recruitment, or mesh selection is not sharp. The minimum size of shrimp in fisheries is subject to considerable variation because the size of recruits varies and the effects of mesh selection do not precisely control the minimum size of shrimp.

(7) Some individual stocks of shrimp occur in waters of more than one coastal state, and thus require concerted international cooperation for effective management.

These are discussed in the following sections and in the final section where proposals are made for dealing with the problems.

3 The biology of shrimp and rate measurements

Growth
Papers containing new data or analysis of growth included those by Lhomme and Garcia, Mathews, Nichols, Parrack, and Pauly. All addressed the seasonal variability of growth, although Parrack noted that seasonal variation was relatively unimportant for offshore *Penaeus aztecus*. Nichols presented a way of considering seasonal variation by examining growth-rate variation. Pauly presented a method for extracting growth curves from length-frequency data.

Discussion centered on evaluating the potential importance of growth variability. Adequate modelling mechanisms appear to be available for dealing with predictable variation. The question arises: are the growth parameters obtained (in any study) biologically meaningful? Clearly, they should be, or the estimation becomes merely a curve fitting exercise that summarizes growth over only a portion of the lifespan, and not a means to 'predict' growth outside the observed range.

The general pattern of shrimp growth seems to be well known, and reasonably consistent from area to area and species to species. For most purposes the available information appears adequate for management purposes. Further study may be needed in special circumstances. For example, the determination of the optimum date to open the fishing season may require particularly accurate knowledge of growth. Again, if a management policy induces large changes in density, knowledge of possible density-dependent growth could be important.

Natural mortality
New estimates of M were reported by Brunenmeister, Lhomme and Garcia, Mathews, Nichols, Parrack, and Ye. The high sensitivity of yield-per-recruit results to typically uncertain estimates of M was reported by Nichols.

In the discussion, the poor precision and accuracy among existing M estimates was stressed. Published estimates appear to include extraordinarily high values of M. However, even the 'reasonable' estimates are highly variable. There is an important distinction to be made between real variation and error variability of estimates. Real variation most certainly occurs with age and size, as the shrimp progress through several different environments during their life history. Variations among years, perhaps in response to variation in abundance of predators, or in occurrence of disease (which might be density-dependent) must also be considered. Improving estimates of M will probably be costly, but may be worth the investment. Detailed investigation of the mechanisms of natural mortality, such as predation or disease, could provide some better understanding of the process of natural mortality, and some indication of the significance of various causes of mortality such as the abundance of predators. Careful re-examination of existing techniques (mark-recapture analysis, *etc*) might also be considered.

The value of comprehensive inter-specific comparisons (such as those made by Pauly) was stressed particularly in providing an objective first approximation in stocks where direct estimates were unavailable. At the same time the danger of certain values of natural mortality being prematurely adopted and the importance of obtaining direct and independent estimates of natural mortality was stressed.

Migration and stock identification
Several papers considered stock-structure explicitly (Brunenmeister, Lhomme and Garcia, Mathews, Parrack, Rothschild and Parrack, and Ye). Recognition of stock structure impacts directly on the validity of the production models, and on development of stock recruitment relationships. A possible latitudinal gradient in migratory behavior was mentioned. The discussion included recognition of the importance of migrations across international boundaries. When such movements occur it is important, in reaching agreement between the countries concerned, for there to be good information on the positions of the main nursery and fishing grounds relative to the national boundaries (this may be accomplished by developing a series of maps showing the location of the main concentrations of each size of shrimp, and of the fishing grounds during each season).

Other biological topics
Latitudinal differences were noted in the relative strength of two seasonal spawning peaks in some species. Apparently there are no examples of a secondary peak declining continuously with time, or with increasing fishing effort. Some 'two-peak' cases may be discrete enough to function as separate stocks in the same general geographic area.

The paper by Penn introduced a behavior-based classification scheme for shrimp species, in which he introduced the concept that some species may have changed from a schooling to a non-schooling behavior as fishing pressure increased. This appears to have occurred in some stocks of *P. merguiensis* in Australia. Possible evidence, pro and con, for similar changes among species that he suggested might show similar behavioral changes was discussed. The variability of aggregative behavior in response to environmental variation, particularly turbidity, was also considered. Understanding these modifications in behavior has a direct impact on understanding (and forecasting) the relationships between fishing effort and fishing mortality.

4 The data base

Catch statistics

The importance of adequate data, comprising at least comprehensive statistics on catch and fishing effort, distinguishing catches of different species of shrimp, and some data on the sizes of shrimp caught was assumed to be generally recognized. The meeting therefore discussed situations where, despite this recognition, the data are still inadequate.

Statistics on total catch are readily available for most if not all the main industrial shrimp fisheries, but several participants expressed concern that significant catches were not recorded at all, *eg*, from sport, subsistence, or artisanal fisheries. Two examples of misreporting (or failure to include a significant part of the fishery) were mentioned for the white shrimp on the U S Gulf Coast (Christmas), and for the Kuwait-Bahrain area (van Zalinge). In the latter exclusion of the increasing catches of the artisanal fishery had resulted in a much greater apparent decline in total catch than had actually occurred.

As complete catch data are basic to many analytical approaches (production modelling, cohort analysis), omissions of potentially large components of the total catch can be a serious problem. The sizes of these unreported components may vary radically over time. Inability to address or even detect such changes could create a very biased picture of the condition of a stock.

The practice of discarding fish in the shrimp fisheries is well known, but it was pointed out that in several fisheries small shrimp are also discarded. Quantitative estimates of both kinds of discards and of the species involved is most important. The meeting noted that FAO was planning to produce a report or manual setting out appropriate and cost-effective methods of estimating the quantity of discards.

Identifying individual shrimp by sex in samples of catches for species with strong growth differences between the sexes is a problem that impacts on any analyses involving length or age composition of catch statistics. Precision of these analyses could also be improved with better resolution of size data.

Effort data

Fishing effort standardizations by vessel and gear characteristics were incorporated in several of the submitted papers: Bowen and Hancock, Brunenmeister, Lhomme and Garcia, Mathews, Penn, and Ye. Brunenmeister's paper presented an extended analysis including vessel characteristics, spatial and temporal variation, and catches of other species. Penn's paper considered differences in catchability generated by differences among activity patterns and schooling behavior among species, and suggested that the utility of CPUE data may be low for species with highly variable catchabilities.

In the discussion, note was made that standardization is not often attempted, and that it is important to anticipate changes that might occur, particularly regarding fishermen's ability to increase fishing power even with controls in place. The effectiveness of effort is strongly mediated by catchability variation, and by spatial and temporal distribution of effort relative to the distribution of shrimp. Because of this variability, prediction of fishing mortality from effort projections may always require monitoring.

The group was reminded that otter trawls are not the only gear in many fisheries. Thus, a broad consideration of fishing effort as it is applied to shrimp would need to take into account various other gear.

For many applications effort standardization can require more than consideration of just fishing power. Catchability variations may occur at several scales: physical aspects of gear performance, reaction of shrimp to gear, general aggregating behavior of shrimp, the larger-scale distribution of stock density and fishing effort over a stock's range, and the distribution of effort in time within the time interval used in analysis. Decisions are required about what types of effort data and auxillary information must be collected. (For example, experience of vessel captains is often not measured).

5 Methods of analysis

Production models

Production model results were presented in the papers by Bowen and Hancock, Brunenmeister, Ehrhardt *et al*, Mathews, Silas *et al*, and Unar and Naamin. Additionally, plots relating catch to effort, without fitting a model were presented by Bowen and Hancock, Penn, and Villegas and Dragovich.

While production models do not take explicit account of the effect of fishing on different size groups (*eg*, juveniles or adults), they demand less data than age- or size-structured models, and are therefore likely to continue to be widely used. It was pointed out that the short life of shrimp meant that annual pairs of observation of catch and fishing effort were more likely to match the equilibrium condition (particularly if recruitment were not affected by fishing) than was the case in longer-lived fish. One serious limitation in using production models is that of determining a suitable measure of effort (or catch per unit effort). In this connection Garcia mentioned a technique developed by Csirke and Caddy at FAO in Rome which relates the catch directly to total mortality (as estimated, for example, from size or age composition), which under the usual simple assumptions gives rise to a parabola with intercept on the x-axis at M, rather than the origin. While this shares several theoretical drawbacks with other production models, and there can be difficulties in obtaining adequate estimates of total mortality, this approach appears to be promising for some situations.

The production models usually showed a curved left-hand limb, sometimes a suggestion of a maximum, but very seldom a declining right-hand limb. Various reasons were suggested. Yield-per-recruit analyses suggest a flat-topped curve so that if recruitment is not affected a flat-topped curve of total yield may be more representative of reality than a parabola. Alternatively, declining total catches, and therefore even faster declines in catch-per-unit-effort could cause the expansion of effort to stop for economic reasons before a declining right-hand limb can be observed. Difficulties in measuring true fishing effort were also mentioned.

Age- or length-structured models

Models of this type, based more or less directly on the yield-per-recruit calculations of Ricker or Beverton and Holt are essential to study the effects of changes in the fishing practice which involve changes in the pattern of distribution of fishing mortality with age. A number of yield-per-recruit analyses were presented (Ehrhardt *et al*, Nichols, Rothschild and Parrack). There were inevitably problems, especially in the estimation of certain parameters (notably natural mortality). This could be somewhat reduced by using figures based on comparisons with related stocks as first approximations.

It was pointed out in Pauly's paper, and in a presentation by Jones during the meeting, that given a growth curve, length composition data could be used directly to carry out several types of analysis, including cohort analysis, without having to estimate age-composition. These techniques which have been described in an FAO Fishery Circular (Jones, 1981) were felt to have wide promise in applications in analyzing shrimp fisheries. At the same time problems and difficulties, *eg*, the need to consider the curve of age as a function of length, rather than the more normal growth curve (length as a function of age) were underlined. One of the advantages of a length-structured analysis was that the critical parameter was usually M/K rather than M, and it was suggested that M/K might be fairly constant within a species group, *eg*, penaeid shrimp.

Stochastic models

Stochastic modelling appears to be a fruitful direction for shrimp research. Possible uses in yield-per-recruit and stock-recruitment were considered. Stochastic modelling may be particularly critical in examining the true nature of 'collapse', and in understanding the relationship of parent stock size and environmental variation in establishing recruitment strength.

There is a distinction between stochastic models in which an intrinsic, real, but random variability of one or more parameters is incorporated, and sensitivity analysis, which investigates the effects of either estimate error or real variability in a deterministic context. Both types of models will have applications in future research.

Stock and recruitment

Stock recruitment shows considerable variations from year to year which are not connected with any obvious changes in adult stock. It is clear that whatever the relation between the size of the adult stock and the average recruitment from the stock, the actual recruitment in any one particular year is determined very largely by environmental conditions in that year. It is therefore better to describe the stock-recruitment relation by a family of curves, each corresponding to a given set of environmental conditions (weather, food supply, state of development of the coastal zone, *etc*). Each curve gives the recruitment that would arise from a given size of spawning stock under the defined environmental conditions (*See Fig. 1*).

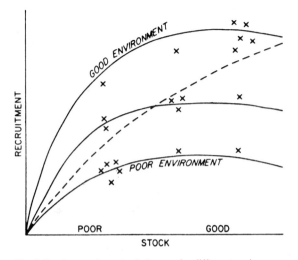

Fig. 1. Stock-recruitment relations under different environmental conditions, and the likely distribution of observed points and empirically fitted curve

At best, the environmental effects add noise to the system and make it difficult to determine in what way the average recruitment changes with changes in adult stock, and in particular to detect at what point a reduction in spawning stock will cause a significant fall in recruitment.

In practice, the existence of environmental effects may bias the estimates of the stock-recruitment relation. The size of the adult stock of shrimp is largely determined by the success of the previous spawning season as modified by the intensity of fishing between recruitment and the spawning season. If there is a significant correlation between environmental conditions in successive years, this can give rise to a correlation between stock and subsequent recruitment which is really due to the influence of recruitment on subsequent stock size. In terms of *Figure 1* it is likely that large spawning stocks will occur during periods of generally good environmental conditions, giving rise to observations on the upper curve of the family of stock recruitment curves. Similarly periods of poor environmental conditions will give rise to small spawning stock and poor recruitment and observations on the lower curve. This is indicated in *Figure 1* where the likely distribution of observations gives rise to an empirically fitted line (broken line)

that suggests a much bigger change in recruitment with changes in adult stock than those corresponding to any fixed set of environmental conditions.

In some cases it is possible to determine what environmental factor is important, and to make allowance for it. In the Gulf of Carpentaria and elsewhere recruitment is strongly affected by rainfall. There has been a downward trend in recruitment over the past few years, which could have been ascribed to the effect of fishing, but when the effect of rainfall (which also shows a trend) is taken into account, there is no trend in the residuals. The possibility of bias, and of observing a spurious relation between stock and recruitment is particularly high for changes in the coastal environment (developing for housing, roads, cutting of mangroves, *etc*) which have significant and (over the short-term at least) irreversible impact on recruitment. It is possible that the strong correlation between stock and recruitment observed in Kuwait waters and perhaps also in adjoining areas is due to a steady degradation of the coastal nursery areas.

Despite the difficulties of interpretation, the increasing number of cases in which lower spawning stock sizes are associated with lower recruitment, and which cannot be immediately explained by environmental factors (natural or man-made) are matters of great concern. If these are indeed cases of low stock causing low recruitment, and fishing on these stocks is maintained at a high level, the risk of a stock collapse ('recruitment overfishing') is very real. The reason for this is well known, and is illustrated in *Figure 2*. *Figure 2* represents the simplified situation, in which

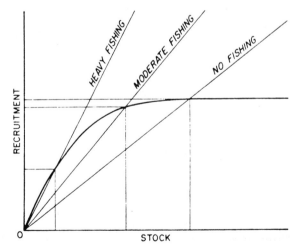

Fig. 2. Equilibrium positions under a given stock-recruitment relation for various levels of fishing effort

16

environmental factors are constant. Under a given pattern of fishing the spawning stock resulting from a given recruitment will be proportional to that recruitment. This is indicated by the straight lines, with slopes increasing with increasing fishing. Where the line cuts the stock-recruitment curve it gives the equilibrium position for any pattern of fishing. If the fishery is on the fairly flat part of the stock-recruitment curve (recruitment is effectively independent of stock), then changes in fishing effort change the equilibrium recruitment little, but if the present fishery is near the left hand part of this curve, even a moderate increase in fishing effort can cause a great reduction in the equilibrium recruitment, and it may not take much increase in effort to cause a complete collapse. The observed relation between stock-and-recruitment where the relation of stock to recruitment is approximately proportional (eg, for the Gulf of Mexico brown shrimp) is therefore very disturbing.

6 Multispecies problems

Several species of shrimp

Most of the papers presented to the meeting and most of the discussions treated the problem and the scientific analysis in terms of a single species (or at least of a single stock) without distinguishing species. In fact, shrimp fisheries with a few exception (eg, the Chinese fishery for P. orientalis) are based on more than one species. Two types of interaction can be distinguished – first when the species are biologically separated in space or time, so that the interaction is only in terms of the operation of the fishery, and second, when there is significant biological interaction.

The second situation presents interesting scientific problems of management, for example, the possibility that the economic damage from the depletion of one species may be reduced by the increase of other species – the situation in the southeast Pacific where sardine stocks have greatly increased following the collapse of the anchovy offers an analogy in a different ecological context. Detailed examination of the life-history patterns of different species suggest that different species of shrimp do not have overlapping requirements, usually being on different grounds, or, if occurring on the same grounds, do so at different seasons. The likelihood of one species benefitting to any significant degree from the depletion of another shrimp stock seems small and there would be no justification in delaying implementation of needed management measures in the hope that such interactions would take place.

The first type (of operational interaction) causes the greatest management problem when the existence of a fishery on one stock allows fishing on another stock to continue even when the economic return from the second species is very low. It has generally been assumed that economic constraints will cause the expansion of fishing effort on a stock to cease before there is serious risk of recruitment overfishing. This assumption is being challenged on general grounds (see Section 5), but the possibilities of such overfishing are greatly increased in multispecies fisheries. For example, the stock of brown tiger prawn in Shark Bay, Western Australia, seems to have been reduced to a very low level because the fishery can continue by fishing principally on other species. The existence of two or more species can also cause difficulties in data collection and analysis, particularly in ascribing correctly fishing effort to one or other species. Usually this difficulty can be resolved by collecting statistical data in sufficient detail in space and time.

Fin fish

Shrimp trawls are not selective, and, usually large quantities of fin fish – up to 95% of the total catch – are taken by shrimp trawlers. (Exceptions are the fishery on schooling banana prawns in the Gulf of Carpentaria, Australia, and the 'mud-bank' fishery of southwest India, in which the catches consist almost entirely of shrimp.)

Most trawl fisheries have bycatches, but the proportion of the bycatch that is discarded is probably higher in shrimp fisheries than in most other fisheries and is due, largely, to the high disparity in price between shrimp and the bycatch species. (This price difference also occurs and leads to a high proportion of the bycatch discarded in tropical trawl fisheries for high valued products such as cephalopods, sea bream, etc, operated by long-range fleets off West Africa and by the larger Thai trawlers in southeast Asia.) Rough estimates of the quantities of discards involved in the major shrimp fisheries are given in Table 1; these are first approximations. Altogether the quantity discarded is probably between one and two million tons, with the greatest quantities occurring in North America, and the northwestern part of South America.

The fate of these fish once they leave the water is highly variable, though this variation is more systematic than the variation in the ratio of shrimp to fish in the catch. At one extreme are the artisanal fisheries, particularly in southeast Asia, operated by small vessels with no refrigeration or even ice. These vessels make trips of only a few hours, and

Table 1
ROUGH ESTIMATES OF BYCATCH AND DISCARDS IN PENAEID SHRIMP FISHERIES[a]

Country	Shrimp catch (1979) tons	Bycatch ratio	Bycatch tons	% Discarded	Quantity discarded	Date of observation and author
China[d]	7 500	8:1–4:1	35–60 000	nil	nil	October 1962
Indonesia (Arafura Sea)	6 000	variable	ca. 100 000	high	ca. 100 000	1970s (Unar)
Indonesia (other areas)	157 000	3:1–1:1 (inshore) 20:1–30:1 (offshore)	115 000	<2%	<2 000	
Australia	21 000	variable[b]	unknown	high	unknown	
Thailand	100 000	variable	750 000[c]	small	small	
India	183 000	4:1	316 000	<2%	5 000	1970s (George)
Kuwait	1 600	10:1	15 800	95%	15 000	1978 (Mathews)
Senegal	5 500	variable	80 000	ca. 50%	40 000	1970s (Garcia)
United States (Atlantic coast)	105 000	2·8:1	37 000	ca. 100%	37 000	1970s
United States (Gulf coast)		9:1	600 000	ca. 100%	600 000	1970s
Mexico (Pacific coast)	46 000	10:1–15:1	400–500 000	>95%	400 000	1970s and 1980s (Ehrhardt)
Brazil and Guyanas	21 500	10:1	215 000	high	200 000	Present (Villegas and Dragovich)
TOTAL	658 600		ca. 2 700 000		1 399 000	All countries above except Australia

Total world landings, all types of shrimp 1 526 000.
Total world landings (less pandalids, sergestids *etc*) 1 238 000.

[a] All estimates are imprecise; the figures are presented here to illustrate the general magnitude of the quantities involved, and the regional variation.
[b] Bycatches are very low on schooling prawns, but can be high in other fisheries. Glaister reported bycatch of 21 700 tons (18 000 discarded) in the eastern Australia fishery that landed 2 500 tons of prawns in 1979.
[c] Taken as equal to the quantity of 'unspecified marine fish' reported by Thailand in the Yearbook of Fishery Statistics.
[d] Yellow Sea fishery only.

usually all their fish are brought ashore. Not all of it may be used for direct human consumption, the rest being used for duck food, fish meal, *etc*. At the other extreme are the specialized shrimp fisheries with relatively large trawlers; these usually freeze their catch at sea, and make long trips. Their specialized interests and limited handling and storage capacity usually result in all or nearly all the incidental catches of fish being discarded. The rate of discards is particularly high in regions (the United States and the Gulf of Carpentaria – Arafura Sea area) where there is no great need or local demand for fish. The type of vessel may also affect the proportion of fish discarded. Off Senegal the trawlers that keep their catch on ice catch a higher ratio of fish than the larger freezer trawlers which work further offshore, though the latter discard a higher proportion of the fish they do catch. Measures have been taken by a number of countries to encourage the landing of more fish. For example,

the government of Guyana has required since 1974 that all shrimp trawlers include 909 kg of fish in each landing (Villegas and Dragovich). Increasing demand for fish is also increasing the proportion kept, *e.g.*, in Senegal.

The failure to use this great quantity of potentially valuable protein food has attracted considerable attention and concern, particularly since much of this wastage occurs not far from places with large populations suffering from shortage of protein. The possibilities for better use of the discards was the subject of an FAO-IDRC Technical Consultation on the Utilization of Fish Bycatch in Shrimp Trawling, held in Georgetown, Guyana in October–November 1981. This aspect was therefore not considered in detail at the present workshop, though it was stressed that this aspect of discards was not just a technological problem of developing an appropriate method of processing the bycatch, but included the economic problem of making it attractive (or indeed even viable) for the specialized shrimp trawlers to adopt these methods.

The present workshop concerned itself with other questions, particularly the following:

(*1*) What is the impact of the shrimp fishery on the fisheries (actual or potential) of the various species of fin fish?

(*2*) Do the discarded fish and small shrimp supply a useful source of food for shrimp, so that reduction of discards might affect future shrimp catches?

(*3*) Do some or all of the species of fish in the bycatch compete with, or prey upon, shrimp to the extent that a reduction in bycatch (through the use of selective trawls) would result in a detectable reduction in future shrimp catches?

No formal assessment of the impact of shrimp trawls on bycatch fish stocks was presented at the workshop – nor was any attempt made to evaluate the potential benefit of decreasing trawling pressure on these stocks. The necessary calculations are straightforward, and similar work has been done for many multispecies fisheries, particularly in the North Sea. In exploring the question, Caddy (in an FAO paper presented to the Fisheries Commission for the Western Central Atlantic, WECAF) found that the fishery potential in the WECAF area that was lost due to such discards was high under any combination of assumptions used.

The question of discards as a food supply for shrimp was addressed by Cushing, and by Sheridan *et al.* Though there are a number of uncertainties remaining – including the degree to which shrimp do in fact feed on dead fish – it does appear that the additional growth of shrimp due to discards either by direct consumption, or through the recycling of nutrients, is at most small.

The impact of fish, as predator or competitor, on shrimp stocks is less clear. Shrimp do not appear to be an important element in the diet of most species of fish that occur in the bycatch. However, they do occur, and in view of the large number of fish compared with shrimp, the impact of predation on shrimp could be significant. It was suggested that most predation takes place on the fringe of the main distribution of shrimp. It thus may be due mainly to species that do not feature largely in the bycatch, the latter being on the whole about the same size as the shrimp.

Pauly analyzed the data of trawl surveys and commercial catches in the Gulf of Thailand and found a strong correlation between the survival of juvenile shrimp between spawning and recruit-

ment, and the abundance of demersal fish. He found that in recent years this survival rate has increased, sufficient to balance the decrease in spawning stock of shrimp due to heavy fishing.

The effects of predation may not be obvious. In the Irish Sea cod eat *Nephrops*, so that one might expect that reducing cod populations might increase the harvest of *Nephrops*. The true picture is more complicated; cod eat small fish which eat juvenile *Nephrops*, so that a reduced cod stock might result in more small fish and in less *Nephrops*.

In conclusion it was felt that there would be no significant disadvantages to the shrimp stocks (and hence the shrimp fisheries) by reducing discards, the bycatch being kept constant. There would be benefits, possibly large to any directed fishery for fin fish if the bycatch could be reduced, whether by gear modifications such as the separator trawl being developed in the United States, the use of a larger mesh size, or a general reduction in the shrimp trawling effort. The impact on shrimp of eliminating bycatch is less clear, and needs more investigation. It might be negative, but the effect is neither so large nor so certain as to argue against reducing bycatch for the other reasons already mentioned.

A final reason for reducing discards which emerged in discussion was that the fish discarded by one vessel could, especially in areas of heavy fishing, be picked up in the trawl of another vessel, and badly affect the keeping qualities of the catch.

7 Environmental aspects

Shrimp are sensitive to the environment in which they live at all stages of their life. The environment also affects the operation of the fishery in many ways. There are therefore many aspects of the interaction between shrimp and shrimp fisheries and the environment which could be studied. The meeting agreed that the important aspects to study were those that can enable us to limit unfavorable changes such as habitat destruction or to predict changes.

In studying the effect of the environment a number of general problems arose. These included:

(*a*) *Lack of experimentation* The data base used is mainly extracted from the fishery and because of the impossibility to do any experimentation, the environmental effects have been studied most of the time by correlation analysis. As a consequence, the cause-effect relationships suggested are to be taken as suggestions only.

(*b*) *The environment parameters available are limited in number and not independent* Temperature, oxygen, and salinity are linked and describe water mass. Turbidity, plankton density, and photo-period are not independent parameters. The same holds often true for depth, sediment texture, organic content, and benthic biomass. Any of these parameters may be taken as an indicator at best, and cause-effect relationship with any one parameter of the set is difficult to ascertain. This is complicated by the fact that the number of parameters usually measured is limited (usually consisting of temperature, salinity, and sometimes dissolved oxygen). Others (turbidity, photo-period, currents, amount of food, *etc*) are often not recorded, so that it is impossible to establish correlations, even when their influence may be important.

(*c*) *Measuring the biological phenomenon* An appropriate index for measuring the phenomenon of interest in the shrimp stock (spawning, migration, recruitment, *etc*) may not be available, or can only be obtained at the cost of considerable research.

(*d*) *The correlation obtained* sometimes between magnitudes oscillating seasonally may be largely spurious, and so the proper lag-time needs to be carefully researched.

(*e*) *The signal/noise ratio* should be taken into consideration when trying to identify a cause – effect relationship. This ratio varies from one region to another and depends on the parameter(s) considered. In temperate and subtropical countries, temperature might be an important triggering factor while in an equatorial-type rainfall, food, turbidity and associated change might be more important. In this connection signal refers to 'long-term' conditions as opposed to short-term ones (considered as noise). At any time-scale the noise can be interpreted as the result of variations at higher frequency than the signal, *eg*, a 'noise' is only non-understood information and when it becomes too important it has to be analyzed. This refers particularly to stock-recruitment relationships and to production models.

Despite these difficulties, a number of environmental influences on shrimp stocks have been established, as reported in the published literature, and at the meeting. These include the following:

(*a*) *Survival* – for larvae and juveniles it is governed by the combined effect of temperature and salinity. A combination of low temperature and low salinity is very unfavorable.

(*b*) *Distribution* – the main parameters governing distribution of shrimps are the following (Garcia and LeReste):
- Temperature – the shrimps often react to strong changes in temperature by migration (geographic or bathymetric).
- Concentrations of shrimp are associated with the presence of estuaries. There are apparent exceptions (Mexico, Ehrhardt *et al*; Arabia, van Zalange). The association tends to be highly variable from species to species.
- Shrimp concentrations are also associated with fine sediments (from sand to mud) and the preferenda are different among species or groups of species.

A clear distinction has been suggested between 'white' and 'brown' shrimps, the 'whites' being littoral species, closely associated with areas of high runoff; low, variable salinity and very muddy bottom. The 'browns' are found on more typically marine areas that are hydrologically more stable, and on sandy or sandy mud bottoms.

It has also been suggested that the different requirements of larvae, sub-adults and adults apparently lead to a decrease in overlap in time and space for the different species, possibly reducing interspecific competition within the shrimp group.

(*c*) *Migration* – considerable information is available on migration (Garcia and LeReste). The suggested triggering factors for the sub-adult migration are marked changes in temperature (cold fronts; Rothschild and Parrack), salinity, currents (linked seasonal changes in river outflow; Walker). Daily cycle and moon phase also seem to be important.

It must be noted that the migration rate out of an estuary is also linked to the amount of shrimp available for migration and therefore to the seasonal pattern of reproduction at sea and environmental conditions in the estuaries a few months before migration starts. The number of migrating shrimp is in fact the result of various superimposed seasonal patterns – reproduction (larval production), coastal hydrography (currents, larval transport), estuarine environment (larval survival) before migration in addition to the environment factors at migration.

The main findings are:

(*1*) The strength of a migrating cohort depends in the conditions prevailing during the estuarine period. There is an optimal time-space window.

(*2*) The size/age at migration varies seasonally between years (Staples *et al*; Garcia and LeReste).

(*3*) The 'normal' migration pattern, corresponding to the temporal pattern of reproduction is distorted by events like floods, cold fronts, *etc*.

(*4*) The shrimp swimming behavior is linked to changes in salinity. This may explain how shrimp orientate themselves in the inshore/offshore salinity gradients when migrating. The generalization of the theory is, however, still to be established.

(*d*) *Spawning* – This aspect of shrimp biology was discussed in the paper by Garcia. The main problems lie in the measure of spawning activity (percentage of gravid females is not enough) and on the limited availability of environmental parameters.

It can be said that in general two spawnings occur – in spring and autumn. The first is generally found to be the more important and stable one.

There are, of course, some differences between species and areas and the amplitude of the seasonal reproductive pattern depends upon the overall stability of the environment.

The literature is rich in statements about factors triggering reproduction. Temperature seems to trigger an increase in percentage of gravid females in some areas while rain is the apparent triggering factor elsewhere. The increase in actual spawning presumably follows the increase in gravid females after some delay period, but has seldom been observed directly. The real effect of some other factors (plankton bloom, food) has not been looked into enough especially in oligotrophic environments.

The most important problem is that generally the autogenic aspect of shrimp reproduction has not been considered (adaptive process) and that environmental aspects of reproduction have been studied more at the individual than at the population level.

In short-lived species like shrimp (limited number of year classes) the close adaptation of the seasonal spawning potential to seasonal change in environment may be the key to their permanence in oscillating environments.

(*e*) Catchability has been shown to vary with turbidity and temperature. The apparent response of the shrimp to these parameters may depend on species behavior (burying or non-burying, nocturnal or diurnal).

(*f*) Growth is certainly affected by temperature and this meeting has provided an interesting set of observations on that aspect (Garcia and LeReste, Nichols, Pauly *et al*). In general, growth increases with temperature for a given size. However, Nichols found that growth was not so linearly correlated with size as implied with the von Bertalanffy growth function. Growth is also slowed down by spawning.

(*g*) Abundance (Staples *et al*, Garcia and LeReste, Ehrhardt *et al*, Walker). Most of the papers presented touch on this problem, several extensively. Abundance (or measured catch rate and catches) is apparently correlated with sunspot activity, temperature, mangrove area, latitude, estuarine/marine interface length, rainfall, river outflow, *etc*.

The relations have been shown to be positive or negative, depending on the area and the species considered, and it has been suggested that the relation with environment may not be linear within the whole range of possible values.

Abundance is linked directly to recruitment and there is evidence that the success of a cohort depends on the environment during its larval/juvenile phase. A number of factors interact, with rainfall and river outflow being particularly mentioned as affecting the year-to-year variations.

Relationships between favorable nursery areas and production have been demonstrated and it has been proposed that the important parameter is the 'ecological volume' defined as the overlapping between a 'static' habitat (favorable depth/area) and a dynamic one (the optimum characteristics of the water mass). This raises the problem of the conservation of the physical habitat (marshland area, estuarine-marine interface length, *etc*) and implies the necessity, in addition to the traditional management measures aiming at optimizing the yield-per-recruit, to strengthen the measures aiming at reducing undue larval mortality by littoral management.

Periodic and aperiodic changes

A distinction between these two types of change is important because the first refers to naturally reversible phenomenons, *eg*, periods of drought, while the second most probably refers to non-reversible ones, such as development of the coastal zones for housing, industry, *etc*. The latter can lead to completely different problems and may require quite different management solutions.

One useful question is: What can a manager do in face of periodic natural variations? It is felt that there is still a need to detect and understand them in order to avoid unnecessary troubles in the fishery (proclaiming for a collapse when it is not the case) and take eventually the necessary steps for reducing the adverse predictable consequences of the variations. In the case of the aperiodic changes, the management advice is straightforward if the changes are the consequences of man's activities even if the implementation of the advised measure may raise some problems.

Predictive models
Their usefulness for management purposes was discussed. It was felt that it was necessary that the important changes be detected and predicted in order to look for appropriate measures of lessening of the effects. Short-term predictions have been made in Louisiana, China, and Australia over a long period of years and are felt to be useful.

In fact the usefulness of a predictive model is inversely related to the unexplained variability. It has been often stated that in general these models very often fail to predict when they are confronted with the test of time, and that they are only able to make useful predictions at the extremes of the range of possible environmental values.

One way of testing the accuracy of the model is to build it using only part of the information available (*eg*, part of the time series) and then predicting later values that can be readily compared with observed ones.

It has been pointed out that more useful research is to be done along these lines, but that the first priority should be given to the development of an 'understanding model' before the mathematical ones are developed.

It was also remarked that relations with the biological environment should also be looked at. For example, the abundance of predators may provide a good indicator of recruitment levels.

When building mathematical models, a progressive procedure, introducing more and more variables in order to explain more and more of the observed variances is useful. However, the number of variables must stay small as compared to the number of data points available.

Attention was finally drawn to the necessity of checking long term changes in catchability before changes in abundance (and recruitment) are inferred.

8 Management

Management objectives
For the purpose of this discussion we might define fisheries management in a broad sense as the manipulation of factors to achieve societal goals from a stock of fish. More specifically this goal is usually quantifiable in terms of societal benefits in the form of food production, gross or net value, employment, the income of individual fishermen, or some combination thereof while maintaining the stock at some high level for sustainable production. The objective is usually to achieve an optimum balance between inputs and various outputs. As the fishery is developed and societal needs and values change, the management goals will change.

The goals and values to be obtained from the fishery are determined by the society and it is the responsibility of a decision-maker (fishery manager) at some level to decide how to obtain these benefits from the fishery. The manager must be able to identify the need for action and be prepared to act promptly. If there is to be a scientific basis for the management program, the manager needs biological, economic, and sociological information to assist in the decision-making process. Also included in influencing the decision process are people whose decisions affect the investment in new or larger vessels, either directly (*eg*, in regional development banks) or indirectly (*eg*, through tax policies). It is not the responsibility of the scientists to formulate the management objectives but to provide the fishery manager with the scientific basis for a range of management options and the ramifications of their implementation.

The scientist should take care that he does not second guess what he believes to be the desires of the manager, but provide him with an appropriate range of options.

Simple bioeconomic models to predict the outcome of fishery management actions are needed to aid the fishery manager in the decision process. However, the models should go beyond catch and effort relationships, should not be overly expensive, and should concentrate on the significant parameters.

After the manager selects his option and implements his program, there are others who influence its effectiveness. Fishermen must be willing to accept and employ the measures, while bankers and investors can influence the development of the fishery by distribution of capital. The shrimp fisheries throughout the world are generally fully exploited, and there is concern in many areas (China, Mexico, Australia, Indonesia) over the impact of the high level of exploitation on the stocks. In nearly all areas the abundance of shrimp, as measured for example by the catch-per-unit-effort, has sharply declined. In some areas (in parts of India and in the Kuwait-Saudi Arabia region) total production has also sharply declined, possibly as a result of heavy fishing. The fisheries in some countries face economic problems resulting from the high energy costs (fuel) in shrimp production (China, United States, Australia). Allocation among user groups of offshore, inshore, and artisanal fishermen is another problem in nearly all countries.

It was the consensus of the Workshop that because of the highly developed nature of the world's shrimp fisheries some form of management is in order for all stocks. This management will almost certainly involve control of the total amount of fishing, and will probably have direct or indirect effects on how this total is distributed. Stock maintenance is of increasing concern, and the precarious position of some stocks may be masked by high gross economic yield.

Various objectives such as adjustment of fishing mortality, fishing capacity, size at first harvest, and allocation among user groups have been sought through a variety of management measures. These are described in the papers and discussions as meeting with varying degrees of success.

Age-specific fishing mortality
Because of the rapid growth of young shrimp – and of the even greater increase in value of the individual shrimp – 'growth overfishing' is likely to occur in shrimp; that is, the total weight and total gross value of the shrimp catch is likely to be increased by shifting fishing mortality from the smallest sizes of shrimp onto the larger sizes. The separation in space and time between the main concentration of small and large shrimp mean that there are several ways of doing this, as discussed below. This separation has also encouraged the growth of distinct fisheries on the two groups – typically artisanal fishermen catching small shrimp with traditional gears in the lagoons, and industrial vessels trawling for large shrimp offshore. Management is often concerned in giving preference to one of these groups. This decision will be based ultimately on overall national policies (*eg*, the decision in Indonesia to ban trawling where there are many small-scale fishermen), but it is important that the managers are supplied with sufficient information on the costs and benefits involved (for example the calculations of Griffin of the differences in value from catching shrimp in the inshore and offshore fisheries of the Ivory Coast). Once the decision is made, the procedures to give priority to one or the other sector are likely to be fairly straightforward,

and were not discussed further. The meeting therefore concentrated on controls on the ages (or sizes) of fish caught within a fishery, particularly the industrial trawl fishery. Several measures have been employed.

Seasonal closures
Seasonal closures can be used effectively to select the size at first harvest in some shrimp stocks where there is a seasonality of recruitment of major portions of the stocks. Mathews discussed closed seasons in the Arabian Gulf off Kuwait to protect young recruits and which by reducing mortality would be expected to protect young recruits and which by reducing mortality would be expected to increase biomass. Winter/Spring spawners would also be directly protected by one of the seasonal closures. He concludes that if relatively high values of Z are assumed to be applicable, then a 3–5 month closed season is clearly useful, and a 5 month season is more likely to increase recruitment and landings than a 3 month closed season.

Poffenberger described the United States' closure of its western Gulf of Mexico to increase the harvest size of juvenile brown shrimp emigrating from estuaries. Data indicate increased yields and a higher value for the larger shrimp taken.

In Australia's Gulf of Carpentaria Kirkwood reports a decline in the sizes caught in 1976 due to a 15 day earlier re-opening of a closed season.

Garcia and LeReste suggested an advantage of closure of sea fishing at the moment of most intense recruitment to avoid exploitation of concentrations of juveniles at a time of rapid growth. They suggested this closure might be coordinated with a closed season in the estuaries if a fishery exists there. They caution that consideration should be given to the economic impact of an idle fleet.

Ye described China's unilateral seasonal closure of a northern portion of the Yellow Sea and the need to expand this closure to afford protection of the *Penaeus orientalis* brood stock in its extensive migration for spawning.

Area closures
The permanent closure of an area which serves as a nursery to juvenile shrimp and contains few individuals of a preferred harvest size has been effectively employed. In the Mexican Pacific fishery managers have closed to trawling two well-identified nursery areas which contain juvenile shrimp throughout most of the year, according to Ehrhardt *et al*.

Van Zalinge reported a somewhat larger average size in the months following the 1980 closed season adopted by all of the west coast countries on the Arabian Gulf.

A permanent closure of the Tortugas Shrimp Sanctuary in the U S eastern Gulf of Mexico was initiated to protect juvenile pink shrimp from growth overfishing. Small shrimp migrate through the area (which extends offshore to about ten fathoms) through the year. Poffenberger discussed the economic gains of such a measure. Such regulation (though it may be effective in increasing catch and value) has socio-economic consequences because it may eliminate small boats that are unable to fish beyond the closed waters.

Closed seasons and areas are used in Australia to protect nursery areas and control size of harvest (Walker).

Minimum size limits

Control of size at first capture by use of minimum landing sizes of shrimp has encouraged the wasteful practice of culling and discarding undersized shrimp. In an open access resource fishermen have little incentive to refrain from fishing mixed stocks when there is a profit to be made from retaining the larger individuals. Therefore, this measure has proven ineffective where employed until recently in the U S Gulf of Mexico. A recently implemented set of regulations in federal offshore waters adjusts harvest size by area closures. Two states, Florida and Texas, have subsequently repealed their landing size restrictions.

Griffin *et al* estimated that if small shrimp previously being culled in the Texas fishery could have been landed, the value of greater profits to vessel owners would range between $18·6 and $27·4 million and to their crews between $4·65 and $6·84 million.

Mesh size regulation

The adjustment of net mesh size may be used to release small individuals of the stock, small shrimp in a mixed stock, and unwanted bycatch. The theoretical dynamics of mesh size selection with reference to shrimp are discussed by Jones. Since the selection range in shrimp selection curves extends over a relatively large part of the exploited length range, assessment methods that permit the rate of exploitation to vary continuously with length are to be preferred to those that assume knife-edge selection and a constant value of fishing mortality for the exploited length groups.

It was agreed, primarily because of the wide selection range, that regulation of mesh size was unlikely to be particularly useful in terms solely of the shrimp catch. However, it could be useful for controlling the fish catch, especially in the many areas (Indonesia, Thailand, *etc*) where the fishery is based on a mixture of shrimp and fish. Cushing reported on this type of application of mesh size in the *Nephrops* fishery of the Northeast Atlantic. In such cases the optimum mesh size would have to be based on a compromise between the large mesh, which would be best for larger species of fish, and the smaller mesh, appropriate for smaller fish and shrimp. This still might be considerably larger than the mesh in use in several fisheries, especially in Asia, for which the use of a larger mesh could be beneficial.

If a mesh size regulation is being considered, a number of factors need to be examined. These include the selection pattern (including the effect of different types of net and twine, and of different sizes of catch and towing times), the assessment of the immediate and long-term effects (for which the length-structured models, such as those discussed in *Section 5*, are likely to be useful), and the determination of effective measures of enforcement (which include controls on the use of chafing gear, double layers of netting, *etc*, as well as a legally acceptable method of measuring the meshes).

Control of the amount of fishing

A variety of methods have been employed to reduce fishing mortality with varying degrees of success. Shorter fishing seasons, less area open to fishing, less efficient fishing methods and gear, quotas, limited entry and limitation of capital are some methods considered.

Because the abundance of stocks may fluctuate greatly as a result of environmental factors, as discussed by Christmas in the U S Gulf of Mexico, the manager must be provided with current predictive information if he is to control fishing pressure to prevent recruitment over-harvest. The manager must also be aware of any change in fishing effort or practices which may affect the total catch. The shift of catch and effort by the industrial and artisanal fisheries of the Arabian Gulf in the 1970s is one example of a changing fishery and the impact on stocks.

The manager must also consider the socio-economic impact of reducing the efficiency of the fishermen particularly during a period of rapidly increasing costs of fishing and processing the catch.

Gear restriction

Catch can be reduced by restricting the efficiency of the fishing unit provided that the fisherman does not compensate by increasing effort. Methods commonly employed include limitation of trawl size (or footrope) or even elimination of the trawl from specific areas. Where several groups of fishermen exploit the same stock controls on the type of gear that can be used can be very effective in discriminating in favor of one group. Thailand has prohibited the use of trawls with motorized boats within 3 000 meters of shore (Srimukda). Unar described the experience in Indonesia to restrict trawling from water where high densities of small shrimp may occur, and where many small-scale fishermen are operating a variety of traditional gears. This action has a great socio-economic impact because it allocates a portion of the resource to an artisanal fishery.

Catch quotas

Though quotas on the total annual catch have been a common method of management for long-lived animals (whales, halibut, cod, *etc*), annual catch quotas are not a suitable measurement for shrimp. Since they are short-lived, an annual quota does not control the fishing mortality, and might in fact encourage intense fishing at the beginning of the season – though this might be dealt with by setting quotas for short periods, *eg*, months.

A daily vessel or trip catch limit has been used to limit mortality. It also affects capacity. This measure applied in some inshore waters of the U S Gulf of Mexico to restrict harvest of juvenile shrimp requires a high level of monitoring for enforcement to be effective.

Controls on fishing effort

While other measures such as catch quotas can achieve the biological objectives, some direct control on the fishing effort (or the capacity of the fleet) is likely to be necessary to realize the significant economic benefits that could come from effective management. These measures are also likely to result in what is, in effect, an allocation of the resource between different user groups.

Criteria to be considered in setting a level of fishing effort includes maintaining stock at a desired level of productivity, keeping costs to a minimum, and obtaining the support of the affected fishermen.

Some possible ways to restrict effort are quotas, limited entry, delegation of fishing rights, and taxation or license fees. Catch quotas, in addition to the disadvantages discussed earlier, require a high level of enforcement to be effective in large fisheries.

Limited entry (or the restricting of the number of fishing units – usually the number of vessels

licensed to fish) does not necessarily result in limiting effort. In the Australian experience of limited entry, which resulted in a restriction on the number of vessels, fishing effort continued to increase.

The objective of such a system should be the maintenance of the resource at the desired level while maintaining an economically viable fishery and industry. This is essentially the objective in Western Australia. Under United States law, limited entry cannot be implemented solely for economic purposes, but must also consider biological, sociological, and other factors.

Limited entry can tend to generate increasing real fishing effort in two ways. First, each fisherman will try to increase his effort within the terms of his permit, for example, the increased size of boat in Australia. Second, the degree to which the measure is successful and generates income for those in the fishery, will stimulate the fishermen to improve their effectiveness still further. In principle this should be controlled by the terms of entry limitation, but the ability of fishermen to outwit regulations is great.

Unar described the Indonesian program as successful to limit effort in Irian Jaya based on the optimum effort suggested by production models. He cautions that fishing mortality is dependent on effort, and in some Indonesian fisheries there is a trend to increase vessel power, net size, and fishing hours.

Fishing rights are usually delegated in small, local fisheries and may have some value in allocating and restricting inshore artisanal fisheries.

Taxation in the form of high license fees is a method of limiting entry but as in limited entry could stimulate increased effort by those authorized to fish. To the extent that the fees would return some of the economic rent to the society, this stimulation will be reduced.

Some of the more broad measures which may affect capacity are import duties and quotas which increase markets for domestically produced shrimp in an importing country. Government subsidy of vessel construction, loan guarantees, or fuel costs would also tend to increase fishing capacity, maintain excessive effort, and generate excess capital.

Habitat modification

Although the maintenance of the quality of the fishery habitat affects natural mortality and recruitment to the fishery, it is not always within the direct control of the manager of the fishery. Habitat may be lost as described by Christmas in the case of the white shrimp in the Northern Gulf of Mexico.

Shrimp habitat can be enhanced by such programs as water management and pollution control.

Enforcement
To be effective, a management measure must be enforceable as well as acceptable to most of the fishermen who are regulated. The cost and level of enforcement necessary to implement regulations should be considered at the onset.

Monitoring program
The fishery managers and scientists should monitor the condition of the fishery and be prepared to take prompt action to revise the management objectives and techniques if the need arises. Advisory boards of fishermen and technical experts in fields closely associated with the fishery can be useful for this purpose.

9 Future work

The Workshop was concerned with the scientific basis for the management of penaeid shrimps. In considering future work it was therefore necessary to have in mind that the purpose of fisheries research is to provide a basis for management decisions. This is not to say that each research program will have a direct management application, but in the final analysis the integration of the research data should provide an understanding of the penaeid shrimps about which management decisions have to be made.

Much research has already been undertaken on the shrimp stocks of the world and it is not the purpose of this Workshop to attempt to provide either a catalogue of that research or a manual of research requirements. Diverse programs of research on the shrimp stocks, the environment in which they live, and on the units exploiting those stocks will continue to be undertaken. However, the Workshop was convened because the research undertaken so far has brought about a number of concerns regarding some of the shrimp fisheries, and the time was opportune to consider those areas of research which appear to require special attention so that the scientific basis for management decisions might be strengthened.

In general, the concerns being expressed are:
(*a*) Most shrimp stocks are now being heavily fished;
(*b*) The abundance of some shrimp stocks appears to have declined and the reasons for the decline are unclear;
(*c*) The heavy fishing pressure in some fisheries may have resulted in a decrease in the abundance of spawning stocks to a level which is resulting in reduced recruitment;
(*d*) In some areas there is a decline in the quality of the juvenile habitat;
(*e*) The cost of operation of some segments of shrimp fisheries is increasing at a rate faster than income;
(*f*) There is conflict among user groups as to area and size at which shrimps are to be harvested. This can be at both the national and international level;
(*g*) The large fish bycatch can damage directed fisheries on the fin fish species, and if used for human consumption, could increase the world's supply of high-grade protein.

Within this framework of concern about the state of many of the world shrimp stocks, the Workshop discussed future research needs, and proposed that special attention be given to the following areas of research:

(i) Stock and recruitment relationship
Concern was expressed that whereas management decisions in the past had mostly been made on the basis that recruitment numbers are independent of parent stock abundance at the levels of exploitation being experienced (with variations being determined by environmental conditions), data from a number of fisheries now indicate that this degree of independence may not hold true for all shrimp fisheries when exploitation rate is high.

In researching stock/recruitment relationships special attention needs to be given to:
(*a*) definition of index of breeding stock abundance;
(*b*) fecundity, with a view to estimating an index of egg production;
(*c*) definition of index of recruitment;
(*d*) recruitment variability due to environmental factors.

Emphasis should be given in establishing causal mechanisms that could affect recruitment, and where feasible identifying density-dependent effects which could generate a stock-recruitment curve.

(ii) Natural mortality
While there have been advances in the determination of estimates of natural mortality, the range of estimates is so wide and the parameter of such importance for management advice, that there is an urgent need for special attention to be given to research which will provide a greater understanding of the value of M. In this regard, the following lines of research were suggested:

(a) Comparative studies using data already available to obtain a greater understanding of the natural mortality of the different shrimp types.

(b) Studies of the underlying causes of mortality – predation, physiological death, diseases.

(c) Further tagging studies with particular attention being given to the degree of tagging mortality.

(d) Life table studies and DeLury type techniques. In this regard, attention was drawn to a recent paper by Chien and Condrey, 1981 manuscript.

(iii) Identification and standardization of effective fishing effort

Many shrimp fisheries consist of an inshore and an offshore segment. The inshore segment involves the use of different types of fishing gear – fixed nets, small boats with nets, and so on. The offshore segment is the industrial fleet of multi-rigged trawl vessels. It is of increasing importance to have a sound understanding of the effective fishing effort, preferably of both segments of the fishing industry but at least of the industrial fleet.

The nature of the inshore fishery may make difficult the calculation of effort estimates of this segment. However, even if this is not possible, data (eg, the numbers of fishermen and the types of gear used) should be gathered to understand whether changes are taking place in the fishing effort of the inshore fishery and the relative strength and direction of these changes.

It is important to establish the unit of effective fishing effort of the industrial fleet when the data are first being collected from that fishery. Failure to do this will result in the data set being of less value when stock analyses are undertaken.

Even though estimates of fishing mortality may not be available for the inshore fishery, managers will be called upon to make decisions about the use of the shrimp resources by different user groups. It is important therefore to attempt to assign a fishing effort figure to all segments of the fishery to aid the decision making process.

Areas of research which should be considered in the future include:

(a) Independent estimates of the stock, eg, by fish locating techniques.

(b) Catchability studies – behavior of the animals and fishing pattern of the fleet.

(c) Gear research – to estimate amount of fishing mortality generated by a particular gear type and to establish the selectivity of nets of different meshes for shrimp and for the main species of fish occurring in the bycatch.

(d) Methods of analyzing length-frequency data, including adaptations of cohort analysis.

(iv) The habitat

The immediate coastal zones are the main nursery areas for shrimps, and it is these zones where several changes may occur. The changes may either occur naturally, eg, variations in river run-off, or be man-made, eg, dredging of the estuarine system or removing mangroves.

As changes in the habitat are likely to have a major influence on shrimp recruitment, it is important to pursue studies on the nursery habitat so that causes in recruitment variation can be better understood. There is very little data available on habitat destruction and its effect on shrimp stock abundance.

Information on habitat changes is not only important in analyzing the stock production data and stock-recruitment data, but also in providing the administrator with information about the likely effect of proposed man-made changes to the nursery habitat. Furthermore, such studies provide an opportunity for advice to be passed to the fishing industry indicating the likely relative abundance of shrimp some months ahead of the fishing season for the industrial fleet.

In summary, research should be undertaken on the life history of shrimp species in relation to the critical environmental influences. Also, a valuable contribution to the development of future research programs would be a global view of types and areas of inshore habitat in relation to shrimp abundance, and including information on habitat changes which have occurred.

(v) Data base

The Workshop identified the establishment and continual updating of a data base as being of critical importance. The data base should be a historical description on an annual basis of the fishery describing:

(a) catch, catch composition and effort data (including bycatch species, discards and estimates of unreported catch);

(b) number and type of fishing units;

(c) number and type of personnel operating the units;

(d) the fishing grounds, eg, artisanal and industrial;

(e) methods of handling the production on the fishing units and in the factories;

(f) the market system;

(g) the value of the product at specified points of sale, and easily obtainable allied economic data;

(h) significant changes which have taken place

in the fishing units, personnel, grounds, marketing;
(*i*) the environment, quantified where possible.

It is recognized that the amount and sophistication of data collection will vary from country to country. However, it was emphasized that whatever data are available should be properly identified so that they are in a form capable of being used by those providing advice as well as by administrators.

(vi) Data integration

Fisheries scientists collect an array of data on shrimp stocks, and some of these data are essential for other studies, such as those undertaken by economists. It is important, therefore, that fisheries scientists understand the requirements of economists, and the sociologists too, so that the data base is capable of being used by all personnel who have a responsibility to provide advice to management. Furthermore, the advice of the fishery scientist will be more meaningful if the scientist has a basic understanding of the work being undertaken by the economist and sociologist.

Future work should be undertaken with greater attention being given to all personnel having an understanding of the method of collection and accuracy of the original data set. The data set will increase in complexity and value as research workers from the various disciplines start to work the data and make more specific their requirements for data collection. An appropriate technique or management information system should be adopted to assist in the integration of the data set, and this integration should include the financial implications.

(vii) Use of models

Production models are of value in gaining an understanding of the effect of fishing on the stocks. However, concern was expressed that too much reliance may be placed on their use in terms of achieving optimum yield on a long term basis. The production model suggests what will happen on the average, but does not provide information early enough on the possibility of a severe reduction in catch if effort is increased significantly.

The assessment obtained with a production model will be more valid the longer the series of data on which it is based. There is danger with assessments based on short periods of confusing variations due to environmental changes with those due to changes in effort. It is important that continuing research be undertaken to examine the causes of the variations.

Age- or size-structured models (of the Ricker/Beverton and Holt type), giving the relation of yield-per-recruit to different fishery patterns are essential in determining the effect of changes in the balance between different sectors of the fishery (inshore/offshore, *etc*). Because of large variations in estimates of some parameters of shrimp stocks, such as natural mortality, thus reducing the applicability of the yield-per-recruit model, attention should be given to providing scientific advice in terms of the probability of the result of taking a particular management decision.

On the development of new models it was pointed out that there were a number of outstanding problems for which adequate models were not available. Examples given were the requirement for bioeconomics, recruitment, decision-making and allocation models. Furthermore, there is a requirement for future work to include a model in conceptual form describing how the fishing fleet might respond to management options being considered and to identify influences such as a rapid increase in fuel price and to describe the effects on the stocks of the fleet response.

(viii) Analysis of the system

The Workshop drew attention to the importance of scientific advice being presented in a manner which integrates the array of data available on the stock, the fishing units, and the environment. In the transfer of information there is a need to develop simple models which draw together the dominating variable factors of the environment, stock parameters and the fishing practice, and describe their integrated effect on stock abundance and fishing industry success. Some of the important elements of the system to be considered are:

(*1*) Fishing section – industrial fleet, inshore fishery

(*2*) Estimates of F, M, K, L_∞

(*3*) Length at entry into the fishery, length at maturity

(*4*) River flow

(*5*) Rainfall

(*6*) Temperature

(*7*) Cultural enroachment on estuarine habitat

(ix) Ecological interactions

In some shrimp fisheries, such as that on banana shrimp in the Gulf of Carpentaria, there is very little bycatch taken with the shrimp. However, in others the capture of shrimp is accompanied by considerable quantities of fish, much of which is discarded; while in others again the fishery is directed at both shrimp and fish.

Research is required on the ecological interactions of the fauna on the shrimp grounds to provide information on the likely consequences in terms of total yield of introducing gear changes such as a shrimp separator trawl.

In relation to this subject, studies are required on the selectivity in feeding habits of the fish, measurement of the relative biomass of the various prey in the stomach of the fish, and a comparison with the relative biomass of these prey items in the prawn environment.

(x) Socio-economics

Future work should include greater emphasis on the role of the socio-economist in providing advice to management. Areas of work needing particular attention are those which:

(*a*) Clarify management objectives for any particular fishery taking into account the existence of an inshore and an offshore fishery. In considering this subject consideration will need to be given to such matters as quantifying trade-offs between net revenue, employment, and individual income.

(*b*) Determine costs and how these might be lowered by variations in the balance of elements of capital, manpower, energy, and in cost structure.

(*c*) Determine the multiplier effect under various management options. For example, it is important to determine whether F for maximum employment is far to the right of F for maximum net revenue or F for optimum individual income. Such a study would assist in the resolution of conflicts between management objectives.

(*d*) Provide information on the mobility in and out of the fishery of labor (especially in rural areas where there are cultural barriers), and of capital (access to loans, indebtedness, *etc*).

(*e*) Provide information on fishermen's earnings.

(*f*) Add to an understanding of the benefits of management options; this should include the collection of information on management schemes used in some traditional fisheries, such as property rights.

(xi) Priority and balance of research programs

Attention needs to be given to developing methodology for determining criteria for the allocation of finance for research. While the management objectives will differ from country to country thus affecting the priorities for research, the methodology will have general application.

An example of a process which could serve as a useful guide to establishing research priorities is set out in *Table 2*. It must be emphasized that the priority ratings given to each factor is by way of an example only, and the research objectives and management objectives (and the priorities given to them) are likely to vary from country to country. Each country therefore needs to determine its own management objectives, the research needed to attain each of these objectives, and the current state of knowledge relevant to each field of research. For example, countries with urgent social problems may find that high priority should be given to research into the economic dynamics of the harvesting and processing sectors. Nevertheless, it is believed that a tabulation of this type would be found useful by most countries in assessing their national research priorities.

Acknowledgements

Many people contributed to the success of this workshop. Dr William Fox, Director of the Southeast Fisheries Center, initiated the efforts which led to the workshop and secured much of its financial support. Mr Charles Lyles, Executive Director of the Gulf States Marine Fisheries Commission, secured the support of the Commission which also provided much of the requisite financial support. The NOAA/University of Miami Cooperative Institute for Marine and Atmospheric Studies (CIMAS) partially supported one of the workshop conveners, Dr Brian J Rothschild, through its Ecosystem Dynamics Program so we wish to thank the former and acting Directors of CIMAS, Dr Eric Kraus and Dr Claes Rooth. In particular we wish to acknowledge the efforts of Mr Larry Simpson and Mrs Virginia Herring of the Gulf States Marine Fisheries Commission who coordinated much of the logistics of the conference; and Ms Dianne Allen (Southeast Fisheries Center) and Mrs Margaret Bean (Gulf States Marine Fisheries Commission) who provided timely typing of the meeting reports. The efforts of these people assured that the objectives of the workshop were met.

Research Objectives	Management Objectives				Total Score	Mgmt. Need Rank	State of Knowledge Rank**	Combined Rank
	Optimum Size	Max. Economic Function	Min. Biol. Risk	Habitat Mgmt.				
1. Growth	1*	3	2	3	9	6·5	H/9	9
2. M	1	3	2	3	9	6·5	L/3	4
3. F	1	1	1	2	5	1	M/4	1
4. S/R	2	2	1	2	7	3·5	L/2	2
5. Inter-species relationships	3	3	1	1	8	5	L/1	3
6. Environmental interactions	2	1	2	1	6	2	H/8	5
7. Harvesting economic dynamics	2	1	3	1	7	3·5	H/7	6
8. Processing economic dynamics	3	2	3	2	10	8	M/6	7·5
9. Market economic dynamics	3	2	3	3	11	9	M/5	7·5

* *Score Description (only 3 of each per objective)*
1 = Essential
2 = Primary supporting information
3 = Secondary supporting information

** *Level of Current Knowledge*
H = Highest
M = Moderate
L = Least

Appendix I

List of participants

Dr W H L Allsopp
Associate Director (Fisheries)
International Development Research Centre
5990 Iona Drive
University of British Columbia
Vancouver, B C, Canada V6T 1L4

Mr Mohsen M H Al-Hosaini
Mariculture and Fisheries Department
Kuwait Institute for Scientific Research
P O Box 1638
Salmiya, Kuwait

Mr Bernard K Bowen
Director, Western Australian Dept. of Fisheries
    and Wildlife
108 Adelaide Terrace
Perth, W Australia 6000

Dr Joan Browder
National Marine Fisheries Service
Southeast Fisheries Center
75 Virginia Beach Drive
Miami, Florida 33149 U S A

Dr Susan Brunenmeister
National Marine Fisheries Service
Southeast Fisheries Center
75 Virginia Beach Drive
Miami, Florida 33149 U S A

Mr W D Chauvin
Shrimp Notes, Incorporated
New Orleans, Louisiana U S A

Mr J Y Christmas
Assistant Director Emeritus for Fisheries
    Research and Management
Gulf Coast Research Laboratory
East Beach Drive
Ocean Springs, Miss. 39564 U S A

Dr Richard Condrey
Center for Wetland Resources
Louisiana State University
Baton Rouge, Louisiana U S A

Dr David H Cushing
198 Yarmouth Road
Lowestoft, Suffolk NR 32 4AB England

Dr William Dall
Director, CSIRO Marine Laboratories
Division of Fisheries and Oceanography
Northeastern Regional Lab
233 Middle Street
Cleveland, Australia QLD 4163

Mr Nelson M Ehrhardt
Research and Integrated Fisheries Development
    Project
Apartado Postal M-10778, Mexico 1, D F

Dr Theodore Ford
Assistant Secretary
Louisiana Department of Wildlife and Fisheries
400 Royal Street
New Orleans, La. 70130 U S A

Dr William W Fox, Jr
Director, National Marine Fisheries Service
Southeast Fisheries Center
75 Virginia Beach Drive
Miami, Florida 33149 U S A

Dr Serge Garcia
Marine Resources Service
Food and Agriculture Organization of the
United Nations
Department of Fisheries
Via delle Terme di Caracalla
00100, Rome, Italy

Dr M J George
Central Marine Fisheries Research Institute
Post Bag 1912
Ernakulam, Cochin – 682018
Kerala State, India

Dr Wade Griffin
Department of Agricultural Economics
Texas A & M University
College Station, Texas 77843 U S A

Dr John A Gulland
Fishery Resources and Environmental Division
Food and Agriculture Organization of the United
    Nations
Via delle Terme di Caracalla
00100, Rome, Italy

Dr Donald A Hancock
Chief Research Officer
Department of Fisheries and Wildlife
W Australian Marine Research Laboratory
P O Box 20, North Beach
Western Australia 6020

Mr Pierre Jacquemin
c/o Mr N M Ehrhardt
Research and Integrated Fisheries Development
    Project
Apartado Postal M-10778
Mexico 1, D F

Mr Rodney Jones
Department of Agriculture and Fisheries for
Scotland
Marine Laboratory
P O Box 101, Victoria Road
Aberdeen, AB9 8DB, Scotland

Dr G. P. Kirkwood
Sr. Research Scientist, CSIRO
P O Box 21
Cronulla, N S W 2230, Australia

Dr Edward F Klima
National Marine Fisheries Service
Galveston Laboratory
4700 Avenue U
Galveston, Texas 77550 U S A

Mr Terrance R Leary
Gulf of Mexico Fishery Management Council
Lincoln Center, Suite 881
5401 West Kennedy Boulevard
Tampa, Florida 33609 U S A

Dr J Lhomme
Centre De Recherches Oceanographiques
B P V 18
Abidjan (Cote d'Ivoire), Ivory Coast

Mr Francisco Javier Magallon
Fisheries Department of Mexico
National Institute of Fisheries
Av. Avaro Obregon 269 – 10 piso,
Mexico F D F

Dr Sergio Martinez
Director, CIDEP
Instituto Nicaraguense de la Pesca,
Apartado Aereo 2020, Managua
Nicaragua

Dr C P Mathews
Fisheries Management Project
Mariculture and Fisheries Dept.
Kuwait Institute for Scientific Research
P O Box 24885
Safat, Kuwait

Dr Scott Nichols
National Marine Fisheries Service
Southeast Fisheries Center
75 Virginia Beach Drive
Miami, Florida 33149 U S A

Mr Jose Ximenez de Mesquita
Research Scientist
SUDEPE/PDP, AV.
W/3 Norte Q. 506 Bloco C,
Ed. da Pesca 70 740
Brasilia, D F, Brazil

Mr Michael Parrack
National Marine Fisheries Service
Southeast Fisheries Center
75 Virginia Beach Drive
Miami, Florida 33149 U S A

Dr Daniel Pauly
International Center for Living Aquatic Res.
    Mgmt.
MCC P O Box 1501, Makati
Metro Manila, Philippines

Dr J W Penn
Senior Research Officer
W Australian Marine Research Laboratory
P O B 20, North Beach
Western Australia 6020

Ms Patricia L Phares
National Marine Fisheries Service
Southeast Fisheries Center
75 Virginia Beach Drive
Miami, Florida 33149 U S A

Mr John Poffenberger
National Marine Fisheries Service
Southeast Fisheries Center
75 Virginia Beach Drive
Miami, Florida 33149 U S A

Dr Joseph E Powers
National Marine Fisheries Service
Southeast Fisheries Center
75 Virginia Beach Drive
Miami, Florida 33149 U S A

Ms M Conception Rodriguez de la Cruz
Fisheries Department of Mexico
National Institute of Fisheries
Av. Alvaro Obregon
269 – 10 piso
Mexico F D F

Dr Brian J Rothschild
Professor, UMCEES
Chesapeake Biological Laboratory
University of Maryland
Box 38, Solomons
Maryland 20688 U S A

Mr Pichit Srimukda
Brackish Water Fishery Division
Department of Fisheries
Ministry of Agriculture and Cooperatives
Rajdamnern Avenue
Bangkok, Thailand

Dr J P Troadec
Food and Agriculture Organization of the United
    Nations
Department of Fisheries
Via delle Terme di Caracalla
00100, Rome, Italy

Dr Lamarr Trott
National Marine Fisheries Service
3300 Whitehaven Street, N W
Page Building Two
Washington, D C 20235 U S A

Mr M Unar
Director,
Central Research Institute for Fisheries – AARD
J L Kerapu 12, Jakarta
Indonesia

Mr Tom Van Devender
Mississippi Bureau of Marine Resources
Long Beach, Mississippi U S A

Mr John Watson
National Marine Fisheries Service
Mississippi Laboratory
P O Drawer 1207
Pascagoula, Miss. 39567 U S A

Dr Ye Chang Cheng
Marine Fishery Research Service of Liaoning
Luta Liaoning Province
The People's Republic of China

Appendix II

List of papers

1. The limited entry prawn fisheries of Western Australia: Research and management, by B K Bowen and D A Hancock.

2. Ecological interactions between penaeid shrimp and associated bottomfish assemblages, by P Sheridan, J Browder and J Powers.

3. Standardization of fishing effort, temporal and spatial trends in catch and effort, and stock production models for brown, white, and pink shrimp stocks fished in U S waters of the Gulf of Mexico, by S Brunenmeister.

4. (a) Do discards affect the production of shrimps in the Gulf of Mexico?
   (b) The *Nephrops* fishery in the northeast Atlantic, by D H Cushing.

5. Catch prediction of the banana prawn, *Penaeus merguiensis*, in the southeastern Gulf of Carpentaria, by D J Staples, W Dall, and D J Vance

6. Biologie et exploitation de la crevette penaeide *Penaeus notialis* (Perez-Farfante, 1967), by F Lhomme and S Garcia.

7. Analysis of management alternatives for the Texas shrimp fishery, by W Griffin.

8. Some principles of mesh selection, with particular reference to shrimps, by R Jones.

9. Modelling of the Gulf of Carpentaria prawn fisheries, by G P Kirkwood

10. Exploitacion de los recursos camaróneros en Nicaragua, by S Martinez.

11. Impacts of variation in growth and mortality rates on management of the white shrimp fishery in the U S Gulf of Mexico, by S Nichols.

12. Application to shrimp stocks of objective methods for the estimation of growth, mortality and recruitment-related parameters from length-frequency data (ELEFAN I and II), by D Pauly.

13. The behavior and catchability of some commercially exploited penaeids and their relationship to stock and recruitment, by J W Penn.

14. The relationship between catch and time of fishing for the pink, brown and white shrimp of the northern Gulf of Mexico, by P L Phares.

15. A review of the shrimp fisheries of India: a scientific basis for management of the resources, by E G Silas, M J George, and T Jacob.

16. A review of the Indonesian shrimp fisheries and its management, by M Unar and N Naamin.

17. A review of the shrimp fisheries in the Gulf between Iran and the Arabian peninsula, by N P van Zalinge.

18. The Guianas-Brazil shrimp fishery, its problems and management aspects, by L C Villegas and A Dragovich.

19. The prawn (*Penaeus orientalis* Kishinouye) in Pohai Sea and their fishery, by Ye Chang Cheng.

20. Australian prawn fisheries, by R H Walker.

21. Introductory guidelines to shrimp management: some further thoughts, by J A Gulland.

22. The impact of environmental factors on Gulf of Mexico shrimp stocks, by J Y Christmas.

23. Review paper on the U S Gulf of Mexico shrimp fishery, by B J Rothschild and M L Parrack.

24. An economic perspective of problems in the management of penaeid shrimp fisheries, by J R Poffenberger.

25. A historical assessment of the white shrimp (*Penaeus setiferus*) fishery in the U S Gulf of Mexico, by S Nichols.

26. Review of eastern king prawn fishery, east coast of Australia, by J P Glaister.

27. A review of the fisheries biology and population dynamics of Kuwait's shrimp fisheries, by C P Mathews.

28. Shrimp fisheries in the Gulf of Thailand, by Pichit Srimukda.

29 Fishing configurations of common shrimp trawl designs employed in the southeastern United States, by J W Watson, Jr., I K Workman, C W Taylor, and A F Serra.

30. A bioeconomic analysis of the Ivory Coast shrimp fishery, by W L Griffin.

31. The Pacific shrimp fishery of Mexico, by N M Ehrhardt, P S Jacquemin, F J Magallon, and M C Rodriguez.

32. Life cycles, dynamics, exploitation and management of coastal penaeid shrimp stocks, by S Garcia and L LeReste.

33. Exploracao do camarao na regiao norte do Brasil, by J X de Mesquita.

34. The relationship between stock and recruitment in the shrimp stocks of Kuwait and Saudi Arabia, by G R Morgan and S Garcia.

35. Growth rates of white shrimp as a function of shrimp size and water temperature, by S Nichols.

Reports on national fisheries

Australian prawn fisheries

R H Walker

Abstract

Production of penaeid shrimp (prawns) in Australia doubled in the decade to 1980, reaching over 20 000 tons. Catches are taken throughout most of Australia, but the main fisheries are off the west coast from Sydney to northern Queensland, in the Gulf of Carpentaria, in South Australia, and Shark Bay and Exmouth Gulf in Western Australia. The biology and population dynamics of the major species in these fisheries is described. The effects of fishing on the stocks has been analyzed using production and yield-per-recruit models. Natural factors, particularly rainfall, have a significant effect on timing and magnitude of recruitment.

Management objectives include biological, economic and social factors. The measures taken include limited entry to control fishing effort and/ or capital inflow, and closed seasons and areas to protect small animals.

Australian prawn fisheries
Statement of the problem

The Australian prawn catch of 27 000 tonnes in 1980–81 is rather more than double that of a decade earlier (*Table 1a*). This dramatic expansion is due to the exploitation of the hitherto unfished grounds off northern Australia. However in recent years the rate of increase of the prawn catch has slowed considerably and shows signs of plateauing. This increase in catch has been accompanied by a rapid increase in boat numbers particularly in the fishery off the east coast. The number of vessels in fisheries of the Gulf of Carpentaria and Arnhem Land (the Declared Management Zone of the Northern Prawn Fishery or DMZ of the NPF) and in the large embayments of Western Australia and South Australia have been limited by management action. However, action to control boat numbers has only recently been taken off the Queensland

Table 1 (a)
AUSTRALIAN PRAWN PRODUCTION 1972/73 – 1980/81 (TONNES OF WHOLE PRAWNS)

State	1972/73	1973/74	1974/75	1975/76	1976/77	1977/78	1978/79	1979/80	1980/81
New South Wales	2 128	2 755	2 075	2 472	2 618	2 430	1 981	2 436	2 736
Queensland	6 892	11 222	4 414	6 647	11 702	8 428	10 044	10 580	14 448
Northern Territory	2 584	3 997	3 346	3 191	1 366	1 774	3 065	3 212	4 259
Victoria	14	4	—	—	—	—	—	—	*
South Australia	1 789	2 921	2 530	2 679	2 831	2 234	2 475	2 445	2 395
Western Australia	3 308	3 100	3 898	4 665	3 047	3 940	3 472	3 387	3 083
Tasmania	—	—	—	—	—	—	—	—	—
Australia	16 445	23 999	16 263	19 652	21 565	18 807	21 036	22 059	26 921

* Not yet available
Source: Australian Bureau of Statistics

Table 1 (b)
AUSTRALIAN PRAWN PRODUCTION 1979/80 (TONNES OF WHOLE PRAWNS)

Species	NSW	QLD	NT	Vic	SA	Tas	WA	Aust.
Banana Prawn	—	2 285	1 039	—	—	—	165	
Tiger Prawn	—	4 532	—	—	—	—	—	—
Western King Prawn	—	—	11	—	2 445	—	1 758	—
Brown Tiger Prawn	—	—	1 600	—	—	—	1 044	—
Eastern King Prawn	2 436*	1 482	—	—	—	—	—	22 059*
Red Spotted Prawn	—	—	—	—	—	—	—	—
Endeavour Prawn	—	1 149	562	—	—	—	291	—
Western School Prawn	—	—	—	—	—	—	14	—
York Prawn	—	—	—	—	—	—	—	—
School Prawn	—	64	—	—	—	—	—	—
Green-tail Prawn	—	—	—	—	—	—	—	—
Rainbow Prawn	—	9	—	—	—	—	—	—
Other Prawns	—	1 029	0·2	—	—	—	115	—

* All species
Source: Australian Bureau of Statistics

Coast and has not yet been taken off New South Wales (*Figure 1*).

Australian prawn fisheries thus present a range of problems. In South Australia the detailed control of fishing effort is being undertaken, in Western Australia there is concern that one species in a two species fishery is being overfished. In the DMZ the biology, population dynamics and status of only one major species has been elucidated and research on the five other species, which are becoming increasingly important to the fishery, is only just beginning. Off the tropical east coast catch statistics are suspect and although there is a fair understanding of the biology of major species off the temperate east coast, stock assessments have as yet only been undertaken on a limited scale.

The scientific inputs which are required to address these problems range from the development and application of bioeconomic models of some fisheries to establishing reliable catch and effort and biological statistics as a prelude to the management of other fisheries. Experience with the management of Australian prawn fisheries has shown that stock assessment is the primary need of management. However, successful management is greatly enhanced by an ability to predict stock abundance. An understanding of behaviour strengthens both assessment and prediction of stock status (Dall, unpublished).

Shrimp fisheries

The biology of the major species
Walker (1975) summarised what was then known of the biology of seven commercially important Australian penaeid species. Since that time studies in south eastern Queensland and New South Wales have added to knowledge of biology of the eastern king prawn, *Penaeus plebejus*; the school prawn, *Metapenaeus macleayi*; the greentail prawn, *Metapenaeus bennettae*; and the brown tiger prawn, *Penaeus esculentus*. Studies in the Gulf of Carpentaria and the Northern Territory have concentrated on the banana prawn *Penaeus merguiensis* but information on the aspects of the life history of eleven other species has also been obtained (Rothlisberg, Jackson and Pendry, 1978; Moriarty and Barclay, 1981; Moriarty 1977). In Western Australia the western king prawn (*Penaeus latisulcatus*) and the brown tiger prawn (*Penaeus esculentus*) have been studied, as has the western king prawn in South Australia.

Fig 1 The Geographical distribution of Australian prawn fisheries. Localities mentioned in the text are also shown.

9.02° S

DECLARED MANAGEMENT ZONE

DARWIN
ARNHEM LAND
GULF OF CARPENTARIA

BROOME
Southern Boundary of Northern Prawn Fishery

Norman River

GREAT BARRIER REEF

Nickol Bay

Exmouth Gulf

NORTHERN TERRITORY

BOWEN
Southern Boundary of Northern Prawn Fishery

WESTERN AUSTRALIA

QUEENSLAND

Hervey Bay

Shark Bay

SOUTH AUSTRALIA

Moreton Bay
BRISBANE

Clarence River

NEW SOUTH WALES

Hunter River

PERTH
Cockburn Sound

Spencer Gulf

West Coast

Gulf
St. Vincent
ADELAIDE

SYDNEY

Investigator Strait

VICTORIA
MELBOURNE

AUSTRALIA

200 0 200 400 600 800
Kilometres

TASMANIA
HOBART

Migration and distribution of life history stages

With the exception of the greentail prawn, which can complete its life cycle in estuarine conditions, Australian penaeids so far studied conform to the usual pattern of a littoral juvenile phase, followed by an offshore migration of adolescents with an adult spawning phase in deeper water. The life cycle is completed by a return migration of larvae to the coastal shallows.

Young (1978) found that although eastern king, brown tiger and greentail juveniles were found throughout Moreton Bay they were most abundant in shallow areas, particularly seagrass beds. Within the general area eastern king prawns were most abundant in areas with strong marine influences, abundance of tiger prawns increased in areas of influences intermediate between marine and riverine and greentail prawns were abundant in intermediate or riverine areas. Post larval school prawns were restricted to a coastal lagoon.

These restrictions in range reflect recruitment patterns (Young and Carpenter 1977) which are closely related to life histories and spawning distributions of adults. Eastern king prawn spawn well out to sea, greentail prawns spawn in the shallow regions of Moreton Bay and adult school prawns spawn near river mouths (Ruello, 1973; Glaister 1978b). Young and Carpenter (1977) suggest that

38

serial recruitment of the eastern king prawn (peak recruitment July–September) brown tiger (recruited January–February) and greentail (recruited March–June) prawns allows the three species to share the same habitat with minimum interspecific competition.

To account for observed differences in seasonal pattern of postlarval emigration and juvenile immigration of banana prawns to and from their nursery areas in rivers in the Gulf of Carpentaria, Staples (1979) has suggested that the basic life history strategy of the banana prawn is a six month generation time and a lifespan of approximately twelve months with spawning peaks occurring in the adult population in spring and autumn. Survival of autumn spawned larvae dominates the northern rivers and spring spawned larvae dominates the south. Staples (1980, a) considers that a combination of passive and active transport is probably responsible for the overall transport of larvae to an estuary. His studies of banana prawns do not support hypotheses that diurnal vertical migration to take advantage of favourable tidal currents or orientation along salinity gradients provide larvae with transport mechanisms. He notes that wind induced currents, rainfall and freshwater runoff all influence recruitment to the Norman River and suggests that larval response to salinity changes during the tidal cycle are worthy of further investigation. Staples (1980, b) found that the time which banana prawns spent in the Norman River varied considerably and suggests that emigration of this species is stimulated directly by effects of rainfall.

In considering the spawning and fecundity of the western king prawn in Shark Bay and Exmouth Gulf, where temperatures permit year round spawning, Penn (1980) points out that multiple spawning by adults and rapid increases in population size after recruitment to the fishery in late summer and autumn is likely to lead to a high level of egg production during the winter with peaks in autumn and spring. In contrast spawning in Cockburn Sound is limited by water temperatures. Juveniles leave the nursery areas and are recruited to the fishery in the late summer and autumn. In Shark Bay Penn and Stalker (1979) indicate that larvae are transported to the nursery areas around the shore of the Bay. The nursery areas of juvenile western king prawns and shallow sand flats between low water and 1 metre whereas brown tigers utilise seagrass beds which are slightly deeper. Juveniles are recruited to the nursery area throughout the year with peaks of recruitment during spring and autumn. Penn (1975) suggests that this recruitment pattern is due by the larvae

taking advantage of favourable tides by means of diurnal vertical migration. King (pers. comm.) has found that larval western king prawns are associated with seagrass beds in Spencer Gulf.

Growth

Lucas (1974) presents data on growth of eastern king prawns in Moreton Bay based on the results of a series of tagging experiments. He estimates von Bertallanfy growth parameters for males at $L_\infty = 40$ mm carapace length and $K_\infty = 0.10 \pm 0.03$ per week and for females at $L_\infty = 49$ mm carapace length and $K = 0.10 \pm 0.03$. Somers (1975) indicates that the value of K changes throughout the life of the eastern king prawn and suggests that this change is related to temperature. Ruello (1975), reported that growth rates of eastern king prawns migrating from New South Wales were similar to those reported by Lucas.

Glaister (1978b) reports that school prawns tagged in northern New South Wales had a mean growth rate of 0.7 ± 0.3 mm carapace length per week. He found no significant differences in the growth rates of males and females of this species. This rate was comparable with growth rates reported for this species by Ruello (1970) for central New South Wales.

Lucas *et al.* (1979) estimated the von Bertallanfy growth parameters for banana prawns in the Gulf of Carpentaria to be $L_\infty = 38.0$ mm carapace length and $K = 0.08$/week. Munro (1975) found that female banana prawns grow larger and continue to grow over a longer period than males. He indicates that males reach their maximum average size of 34 mm carapace length in about 8 months and females reach 39 mm carapace length in about 12 months. He suggests that some females continue to grow and enter a second year of life. Staples (1980b) found that juvenile banana prawns grew very rapidly and that this growth was temperature dependent. Penn and Stalker (1979) give growth curves for western king prawns and brown tiger prawns in Shark Bay.

Natural mortality

Lucas (1974) undertook a series of tagging experiments to estimate the population parameters of eastern king prawn stocks off southeastern Queensland. He discusses the difficulties of estimating natural mortalities from tagging experiments. He points out that there is a rapid migration through the fishery on young prawns in Moreton Bay (E = 0.17/week) to the fishery for adults offshore and suggests a natural mortality coefficient of 0·11/week with an upper limit of 0·22/week for prawns in Moreton Bay. For the offshore population he estimated a natural mortality of 0·05/week.

Lucas et al (1979) used an estimate of the total mortality in the early years of the banana prawn fishery at Weipa when fishing intensity was low as an estimator of the natural mortality coefficient for banana prawns in the Gulf of Carpentaria. They comment that this value (M = 0·05/week) overestimates the value of the coefficient of emigration.

Penn (1976) reports the results of a series of tagging experiments in Cockburn Sound. He found that the single census (Petersen) method was the most appropriate means of estimating population size from his tag recapture information. He used Ricker's method to estimate mortality rates and found M to have a value of 0·015 − 0·035/week. He points out that variations in catchability can cause inconsistencies in the catch/effort relationship of a penaeid stock and considers that an understanding of the behavioural and physiological factors relating to these catchability changes are fundamental to the estimation of population parameters in penaeid stocks.

Other aspects of biology relating to management
The results of tagging experiments carried out by Ruello (1975) and Montgomery (1981) suggest that there is a single stock of eastern king prawns off southeastern Australia with breeding occurring at the northern end of the distribution off northern New South Wales and S.E. Queensland.

In Eastern Australia movements of tagged school prawns were generally northward from river mouths but Glaister (1978, b) suggests that such migrations are limited in extent.

Tagging experiments in the Gulf of Carpentaria (Lucas et al, 1979) suggest that banana prawns do not undertake extensive directed migrations and there appears to be little genetic differentiation of the stock.

On the other hand tagging in Shark Bay in Western Australia and Spencer Gulf and Gulf St Vincent in South Australia suggest each embayment supports a separate population of eastern king prawns.

Both short and long term variations in catchability have been reported in Australian penaeids. Racek (1959) demonstrated a lunar rhythm in the catchability of school and eastern king prawn. White (1975) has demonstrated diurnal and lunar rhythms in brown tiger prawns in Exmouth Gulf. Staples and Vance (1979) have shown a tidal rhythm in the catchability of juvenile and adolescent banana prawns in the Norman River with juveniles being more catchable during low tide, in contrast, adolescents in deeper water were more catchable at high tide. Young and Carpenter (1977) have shown a diurnal periodicity in recruitment of eastern king prawns to nursery areas in south east Queensland.

There may have been species replacement in some Australian fisheries. In the early years of Moreton Bay fishery quite large catches of banana prawns were taken and the greentail prawn (*M. bennettae*) was the most abundant species. Now banana prawns are uncommon and hardback prawns (*Trachypenaeus* spp.) are often the most abundant.

Penn (this seminar) has suggested that the disappearance of banana prawns from Exmouth Gulf may be largely due to heavy fishing. However Kirkwood (this seminar), citing work in preparation by Vance et al, indicates there has been no stock related trends in recruitment discernible for banana prawns in the Gulf of Carpentaria.

The fisheries
Ruello (1975) and Walker (1975) reviewed the history and status of Australian prawn fisheries in the early 1970's. Although a variety of set nets and beam trawls are used in estuaries and coastal lagoons on the east coast of the continent, otter trawlers produce the majority of the Australian catch. During the 1970's the expansion of the Australian fishery into tropical waters off northern Australia resulted in a rapid increase in catch from about 10 000 tonnes in the late 1960's to a peak of 24 000 tonnes in 1973/74. Since then annual production has been about 20 000 tonnes. (*Table 1*). The 1974 catch was largely due to an exceptional catch in the Gulf of Carpentaria.

The entry of vessels to important prawn fisheries in Western Australia in Shark Bay, Exmouth Gulf, and Nickol Bay, (see Bowen and Hancock (this volume) *Tables 1 to 4*) and South Australia in Spencer Gulf, St Vincent Gulf, Investigator Strait and the West Coast (*Table 2*) has been controlled since the early days of fishery development.

There was no control on entry to fisheries off the east coast and in northern Australia until in 1977 the further entry of vessels to the fisheries in the Gulf of Carpentaria and Arafura Sea (DMZ of the NPF) was limited. More recently the entry of vessels to the fishery off the Queensland coast was controlled.

Although the number of vessels in all Australian prawn fisheries has risen during recent years the major increase in fleet size has occurred off the

Table 2
NUMBER OF AUTHORITIES AND PERMITS ON ISSUE[1] AND NUMBER OF SALES OF AUTHORISED VESSELS, 1968–69 TO 1978-79

Year	West Coast and Spencer Gulf			Gulf St. Vincent			Investigator Strait			All zones		
	Auth.	Sales	Permits	Auth.	Sales	Permits	Auth.	Sales	Permits	Auth.	Sales	Permits
1968–69	26			5						32		
1969–70	30	1		5						35	1	
1970–71	37	2		10						47	2	
1971–72	36	3		10	1					46	3	1
1972–73	36	2	1	10	1					46	3	2
1973–74	34	2	1	10	1					44	3	2
1974–75	39	1	2	12	1				2	51	2	6
1975–76	39	1	3	12	2				5	51	3	10
1976–77	39	4	3	12	2				5	53	6	10
1977–78	39	2	3	14	1				8	53	3	11
1978–79	39	2	1	14	1				8	53	3	10

Source: South Australia, Department of Fisheries
[1]To avoid double counting of reissued authorities or permits, the number shown is the maximum number of concurrent authorities on issue during a given year.
From Byrne

Queensland east coast where the number of trawlers increased from about 700 in the mid 1970's to 1400 in 1981 (*Figure 2*). A freeze on boat numbers in late 1980 has since stabilised numbers at this level.

In all fisheries save where there are rigid replacement rules boat sizes have tended to increase. In remote areas small wooden vessels with ice or brine storage facilities have been replaced by larger steel vessels with dry refrigeration. Electronic navigation and fish finding aids are universal and except where regulations prevent it almost all Australian prawn trawlers are double rigged. In the northern prawn fishery multiple rigs *ie* towing four or occasionally six nets in pairs are common. Bryrne (1982) estimated the relative fishing power of vessels in the South Australian fleet and found that 1% increases in vessel length and continuous rated brake horsepower lead to 0·53% and 0·18% increases in fishing power respectively and that switching from single to double rig increases fishing power by 30%. In the DMZ of the NPF a government subsidy on the construction of vessels longer than 21 metres has had a profound effect on the size composition of the fleet (*Figure 3*).

Rising fuel prices have lead to the adoption of fuel saving equipment such as Kort nozzles, the adoption of minimal drag net designs and more fuel efficient motors.

The data base
Production statistics in terms of the total landed weight and value are collected on a State by State basis for national accounting purposes (*Table 1a* and *b*). In some States (*eg* W.A. and S.A.) this is collected by means of a monthly catch return provided by fishermen. In others (*eg* Queensland) fish

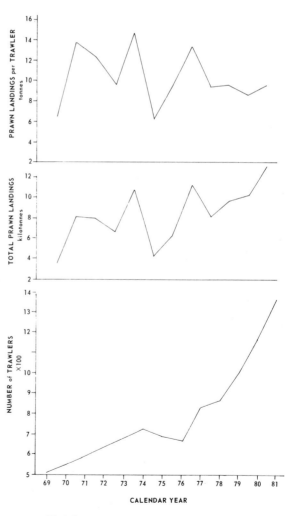

Fig 2 Queensland prawn catches 1969–81.

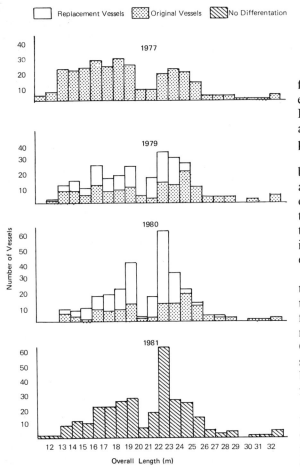

Fig 3 Vessels fishing in the declared management zone 1977–1981 by length class.

processors and markets are the primary source of these data.

In closely managed fisheries such as those in the Western Australian and South Australian Gulfs and the DMZ of the NPF considerable quantities of data are collected. Categories of data collected for the Western Australian fisheries are summarised in Bowen and Hancock (this volume). Data held for the NPF include:

- Processors returns incorporating catches supplied and processing output
- Administrative data including boat and gear descriptions, licensing and ownership history *etc*
- Economic data on costs, earnings and profitability
- Catch/effort and locality data from logbooks together with data on boat and gear factors likely to modify fishing effort
- Research data including biological, oceanographic and meteorological data
- Fishery monitoring data including data on operations, catch make-up and size composition, surveys of nursery areas.

Data for other fisheries such as the limited entry fishery on the Queensland east coast and the northern New South Wales fishery are less complete. But catch/effort, economic data and research data are available for smaller areas and defined time periods.

Extensive data collection requires staff and can be difficult in remote areas but appropriate arrangements such as the employment of part time operators and the use of surveys rather than continuous monitoring can minimise costs. Good relations between field staff and industry, and effective interchange between industry and government are essential for effective data collection.

If catch and effort data is to be collected effectively by means of a logbook system it is essential that books are designed so that they are easy for fishermen to use. Logbooks should collect the necessary information without unnecessary detail. Generally voluntary rather than compulsory provision of logbook data is used in Australia on the basis that unwilling providers are more likely to falsify information.

A particular problem in the DMZ of the NPF is the occurrence of species in the catch which are not distinguished commercially. There are two tiger prawns (*P. esculentus* and *P. semisulcatus*) and two endeavour prawns (*Metapenaeus endeavouri* and *M. ensis*) which occur in the fishery on the same grounds. Regular scientific sampling of the catch is required to adjust logbook data for this fishery.

The effect of fishing on shrimp stocks

There have been suggestions that some stocks of Australian prawns have been affected by fishing. Penn (this volume) reports the disappearance of banana prawns from Exmouth Gulf. As previously indicated the banana prawn is now rare in Moreton Bay. Both Exmouth Gulf and Moreton Bay are at the southern limit of the banana prawn's Australian distribution. In contrast Ruello (1975) reports that a commercial prawn fishery began in the first or second decade of the 19th century in Port Jackson (Sydney) which continues to support a prawn fishery to this day.

Lucas (1974) in study of the eastern king prawn stocks of the Moreton Bay area indicated that an increase in fishing intensity both in the Bay and offshore would correspondingly increase catch with only a small reduction in catch per boat. He pointed out that rapid migration of young prawns through the Bay meant that changes in fishing intensity there would only have a small effect on the offshore fishery.

Lucas *et al* (1979) estimated the yield per recruit for the banana prawn fishery in the Gulf of Carpentaria (*Figure 4*). They pointed out that in this very intense fishery fishing mortality levels had been sufficient to produce yields per recruit within 10% of its maximum value since 1974. More recently it has been suggested that this fishery has been fully exploited since 1971 (*Table 3*). Lucas *et al* (1979) suggest that substantial fluctuations in catch would continue due to fluctuations in recruitment.

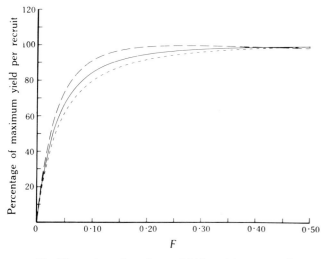

Fig 4 Percentage of maximum yield in weight per recruit as a function of fishing mortality for banana prawns in the Gulf of Carpentaria. The three curves correspond to pairs of values of M/K and Lc/L∞ of 0·40, 0·77 (------); 0·63, 0·82 (———); 0·80, 0·87 (_ _ _ _ _ _). From Lucas *et al* (1979).

In Western Australia, production models have been used to assess the status of prawn stocks in Shark Bay and Exmouth Gulf. Bowen and Hancock (this volume) suggest that the generalised production model gives an acceptable representation for both brown tiger and western king prawns.

They suggest that the effort on western king prawns in Shark Bay is about 10% in excess of optimum, whereas in 1979 effort on brown tiger prawns was well in excess of this optimum. They say that availability of western king prawns on the grounds allowed continuing exploitation of brown tigers after their catch rate had fallen to levels otherwise sufficient to discourage further fishing.

There have been no assessments of other Australian prawn stocks published although Byrne (1982) says that the yields from Spencer Gulf and Gulf St Vincent are near maximum sustainable levels (*Table 4*). Catches of the mixed species fishery for tiger, endeavour and western king prawns are still rising in the Gulf of Carpentaria and Arafura Sea (*Table 5*). Catches in north eastern Queensland and Torres Strait have risen steeply since 1975 and catches of eastern king prawns off south eastern Queensland have been fairly stable.

The fisheries in Gulf of Carpentaria, Nicol Bay, Exmouth Gulf, Shark Bay, St Vincent Gulf and Spencer Gulf are directed at adult prawns and a variety of closed areas and seasons are applied to protect nursery areas and control size of first capture. According to Byrne (1982) several closures are also used to control fishing effort in the South Australia Gulfs (*Table 6*).

Table 3
ANNUAL CATCH, EFFORT AND CATCH PER UNIT EFFORT (CPUE) FOR BANANA PRAWN FISHERY OF THE DMZ (*1968–80*)

Year	Vessels	Total catch (tonnes)	Banana prawn fishery catch[a] (tonnes)	CPUE[b] (tonnes/ vessel-day)	Effort (vessel-day)	Days per vessel per year
1968	19	1 978	—	—	—	—
1969	115	1 070	—	—	—	—
1970	142	1 704	1 527	—	—	—
1971	165	7 365	6 597	1·313	5 024	27
1972	154	4 804	4 131	1·070	3 861	27
1973	166	4 226	3 524	0·711	4 956	32
1974	165	12 711	12 291	1·777	6 916	42
1975	105	2 980	2 855	0·609	4 690	45
1976	145	4 436	4 164	0·647	6 427	44
1977	175	6 216	5 956	0·911	6 535	37
1978	193	2 535	2 263	0·455	4 977	33
1979	199	4 775	4 335	0·662	6 549	33
1980	231	2 681	2 092	0·306	6 826	30

[a] Total catch of banana prawns excluding those caught as an incidental component of the trawl fishery catches.
[b] Catch per unit effort based on logbook records pertaining to banana prawn fishing only.

Table 4
CATCH AND VALUE FOR YEARS 1968–69 TO 1978–79

	West Coast		Spencer Gulf		G. St. Vincent		Investigator Strait		Total	
	Catch	Value	Catch	Value	Catch	Value	Catch	Value	Catch	Value
	'000 kg	$'000	'000 kg	$'000	'000 kg	$'000	'000 kg	$'000	'000 kg	$'000
1968–69	17	n.a. [1]	507	n.a.	16	n.a.			540	589
1969–70	226	n.a.	932	n.a.	101	n.a.			1 259	1 586
1970–71	151	n.a.	909	n.a.	127	n.a.			1 187	1 496
1971–72	276	n.a.	1 001	n.a.	211	n.a.			2 675	4 013
1972–73	243	n.a.	243	n.a.	229	n.a.			1 715	2 881
1973–74	262	n.a.	2 287	n.a.	354	n.a.			2 903	3 774
1974–75	179	n.a.	2 052	n.a.	342	n.a.			2 573	3 860
1975–76	103	n.a.	1 955	n.a.	451	n.a.	106	n.a.	2 615	7 584
1976–77	30	98	2 222	7 028	447	1 844	143	544	2 842	9 513
1977–78	20	46	1 674	4 022	397	1 593	159	652	2 278	6 314
1978–79	16[2]	75[2]	1 837[2]	8 555[2]	275[2]	1 320[2]	180[2]	864[2]	2 308[2]	10 814[2]

Source: South Australia, Department of Fisheries
[1] Prior to 1976–77, regional prices are not available, hence it was not possible to calculate the value of the catch on a regional basis.
[2] Provisional.
From Byrne

Year	Vessels	Total catch (tonnes)	Catch per vessel (tonnes)	CPUE (tonnes/ vessel-day)	Effort (vessel-day)	Days per vessel per year
1968	65	170	1·95	—	—	—
1969	144	845	5·87	—	—	—
1970	191	1 555	8·14	0·262	5 939	31
1971	169	1 583	9·37	0·205	7 725	46
1972	180	1 854	10·30	0·230	8 060	45
1973	217	2 266	10·44	0·294	7 701	36
1974	196	1 153	6·56	0·382	3 016	15
1975	107	1 425	13·24	0·245	5 817	54
1976	145	1 805	12·48	0·261	6 915	48
1977	193	4 071	20·85	0·350	11 637	60
1978	237	4 937	20·07	0·263	19 746	79
1979	240	5 576	22·63	0·299	18 618	78
1980	268	6 543	24·41	0·227	28 858	108

In the long established fisheries of eastern Australia prawns are exploited as juveniles and adults. However, apart from the Lucas' (1974) assessment of the Moreton Bay eastern king prawn fishery no accounts of variations of fishing mortality have been published. However closures and gear regulations of various sorts (Glaister 1978a; Walker, 1975) are applied for a variety of reasons. In some estuaries and coastal lagoons the amateur catch is estimated as of the same order (100 tonnes a year) as the catch of professional fishermen although no statistics are kept (Ruello pers. com.).

Other species

There have been very few Australian studies in the bycatches of shrimp trawlers. Incidental catches of fish taken during experimental or exploratory work have been described by Penn and in the Kappala Reports of New South Wales State Fisheries.

Stocks of the saucer scallop (*Amusium ballotti*) were discovered by prawn fishermen in Hervey Bay (Queensland) and Shark Bay and these are now caught using modified prawn trawls. In Hervey Bay boats tend to move from the scallop fishery to the prawn fishery and back as economic circumstances dictate rather than fish for both animals simultaneously (Dredge, pers. comm.).

In Australian shrimp fisheries high unit value animals such as squid, crabs and slipper lobsters may be retained but fish are very seldom kept. Off

44

Table 6
CLOSURES FOR SPENCER GULF ZONE, 1979 AND 1980

Period	Area
Permanent	Area north of line joining Point Lowly and Port Germein (nursery area)
15 Jan – 28 Feb 1979	Total gulf
1 Mar – 31 Mar 1979	Area north of line joining Webling Point and Shoalwater Point Light (fishing area adjacent to northern nursery area).
1 July – 30 Sept 1979	Total gulf.
22 Dec 1979 – 15 Feb 1980	Total gulf.
16 Feb – 31 Mar 1980	Area north of line joining Webling Point and Shoalwater Point Light (fishing area adjacent to northern nursery area).
1 April – 30 June 1980	Thirteen mile strip parallel to the coast from Cowell to Cape Driver (fishing area adjacent to Cowell nursery area.
1 July – 30 Sept 1980	Total gulf.

Source: South Australia, Department of Fisheries.
From Byrne

central New South Wales however gear regulations applying to the fish trawl fishery takes account of simultaneous captures and retention of fish and shrimp.

Environmental aspects

Natural variations

Staples (1980) says that a comparison with other species indicates that rainfall plays a much larger role in stimulating emigration of juvenile prawns from Australian estuaries than in other countries.

Ruello (1973) has shown both short and long term responses of school prawns to rainfall. He suggests that the emigration of school prawns from the Hunter River in response to flooding was because of increased river flow and the subsequent disturbance of bottom sediments. Glaister (1978a) found a significant correlation between catches of this species in the Clarence River estuary and rainfall in the previous month and that the annual production of prawns offshore was related to river discharge for the same year. However, Ruello (1973) found that prawn catch off the Hunter River was strongly correlated with rainfall in the previous year. He suggested that this could be due to the cumulative effect of rainfall on reproductive success, recruitment of young to the estuary and the growth and survival of all life history stages.

Staples (1981a and b) has shown that rainfall directly effects both emigration and immigration of banana prawns in the Norman River estuary. Rainfall in this estuary is highly seasonal and he suggests that high river discharge rates prevent immigration of post larvae during much of the wet season. He suggests that emigration is directly related to rainy periods but that rainfall associated factors other than a simple salinity change are responsible for triggering it.

Staples *et al* (this volume) have shown that catches of banana prawns in the south eastern Gulf of Carpentaria are best correlated with summer and autumn rainfall. However the fishing season begins in autumn so rainfall in the autumn months cannot be used for prediction. But spring and summer rainfall gives an acceptable prediction of the catch and gives promise of providing a useful model for predictive purposes but further testing of the model is necessary.

Human activities other than fishing

Ruello (1973) expresses the view that the erection of a major water storage on the Hunter River has probably had little adverse effect on the school prawn in the Hunter River estuary. He does point out, however, that the construction of a rock weir on a tributary closed off a nursery area.

Young (1978) points out that all littoral banks in Moreton Bay constituted important nursery areas for various species of prawns. He says that any loss of these areas will result in a corresponding loss to the commercial fishery. He also says that the abundance of juveniles is highest in seagrass meadows and if these were encouraged to increase by reducing water turbidity and by ceasing dredging a corresponding increase in recruitment to the fishery of the species studied could be expected.

Management

Management objectives

The objectives of Australian fisheries managers are usually stated in the legislation under which they operate and are expressed in general terms such as 'seeking optimum utilisation of the resource'.

Both government and industry are united in the need to prevent any stocks being harmed biologically. This 'bottom line' is usually modified by some other aim either explicit or implicit. For example the aims of management in the DMZ of the NPF have been defined as:

- encouraging the establishment of shore based processing plants and other facilities necessary for servicing fishing fleets
- maximising employment opportunities
- implementing management strategies that whilst providing adequate protection for prawn resources, are sufficiently flexible to allow further development of resources where development potential exists.

The South Australian Government has made considerable efforts to control increases in fishing efforts and fishing power in order to maximise the economic return from the fisheries under their control. Byrne (1982) has observed that there have been significant unplanned increases in effort despite controls and notes that there is scope for this to continue even under the existing restrictions (*Table 7*). He suggests that an industry financed buyback scheme would be one way of achieving the desired level effort at the lowest possible cost and so obtain the maximum benefit from the limited entry program.

Bowen and Hancock (this volume) point out that limited entry provides partial control of exploitation rate and capital input but the price for resource security and high profitability is restriction of the personal liberties and freedom of choice of the fishermen. They point out that close contact between industry and scientists and the administrator is essential for the success of a limited entry regime.

Experience with the limited entry regime in the DMZ of the NPF had also demonstrated the difficulty of containing effort and capital input into a limited entry fishery. In practice it proved to be difficult to enforce a rigid boat replacement policy and the size of vessels tended to increase (*Figure 3*). Additionally, a government policy which subsidises the construction of boats larger than 21 metres in length to support the shipbuilding industry has had a marked effect on the size composition of the fleet (*Figure 3*). The sale of entry entitlements by fishermen has allowed the vertical integration of catching and processing sectors. This probably increased the efficiency of the industry but has led to social difficulties. The sale of entitlements probably increased boat numbers in the east coast fishery in two ways; former owners built new boats for the east coast fishery using the sale proceeds and old vessels which had left the DMZ after replacement also operated on the east coast.

An open entry policy on the east coast of Queensland has led to doubling of the numbers of prawn trawlers operating there since 1977 (*Figure 2*). The decline in catch per trawler together with increased operating costs and unwise investment decisions has led a large number of trawler operators into financial difficulty. This situation is probably the most difficult question in Australian prawn fishery management at the moment. The Queensland Government has frozen the number of vessels entitled to operate there but resolution of the problem will be difficult.

In summary the Australian experience has shown that the introduction of a limited entry regime late in a fishery's development tends to increase fishing in the short term as fishermen take advantage of this right to participate in the fishery and in the longer term they seek to maximise their share of the catch by increasing their efficiency. In the Australian context it has proved difficult to devise management measures which closely restrict the capitalisation of licence values and the

Table 7
RESTRICTIONS IMPOSED ON AUTHORISED PRAWN VESSELS FOR 1970 TO 1979

Characteristic	Zone	Restriction
Vessel length	Sp.	1970 onwards. Maximum of 17 metres for new entrant and replacement vessels.
	Sp.	1974–1979. No greater than that of vessel being replaced.
	St.V.	1970 onwards. Maximum of 14 metres.
	St.V.	1970–1979. No greater than that of vessel being replaced.
Rig	Sp.	1971–1974. No changing from single to double rig.
	St.V.	1970–1974, 1976 onwards. No use of double rig.
Net headline	Sp. and St. V.	1970 onwards. Maximum headline (in metres) = $17 + \cdot 882 \times$ vessel length (in metres).
Engine power	Sp.	1979 onwards. Maximum of 300 cBHP for all new entrant and replacement vessels and replacement engines for present vessels of length 17 metres or less.
	St.V.	1979 onwards. Maximum of 250 cBMP for all vessels.

Source: South Australia, Department of Fisheries.
From Byrne

input of capital and fishing power. It appears that even in limited entry regimes there is a tendency to fish prawn resources to their biological maximum and beyond. However, industry is receptive to measures designed to increase the value of the catch such as those designed to prevent the capture of small prawns or designed to protect the viability of stocks such as the closure of nursery areas. They are also receptive to regimes designed to prevent social conflict such as those which provide for rotational access to favoured estuarine fishing sites.

Given that industry and administration have a common desire to protect the stocks, further control over participation in a fishery is compromise between fishery management theory and what is socially and politically acceptable.

Management techniques

Walker (1975) and Ruello (1975) have reported on management techniques used prior to the early 1970s. Limitation of entry to the fishery has been favoured as basic management technique by Australian prawn fishery managers. Almost the only significant fishery left where entry is not limited is the New South Wales prawn fishery. Limited entry has generally been imposed to control fishing effort or capital inflow or both. When introduced early in the fishery's history it has generally slowed down, but not prevented, excessive increases in fishery power and capital investment. When introduced later in a fishery's history it can lead to rapid increases in fishing effort as fishermen take up their entry options and, in the absence of a policy to the contrary, attempts by processing companies to gain a share of the catch by the direct ownership of access entitlements.

Close seasons and areas are used to protect nursery areas and control size of first capture. They are also, in South Australia, used to control effort (*Table 6*). Gear restrictions on total headrope length and the number of nets which may be towed are also used to control effort. In some fisheries boat length and engine power have also been restricted to control fishing power (*Table 7*) (Byrne, 1982).

Where there are significant amateur fisheries the operations of commercial fishermen may be restricted on an area or time basis, *eg* Moreton Bay is closed to prawn trawling during weekends to avoid conflict with amateur anglers.

Preparation and evaluation of management advice

In Australia fisheries management is the responsibility of government. Within their territorial waters the States are solely responsible for management. Although the Federal Government is responsible for management beyond territorial limits it operates in consultation and with the agreement of the States. A series of regional committees has been established through which this consultation takes place. The procedures for developing, evaluating and implementing advice are similar at both levels.

When a management action is contemplated by Government or requested by industry it is usual to establish an expert group, which generally contains both fishery scientists and economists. This group is tasked to review the state of the fishery and the extent of the problem requiring management action and asked to suggest possible management actions. It is usual for the group to recommend requirements for further research should the existing data base be inadequate.

The recommendations of the group and the accompanying argument are placed before the managers. After the management group reviews the recommendations industry is consulted on possible courses of action either by means of standing consultative groups or, if the action contemplated is particularly important, at meetings in the fishing ports. The management group then takes into account the views of industry and the expert group and develops a unified proposal. Industry views may be sought once more on this new proposal if it has far reaching implications but more usually the proposal is passed to Government for approval or rejection by the Minister or Ministers responsible for fisheries.

Management actions the implications of which are not clear cut or are difficult to predict are often introduced on an interim basis for a defined period and data is collected during that period to allow evaluation of the effects of management.

The status of fisheries is formally reviewed at intervals by expert and management groups and recommendations for future research and management forwarded to Government.

Future work

The collection of catch and effort and biological data from fisheries on the east coast is a pressing problem. Studies of biology and population dynamics of the components of the mixed species fishery off northern and north eastern Australia are required. The initial need is for an assessment of state of stocks but future management is likely to require the identification of nursery areas and the elucidation of recruitment processes together with a knowledge of stock identity and the distribution and habitat of the fishable populations.

A better understanding of the economic relationship between the primary and secondary sectors of the industry appears to be necessary.

Although it appears likely that most prawn fisheries on the Australian continental shelf are now being fished, exploration for stocks in remote areas and in deeper water may lead to the discovery of additional resources eg an active deepwater exploration programme of NSW has led to the establishment at a fishery for royal red prawns off central N.S.W. in recent years.

References

BOWEN, B K and HANCOCK, D N. The limited entry prawn
1984   fisheries of Western Australia: research and management. This volume.

BYRNE, J L. The South Australian prawn fishery: a case study in
1982   license limitation. Policy and Practice in Fisheries Management. (AGPS: Canberra).

DALL, W (unpublished). Management oriented research: Penaeid prawns. Managing Queensland Fisheries Conference: Brisbane, August, 1981.

GLAISTER, J P. The impact of river discharge on distribution and
1978a   production of the school prawn, *Metapenaeus macleayi* (Haswell) (Crustacea: Penaeidae) in the Clarence River region, northern New South Wales. *Aust. J. Mar. Freshwater Res.*, 29, 311–23.

GLAISTER, J P C. Movement and growth of tagged school
1978b   prawn, *Metapenaeus macleayi* (Haswell) (Crustacea: Penaeidae), in the Clarence River region of northern New South Wales. *Aust. J. Mar. Freshwater Res.*, 29, 645–57.

KIRKWOOD, G P. Modelling of the Gulf of Carpentaria prawn
1984   fisheries. This volume.

LUCAS, C. Preliminary estimates of stocks of the king prawn,
1974   *Penaeus plebejus*, in south-east Queensland. *Aust. J. Mar. Freshwater Res.*, 25, 35–47.

LUCAS, C, KIRKWOOD, G and SOMERS, I. An assessment of the
1979   stocks of the banana prawn, *Penaeus merguiensis*, in the Gulf of Carpentaria. *Aust. J. Mar. Freshwater Res.*, 30, 639–652.

MONTGOMERY, S S. Tagging studies on juvenile eastern king
1981   prawns reveal record migration. *Aust. Fish.*, 40(9), 13–14.

MORIARTY, D J W. Quantification of carbon, nitrogen and
1977   bacterial biomass in the food of some penaeid prawns. *Aust. J. Mar. Freshwater Res.*, 28, 113–118.

MORIARTY, D J W and BARCLAY, M C. Carbon and nitrogen
1981   content in food and assimilation efficiencies of penaeid prawns in the Gulf of Carpentaria. *Aust. J. Mar. Freshwater Res.*, 32, 245–251.

MUNRO, I S R. Biology of the banana prawn, *Penaeus merguier-
1975   sis* in the south-east corner of the Gulf of Carpentaria. First Australian National Prawn Seminar (AGPS: Canberra).

PENN, J W. The influence of tidal cycles on the distributional
1975   pathway of *Penaeus latisulcatus* Kishinouye in Shark Bay, Western Australia. *Aust. J. Mar. Freshwater Res.*, 26, 93–102.

PENN, J W. Tagging experiments with the western king prawn,
1976   *Penaeus latisulcatus* Kishinouye II. Estimation of population parameters. *Aust. J. Mar. Freshwater Res.*, 27, 230–250.

PENN, J W. The behaviour and catchability of some commer-
1984   cially exploited penaeids and their relationship to stock and recruitment. This volume.

PENN, J W. Spawning and fecundity of the western king prawn
1980   *Penaeus latisulcatus* Kishinouye in Western Australian waters. *Aust. J. Mar. Freshwater Res.*, 31, 21–35.

PENN, J W and STALKER, R W. The Shark Bay prawn fishery
1979   (1970–1976). *West. Aust. Dep. Fish. Wildl. Rep.*, 38, 1–38.

RACEK, A A. Prawn investigations in eastern Australia. *Res.
1959   Bull. State Fish. NSW*, 6, 1–57.

RUELLO, N V. Prawn tagging experiments in New South Wales.
1970   *Proc. Linn. Soc. NSW*, 94(3), 277–87.

RUELLO, N V. The influence of rainfall on the distribution and
1973   abundance of the school prawn *Metapenaeus macleayi* in the Hunter River region (Australia). *Mar. Biol.*, 23, 221–8.

RUELLO, N V. Geographical distribution, growth and breeding
1975   migration of the eastern Australian king prawn *Penaeus plebejus* Hess. *Aust. J. Mar. Freshwater Res.*, 26, 343–54.

RUELLO, N V. An historical review and annotated bibliography
1975   of prawns and the prawning industry in Australia. First Australian National Prawn Seminar (AGPS: Canberra).

ROTHLISBERG, P C, JACKSON, C and PENDRY, R C. Larvae of
1978   Gulf and Carpentaria Prawns reared at sea. *Aust. Fish.*, 37, 8–12.

STAPLES, D J. Seasonal migration patterns and juvenile banana
1979   prawns, *Penaeus merguiensis* de Mau, in the major rivers of the Gulf of Carpentaria, Australia. *Aust. J. Mar. Freshwater Res.*, 30, 143–157.

STAPLES, D J. Ecology of juvenile and adolescent banana
1980a   prawns, *Penaeus merguiensis*, in a mangrove estuary and adjacent offshore area of the Gulf of Carpentaria. I. Immigration and settlement of postlarvae. *Aust. J. Mar. Freshwater Res.*, 31, 635–652.

STAPLES, D J. Ecology of juvenile and adolescent banana
1980b   prawns, *Penaeus merguiensis*, in a mangrove estuary and adjacent offshore area of the Gulf of Carpentaria. II. Emigration, population structure and growth of juveniles. *Aust. J. Mar. Freshwater Res.*, 31, 653–665.

STAPLES, D J and VANCE, D J. Effects of changes in catchability
1979   on sampling of juvenile and adolescent banana prawns, *Penaeus merguiensis* de man. *Aust. J. Mar. Freshwater Res.*, 30, 511–519.

YOUNG, P C. Moreton Bay, Queensland: a nursery area for
1978   juvenile penaeid prawns. *Aust. J. Mar. Freshwater Res.*, 29, 55–75.

YOUNG, P C and CARPENTER, S M. Recruitment of post larval
1977   penaeid prawns to nursery areas in Moreton Bay, Queensland. *Aust. J. Mar. Freshwater Res.*, 28, 745–73.

WALKER, R H. Australian prawn fisheries. First Australia
1975   National Prawn Seminar. (AGPS: Canberra).

WHITE, T F. Factors affecting the catchability of a penaeid
1975   shrimp, *Penaeus esculentus*. First Australian National Prawn Seminar. (AGPS: Canberra).

The prawn (*Penaeus orientalis* Kishinouye) in Pohai Sea and their fishery

Ye Chang Cheng

Abstract

The fleshy prawn (*Penaeus orientalis* or *P. chinensis**) supports a substantial fishery in the Pohai Sea. It makes extensive migrations of some 1 000 km from its nursery grounds in the inner Pohai Sea to wintering grounds in the deeper waters of the Yellow Sea, returning to spawn in the Pohai Sea in April – May. This gives rise to three distinct fishery seasons; in autumn and spring in the Pohai Sea, and in winter – spring in the Yellow Sea. Most catches are now taken in the autumn, using set nets from small vessels, and trawls, mostly by larger vessels. Catches have fluctuated between 10 000 and 90 000 tons, mainly due to variation in recruitment. There is some indication that the strength of recruitment is related to the abundance of the parent stock.

Currently management measures are in force to protect the spawning stock, including the prawns migrating from the wintering grounds. A bio-economic model, which includes the effects of growth and natural mortality of the prawns, and the costs of fishing is used to determine the optimum fishing strategy.

1 Introduction

The prawn in Pohai Sea (*Penaeus orientalis* Kishinouye) is a long distance migration king prawn with high economic value, being the main object of fishing in three provinces and one municipality (*ie* Liaoning, Hebei, and Shandong Provinces and Tianjin Municipality) in north China. The bumper or poor harvest of prawn resource exercises direct influence upon the fishery production, the livelihood of fisherman, and the foreign exchange income of this country.

According to the recapture record of tagging experiments by Mako (1966), Kim (1973), and the Research Institute of Marine Fisheries of the Yellow Sea, it is believed that there are two different geographical populations of prawns present; a small Korean population in the west coast of Korea, and a larger Chinese population in the Yellow Sea and Pohai Sea. The spawning ground of the latter population is widely distributed from

---

*The Latin name commonly used is *P. orientalis* but in his review of the shrimps and prawns of the world, Holthuis (FAO Species Synopsis 125, Vol. 1, 1980) points out that the correct name should be *P. chinensis*.

the estuary of Yalu River (in Jilin and Liaoning Province, northeast China) to the coast of Haizhou Bay (in Jiangsu Province, east China), and the main spawning ground is in the vicinity of the estuaries along the coast of Pohai Sea.

We are only interested in the Chinese population, especially its main stock coming into the Pohai Sea.

At present the main problems in the fishery are heavy fishing effort, poor economic benefit, and extravagant power consumption, which will eventually bring about the fluctuation of the population.

Distribution and Migration

The prawn is an annual animal, and during its life span it migrates along a route about 1 000 km long (*Figure 1*).

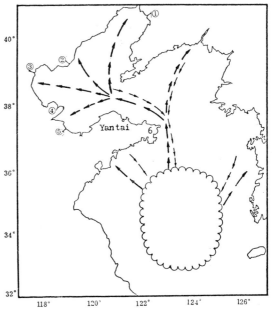

Fig 1 Sketch map of distribution and migration of the prawn. 1 Liao Ho – entering the Liaodong Bay; 2 Luan Ho; 3 Hai Ho – entering the Pohai Bay; 4 Yellow River and Xiaoqing Ho – entering the Laizhou Bay; 6 Chengshantou

From this figure we can see that the wintering ground of prawns covers a large area of sea from 33°00′ to 36°00′ North latitude. In the second half of March every year, it begins to move northwards in groups with the northward movement of the isotherm of 6°C. During the migration, two subgroups separate from the main stock, one moves towards the Haizhou Bay and Jiaozhou Bay at the south coast of Shandong Peninsula, and the other towards the Haiyangdao fishing ground and the sea area near the estuary of Yalu River in Liaodong Peninsula for spawning. The main stock rounds Chengshantou (in Shandong Province) and comes into the Pohai Sea through Pohai Straits along the isotherm of 5°C and in the last ten-day period of April it arrives at the vicinity of the estuaries and river-mouths for spawning. Deng (1980) has reported the amount and distribution of eggs and young prawns in the spawning ground, as well as the environmental conditions such as the water temperature, salinity, and the bottom condition. When the water temperature rises over 13°C, on the 2nd May at the earliest and May 18th at the latest, it begins spawning. The salinity of the spawning ground ranges from 23·0 to 30·0‰. The fecundity is 0·5 – 1 million, and most brood prawns die after spawning. In the last ten-day period of June, the young prawns gather in the coastal shallows looking for food and growing there. By the last ten-day period of August, the body length of young prawns runs up to 60 – 100 mm on an average, and they start to move to deeper water. In September, the body length of prawns runs up to 150 mm or so on an average, and they move in batches from the coast to deep water area over 20 m for food, forming the productive autumn fishing ground. In the second half of October, after the last moult, the female prawns are 170 – 180 mm in body length on an average, and begin to mate. The male prawns are about 150 mm long on an average, and a part of male prawns die after mating. In the second half of November, when the cold wave comes, the water temperature of the bottom in Pohai Sea drops to 12° – 13°C, the prawns move in groups out of Pohai Sea Straits, the females leading the way and the males following. By the end of November, they pass Chengshantou, and, in the second half of January of the next year, reach the wintering grounds where they disperse. In the second half of March, the air and water temperature are going up, and the prawns with sexual glands developed begin to migrate northwards for propagation.

2 General situation of the prawn fishery

2.1 Fishing seasons
There are the autumn fishing season, winter – spring fishing season and spring fishing season in a year for prawning operation. The Japanese fishery is carried out in winter – spring. In China the oper-

ation is carried out mainly in autumn and spring. Before 1963, it was mainly carried out in spring; from 1962 on, it is mainly carried out in autumn, during which more than 90 percent of the annual yield is taken. Due to the adoption of the policy of protecting brood prawns, the spring yield will go on decreasing.

2.2 Fishing boats and fishing gears
There are two kinds of fishing nets. One is the set net, including fyke net, hanging net, and drift net, etc. The fishing boats using this kind of nets are rather small, being small-sized wooden boats of less than 40 hp. They are suitable for use in shallow water operations. The current fisheries management regulation allow the use of these nets from 5th September. The other is the pair trawl, including the motorized fishing boat pair trawl and the power/sail ship pair trawl. The power/sail ship is rather small in motive power, being 135, 100 or 80 hp; they can start fishing on 15th September. The motorized fishing boat is a large-sized fishing vessel with more than 200 hp, and they can start fishing on 5th October.

2.3 Fishing effort
The prawn fishery is a large-scale complex fishery. Therefore, it is quite difficult to accurately estimate the total fishing effort. We take the pair trawl of the power/sail ship as a standard unit of fishing effort. The average yield per pair of power/sail ship in the autumn is used as yield per effort. Dividing the total catch in the entire autumn by this value, we obtain the standard fishing effort in the autumn fishing season. The data are shown in *Figure 2*. From this diagram it is seen that from 1973 on the fishing effort has been increasing rapidly at an average rate of 17 percent a year.

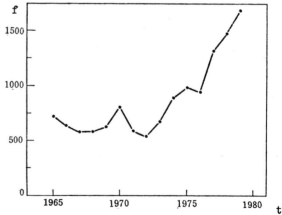

Fig 2 Diagram of prawn fishing effort in autumn in Pohai Sea, 1965–79

50

2.4 Variations in catch

The data for the yield of year class in the 14 years from 1965 to 1978 are shown in *Figure 3*. It indicates the trend of variation in catch in the entire prawn fishery. In addition, we use the yield and yield per effort in autumn in Pohai Sea to indicate the dynamics of prawns stock in Pohai Sea. In *Figure 4* there are two curves, one is the curve of yield (*y*) in autumn in Pohai Sea, and the other is the curve of yield per effort (*y*/*f*) in autumn in Pohai Sea. By and large, these two curves coincide in tendency.

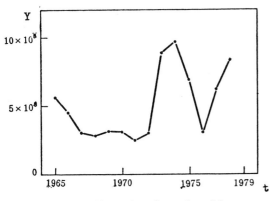

Fig 3 The yield in number of year-class of the prawn

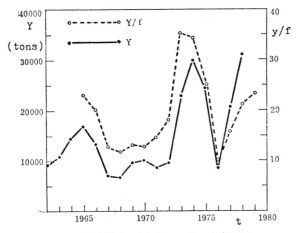

Fig 4 Curves of yield (*y*) and yield per effort (*y*/*f*) of the prawn in autumn in Pohai Sea, 1962–78

Figure 5 shows the estimates of the relative abundance of young prawns in Pohai Sea in the first ten-day period of August in the 15 years from 1965–79. (Unit: number per hour. See Section 4 'Forecast'.) These three kinds of data mentioned above illustrate the fluctuation in amount of the prawn in Pohai Sea. During the period of observation, the 1968 generation was the least abundant while the 1979 generation was the most abundant,

the latter being approximately 5·7 times the former. The dynamics of prawn stock is related mainly to the amount of parental prawns and the environmental conditions of spawning grounds. Based on primary investigations, we see that the amount of prawns is in direct correlation with the rainfall in Pohai Sea and the runoff of the Huanghe River (Yellow River); and is in inverse correlation with the salinity of the sea water.

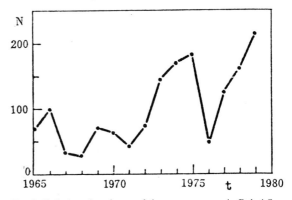

Fig 5 Relative abundance of the young prawn in Pohai Sea, 1964–79

3 Growth and mortality

3.1 Growth

The male and female prawn individuals are conspicuously different in size. When the sexually mature prawns begin to pair, the female may be about 30 mm larger than the male in body length.

Jingyao (1981) studied the data obtained in the five years from 1959 to 1964 and the average relations between the body weight and body length to be:

$$W\male = 11 \cdot 3 \times 10^{-6} \, l^{2 \cdot 9987}$$

$$W\female = 11 \cdot 0 \times 10^{-6} \, l^{3 \cdot 0014}$$

The growth of prawns may be described by von Bertalanffy's equation. Assuming $t = 0$ on the 25th May, according to the data of the growth of prawns each five days, Jingyao obtained the value of the parameters as follows:

female: $W_\infty = 91 \cdot 8$ g; $L_\infty = 201 \cdot 3$ mm; $K = 0 \cdot 018$; $t_o = 25$ days

male: $W_\infty = 49 \cdot 1$ g; $L_\infty = 163 \cdot 5$ mm; $K = 0 \cdot 0168$; $t_o = 29$ days.

3.2 Mortality

(*i*) The mating mortality of males

After mating, the males die in large numbers.

51

Certainly, there is some evidence that a part of males die immediately after mating. Firstly, the abundance of males decreases in the catch after the mating season, the sex ratio being about 75% females by the end of November. Secondly, study of the statistics of the market commercial landings and the results of survey in the following year, show that the sex ratio in the spring fishing season is similar to that at the end of November, because the fisherman and market officials can separate the males from the females according to the external features and colours. Therefore, the record in reliable. Finally, the observation results by Xu (1980) confirmed that a large part of males die after mating season. This experiment is performed in a pond $(2.7 \times 1.6 \times 1.2$ m) in the laboratory. Changcheng (1981a) derived a formula to roughly estimate the mating mortality of males, by assuming that males and females are equal in number before mating, the males and females are equally likely to be caught, and except for the mortality of males due to mating, the natural mortalities of the male and female individuals are equal. The formula is as follows:

$$\sum M_m = \ln(P_F/P_m) \qquad (1)$$

where P_m and P_F are the proportions of males and females in the catches and M_m is the additional mortality each month due to mating. The data of sex ratio in 1973-75 and the estimated values of M_m are listed in *Table 1*.

(ii) Estimate of total mortality (Z)

Under common assumptions the logarithm of the catch in number of a generation in successive time intervals is the linear function of time t. By this method we have estimated the total mortality of the prawn in Pohai Sea. *Table 2* lists the data of the catches in number in autumn fishing season in Pohai Sea over the ten years from 1970–79, the interval of times of sampling being one month.

Based on these data, by using the method of linear regression, we have estimated the value of Z

Table 2
THE CATCHES IN NUMBER OF THE PRAWN IN POHAI SEA, 1970–79

| | (Number $\times 10^6$) | | | (Number $\times 10^6$) | | |
Month	Female	Male	Total	Female	Male	Total
	1970			1971		
September	42·49	37·37	79·86	38·01	34·11	72·12
October	49·44	22·74	72·18	38·44	27·84	66·28
November	20·12	12·71	32·83	10·41	7·27	17·68
December	7·36	1·20	8·56	3·41	2·31	5·72
	1972			1973		
September	60·52	54·32	114·84	124·68	115·09	239·77
October	41·61	21·06	62·67	81·24	47·71	128·95
November	24·42	8·94	33·36	52·08	21·27	73·35
December	6·98	2·56	9·54	23·66	8·75	32·41
	1974			1975		
September	198·35	186·79	385·14	156·89	153·78	310·67
October	174·91	95·44	270·35	86·78	57·13	143·91
November	59·56	22·03	81·59	20·06	10·19	30·25
December	20·18	6·37	26·55	8·86	3·11	11·97
	1976			1977		
September	61·06	54·15	115·21	109·73	97·31	207·04
October	41·02	24·09	65·11	117·82	69·20	187·02
November	14·55	7·80	22·35	32·97	17·68	50·65
December	4·82	2·05	6·87	12·32	5·23	17·55
	1978			1979		
September	181·10	160·59	341·69	265·83	235·73	501·56
October	130·56	76·67	207·23	165·79	97·37	263·16
November	58·03	31·11	89·13	58·20	31·20	89·40
December	39·86	16·92	56·78	30·29	12·86	43·15

(total mortality per month) of either sex and listed them in *Table 3*.

Over the ten years, the average mortality per month of the female prawn was 0·746, and accordingly the mortality per ten-day period was roughly 0·25.

In addition, with Y/f per ten day period as abundance index, the total mortality rate per ten day period was estimated and the results are 0·214 for female and 0·340 for male in average of 16 years from 1963 to 1979 among which the data of 1973 is absent. With Cohort Analysis we obtained the abundance of male prawns. By early October there

Table 1
SEX RATIO OF THE PRAWN IN POHAI SEA AND INSTANTANEOUS NATURAL MORTALITY OF THE MALE DUE TO MATING 1973–75

| | 1973 | | 1974 | | 1975 | | Average | | | |
Month	Female	Male	Female	Male	Female	Male	Female	Male	$\sum M_m$	M_m
September	52·0	48·0	51·0	49·0	50·3	49·7	51·1	48·9	0·044	0·044
October	62·7	37·3	64·7	35·3	60·3	39·7	62·6	37·4	0·515	0·471
November	71·0	29·0	73·0	27·0	66·3	33·7	70·1	29·9	0·852	0·337
December	73·0	27·0	76·0	24·0	74·0	26·0	74·3	25·7	1·062	0·210

is no mating for Pohai Sea prawn, the abundance of male is equal to that of female, thus the mortality of female may be estimated, being about 0·31. As the total mortality estimated from a variety of sources is similar, we confirm that the results are available.

Table 3
THE INSTANTANEOUS TOTAL MORTALITY OF THE PRAWN IN POHAI SEA, 1970–79

	Female	Male	Total	Female	Male	Total
		1970			1971	
Z	0·616	1·090	0·749	0·863	0·942	0·898
r	0·913	0·924	0·937	0·945	0·967	0·956
		1972			1973	
Z	0·701	1·002	0·908	0·542	0·853	0·656
r	0·960	0·967	0·982	0·988	0·999	0·997
		1974			1975	
Z	0·793	1·160	0·922	1·009	1·342	1·133
r	0·959	0·990	0·978	0·988	0·995	0·992
		1976			1977	
Z	0·865	1·095	0·953	0·783	1·013	0·871
r	0·982	0·994	0·989	0·939	0·976	0·957
		1978			1979	
Z	0·535	0·765	0·623	0·756	0·986	0·844
r	0·985	0·998	0·993	0·991	0·999	0·995

$n = 4$, r = correlation coefficient

We can draw the *Figure 6* to find the relationship between the total mortality of the female and the fishing effort (f).

From this figure it is seen that when f is within the range from 500 to 1 000, the values of Z tend to vary with f, and when f is greater than 1 000, the increase of f exercises little influence on the value of Z.

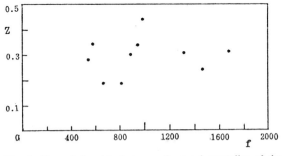

Fig 6 The relationship between the total mortality of the female prawn (Z) and the fishing effort (f)

According to this analysis, when the effort is less than 1 000, the natural mortality (M) may be estimated from the relation $Z = qf + M$. The correla-

tion coefficient is about 0·619. A natural mortality of female prawns in Pohai Sea of 0·043 per ten day period, corresponds to survival rate about 39% from 11 September to the following year 21 April in the absence of fishing. Before 1962 the Chinese boats were operated in spring fishing season. According to the production at that time we infer that the figure of 0·043 may include an effect of the spring fishery, and the natural mortality may be smaller.

4 Forecasts

Before 1962, we predicted the catch of the prawn in Pohai Sea in the coming spring fishing season based on the records of catches by powered fishing boat trawls in wintering ground, and got fine results. In 1962, when the operation was switched over from fishing in spring to fishing in autumn, we tried to predict the catch in the coming autumn fishing season based on the runoff of the Yellow River and the rainfall in the Pohai Sea coast, but the results were not so satisfactory. From 1962 on, in the first ten-day period of August every year, we established stations in the young prawn distribution areas in the three bays (i.e., Liaodong Bay, Laizhou Bay and Pohai Bay) of the Pohai Sea respectively, and used 60 hp vessels with trawl nets to carry out trial fishing in each station for half an hour, and calculated the relative abundance in each bay with number per hour as unit. During the period of 1963–73, the forecast of autumn catch was based on the relative abundance in each bay. We also took other factors into consideration, including runoff, strong wind, and brood stock abundance, etc., and interviewed fishermen to listen to their opinions. Summing up these factors and comparing with the situations in the past years, we predicted the catch in the coming autumn fishing season. In the years when the fluctuation in amount was rather great, the forecasting of catch stood in comparatively great error.

In 1973, Chuanzhen, Jungi, and Weixi (1981) found the weights of relative abundance in each bay; they are: Liaodong Bay, 0·175; Laizhou Bay, 0·400; and Pohai Bay, 0·425. Based on these data, we are able to calculate the relative abundance of the young prawn in Pohai Sea. These data and the catch in Pohai Sea in the autumn fishing season from 1965–79 are given in *Table 4*.

With this table, we can forecast the coming autumn yield by linear regression. That is

$$\hat{y} = a + bx.$$

The values of the parameters a and b, the correla-

Table 4		
THE AUTUMN YIELD AND THE RELATIVE ABUNDANCE IN THE POHAI SEA, 1965–79		
Year	Autumn yield (y)	Relative abundance (x)
	t	
1965	13 931	70·1
1966	13 342	98·3
1967	7 014	34·1
1968	6 889	29·7
1969	9 756	70·0
1970	10 337	64·6
1971	8 647	42·5
1972	9 505	74·5
1973	23 104	145·3
1974	30 573	170·1
1975	24 776	181·1
1976	8 525	48·8
1977	20 703	124·1
1978	31 224	161·9
1979		216·6

tion coefficient r, the standard deviation s, and the statistical figures over the years from 1973 to 1979 are listed in *Table 5*. And based on the data of *Table 4* we can draw *Figure 7*.

Table 5					
THE VALUES OF PARAMETERS AND STATISTICAL FIGURES, 1973–79					
Year	b	a	r	s	n
1973	94·7	4236	0·849	1 483	8
1974	132·5	2 164	0·944	1 767	9
1975	160·0	549	0·963	2 186	10
1976	144·7	1 484	0·959	2 408	11
1977	144·7	1 480	0·961	2 286	12
1978	144·7	1 450	0·962	2 206	13
1979	157·6	814	0·957	2 891	14

Notes:
a, b: parameters in the equation $\hat{y} = a + bx$
r: correlation coefficient
s: standard deviation
n: number of years used

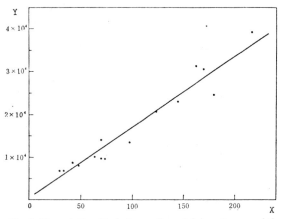

Fig 7 The relationship between the catch in autumn and the relative abundance

The significance of correlation coefficient may be examined by t; and through examination we see that the correlation is quite significant. The forecasting value may be estimated by $\hat{y} \pm 2s$. The fishery administration offices and fishermen are extraordinarily interested in the forecast of the catch of prawns in the coming autumn fishing season. For the purpose of enhancing the accuracy of forecasting, we are now engaged in the research work to further improve the method of forecasting.

5 Stock and recruitment

Changcheng (1980) has reported that it is advisable to use the Beverton's model (1957) (formula(3)) and Ricker's model (1958) (formula (2)) to describe the relationship between the parental abundance and the recruitment, that is

$$R = aAe^{-bA} \qquad (2)$$
$$R = A(aA + b)^{-1} \qquad (3)$$

where A is the brood stock abundance, R is the recruitment, and a and b are two parameters to be determined.

5.1 *Data and results*

The prawn is an annual animal, and the fishing effort is great. Therefore, the annual yield is equal to the generation yield and would approach the recruitment, and may serve as a relative abundance of recruitment. The catch of the spawning prawns in spring may be approximately regarded as a relative abundance of spawning parental prawns. The data are listed in *Table 6*, and illustrated in *Figure 8*. In the figure, the curve A is drawn according to Beverton's model, while the curve B to Ricker's model. *Table 7* lists the values of the two parameters calculated according to the data in *Table 6*, the correlation coefficient r, and the correlation ratio G. We see that the correlation is moderately high. *Table 7* also lists the calculated values of maximum recruitment (R_{max}) and the parental abundance necessary for the maximum recruitment (A_{max}).

5.2 *Determination of appropriate parental abundance*

In Changcheng, Chuanzhen and Peijun's (1980) opinion, Beverton's model would be somewhat better than Ricker's model in describing the relationship between parent stock and recruitment.

From *Table 6*, the average number of parent stocks from 1961 to 1976 was $2·7 \times 10^7$. This is far smaller than that giving maximum recruitment in

Table 6
THE DATA OF PARENTAL ABUNDANCE AND RECRUITMENT, 1961–76

Generation (year)	Parental abundance (in number)	Recruitment (in number)
1961	$2\cdot459 \times 10^7$	$3\cdot165 \times 10^8$
1962	$2\cdot927 \times 10^7$	$3\cdot478 \times 10^8$
1963	$3\cdot513 \times 10^7$	$3\cdot672 \times 10^8$
1964	$2\cdot994 \times 10^7$	$5\cdot365 \times 10^8$
1965	$4\cdot581 \times 10^7$	$5\cdot249 \times 10^8$
1966	$3\cdot838 \times 10^7$	$4\cdot205 \times 10^8$
1967	$2\cdot740 \times 10^7$	$2\cdot988 \times 10^8$
1968	$1\cdot085 \times 10^7$	$2\cdot643 \times 10^8$
1969	$1\cdot643 \times 10^7$	$3\cdot020 \times 10^8$
1970	$1\cdot302 \times 10^7$	$2\cdot981 \times 10^8$
1971	$1\cdot230 \times 10^7$	$2\cdot373 \times 10^8$
1972	$0\cdot774 \times 10^7$	$2\cdot706 \times 10^8$
1973	$2\cdot385 \times 10^7$	$8\cdot233 \times 10^8$
1974	$6\cdot649 \times 10^7$	$9\cdot351 \times 10^8$
1975	$4\cdot084 \times 10^7$	$6\cdot258 \times 10^8$
1976	$1\cdot723 \times 10^7$	$2\cdot363 \times 10^8$

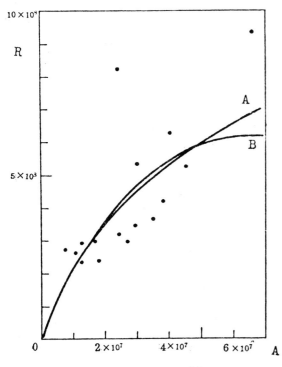

Fig. 8. Reproduction curves of the prawn

Table 7. To increase recruitment, or to make it fluctuate around a higher level, it is necessary to increase the spawning stock to more than 6×10^7. If the amount of parental prawns can be kept at this level, under general environmental conditions, the generation yield may fluctuate around 60×10^7 in number, or 30 000 t in weight. Theoretically, if the various conditions can be satisfied, the generation yield will possibly run up to 120×10^7 in number, or 60 000 t.

5.3 *Fishery management measures for conserving brood prawns*

It is necessary to forbid catching brood prawns in spawning grounds or on the way of reproductive migration from wintering grounds during the period from the beginning of March to the end of June. Without this measure, the parent abundance may decrease, resulting in the decrease of the amount of prawns. At present, the People's Republic of China has unilaterally forbidden catching spawning brood prawns in the Pohai Sea area north of Chengshantou of Shandong Province.

6 The yield per recruit (Y/R)

The purpose of calculating the yield per unit recruitment is to decide on the optimum conditions. We may calculate the prawn yield per unit recruitment (Y/R) by the usual equation

$$Y = \sum_{i=1}^{n} \frac{N_i(1 - e^{-Z_i})}{Z_i} F_i W_i \qquad (4)$$

where

$$N_i = N_{i-1} e^{-Z_{i-1}}. \qquad (5)$$

When $i = 1$, $Z_{i-1} = 0$, then $N_1 = R$, the number recruiting at the beginning of fishing in the autumn. W_i is the average individual weight in the period i. Substituting formula (5) into formula (4), we have

$$Y/R = \sum_{i=1}^{n} \frac{(1 - e^{-Z_i})}{-Z_i} F_i W_i e \left(-\sum_{j=1}^{i} Z_{j-1} \right). \qquad (6)$$

Formula (6) gives the yield per unit recruit in weight. There are two controllable factors in the

Table 7
THE VALUES OF CORRELATION COEFFICIENT (r), CORRELATION RATIO (G), PARAMETERS (a, b), MAXIMUM RECRUITMENT (R_{max}), AND PARENTAL ABUNDANCE NECESSARY FOR THE MAXIMUM RECRUITMENT (A_{max})

Model	r	G	a	b	A_{max}	R_{max}
Ricker	0·57	0·67	24·43	$1\cdot450 \times 10^{-8}$	$6\cdot7 \times 10^7$	62×10^7
Beverton	0·57	0·69	$8\cdot205 \times 10^{-10}$	$4\cdot256 \times 10^{-2}$	no solution	122×10^7

prawn fishery as well as in other fisheries, one is the fishing effort, and the other is the date to allow to catch. In formula (6), the date to allow to catch is represented by the time i, which is arbitrarily selectable. We take a ten-day period as the unit of time in prawn fishery. (In China, a month is divided into three ten-day periods: first, second, and last.) To illustrate, if the fishing operation begins on the 11th September, $i = 1$. Towards the end of November or the beginning of December, most of the power/sail ships have been transferred to other fishing grounds to catch finfish, while some of the motorized fishing boats continue to catch prawns until the first ten-day period of December. Therefore, if the operation begins on the 11th September, $n = 9$; and if it begins on the 21st September, $n = 8$. Due to the difference in growth characteristics between the male and female individuals, we should calculate $(Y/R)m$ and $(Y/R)f$ respectively, then add up to Y/R. The growth data and the values of natural mortality of male prawns are listed in *Table 8*.

Table 8

THE AVERAGE BODY WEIGHT OF MALES AND FEMALES AND THE NATURAL MORTALITY OF MALE DUE TO MATING, 1973–75

Time		Average weight		Natural mortality of male as to pair Mm
		Female	Male	
September	11th	44·7	31·2	0·011
	21st	49·0	31·6	0·011
October	1st	56·3	33·0	0·157
	11th	58·8	33·1	0·157
	21st	61·1	33·2	0·157
November	1st	63·8	33·3	0·112
	11th	68·4	33·5	0·112
	21st	71·8	33·5	0·112
December	1st	72·3	33·5	0·105

In this table, the values of natural mortality are obtained by converting the mortality per month (see *Table 1*) into the average mortality per ten-day period. (In September and December, it is converted into the average mortality per twenty days). This may lead to some errors. On calculating Y/R, we selected three different opening dates, the 11th September, the 21st September, and the 1st October. The results of calculation are listed in *Table 9*.

These data illustrate the effect of fishing mortality and opening date on the yield per unit recruit. We should now follow the old routine to determine the maximum yield per unit recruitment $(Y/R)_{max}$ and its corresponding optimum control variables. As mentioned in Section 3, the average value of mortality per ten-day period over the ten years (1970–79) equals 0.25, and when f is greater than 1 000, the increase of fishing effort exercises little influence on mortality. This fact shows that there is a certain limit to the fishing mortality with regard to the prawn fishery in the autumn fishing season, being about 0·30 per ten-day period, and it cannot increase limitlessly. This limit condition must be taken into consideration. The values of Y/R of $F > 0·30$ in *Table 9* can by no means be taken as feasible plans for consideration. Therefore, the optimum solution is

the opening date of fishing to allow to catch on the 21st September and $F_{\text{ten-day}} = 0·30$

The natural mortality with exception of male mating mortality was not included in above computation. If the natural mortality (0·043) was taken account of, the optimum opening date of fishing would be changed to 11th September.

7 Bioeconomic model

We are very much interested in the goals of fishery management of best economic result, least power consumption, and most numerous employment opportunities. In this respect, bioeconomic models can provide valuable reference data.

Table 9

THE EFFECT OF FISHING MORTALITY AND OPENING DATE OF THE FISHING SEASON ON YIELD PER UNIT RECRUIT (Y/R)

Fishing mortality F	Opening dates								
	September 11th			September 21st			October 1st		
	Male	Female	Total	Male	Female	Total	Male	Female	Total
0·10	14·26	34·64	48·90	12·79	33·58	46·37	11·09	31·96	43·05
0·20	21·30	46·89	68·19	19·62	47·37	66·99	17·42	47·01	64·43
0·30	25·00	50·57	75·57	23·83	52·61	76·44	21·19	53·88	75·07
0·35	26·18	51·05	77·23	24·74	53·72	78·46	22·50	55·70	78·20
0·40	27·07	51·11	78·19	25·76	54·27	80·03	23·56	56·86	80·42
0·50	28·30	50·62	78·92	27·22	54·46	81·68	25·13	58·01	83·14
0·75	29·67	48·78	78·45	29·16	53·35	82·51	27·38	58·21	85·59
1·00	30·28	47·46	77·74	30·08	52·00	82·28	28·58	57·60	86·18

7.1 Model

Ricker's (1958) model is expressed as follows:

$$Y = \sum_{i=1}^{\lambda} \overline{B}_i F_i \qquad (7)$$

where Y is yield, and $\overline{B}$ is mean biomass. Taking

$$\overline{B}_i = \tfrac{1}{2}(B_{i-1} + B_i),$$

where $i = 1, 2, \ldots, \lambda$, and B is biomass, we have

$$\sum_{i=1}^{\lambda} \overline{B}_i = \sum_{i=1}^{\lambda-1} B_i + \tfrac{1}{2}(B_0 + B_\lambda)$$

But

$$B_i = B_0 e^{(G - M - F)i}$$

where G is growth coefficient $= \ln W_t / W_{t-1}$ it follows that

$$Y = \sum_{i=1}^{\lambda-1} [B_i + \tfrac{1}{2}(B_0 + B_\lambda)] qf \qquad (8)$$

Let v'=unit price of prawn per t, and v = total revenue of enterprise, then $v = v'Y$; let J'= consumption per unit fishing effort, and J =cost of enterprise, then $J = J'f$; and let u = profit of enterprise, we obtain

$$u = v - J = \left(qv' \sum_{i=1}^{\lambda-1} B_i + qv' \tfrac{1}{2}(B_0 + B_\lambda) - J' \right) f \qquad (9)$$

This formula can be applied after estimating parameters based on statistical data of fishery.

In order to find out the maximum profit of fishery, we may find the derivative of formula (9) with respect to f, then let $du/df = 0$, and solve it for $f_{e.op}$ (optimum economic effort).

That is

$$f_{e.op} = \frac{qv' \sum_{i=1}^{\lambda-1} B_i + qv' \tfrac{1}{2}(B_0 + B_\lambda) - J'}{q^2 v'[\sum_{i=1}^{\lambda-1} iB_i + (\lambda/2)B_\lambda]} \qquad (10)$$

Substituting formula (10) into formula (9), we get

$$u_{max} = \left(qv' \sum_{i=1}^{\lambda-1} B_i + qv' \tfrac{1}{2}(B_0 + B_\lambda) - J' \right) f_{e.op} \qquad (11)$$

7.2 Estimation of parameters

(a) The value of M

The M in economic model refers to the natural mortality of both male and female individuals. What we estimate is the additional mortality of males individuals when pairing. We take approximately $M = \tfrac{1}{2}Mm$ (see *Table 8*). It may lead some error due to the difference in amount between the male and female individuals.

(b) q, i and λ

It is very difficult to accurately estimate the catchability, being fluctuation from year to year. Only we can roughly estimate the catchability for Pohai Sea prawn fishery. In the ten years from 1970–79 the mortality of female prawns per ten-day period $F_{ten-day} = 0.25$, and the fishing effort f is around 800 pairs of standard power sail ships. Therefore, the catchability

$$q = 3.125 \times 10^{-4}$$

The i and λ are equivalent to n in formula (6). If the date to allow to catch is the 11th September, then $\lambda = 8$.

(c) The value of G

The growth coefficient G is based on the weighted mean of the male and female sizes. The values of G in the three years from 1973–75 are listed in *Table 10*.

(d) The value of B_0

The value of B_0 refers to the biomass of prawns recruiting at the beginning of fishing season. Based on the statistical data of prawn fishery, we give B_0 in advance as 10 000; 20 000; 30 000 and 40 000 t. The method for estimating B_0 in advance is discussed later in Section 7.4.

(e) Economic parameters

We use the values $v' = 3\,000$ and $J' = 30\,000$.

7.3 Results

(a) The optimum fishing effort in prawn fishery $f_{e.op}$

We have determined the optimum economic effort in the prawn fishery by formula (10) and drawn *Figure 9*.

This figure illustrates the variation of $f_{e.op}$ with the variation of B_0. In recent years, the biomass of prawns was about 20 000 – 30 000 t, and the corresponding $f_{e.op}$ being 680–840 pairs of standard power/sail ships. Energy economization is a matter of interest to everybody. It is comparatively simple to estimate the optimum effort in terms of oil consumption (0_{opt}) directly from $f_{e.op}$ from the relation

$$0_{opt} = af_{e.op}$$

where a is by and large a constant, being the oil consumption per unit fishing effort, which can be estimated with readiness. With regard to the fishing operation in the autumn fishing season in Pohai Sea, $a = 19.2$ units. Therefore, in correspondence with $B_0 = 20\,000 - 30\,000$, $0_{opt} = 13\,056 - 16\,128$ units. The fishing effort now available in this fishery is 1 600 pairs of power sail ships with oil

Table 10
THE VALUES OF GROWTH COEFFICIENT G, 1973–75

Date		i	Weight		Sex ratio (%)		Average weight	G_i
			Female	Male	Female	Male		
September	11th		44·7	31·2	51·0	49·0	38·1	
		1						0·0636
	21st		49·0	31·6	52·0	48·0	40·6	
		2						0·1400
October	1st		56·3	33·0	58·7	41·3	46·7	
		3						0·0542
	11th		58·8	33·1	63·0	37·0	49·3	
		4						0·0456
	21st		61·1	33·2	66·0	34·0	51·6	
		5						0·0492
November	1st		63·8	33·3	68·7	31·3	54·2	
		6						0·0643
	11th		68·4	33·5	69·7	30·3	57·8	
		7						0·0555
	21st		71·8	33·5	72·0	28·0	61·1	
		8						0·0194
December	1st		72·3	33.5	74·3	25·7	62·3	

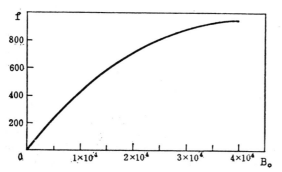

Fig 9 The curve of variation of $f_{e.op}$ with the variation of B_0

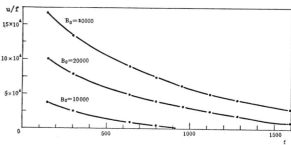

Fig 10 The curve of variation of u/f (profit per vessel) with variation of f

consumption of 30 720 units. If we manage the autumn prawn fishing operation in the light of economic model, we can cut down the power consumption by 17 664 – 14 592 units.

(b) The profit per unit fishing effort (u/f)

Figure 10 is the curve of variation of u/f with the variation of f for different values of B_0. It shows that, for a given B_0, u/f decreases with the increase of f. If we are interested in increasing employment, we may use these data to control the profit per unit fishing effort and determine the number of fishing boats. To illustrate, when $B_0 = 20\,000$, $f_{e.op} = 680$, $u/f = 4·70$, $0_{opt} = 13\,000$ units. If we control $u/f = 3·0$, then the fishing effort increases to 960, or in other words, increases by 280 units, at the expense of increasing power consumption by about 40 percent, and the fishery profit decreases by 7 percent.

(c) Profit isopleth diagram

Figure 11 is the profit isopleth diagram of prawn fishery, which is drawn according to economic model. The dotted line in this figure is the optimum economic fishing effort corresponding to each B_0. This curve intersects with the isoplethic curve at the point u, and this value of u means the u_{max} corresponding to certain B_o. The values of u on both the right and left sides of this curve are all smaller than u_{max}. The zero isoplethic curve implies no profit in fishery, and the area under the zero isoplethic curve gives the loss in fishery. At present, the fishing effort in prawn fishery is 1 600, which is located on the right side of the figure. This isopleth diagram reflects the conditions of input, output and resources in a comprehensive way, and it is used to advantage for considering and studying the economic benefit in fishery.

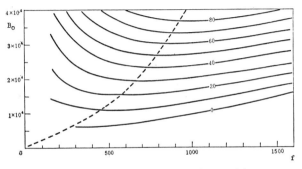

Fig 11 The profit isopleth diagram of prawn fishery

7.4 *Determination of the value of* B_0

B_0 is an undetermined factor. Under general conditions it cannot be estimated. On this account, we are handicapped in applying the economic model based on Ricker's model.

There is a yield forecast before the beginning of the fishing season in Pohai Sea. (See Section 4.) Taking advantage of this estimated value of yield in the forecast, we may roughly estimate the initial abundance B_0.

Since

$$Y = \overline{B}F,$$

and taking

$$\overline{B} = B_0(e^{G-Z} - 1)/(G - Z)$$

we have

$$Y = \frac{B_0(e^{G-Z} - 1)}{G - Z} F$$

And taking the difference in growth between the male and female individuals into consideration, the above formula becomes

$$Y = \frac{0{\cdot}59B_0(e^{G_F - Z_F} - 1)}{G_F - Z_F} F_{\female}$$
$$+ \frac{0{\cdot}41B_0(e^{G_M - Z_M} - 1)}{G_M - Z_M} F_{\male}$$

where 0·59 and 0·41 are the proportions of the female and male prawns in the initial biomass respectively, which are calculated according to the average weight of body on the date to allow to catch (the 21st September) over the three years of 1973–75. The other parameters have been discussed in the sections above. Through calculation we find

$$Y = 0.8B_0$$

in which Y may be substituted by the forecasting value $\hat{Y}$, and B_0 is approximately $\hat{Y}/0{\cdot}80$.

8 Management strategy

This section is devoted to a brief account of management strategy in prawn fishery based on the above-mentioned data and calculating results.

Generally speaking, the prawn fishery management falls into two parts, which are different from each other in character; one is the protection of brood and young prawns, and the other is the fishery management in the autumn fishing season.

8.1 *The protection of brood and young prawns*

The protection of brood and young prawns is an important management strategy to make the dynamics of prawn stock around a high level, as well as an effective measure for increasing the amount of prawns.

(a) The protection of brood prawns

The protection of brood prawns touches upon international fishery relations. At present, China has unilaterally closed the fishery on brood prawns in the spring fishing season in the sea area north of Chengshantou of Shandong Province. But this measure is by no means adequate because in the spring fishing season the remaining spawning prawns which pass Chengshantou are quite small in number. If we do not expand the protection sea area, the amount of prawns will possibly decrease by a big margin due to the insufficiency of brood prawns in number. For this reason, it is imperative to expand the closed fishing zone to 35° north latitude, and the closed fishing period should begin early in March when the adult prawns are migrating for propagation. This is a management measure necessary for protecting spawning parental prawns.

(b) The protection of young prawns

Taking in water by saltworks and unlawful fishing may cause the loss of young prawns in large numbers. The Regulations have made clear stipulations on various nettings in the closed fishing period and closed fishing zone, and forbid catching young prawns, but there is no prohibition against damaging young prawns when saltworks take in water. According to investigations, the young prawns damaged by intake ranges from a few hundred million to more than one thousand million in number, and in some years the number of young prawns damaged exceeds the number of prawns caught. In our opinion, the Regulations should stipulate that saltworks must be prevented from damaging young prawns when taking in water.

8.2 *Fishery management in the autumn season*

The regulations in force stipulate that the opening

59

date for the autumn fishing season is the 15th September, but the amount of fishing boats is not strictly controlled. The present study shows that it is fundamentally reasonable to set the date to allow to catch at the 15th September. The problem is how to adjust the finishing effort, which is related to the goals of management. We have put forward three goals of management and their corresponding fishing efforts. These goals of management are in conflict with each other, and their corresponding optimum efforts also differ greatly from each other. If we intend to obtain the optimum economic result in the fishery, it is necessary to cut down the fishing effort. But if we intend to increase employment opportunities, we have to sacrifice the economic benefit and increase power consumption. We cannot have both at the same time. In essence, the prawn fishery management involves the distribution of yields, economic results and resources, and many politically and socially complicated factors. This article provides some scientific data and optimum conditions for different goals of management as well as their corresponding results. Based on these factors and optimum conditions, the policy-maker may weigh the pros and cons and work out a feasible and satisfactory management plan.

References

BEVERTON, R J H and HOLT, S J. On the dynamics of exploited
1957 fish populations. *Fish. Invest. Minist. Agric. Fish. Food G.B. (2 Sea Fish.)*, 19: 533 p.
CHANGCHENG, YE. The mortality of the Pohai sea prawn,
1981 *Penaeus orientalis. Chinese J. Zool.*, 4: 22–4 (in Chinese)

CHANGCHENG, YE. The management of prawn, *Penaeus orien-*
1981a *talis*. Kishinouye fishery in autumn, Pohai sea. Draft (in Chinese)
CHANGCHENG, YE, CHUANZHEN, LIU and PEIJUN, LI. A study
1980 on the relation between adult stock and recruitment of prawn, *Penaeus orientalis* Kish. from Po Hai. *J. Fish. China*, 4(1): 1–7 (in Chinese, English summary)
CHUANZHEN, LIU, JUNGI, YAN and WEIXI, CIU. A study on the
1981 method of prediction for the autumnal prawn catches in Po Hai. *J. Fish. China*, 5(1): 65–74 (in Chinese, English summary)
JINGYAO, DENG. Distribution of eggs and larvae of penaeid
1980 shrimp *Penaeus orientalis* in Po Hai Bay and its relation to natural environment. *Mar. Fish. Res.*, 1: 17–25 (in Chinese, English summary)
JINGYAO, DENG. Studies on the growth of penaeid shrimp
1981 (*Penaeus orientalis* Kishinouye) in the gulf of Po-Hai. *Mar. Fish. Res.*, 2: 85–93 (in Chinese, English summary)
KIM, BONG-AN. Studies on the distribution and migration of
1973 Korean shrimp, *Penaeus orientalis* Kishinouye, in the Yellow Sea. *Bull. Fish. Res. Dev. Agency, Busan* (11): 7–23 (in Korean, English summary)
MAKO, H, NAKASHIMA, K and TAKAWA, K. Variation of length
1966 composition of Chinese prawn, *Penaeus orientalis* Kishinouye *Bull. Seikai Reg. Fish. Res. Lab.* (34): 1–10 (in Japanese)
RICKER, W E. Handbook of computations for biological statis-
1958 tics of fish populations. *Bull. Fish. Res. Board Can.* (119): 300 p.
RICKER, W E. Computation and interpretation of biological
1975 statistics of fish populations. *Bull. Fish. Res. Board Can.* (191): 382 p.
XU, GAO HONG. The observations on the mating of Chinese
1980 prawn, *Penaeus orientalis. Mar. Sci.* 3: 5–7 (in Chinese)

Acknowledgements

The author wishes to thank Prof. Li Guanguo, of Shandong College of Oceanology, and Mr Zhu De-Shan, of Yellow Sea Fisheries Research Institute, for valuable advice and comments. I am deeply grateful to Mr Deng Jingyon, also of Yellow Sea Fisheries Research Institute, for drafting Section 1 and a part of Section 3.

The Guianas – Brazil shrimp fishery, its problems and management aspects

L Villegas and A Dragovich

Abstract

A large fishery has developed since 1960 off the northeastern coast of South America between the eastern border of Venezuela and a little to the eastward of the mouth of the Amazon. Catches increased rapidly until 1969 and have since fluctuated between 15 and 20 thousand tons (live weight). Most of the catches are taken by industrial-scale fisheries, including many vessels from countries outside the region (particularly Japan and the U S A). Since the establishment of EEZs the activities of these vessels are controlled by the coastal state under some form of licensing.

Four main species of shrimp are taken in the fishery, but it is not known to what extent there may be separate stocks within one or other species. It is likely that there is significant movement of shrimp across the boundaries between adjacent EEZs. This underlines the importance of a regional approach to the assessment and management of the resources. By-catches of fin-fish are considerable, and while some marketable fish are brought ashore, large quantities of marketable and trash fish are discarded.

The paper reviews the history of the fishery, the biology of the shrimp, and the assessment of the stocks, and discusses the needs for management and the possible measures that could be implemented.

1 Introduction

In the early sixties foreign fleets started to exploit the offshore shrimp resource off the Guianas and northeastern Brazil. Landings rose rapidly until 1969. From 1970 onward they have fluctuated between 15 000–20 000 t. The number of boats also increased rapidly up to a record of 658 in 1977, followed by a decline as a result of the extension of fishery jurisdictions by coastal countries.

The offshore fishery is based on adults of four species: brown shrimp (*Penaeus subtilis*), pink-spotted shrimp (*P. brasiliensis*), pink shrimp (*P. notialis*) and white shrimp (*P. schmitti*).

In addition to the modern offshore highly-capitalized fishery there are artisanal fisheries in each of the countries of the region that are catching juvenile forms of the same species as the offshore fishery. There is some information (Dragovich and Villegas, 1981; Chakallal and Dragovich, 1981) indicating that the only artisanal fishery of some importance directed towards shrimp is carried out in the estuarine areas of Brazil.

The present report discusses the main problems related to the management of the Guianas–Brazil shrimp fishery in view of the existing knowledge on the fishery and the socio-economic conditions prevalent in the region.

2 The fishery

Shrimps are fished from Trinidad to Tutoia (Brazil) (ca 42°W) by artisanal and industrial fleets. Offshore adult shrimps are caught by commercial fleets representing several countries. Unknown amounts of juvenile shrimps are caught by artisanal fishermen from the coastal countries.

Artisanal fishermen use small trawls, cast nets and passive gears such as Chinese seine nets in the estuarine areas. By far the most important artisanal fishery is carried out in the States of Para, Maranhão and Piaui (Brazil). An average of 800 t (live weight) per year has been estimated to be landed in the State of Pará during the period 1976–79 (SUDEPE/PDP, 1981). Species caught by artisanal fisheries are: seabob (*Xiphopenaeus kroyeri*), fished in all the region; white shrimp,

reported from Brazil and Venezuela; brown shrimp, fished in Brazil and French Guiana and pink-spotted shrimp, caught at times in Suriname only.

Commercial offshore shrimping started in 1959, after Government-sponsored surveys (Higman, 1959; Bullis and Thompson, 1959) attracted the interest of fishermen and the industry. Foreign and local trawlers fished first off Guyana and later their operation extended to cover the entire fishery down to Tutoia (Brazil).

The offshore fisheries grew rapidly. Landings increased from 2 800 t (live weight) in 1960 to stabilize around 15 000–20 000 t after 1966. Annual landings in excess of 19 000 t were recorded in 1968–70 and again in 1973 and 1977. The number of vessels also increased steadily from about 100 at the beginning of the sixties to more than 500 in the period 1973–78 (*Table 1*). The general extension of jurisdictional waters in 1977 ended the increase of fleet sizes and has been followed by a gradual decline in the number of trawlers. In April 1981 there were at least 442 boats operating in the Guianas–Brazil area (Dragovich, 1981).

Table 1
NUMBER OF VESSELS, LANDINGS (T. LIVE WEIGHT) AND CATCH PER VESSEL

Year	No. of vessels	Landings	T/vessel
1960	—	2 785	—
1961	100	3 095	30·9
1962	96	4 371	45·5
1963	147	7 430	50·5
1964	187	9 262	49·5
1965	203	11 230	55·3
1966	281	15 475	55·1
1967	342	17 222	50·3
1968	362	19 259	53·2
1969	403	19 136	32·6
1970	421	19 081	45·3
1971	346	15 500	44·8
1972	370	16 126	43·6
1973	523	19 606	37·5
1974	562	18 136	32·3
1975	591	15 581	26·4
1976	586	16 926	28·9
1977	658	19 615	29·8
1978	514	15 447	30·0
1979	481	21 514	44·7

From the beginning of the fishery until 1966 there were increases in gear dimensions and vessel sizes and power but, since that time, the fishing power of vessels has remained fairly stable.

The fishing fleet consists chiefly of Florida-type trawlers, fairly modern and uniform in size and fishing gear. Most of the trawlers are equipped with refrigeration systems. The vessels usually range from 21–23 m (70–75 ft) in length and they are usually rigged to fish two trawls 12–14 m (40–55 ft) in length with 2–3 m (8–10 ft) doors.

Present fleets operate out of Cumaná, Guiria (Venezuela), Georgetown (Guyana), Cayenne (French Guiana), Paramaribo (Suriname) and Belém (Brazil). In the earlier days trawlers operated also out of Bridgetown (Barbados), Port-of-Spain (Trinidad and Tobago) and St. Laurent (French Guiana).

The offshore fishery exploits primarily brown, pink-spotted and pink shrimp; white shrimp is the least abundant.

Fleets of many nations participate in this fishery. Recently, U S A and Brazil fleets have been the most prominent (*Table 2*). Certain fleets (Barbados, Trinidad and Tobago and Cuba) are not currently engaged in shrimping off Guianas and Brazil.

Except for the Brazilian fleet and to some extent of the Venezuelan fleet (which has been fishing from Trinidad to Guyana) other fishing fleets operated without restrictions throughout the region and during every month of the year until 1970, when Brazil extended its jurisdiction to 200 nautical miles.

From 1972 through December 1977 Brasil required foreign vessels fishing her waters to have permits issued under bilateral agreements; the numbers of permits was limited. There were also seasonal and other types of restrictions. In December 1977 these agreements between Brazil and foreign governments expired and in the following years no foreign flag vessels have fished for shrimp in Brazilian waters. In 1977, Guyana, Suriname and French Guiana initiated licensing systems. These restrictions forced fleets to trawl only off their respective countries of operation.

At present there is a reciprocal fishing agreement between Suriname and French Guiana that permits the operation of a few shrimpers from each country in the other's territorial waters.

The impact of the extension of jurisdictional waters to 200 nautical miles has apparently had a minimal effect upon the participation of foreign fleets in this fishery, as boats have continued to operate through fishing licences obtained from coastal countries. In 1976 foreign vessels represented 66 percent of the fleet as compared with 64% in 1981. Extended jurisdictions have modified the sharing of total landing amongst participating fleets. U S A landings, which represented nearly 50% of the total in 1973 have decreased to 39% in 1979.

Even though an opportunistic type of fishing prevailed prior to the establishment of national fishing jurisdictions by each coastal country many U S and other than U S vessels fished Brazilian grounds at the beginning of the year, followed by French Guiana and Suriname grounds by the middle of the year and Guyanese grounds in the latter months of the year. Other vessels fished in the same area throughout the year without a particular pattern. It is likely that separate fleets have been exploiting different portions of the total resource because they fished different areas, caught different species and different sizes.

Only Brazilian and, in a lesser degree, Venezuelan boats have been fishing in their own waters. While Japanese activities have been concentrated off Guyana and French Guiana, most of the U S catches were obtained (until 1977) from French Guiana and Brazil. After 1977 most of the U S catches have come from Guyana and French Guiana.

Brazilian catches consisted chiefly of brown shrimp (90% in weight in 1976–78); U S landings

Table 2
NUMBER OF SHRIMPERS BY FLAG COUNTRY OPERATING IN THE GUIANAS – BRAZIL FISHERY

Year	Brazil	Cuba	Barbados	Japan	Korea	Guyana	U S A	Trinidad	Venezuela	Suriname
1972	25	0	0	65	10		153[1]		0	
1973	37	0	6	102	21		198		107	
1974	55	11	21	128	55		207		66	
1975	25	14	20	123	82		157	56	88	
1976	38	6	20	70	110		134	63	90	
1977	48	13	0	67	130	47	141	0	129	20
1978	52	13	0				122	0	58	
1979	79	13	0					0	24	
1980	139	0	0			25		0	24	
1981[2]	95[3]	0	0	64	67	38	152	0		18

[1] Half year
[2] April 1981 (Dragovich, 1981b)
[3] Includes 69 Korean, Japanese owned boats operating under Brazilian flag.

(1972–78) also consisted chiefly of brown shrimp (50–70%), followed by pink-spotted shrimp (20–40%) and pink shrimp (less than 10%) and Japanese landings (1970–77) were 60–80% pink-spotted shrimp. Venezuelan landings have a higher percentage of white shrimp (40%) than landings from other countries, due to the fact that their fishery to a large extent exploits shallow near shore waters, where white shrimp are usually found.

Size compositions of the catches of the two major fishing fleets, U S and Japanese, during the period 1975–77 were slightly different, Japanese catches having a higher proportion of small sized shrimps. Such differences are probably due to differences in the fishing grounds exploited by the two fleets.

Information from log-books indicates that the higher catches in the area from Guyana to Brazil are obtained from March to April and that most of the effort in that region was concentrated on intermediate depths, between 37–64 m (21–35 fm) (Dragovich, 1981a).

3 Other demersal species

3.1 Surveys

Since 1945 numerous experimental and exploratory fishery surveys have revealed the existence of ground-fish stocks in commercial quantities (Dragovich, 1980).

UNDP/FAO and national surveys suggest an existing resource (standing stock) of 350 000 t. While catch rates for these surveys vary with season, depth, time of day and fishing power of vessels, results indicate that:

(1) standing stock is consistently highest in inshore (<40 m) areas and increasing depth is generally associated with reduced trawl catches;

(2) for trawl surveys, daytime catches average higher than night-time;

(3) riverine discharges are concurrent with fish abundance in near shore areas;

(4) coastal waters off Suriname yield consistently higher catches than other areas;

(5) species compositions vary throughout the area but distinct groups can be defined and related to benthic features; inshore trawl catches tend to be marketable species and species that could be used for fish-meal dominate offshore.

3.2 Commercial exploitation

Over 200 000 t of demersal finfish are conservatively estimated to be lost as by-catch in the reg-

ional shrimp fishery each year (Jones and Villegas, 1980).

Shrimpers operating out of Guyana are responsible for approximately 80 000 t of by-catch. The most notable feature in the available information on by-catch is the large amount of marketable fish that are recovered from shrimpers' catches (24–69 percent of the total by-catch). The most common species are sea trout (*Cynoscion virescens*), bangamary (*Macrodon ancylodon*), croaker (*Micropogonias furnieri*) and snappers (*Lutjanus* spp.). Since 1974 the Guyanese Government made mandatory for each shrimp vessel to deliver 909 kg (2 000 lb) of fish per each fishing trip. Four Venezuela stern trawlers also fished in Guyanese waters. Their operation concentrated on demersal fish (especially *Cynoscion* sp.). Cuban vessels while operating in Guyanese waters received fish from shrimpers.

Information of shrimp by-catch in Suriname is practically nonexistent. Data from two research vessels that operated in 1962–65 and 1967–68 indicated that 21–31 percent of the total by-catch was marketable fish. The most common species were sea trout and croaker. There are 6–10 semi-commercial Suriname flag vessels that land an estimated 1 t of fish daily. Their catch consists principally of sea trout, croaker, dog trout (*M. ancylodon*) and butterfish (*Nebris microps*). Of the estimated 1 200 t of marine fish sold in Suriname in 1976, approximately 80 percent of it was sold directly by the shrimpers (Young, 1979).

Off French Guiana no commercial trawling for fish is practised and practically no by-catch is landed by the trawlers based in Cayenne.

A trawling survey was carried out by the R/V RIOBALDO in 1978 in the region Amapá-Pará. Fishing took place in the 10–90 m depth interval. 44–50 percent of the fish caught was marketable. Croakers, weak fish (*M. ancylodon*) and catfishes (*Trachysurus* spp.) were the most common species caught (Oliveira, 1980). About 70 vessels are engaged in pair trawling for catfish (Piramutaba, *Brachyplatystoma vaillantii*) at the mouth of the Amazon river over shallow waters (<15 m). The catfish fleet landed 17 000 t of marketable fish (80 percent catfish) in 1979. It is estimated that during the same period the fleet discarded 20 000 t of fish. Species discarded include small 'piramutaba' and dolphinfish, other catfishes and croakers. No shrimp is reported caught by the catfish fishery.

4 Biological information on commercial species of shrimp

Most of the species caught occur throughout the

region but, based on available information, differences in their bathymetric and geographical distribution have been established.

Adult brown and pink-spotted shrimps are found at various levels of abundance mainly at 27–82 m (15–45 fathoms) throughout the region; pink shrimp has only been recorded off western French Guiana, Suriname and Guyana (Dragovich, 1981a). Brown shrimp is predominant off Brazil and eastern French Guiana and pink-spotted off western French Guiana, Suriname and Guyana (Dragovich and Coleman, 1980). White shrimp are also found over the entire region but in shallower waters—less than 37 m (20 fathoms). Commercial concentrations of white shrimp are found only off Trinidad, Venezuela, Guyana and off Pará-Maranhão (Brazil). It is possible that two separate stocks of white shrimp exist in the region.

Juvenile shrimp are found in the estuarine areas and lagoons of the region. Juvenile brown shrimp are taken throughout the year by the artisanal fishermen in Pará-Maranhão. Based on data from U S trawlers small sizes of brown shrimp occur frequently off Amapá (Brazil) and off eastern French Guiana, especially in March and April and also off Guyana. Biologists of the Division of Fisheries observed large concentrations of juvenile pink-spotted shrimp in the estuarine areas of eastern Suriname from April to June. Fishing companies in Paramaribo report also that Suriname is the only area in the Guianas–Brazil where large quantities of juvenile pink-spotted shrimp are caught. These observations suggest that young brown and pink-spotted shrimp are being recruited onto the fishing grounds principally from these areas.

The presence of extensive estuarine areas in Pará-Maranhão (the biggest in the region) suggest that this area is an important source of recruitment of brown shrimp to the offshore fishery of Brazil and probably eastern French Guiana also.

There is no information available on growth, mortality or migration of postlarval, juvenile or adult shrimp, except that postlarvae of brown shrimp enter a coastal lagoon near Cayenne in February–March at a total length of 1 cm and migrate to the sea after two–three months at about 7 cm total length and that some brown shrimp tagged in the Mana estuary (close to the French Guiana/Suriname border) were recovered off Brazil (between 4° and 5°N) and off Suriname.

As in other studies of shrimp, U S and Brazilian studies have shown that female brown, pink-spotted and pink shrimp reach larger sizes than males. Whereas male brown and pink-spotted shrimp were more abundant than females in landings at Georgetown (Guyana), pink shrimp females were more abundant than males.

Based on data from U S port sampling carried out at Georgetown from 1976 to 1978 all four stages of maturity for each species (females) occurred during each month of the year suggesting year-round spawning.

5 Environmental factors

Environmental factors determine to a large extent the abundance and distribution of shrimp populations. White and brown shrimp are associated with mud substrate with a relatively high organic content, and pink with sand substrate. Annual temperature variations are relatively small in this region (approximately 7° C for surface waters and 6° C for bottom waters). Nevertheless, seasonal variations in temperature may have effects on growth rate that are important for population analysis.

The Guyana Current that flows along the coast from east to west and covers an area as wide as 200 miles is also a factor that presumably influences shrimp distribution. The Current, with a major contribution from the Amazon river, separates from the coast near Cayenne to follow the continental break off Guyana, and its influence is rather weak westward of the Maroni river. The inshore–offshore water movement on the shelf off Suriname implies the existence of a certain degree of upwelling at some distance from the coast. Very little information is available on the presence of countercurrents or eddies in this current system.

6 Stock assessment

6.1 *Data Base*

In the Guianas–Brazil region in addition to the exploratory surveys prior to the beginning of this fishery, numerous research surveys have been carried out by different countries and international organizations (Dragovich, 1980).

Under the terms of the U S A–Brazil Fishery Agreement, the U S National Marine Fisheries Service (NMFS) began to collect catch and effort statistics in 1972 from their trawlers fishing in the Agreement Area on a newly introduced log-book. The collection of catch and effort statistics included data from U S shrimpers fishing off Guianas–Brazil. Information was collected from 13 statistical zones. In 1976, NMFS initiated a port sampling programme of shrimp landings to obtain more detailed biological information at the species level. This sampling programme, which included

Georgetown, Paramaribo, Cayenne and Port-of-Spain was and is successfully carried out only in Georgetown.

Operating out of Belém, the National Superintendency for Fishery Development—SUDEPE—began in 1969 to collect catch and effort statistics. Log-books were introduced in 1974 without much success. A biological sampling programme of landings of shrimpers based at Belém was introduced in 1975. A similar programme for juvenile brown shrimp caught by artisanal fishermen in the State of Pará was initiated in 1976.

Japanese Far Seas Fisheries Laboratory started in 1970 and continues to collect catch and effort information from Japanese trawlers operating in this fishery. Japanese log-books are very elaborate and include information related to fishing zones on catch composition by species and commercial size categories.

Information on catch and effort statistics by Cuban (from 1974), Korean (from 1969) and Venezuelan (from 1973) trawlers is being collected by their respective Departments of Fisheries.

6.2 *State of exploitation*

There are perhaps several shrimp stocks in the Guianas–Brazil fishery, but present information is not sufficient to substantiate this supposition. At the meeting in April 1979 of the Working Group on Shrimp Fisheries of the Northeastern South America convened by the WECAF Project, all the species caught were considered to be part of a single stock (Jones and Villegas, 1980).

The Working Group did not make estimates of MSY as such estimates based on the available data would not be realistic because there was a lack of information on:

(1) the magnitude of the juvenile fishery on nursery grounds;

(2) temporal changes in species composition of offshore catches during the period under study; and

(3) species-directed effort necessary to develop accurate measures of cpue indices of abundance for the various species that are consistent throughout the period.

The period from 1960 to 1968 can be considered as the expansion period for this fishery as catch per boat increased from year to year (*Table 1*). From 1969 onward the total landings have levelled off between 15 000 to 20 000 t (live weight) whereas the numbers of vessels increased from 400 in 1969 to a maximum of 660 in 1977.

A plot of the number of vessels and the recorded landings indicates that since 1968 the total landings

have not increased despite the increase of the fleet size (*Figure 1*), and the catch per boat has decreased steadily during most of the seventies (*Figure 2*). These statistics do not take account of:

(1) possible changes in the estuarine areas that could affect recruitment to offshore areas;

(2) possible increases of juvenile catches resulting in a depressed yield per recruit and, thus, decreased long-term yield;

(3) concomitant changes in the abundance by area and/or by species which may have been offset by fleet mobility in the past.

The statistics of several national and area-specific offshore fleets exhibit trends which differ in detail (*Figure 2*). This may be attributed to the fact that these fleets are, at least in part, exploiting different segments of the total resource.

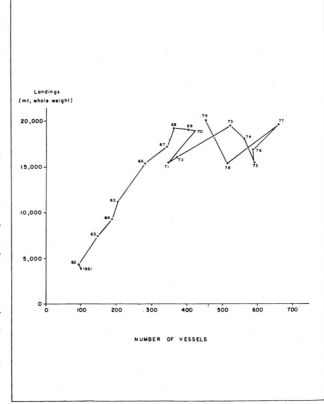

Fig 1 Relationship between total landings and total number of vessels in the Guianas–Brazil shrimp fishery

The Brazilian fleet operates in the inshore area, south of the Amazon River, and the offshore area along the entire coast, north and south of the Amazon. Their catches are mainly brown shrimp. The fleet fished north of the Amazon River from 1969 to 1973. In 1974 the catch per day in that area

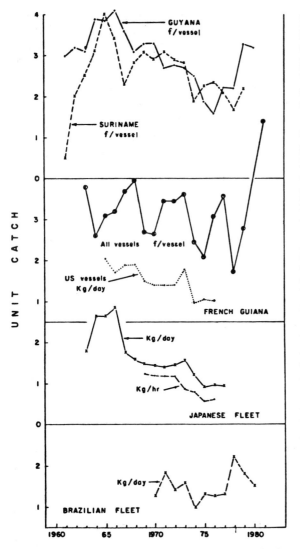

Fig 2 Catch per unit of effort tends for shrimp fleets operating in the region (live weight)

prior to 1977 indicates that the catch per unit effort generally decreased from 1966 onward (*Figure 2*), even though the fleet shifted fishing grounds to prevent a decline in catch rates. The long-term trend suggests a decline in the abundance of pink-spotted shrimp in the area Guyana – French Guiana.

Prior to 1977, the U S fleets based in French Guiana, Suriname and Guyana fished the entire shelf off these countries and Brazil (Agreement Zone). Most of the catches were brown shrimp (50 – 70 percent) and pink-spotted shrimp (20 – 40 percent). The overall drop in catch per day fished from French Guiana-based vessels, concomitant with decreasing effort, recorded during the period 1965 – 75 could reflect the drop in abundance of pink-spotted shrimp as indicated by Japanese cpue and a general decrease in overall offshore abundance of brown shrimp. That trend is corroborated by catch per vessel figures for the Guyana and Suriname-based fleets (*Figure 2*).

The consequence of the extension of jurisdictional waters in 1977 has been that fleets operated exclusively in the waters of the country where they were based. This concentration of effort has been particularly marked off Guyana, Suriname and French Guiana. The decrease in fishing activities that took place off Brazil during 1978 – 79 (*Table 3*) has been followed by an increase in the catch per vessel, specially for the French Guiana-based fleet. Unfortunately, a lack of information on catch rates at the species level precludes the determination of any relationship between these phenomena.

7 Management

7.1 *Possible measures*
Presently there is no management scheme for the entire fishery. Each country, from the Amazon to the Orinoco Rivers, is exploiting its own piece of

reached its lowest level. As a consequence, the fleet started to operate also south of the Amazon (in the area of Tutoia) in mid 1974. Since that time, the fleet has fished in both areas to maximize catch rates. A long term positive trend in the values of catch per day has been recorded from 1975 onward in spite of an increase in the fleet size. Such figures indicate that the overall abundance of brown shrimp in the areas fished by the Brazilian fleet has not decreased.

Japanese landings are dominated by pink-spotted shrimp (60 – 80 percent). The Japanese fleet has mainly operated from Guyana to French Guiana although the areas where most of the shrimp have been caught have changed considerably during the period under review. Information

Table 3
ESTIMATED NUMBER OF SHRIMPERS OPERATING OFF BRAZIL

| Year | Flag of the Vessels | | |
	U S A*	Brazil	Total
1972[1]	153	25	178
1973	193	37	230
1974	214	55	269
1975	92	25	117
1976	99	38	137
1977	90	48	138
1978	0	52	52
1979	0	79	79
1980	0	139	139

Notes: *Licensed vessels (from Dragovich, 1981)
[1] Half year.

66

adjoining shelf without consideration of regulations enforced by other countries sharing the common shrimp resource. The existing fishery regulations by each country are made strictly for the waters under national jurisdiction and they chiefly consist of spelling out the licensing requirements and delimiting closed areas for the offshore fleets.

Some countries (Brazil, French Guiana and Suriname) are attempting to limit total fishing effort through catch and/or vessel quotas. French Guiana has established closed areas for part of the year as a means to protect small shrimp. In general, regulations which cannot achieve their purpose (e.g., the application of size limits to an offshore fishery) have not been applied. For proper management of the resources, in view of the general lack of data and means to support the implementation of an elaborate management plan this approach appears to be a very reasonable one. At the same time taking independent actions at the country level, especially in regard to the number of boat quotas, could well result in excessive fishing pressure and in oversized fleets operating uneconomically.

Boerema (1980) pointed out that, despite the fact that the knowledge of the shrimp fishery off the Guianas and northern Brazil was still very incomplete, there was a need to consider management measures. He based his opinion on the fact that it was likely that some segments of the fishery were already heavily exploited.

A rational objective in managing a fishery is to ensure its best utilization on a continuing basis. Best utilization implies to take into account the biological characteristics of the resources supporting the fishery and the socio-economic conditions prevalent in the countries having jurisdiction over it. Thus the specific objectives for the management of this fishery could be:

(1) maximum yield of shrimp over time;
(2) maximum yield of fish over time;
(3) maximum economic return;
(4) maximum employment, or
(5) a combination of the other four objectives.

Effective management of this resource will require that the stocks of shrimp be managed throughout their range. Since the stocks of shrimp extend along the coasts of several countries and the fishery is also distributed along the entire coast, fishing activities in waters under the jurisdiction of one country will affect those parts of the stocks living in waters of the remaining countries.

The adoption of any management scheme remains basically the responsibility of the coastal countries concerned. As their management objectives could well differ from each other, it is urgent and imperative that such objectives be defined in each country and common management measures be agreed upon, implemented and monitored throughout the area where the resource is exploited. The present way of granting licences by each country, without taking into consideration the number of licences given in other countries, gives support to the need for regional management.

Basically, management of shrimp fisheries can be grouped into two types of regulatory measures:

(1) protection of young shrimp: increase in the mesh size of nets, establishment of size limits for landed shrimp, establishment of closed seasons during the recruitment period, and establishment of closed areas where small shrimp predominate;

(2) amount of fishing: limiting the amount to be caught, establishment of closed seasons, limitations of the number of boats fishing and the types of gears used.

Increased mesh size of nets has been considered by some authors as of limited value as a management tool, as fairly large meshes still retain a substantial amount of small shrimp but, as García and L'Homme (1977) (in García and Le Reste, 1981) pointed out, increased mesh sizes of nets used by shrimpers will result in:

– an increase in the average sizes of shrimps caught and possible losses in the weight caught will be compensated by increases in the values of landings;

– a reduction in the amount of by-catch, which is generally made up of juveniles of commercial species. Part of these juvenile fish could grow to bigger sizes and guarantee better prices.

The establishment of minimum size-limits often results in the discard of undersized shrimp, without bringing any positive benefit to the resource.

Closed seasons at the peak of the recruitment period benefit the juvenile shrimp and, at the same time, reduce fishing effort. Closed seasons during other periods of the year also reduce fishing effort but seasons need to be carefully chosen, taking into account the combined effect of recruitment and growth and their regional variations.

Closed seasons at the peak of the recruitment period and, especially, the closure of estuarine areas to fishing will greatly affect the well being of artisanal fishermen, a group already living under difficult conditions.

The protection of nursery grounds can be achieved through limitation of the amount of fishing in estuarine areas, by controlling the sources of pollution (industrial or agriculture related pesticides and fertilizers) and the eventual destruction of such grounds resulting from engineering works (dikes, dams).

Catch quotas for short-lived animals with large annual variations such as shrimp have little practical value, as the amount of data needed to set appropriate annual quotas is excessive. A more direct way to control the amount of fishing on the stock is to regulate the number and type of boats participating in the fishery. The determination of the permitted fishing effort needs data which are less difficult to obtain than those for catch quotas, and this could be the measure to be considered initially. This type of measure is also appropriate for the characteristics of the Fisheries Administrations in the region.

The number and type of boats authorized to fish inside a particular zone would require the introduction in the country of a system for licensing fishing vessels.

Any management scheme to be adopted by coastal countries to regulate the shrimp fishery in the Guianas–Brazil area needs to take into account the various management objectives chosen by the countries, the local capabilities to prepare, implement and enforce a management scheme and the effect of estuarine shrimp fisheries on the total yield of the fishery. The effect of future development of trawl fisheries for finfish also needs to be taken into account.

At the present time, there are great differences in the capabilities of the countries to carry out an effective surveillance of offshore fishing in their waters. The staff of local Fisheries Services are small in numbers and need to be supported by properly trained people if a management scheme is going to be implemented.

The effects of the estuarine fisheries need to be determined. If these effects are considered to be significant, then some control over the estuarine fisheries may be necessary, but management measures affecting such fisheries will have a bigger impact on artisanal fishermen living in estuarine areas, especially if their activities cannot be shifted to other productive work. This problem is especially delicate in some areas where unemployment is one of the biggest constraints. It could be that too many restrictions on artisanal fisheries would not be possible for socio-economic reasons.

7.2 Research needs and future research programmes with regard to management

7.2.1. *Research needs*   It is obvious that a good deal of information needed for a proper analysis to be made of the state of exploitation of the resource is lacking. There is also no information on the economics of the fishery and very limited information on users of the fishery, especially about artisanal fishermen and their activities.

Due to the inadequate information on species composition, the shrimp resource has been treated up to now as a single stock, although there are four different species in the catches. Adequate information on species composition of the landings is essential for better evaluation of the fishery. Participating countries should take steps to collect data. Various fleets participating in the fishery do not have the same relative efficiency on the same level of fishing power per unit vessel, so that it will equally be necessary to standardize fishing effort in terms of a certain class of vessel.

A surplus production model needs relatively simple data, *viz*, catch by species and standardized estimates of fishing effort. It provides estimates of MSY and the appropriate level of fishing. It can be used for offshore (industrial) fisheries when the level of artisanal (inshore) fishing does not change greatly. The yield per recruit model needs more sophisticated data (growth and mortality) and provides information, for each species, of the optimal size of the animal to be harvested and on the fishing effort that will optimize the yield. The surplus production model might be adequate for the level of information available at present but a yield per recruit model would permit better management of the fishery. The additional data requirements of a yield per recruit model need to be taken into account when developing future research programmes.

7.2.2 *Future research programmes*   The Working Group on Shrimp Fisheries of the Guianas–Brazil Region that met in April 1979 in Panama agreed to focus the research on the following:

(*1*) data collection for catch and effort statistics;
(*2*) data processing for catch and effort statistics;
(*3*) tagging operations; and
(*4*) surveying of nursery grounds and artisanal fisheries.

(*1*) *Collection of catch and effort statistics*   The collection of catch and effort statistics will provide the basic input for the surplus production model. The collection of catch data should be the full responsibility of each country, in accordance with the resolution taken by the Fisheries Commission of the Western Central Atlantic (WECAF Commission), but the report on fishing effort by statistical areas should be the responsibility of the flag country of a particular vessel. Future collection of catch and effort statistics must also include information from artisanal fisheries for shrimp and from fisheries for other species that might take shrimp as a by-catch.

To facilitate data collection, processing and analysis there is a need to develop a uniform statistical reporting system to be used by all participating fleets within a reasonable time period. Presently, several countries participating in the fishery have different reporting systems.

(2) *Central data processing system for catch and effort statistics* Whilst most of the countries participating in the fishery are attempting to collect common data through the use of log-books and a trip-reporting system, it is considered that the shrimp fishery cannot be assessed at the national level as geographic limits of shrimp stocks do not necessarily follow national boundaries. Assessment of the resource requires a regional approach which necessitates regional data. A regional data bank under the responsibility of a body such as the WECAF Commission will reduce the time lag between the data collection in the field and its subsequent analysis.

(3) *Tagging operations* Mark-recapture studies are extremely valuable to determine stock boundaries, migration patterns, sources of recruitment to the offshore fisheries, growth and mortality.

A cooperative shrimp marking programme has been started in August 1981 with the release of stained and ribbon tagged shrimp in the estuarine areas of the State of Pará. This operation will be followed by an offshore tagging in the area Guianas - Brazil that will take place in 1982.

(4) *Survey of nurseries and artisanal fisheries*

As there is a great need for more information on this subject, field surveys of the region are in progress to identify potential nursery areas of shrimp and artisanal fishing communities. The results of these surveys will be of help in the planning of future tagging operations and in the development of a sampling scheme to obtain information on catch and effort from artisanal fisheries.

Among other potential fields of research not mentioned by the Group are:

(a) mesh selectivity studies;
(b) socio-economic studies.

(a) *Mesh selectivity studies* The use of trawls with bigger mesh sizes could eventually permit the release of juveniles of commercial fishes without greatly affecting the economic returns from shrimp catches. Such action could help lower the vast amount of fish that is lost through shrimping oper-

ations in its present form. Finfish spared by shrimpers will have an opportunity to grow to bigger sizes and could be the basis for eventual finfish-directed fisheries.

(b) *Socio-economic studies* A management scheme to be developed for the Guianas - Brazil shrimp fishery must take into consideration certain problem areas regarding management objectives. Studies of cost and earnings for the various segments of the total fleet, of employment, users and consumers of the resources, etc., will help greatly in assessing the impact of the various alternatives proposed for management.

8 Conclusions

The analysis of the available information on the fishery has shown that the level of catches have fluctuated around 15 000 – 20 000 t (live weight) from 1970 onwards. This catch level has been obtained with a fleet ranging in size from 300 to 660 boats. Despite the fact that increased numbers of boats have not resulted in a decrease in total catches, the decline in overall catch per boat recorded from 1969 and in catch rates for some fleets from 1966 – 67 indicates that the amount of fishing has been perhaps excessive during the past ten years, as it could have provoked overexploitation, at least in some segments of the resource.

Decline in the catch rates and increase in the oil prices have affected the rate of returns to the fishing industry. If one considers that before the extension of the economic zones vessels shifted grounds more or less continuously to maximize returns, the imposed territorial limitations on the free fishing resulting from extended jurisdictions have increased the economic problems of the industry.

The resource is made up of several species, each one with a particular geographic distribution not confined to the jurisdictional waters of any particular coastal country. It is also highly possible that nurseries situated in one country are feeding the offshore grounds in another country. In such a situation, activities affecting the resource in any of the countries of the Guianas – Brazil region will have consequences outside their borders.

The fleet size declined, following the extension of the jurisdictional waters but the present policy followed by all coastal countries of granting licences regardless of the amount of licences granted in other countries of the region could result in an oversized total fleet. Such a situation could aggravate even more the decline in the relative abundance detected for some species and in the economic yield from the fishery. It will make more evident the socioeconomic and political problems related to overcapacity of the fishing industry.

The increase in agricultural activities at the expense of nursery areas and other human activities that could affect these areas, such as pollution, artisanal fisheries and engineering works, will add more pressure on the resource. It is known that there are plans to start large-scale rice cultivation in some of the shrimp nurseries. Eventual plans to develop artisanal fisheries in Brazil would result in increased fishing pressure on juvenile shrimps.

Another eventual development that could add indirectly more fishing pressure to the resource would be to promote the introduction of trawlers fishing for finfish.

The possibility of actions to be taken at the national level that will affect the resource makes the adoption of a regional management scheme very urgent in order to avoid further damage to the resource and to the groups of people that are involved in it. The regional approach will need the establishment of a central body representing coastal countries to develop a common management scheme, to implement it, to enforce management measures adopted and to monitor their effects on the resource and their users.

At present, the scientific advice that can be provided for the management of the fishery is rather limited. Studies to determine stock boundaries (including nurseries vs offshore fisheries), growth and mortality rates and better statistics on catch and effort (including artisanal fishery data) will help in obtaining better estimates of the potential of the resource. Information from all coastal countries on socioeconomic aspects, such as the value of the various segments of the fishery, rate of return to the fishing industry, employment and users of the resource will help to decide on management objectives and strategies. Any future research programme needs to consider data collection on the above mentioned aspects.

An important point to be taken into consideration in the preparation of a management plan and its implementation is the limited capacity in terms of people and means, of some coastal countries. To establish a management scheme will need international technical assistance to help coastal countries overcome their limitations and enable them subsequently to continue working on their own.

Technical assistance given in the past to a group of countries has many times failed to reach its objectives as differences in mentalities, political regimes, economies, *etc* and lack of experience in the field of regional cooperation amongst the countries concerned have not been taken into account in the determination of the period of time allotted for the assistance. As a result, the duration of the technical assistance has been too short to arouse the support of local governments and create a nucleus of people in each country determined to pursue the lines of action initiated by the technical assistance programme. This possibility of failure should be taken into consideration in the planning of the assistance programme. Technical assistance must be continued long enough to give time to countries to learn and act together and build their own capabilities.

References

BOEREMA, L K. Expected effects of possible regulatory meas-
1980   ures in the shrimp fishery with special reference to fisheries of the Guianas and Northern Brazil. *WECAF Rep.* (28): 144–51.

BULLIS, H R and THOMPSON, J R. Shrimp exploration by the
1959   M/V OREGON along the northeast coast of South America. *Commer. Fish. Rev.* 21 (11): 1–19.

CHACKALLAL, B AND DRAGOVICH, A. The artisanal fishery of
1981   Guyana. Southeast Fisheries Center, Miami, U S A (Ms.), 27 p.

DRAGOVICH, A. National Report—U S A. *WECAF Rep.*, (27):
1980   59–75.

DRAGOVICH, A and COLEMAN, E M. The United States shrimp
1980   fishery off the coast of Northeastern Brazil, French Guiana, Suriname and Guyana (1975–77). *WECAF Rep.*, (28): 77–115.

DRAGOVICH, A. Guianas–Brazil shrimp fishery and related U S
1981a   research activity. *Man. Fish. Rev.*, 43 (2): 9–18.

——   Report on site visits to the fisheries research centres,
1981b   shrimp companies and processing plants in Guyana, Suriname, French Guiana and Northeastern Brazil. April 4–25, 1981 (Ms.) 21 p.

DRAGOVICH, A and VILLEGAS, L. The small-scale fisheries of
1981   Northeastern Brazil (Pará), French Guiana, Suriname and Guyana, Southeastern Fisheries Center, Miami, U S A (Ms.) 20 p.

GARCIA, S AND LHOMME, F. La crevette rose Penaeus duorarum
1977   notialis de la côte ouest africaine: évaluation des potentialités de capture. *FAO Circ. Pêches*, 703, 28 pp.

GARCIA, S and LE RESTE, L. Cycles vitaux, dynamique, exploita-
1981   tion et aménagement des stocks de crevettes penaeides côtières. *FAO Doc. Tech. Pêches*, (203): 210 p.

HIGHAM, J B. Suriname fishery exploitations, 11 May–31 July
1959   1957. *Commer. Fish. Rev.*, 21 (9): 8–15.

JONES, A C and VILLEGAS, L (Eds.). Report of the Working
1980   Group on Shrimp Fisheries of Northeastern South America, Panama City, Panama, 23–27 April 1979. *WECAF Rep.*, (27): 3–31.

OLIVEIRA, DE G M. Resultados de la pesca de arrastre en la costa
1980   del Estado de Pará y el territoria de Amapa con el barco pesquero RIOBALDO Proyecto WECAF, Panamá, Panamá, (Ms.) 40 p.

SUDEPE/PDP. Custo de captura da frota-camaroneira e
1981   piramutabeira—Estado de Pará. SUDEPE (Brazil), Internal Report, 103 p.

YOUNG, R H. The status of shrimp by-catch. Utilization in some
1979   countries of the WECAF region. Report on an FAO consultancy (Ms.) 62 p.

The shrimp fisheries in the Gulf between Iran and the Arabian Peninsula

N P van Zalinge

Abstract

Catches of shrimp in the Gulf area are dominated by a single species, *Penaeus semisulcatus,* whose biology and distribution is described. Production by large-scale industrial fleets expanded rapidly in the 1960s, reaching a peak of over 7 000 tons in the 1973–74 season. Some of the decline since then may have been balanced by an increase in catch by the artisanal fleet, but in the 1978/79 season, and subsequently, there was a large fall in recruitment and in total catch. The work done in data collection and analysis, and in stock assessment through national and regional activities is reviewed. The possible reasons for the fall in recruitment, pollution, land-reclamation and recruitment over-fishing, are discussed. The paper outlines the possible management measures. Most countries have introduced a closure of varying lengths at the beginning of the fishing season in the period February–June.

1 Statement of the problem

The Gulf shrimp fisheries have generally been described as overexploited, *eg* IOFC 1973, 1977, 1979; FAO 1978, on the basis of declining trends in industrial shrimp catches, seemingly culminating in the 1979 closure of three industrial companies working on the Gulf's west coast.

Even though the statistics remain poor, the information that has become available in recent years indicates that initially the catches of the artisanal shrimp fishery, that developed well after the establishment of the industrial fishery, increased sufficiently to compensate for the diminishing industrial catches. Nevertheless overall catches started to drop sharply after 1976, particularly in Saudi Arabia/Bahrain (*Table 2*). This motivated the Gulf countries to take management measures aimed at the reduction of fishing effort, the underlying assumption being that excessive fishing was the main cause (Boerema, 1979; IOFC, 1979; Van Zalinge, 1980).

However it was also suggested that the significant interannual variations in the catches (Van Zalinge, 1980), as well as the apparent decline in recruitment over the years, are unrelated to fishing and if caused by land reclamation or other human interventions, may well turn out to be irreversible (Morgan and Garcia, 1982).

To address these problems improved catch and effort statistics as well as biological information in its environmental context are a prerequisite.

2 Shrimp biology

2.1 *Species composition*

Penaeus semisulcatus is the only major species, and accounts for more than 90 percent of the landings of the industrial fishery in Iran, Kuwait, Saudi Arabia and Bahrain (Boerema, 1969; Price and Jones, 1975; IOFC, 1977; Van Zalinge, *et al* 1979; FAO, 1980), although recently some doubt was cast on the validity of this statement for Kuwait (Mathews, 1981a). It also dominates artisanal landings in Abu Dhabi, Qatar, Bahrain and Saudi Arabia, but is quantitatively exceeded in Kuwait by *Metapenaeus affinis* and *Parapenaeopsis stylifera*, which form approximately 60 percent of the artisanal landings. The latter species that are distributed throughout the head of the Gulf and along the north-eastern coast (Drobysheva and Pomazanova, 1974) probably also prevail in the Iraq and Iran landings from the estuarine areas in the northern Gulf. The shrimp fishery off Bandar Abbas on the southern Gulf coast of Iran seems to depend on *P. merguiensis* (Boerema and Job, 1968; Longhurst, 1969; Van Zalinge, travel report).

In the Gulf of Masira (Oman) *P. semisulcatus*, *P. indicus* and *M. monoceros* are found.

Other species regularly occurring in the landings in varying though usually small quantities are: *Penaeus japonicus, P. latisulcatus, Metapenaeus stebbingi, Metapenaeopsis stridulans, Trachypenaeus curvirostris* and *Solenocera crassicornis* (Mohamed *et al*, 1981). The available literature on Gulf shrimp, however, pertains mostly to *P. semisulcatus*.

2.2 *Life cycles*

Along the west coast of the Gulf *P. semisulcatus* lifecycle patterns appear to be generally in good agreement (FAO/UNDP, 1982).

The spawning pattern, as measured by the variations in percentage of mature females (FAO, 1978; Mohamed *et al*, 1979b, 1981; Al Hossaini, 1981) and by post larval occurrence (Price, 1979a; Al Attar, 1981), shows two peaks per year, one in spring and one in autumn, respectively corresponding to periods of rising and falling temperatures and of high plankton abundance. These peaks vary in relative importance from year to year, although generally the spring peak is the more important one. Recruitment in autumn apparently stems from spring spawning, while the spring recruitment is produced by the autumn spawning (Mathews, 1981b). Age at first spawning is thus about 12 months for the majority of the shrimps. Al Attar (1981) reports that spawning occurs near shore on muddy sand bottoms.

Recruitment to the artisanal fishery is in April–June and September–November, while catchrates show one major peak in May–June. The subsequent decrease in these catchrates reflects recruitment to the industrial sector with a delay of about 1 to 2 months.

At low effort levels, industrial catchrates show a maximum in October–December, which shifts with increasing fishing effort towards the main period of recruitment. The patterns found for the Saudi Arabia/Bahrain and the Kuwait stocks indicate respectively that prior to 1979 these stocks were much more intensively exploited in Saudi Arabia/Bahrain waters than in Kuwait waters (FAO/UNDP, 1982).

2.3 Migration and distribution.
P. semisulcatus populations are thought to be strung along both the east and west coasts of the Gulf in series of more or less independent units, although overlapping and partial mixing particularly on the larval level cannot be excluded (Price and Jones, 1975; Basson *et al*, 1977; Van Zalinge *et al*, 1981). This is borne out by tagging experiments in Kuwait waters that showed adult migration to be generally limited. The maximum recorded distance was 85 km (Mohamed *et al*, 1979a; Farmer and Al Attar, 1981), but most movements were of much shorter distance and no marked shrimps were reported from Saudi Arabia or other countries (Mohamed *et al*, 1979a). The only clear case of a shared stock seems to be in the southern Saudi Arabia (Tarut Bay)/Bahrain area, although a comparable situation involving other species probably exists in the northern estuary (Farmer and Ukawa, 1980).

Kuwait Bay is one of the important nursery areas on the west coast and seems to be shared by two species: *M. affinis* and *P. semisulcatus*, at different times of the year. Mohamed *et al* (1981) and Jones and Al Attar (1981) report that sea weed (*Sargassum sp.*) affixed to the bottom as well as floating, forms an important habitat for post larval

and juvenile *P. semisulcatus*. Basson *et al* (1977), working in Saudi Arabia found that *P. semisulcatus* in Tarut Bay is ecologically dependent on algal and seagrass beds during the post-larval and juvenile stages, requiring such a synchronization of its cycle that the post-larval stage coincides with maximum algal development and the juvenile stage with the upcoming grass beds. After leaving the grass beds, *P. semisulcatus* is reported to prefer joining what Basson *et al* call the Murex/Cardium community—a community with a relatively high animal diversity—on a mud bottom substrate in 6–15 m deep water, although *P. semisulcatus* is also reported to occur in the subtidal sand biotope.

2.4 Growth
Information on growth is largely limited to *P. semisulcatus*. Monthly carapace length increases of adult shrimp were estimated to be 1·7 mm for males and 2.4 mm for females in Bahrain waters (FAO, 1978) and 2·2 and 3·0 mm respectively in Kuwait waters (FAO, 1980). Only for Kuwait are estimates of K and (carapace) L_∞ available, of which those of Van Zalinge *et al* 1979b and Mathews 1981b are given below.

	K (annual)	L_∞ (mm)
Males	2·11–2·15	33·9–35·5
Females	1·44–1·56	54·9–56·3

2.5 Mortality
Preliminary mortality estimates for *P. semisulcatus* from Kuwait waters are given by Jones and Van Zalinge, 1981; Van Zalinge *et al*, 1979b; and Mathews, 1981b.

Estimates were also derived from the two tagging experiments conducted by Mohamed *et al* (1979a) and are shown below. (Range in parenthesis).

	Z annual	F annual
First experiment		
(Febr.–June 1979)	17·8 (16–22)	4·9 (3·5–6·5)
Second experiment		
(July–Sept. 1979)	24·6 (19–28)	2·7 (2·2–3·1)

First experiment values pertain mainly to the industrial fishery and second experiment values to the artisanal fishery. The apparent mortality (Z–F) is much too high to be realistic, while active searching for tagged shrimp particularly in the first experiment, when a very high reward was paid for each returned shrimp, may have caused F to be higher than normal.

2.6 Behaviour and catchability
Schooling of shrimp is according to fishermen a phenomenon that regularly occurs in most years, although associated more with years of higher

catches in particular during the early years of the fishery (Kristjonsson, 1968; FAO, 1970; Drobysheva, 1971; Van Zalinge *et al*, 1981; Mathews, 1981a; Hamdan *et al*, 1981). The following table is derived from Hamdan *et al*, 1981 and gives the range of catch-rates associated with schooling and the months in which this behaviour was observed.

Species	Kg/hr	Months
P. semisulcatus	1 100 – 4 000	August
M. affinis	120 – 2 500	March, May, July, November
P. stylifera	225 – 660	March

P. semisulcatus apparently schools at the time of major recruitment to the fishery.

3 Description of the fishery

The fishery has two components, an industrial one that was the first to exploit shrimp on a large scale and an artisanal one that became only seriously involved in shrimp fishing after the modern method of trawling had been adapted for artisanal application, which occurred well after the establishment of the industrial sector.

3.1 *The industrial fishery*
Initially it seems fairly small boats (12 – 18 m) were used that delivered their catch to motherships and later to land-based processing plants (Kristjonsson, 1968). Later additions to the fleet were larger (21 – 55 m) and most had modern freezing equipment onboard for immediate processing of the catch (Egenaes, 1965). All vessels are fitted for Gulf of Mexico type double rig trawling, using flat trawls (12 – 34 m), semi-balloon and balloon trawls (16 – 23 m) (Kristjonsson, 1968). Mostly two layers of netting material of a stretched mesh of 30 – 40 mm are used in the codend, often covered with a protective layer (chafer) of 110 mm mesh (Van Zalinge *et al*, 1981). The use of two layers of already small meshed netting means a further reduction in the effective mesh opening and hence discarding of small shrimp is reported. (Kristjonsson, 1968; Van Zalinge *et al*, 1979a, 1981). Although ownership in the industrial sector in most cases belongs for 100 percent to local interests, the day-to-day management and execution of the fishing activities is exclusively carried out by expatriates.

3.1.1 *Industrial expansion: 1959 – 69* This was the period of the shrimp fishing boom, characterized by high catchrates, increasing landings, the appear-

ance of new companies and fleet expansion. Even foreign fleets (*eg* from the USSR, Drobysheva, 1971) were taking part in the bonanza.

The richest shrimp resources are found on the Iran side of the Gulf and it was here that industrial shrimp fishing started around 1959. First on the Bandar Abbas (Hormuz Strait) grounds, possibly as a result of the FAO exploratory survey for shrimp in that area in 1955 (FAO, 1957) and later on the more northern grounds of the Bushehr-Ras Al Motaf area.

In 1961 a Kuwait company began exploitation of the Iran stocks and maybe also the Kuwait stocks, although data pertaining to the latter grounds were not available until 1965, when two other Kuwait companies without fishing concessions for Iran appeared on the scene.

Shrimp fishing in Saudi Arabian waters began in 1963 and in Bahrain in 1966.

At the beginning of 1969 there were altogether eleven companies, four in Kuwait, two in Iran, two in Saudi Arabia, one in Bahrain, one in Qatar and one in Iraq, operating a fleet of at least 125 trawlers (Boerema, 1969) although a FAO (1970) report mentions that in 1969 the Kuwait fleet alone already consisted of 205 catcher boats, of which 200 worked in the Gulf, and the others were employed as far away as New Guinea.

In Iran landings reached a top of 9600 t in 1964 – 65 and in Kuwait a top of 3325 t in 1966 – 67. Also industrial landings from Saudi Arabian/Bahrain waters were generally very high in this period, even though the top (7400 t) occurred in 1973 – 74 (*Tables 1, 2, 3*).

3.1.2 *Industrial decline: 1969 – 1979*
Iran In October 1969 the Kuwait concession for Iran waters, which permitted the operation of a fleet of approximately 70 catcher boats and a number of motherships, expired and was not prolonged possibly as a result of the extreme poor start of the 1969 – 70 season in Iran (Boerema, 1969) and because an Iranian company intended to enter shrimp fishing activities with a fleet of its own. This caused a temporary slump in total production from Iran waters. (*Table 3*) (Bromiley, 1972).

Even though a large Kuwait company managed to obtain again a concession for Iran in 1973, it is not fully known what happened on the Iran scene, although increasing landings indicate that the general situation may not have been unfavourable. In 1978 Keyvan *et al* reported that the Iran fleet consisted of 30 double rigged shrimpers, 3 stern trawlers and 10 collectors, while in addition 6 shrimpers were under construction.

Table 1

DATA ON THE KUWAIT SHRIMP FISHERIES. LANDINGS IN METRIC TONS OF WHOLE SHRIMP[1]. INDUSTRIAL CATCHRATES[2] IN KG PER BOAT DAY. DATA ADAPTED FROM FAO/UNDP 1982

| | Landings | | | | Estimated effort |
Year	Industrial	Artisanal	Total	Catchrate	in industrial boatdays
1965–66	2 900		2 900	915	3 170
1966–67	3 335		3 335	1 005	3 320
1967–68	2 850		2 850	776	3 670
1968–69	2 490		2 490	820	3 035
1969–70	1 810	60	1 870	465	4 020
1970–71	1 050	115	1 165	412	2 830
1971–72	1 670	225	1 895	552	3 435
1972–73	1 590	485	2 075	438	4 735
1973–74	700	785	1 485	597	2 485
1974–75	900	755	1 655	448	3 695
1975–76	375	665	1 040	445	2 335
1976–77	1 190	855	2 045	526	3 890
1977–78	450	670	1 120	339	3 305
1978–79	650	750	1 400	221	6 335
1979–80	645	665	1 310	188	6 970[3]
1980–81	1 000	525	1 525	457	3 335[4]

[1] Whole weight × 0·6 = tail weight
[2] yearly catch divided by yearly effort
[3] closed season in April, May, June 1980
[4] closed season from February through June 1981

Table 2

DATA ON THE SHRIMP FISHERY IN THE SAUDI ARABIA/BAHRAIN AREA. LANDINGS IN METRIC TONS OF WHOLE SHRIMP[1]. CATCHRATES[2] FOR THE TWO MOST CONSISTENT GROUPS OF INDUSTRIAL VESSELS. GROUP I TENDED TO FISH MORE THE NORTHERN AND GROUP II THE SOUTHERN SAUDI ARABIAN WATERS. DATA PARTLY FROM FAO/UNDP 1982

| | Total landings (tons whole) | | | Catch (kg) per boatdays | | Estimated effort in boatdays | |
Year	Industrial	Artisanal	Total	Group I	Group II	Group I	Group II
1962–63	30		30				
1963–64	300		300				
1964–65	400		400				
1965–66	1 900		1 900	(1 308)		1 453	
1966–67	2 570		2 570	(874)		2 941	
1967–68	6 450		6 450	(1 112)	(800)	5 800	8 063
1968–69	6 510		6 510	1 166	(690)	5 583	9 435
1969–70	6 490		6 490	467	1 000	13 897	6 490
1970–71	6 050		6 050	418	767	14 473	7 835
1971–72	5 330	350	5 680	423	575	13 428	9 878
1972–73	6 410	1 000	7 410	695	807	10 662	9 182
1973–74	7 400	3 300	10 700	778	882	13 753	12 131
1974–75	4 500	2 250	6 750	399	857	16 917	7 876
1975–76	4 720	3 250	7 970	441	450	18 073	17 711
1976–77	4 910	5 150	10 060	677	508	14 860	19 803
1977–78	5 050	2 350	7 400	571	570	12 960	12 982
1978–79	2 900	1 450	4 350	428	305	10 164	14 131
1979–80	550[3]	700[3]	1 250[3]				

[1] whole weight × 0·6 = tail weight
[2] yearly catch divided by yearly effort; figures in brackets are estimates
[3] involving preliminary estimates.

Table 3

Industrial landings (in metric tonnes of whole shrimp)[1] from Iran waters, adapted from data in IOFC, 1977, Van Zalinge et al, 1979B and FAO/UNDP 1982. No data for Iranian companies were available after 1975–76. Catchrates[2] (kg per boatday) of Kuwait companies fishing in Iran waters

Year	Industrial landings	Catchrates
1961–62	300	
1962–63	1 700	
1963–64	2 700	
1964–65	9 600	
1965–66	8 100	1 735
1966–67	7 100	835
1967–68	7 900	630
1968–69	7 400	660
1969–70	2 700	340
1970–71	2 900	(682)
1971–72	2 900	692
1972–73	2 700	
1973–74	4 500	
1974–75	4 000	483
1975–76	6 500	735
1976–77		1 141
1977–78		434
1978–79		711
1979–80		549

[1] whole weight × 0·6 = tail weight
[2] yearly catch divided by yearly effort. The effort data prior to 1971–72 were in boat-months and have been changed to boat-days assuming 20 days fishing per month. The figure in brackets is an estimate.

Kuwait Most of the Kuwait boats evicted from Iran in 1969 were added to the fleets already working in Kuwait and—on payment of royalties—in Saudi Arabian waters, causing a jump in effort, and a drop in landings (*Fig. 2*) and in the profitability of the companies (Bromiley, 1972; IOFC, 1977). This led to a merger of the three major companies in 1972, and a further diversification into fishing outside the Gulf or for fish. The overall number of vessels in operation declined gradually, even though the second remaining company in Kuwait doubled its fleet from 6 to 12 vessels (Van Zalinge et al, 1979b).

Saudi Arabia/Bahrain While some small companies disappeared and the larger Saudi Arabian based company gradually reduced its fleet, the Bahrain and Qatar companies operating in Saudi Arabian/Bahrain waters maintained their fleets.

The decline of the industrial fishery on the west coast of the Gulf has mainly been attributed to the successful competition from the expanding artisanal trawl fishery, the development of which is described below, increasing total shrimp landings to unprecedented heights of over 12 000 t in 1973–74 and 1976–77 (*Table 2*), (Van Zalinge, 1980).

3.1.3 *Industrial fishery: 1979–present*

Saudi Arabia Probably because of the sharp decrease in industrial landings and catchrates in 1978–79 and early 1979–80, Saudi Arabia decided in 1979 to close its waters for foreign industrial vessels. The relative vacuum thus created was rapidly filled by some 4 emergent companies in addition to the already existing one, which were finally all brought under the umbrella of a newly formed Saudi Arabian company in 1981. The Saudi Arabian shrimp fleet consisted of 19 rented vessels in July 1981 with four new ones on order. Even the artisanal fleet underwent a reduction (section 7.1) (Van Zalinge, internal FAO reports).

Kuwait As a consequence of the Saudi Arabian measure one of the two Kuwait companies closed, reducing the industrial fleet to about 30 trawlers. However in 1981 when catches had improved, the defunct company came to life again.

Bahrain and Qatar Also the companies based in Bahrain and Qatar closed, leaving only 2–5 independent industrial vessels working in Bahrain (Abdul Qader, 1982).

3.2 *The artisanal fishery*

Dhows, the ancient Arab craft, have probably always been used for fishing purposes. It is however likely that shrimp fishing activities were limited before the introduction of trawling, as only the occasional use of large scoop nets for catching shrimp in very shallow waters of Kuwait has been recorded (Selim, 1968; FAO, 1970; Van Zalinge et al, 1981).

Following the example of the industrial fishery, trawl nets were adapted for (stern) use in motorized dhows. In the FAO/UNDP (1968) and Boerema (1969) reports it is mentioned that primitive trawl nets were used by the local fishermen in Bandar Abbas (Iran) probably as early as 1965. A gradual diffusion of the technique seems to have followed, as according to local sources the first trawl for a dhow was used in Kuwait in 1969 (Van Zalinge et al, 1981) and in Bahrain and Abu Dhabi in 1971 (Abdul Qader, 1982; Van Zalinge, travel report).

Stimulated by an increasing local demand for shrimp and sometimes favoured by local regulations like the closure of Kuwait Bay to industrial fishing, artisanal shrimp catches increased (*Tables 1 and 2*) and formed between one-third and one-half of the total west coast catches since 1973–74 (Van Zalinge, 1980).

An artisanal shrimp trawler in Kuwait measures on the average 16 m in length and is powered with an inboard diesel engine of about 94 hp. The Gulf of Mexico type 4 seam net often consists of a single mesh size throughout (between 27 and 33 mm stretched), and has a double layered codend, retaining virtually every shrimp that enters the net (Van Zalinge *et al*, 1981). More detailed reports on the Iran, Bahrain and Qatar artisanal fisheries are those of Abbes and Farrugio 1977, Davidson Thomas, 1978 and Kristjonsson, 1978. Temporary regulations for catching fish in Iran waters (Southern Fisheries Company of Iran, 1977) stipulate a minimum (unspecified) mesh size of 20 mm in the bag of local shrimp trawl nets.

Van Zalinge *et al* (1979 b) report that they had counted 123 shrimp dhows registered in Kuwait, while Davidson Thomas (1978) mentions a number of 26 for Sitra, the main base of operation in Bahrain. Shimura (1978) gives numbers of artisanal fishing boats for the Gulf countries, although not always specifying their particular denominations. He reported for Iran 400 boats in Bandar Abbas, 342 in Bushehr (Abbes and Farrugio, 1977) counted 451 boats in 44 ports between Dilam Bay and Bandar Abbas) and an unknown number in Khorramshahr, for the Saudi Arabian Gulf coast 230 (Price and Jones, 1975, estimated approximately 300) and for Qatar 35, Van Zalinge (travel reports) mentions 48 artisanal craft in Iraq and between 10–25 shrimp dhows in Abu Dhabi. Altogether, the number amounts to well over a thousand shrimp dhows in the Gulf.

According to Mathews (1981b) one industrial unit of effort equals four artisanal units of effort. The entire artisanal fleet in the Gulf appears then to be potentially of a similar strength as a fleet of at least 250 industrial trawlers, that is a fleet exceeding the size of the industrial fleet at its maximum expansion around 1969.

3.3 *Data base*
3.3.1 *Industrial data* Statistics have usually been collected on an *ad hoc* basis and were provided by courtesy of the companies. The only exceptions are in Kuwait where FAO organised a routine collection in 1978, and in Iraq.

Catch and effort statistics come usually by country of origin of the catches in terms of landed (size-graded) weight and days fishing/days-at-sea (*Tables 1, 2* and *3*). Problems are the multispecies composition of the catches *eg* in Kuwait (Van Zalinge *et al*, 1979b; Mathews, 1981) and discarding (Kristjonsson, 1968), while large differences in vessel efficiency, even within one company, limit the value of the effort data (Van Zalinge, 1980).

There is also doubt about the validity of the older (pre-1969) data particularly with respect to the origin of the landings made by some of the Kuwait companies (FAO, 1970; Van Zalinge *et al*, 1981).

Data from the Bahrain, Kuwait and Qatar companies are generally complete contrary to those from the Saudi Arabian and Iran companies.

3.3.2 *Artisanal data*
Iran and Iraq Information on the artisanal fisheries is very limited and is not collected routinely.

Kuwait A programme for the collection of detailed catch and effort data was set up by FAO in 1977 and is continued by the Kuwait Institute for Scientific Research (Van Zalinge, 1979; Mathews, 1981a). Market data is collected by the Ministry of Planning since 1972, but does not cover all shrimp supplied by the artisanal fishery and is not separated by species (Van Zalinge *et al*, 1979b, 1981).

Saudi Arabia Port custom authorities register landings of shrimp and fish probably since 1971. Unfortunately there is considerable doubt about the reliability of the data from some ports (Hull, 1978).

Bahrain A catch and effort collection scheme was set up by FAO in 1976 and is running continuously since 1979 (FAO, 1978; Abdul Qader, 1982; Bazigos, 1982). Collection of market data started in 1981.

Qatar and the United Arab Emirates Preparations for routine market data collection are underway.

4 Effect of fishing on the shrimp stocks

4.1 *History of stock assessment in the Gulf*
Boerema (1969) concluded that the fishery off Iran and probably also off the west coast of the Gulf (corroborated by Lewis *et al*, 1973, for Saudi Arabian waters) had already reached a stage in 1968–69 where a further increase in fishing effort would possibly result in a decrease in total landings.

The IOFC 1973 report agreed that fishing seemed to have exerted a substantial albeit from area to area varying effect. Reasons were the serious drop in catch-rates in some areas between 1968–69 and 1969–70, accompanied by an increase of small shrimp in the catches. It was provisionally estimated that between 98–123 industrial vessels would perhaps be required to obtain the maximum yield in the Gulf.

The IOFC 1977 report considered that fishing was the main cause of changes in catch. It was noted that depleted stocks did not recover when (industrial) fishing decreased and it was suggested that heavy fishing had affected the reproductive potential of the stocks, although the possibility of a relationship with environmental changes, especially in the Kuwait area was not excluded. This analysis and conclusion were based on series of somewhat incomplete industrial data alone, and ignored, through lack of data, the already substantial artisanal fishery, of which the performance most likely accounts for the observed failure of the industrial landings to recover.

Van Zalinge (1980) considering only the west coast stocks in their totality and taking account of all the available, albeit still scanty and approximate data, on the artisanal fisheries, found that the average level of the combined landings had probably not declined, even though the industrial share in the total landings showed a decreasing trend. Considering the great year-to-year fluctuations in the landings that appeared to occur in the 1970s, it was proposed that, due to the competition of the artisanal and industrial fisheries, overall fishing effort probably had risen to unprecedented high levels, thereby inducing an increased risk of recruitment overfishing among the shrimp populations which under unfavourable environmental conditions could lead to a serious decline in shrimp yields, as occurred in 1978–79 and 1979–80.

Mathews (1981a and b) evaluating the Kuwait stocks, tentatively concluded, that a reduction of effort on *P. semisulcatus* would increase catches. This could be done *eg* through a closed season. Alternatively it was recommended to consider closing Kuwait Bay for protection of *P. semisulcatus*, as the artisanal fishery that is permitted to operate in the Bay, contrary to the industrial fishery, could have an effect on the industrial catches. On the other hand it was thought that *M. affinis* was not in need of protection.

4.2 *Present situation*

The most up-to-date assessment of the state of the shrimp stocks in the Gulf was made at a workshop in Kuwait (FAO/UNDP, 1982).

Kuwait In Kuwait (*Figure 1*) the combined effort expenditure in the industrial and artisanal sectors in most years appeared to have been fairly stable between 3 000 and 4 000 boat days per year, roughly giving the optimum catch. However catch and catchrates have been fluctuating a lot particularly when the years before 1969–70 are taken into account.

Fishing did not appear to be the cause of these variations, which was supported by the low exploitation aspect of the seasonal catchrate patterns (see section 2.2). It was assumed that recruitment

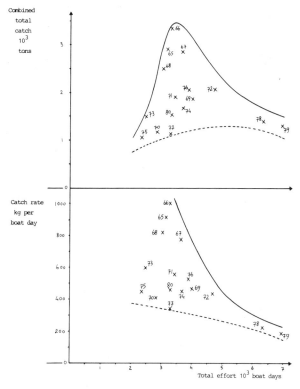

Fig 1 Kuwait. Production model for the combined industrial and artisanal fisheries. Adapted from FAO/UNDP, 1982

was the most probable cause, although errors in the catch and effort data and changes in annual catchability were mentioned.

The closure of the Saudi Arabian waters to foreign shrimp fishing in 1979 led predictably to an excess in fishing effort expenditure in Kuwait waters in 1978–79 and 1979–80. This was further aggravated by the outbreak of the Iraq-Iran war in 1980, which prevented a large part of the industrial fleet from fishing in Iran waters, although some compensation came from the reducing effect of the war on the artisanal fleetsize. The early onset of the closed season for shrimp fishing in February 1981 limited effort in 1980–81 to a pre-1978 level (fig. 1).

The possible effects of a February–June and an April–June closed season on the Kuwait shrimp stocks were analysed using a Ricker yield per recruit model for different sets of values. The more likely outcome (with $M = 3$) indicated that a very small decrease in yield per recruit, but a significant increase in spawning biomass would occur for both closed periods; the effect becoming more pronounced with increasing duration. Similar conclusions were reached by Garcia and Van Zalinge (1982) using a different approach involving the swept area method. It was pointed out though, that a closed season lasting through June can partly deprive the artisanal fishery of the new recruitment catch, as the bulk of the shrimp may have migrated away from the artisanally to the industrially exploited grounds by the time fishing is allowed again.

Saudi Arabia and Bahrain The data base being rather weak, it was not possible to separate catches and effort into their Saudi Arabian and Bahrain components.

The picture emerging from this combined data (*Table 2*) is typical for a heavy exploited fishery (*Fig. 2*). After reaching a level of about 6 000 tons the average catch appeared to plateau more or less, rising only slightly relative to the huge increase in effort, while variability increased markedly.

This situation changed in 1979 when effort was dramatically reduced by banning foreign fleets from Saudi Arabian waters. The resulting very low level of effort expenditure was apparently maintained in the following years, although precise data are lacking.

Abdul Qader (1982) mentions that in Bahrain the 1980–81 catchrates had improved considerably after the effort reduction through the demise of the industrial company and the application of a closed season, although this improvement did not continue in the following year.

Iran It is uncertain what happened in Iran due to the complete lack of artisanal fishery data. Moreover the only information on the industrial fishery after 1975 is from the Kuwait boats operating in Iran waters. The catch per unit of effort in successive years (*Table 3*) indicating changes in stock size showed a similar, although somewhat bolder, pattern of fluctuations as was found for the west coast.

4.3 *Stock-recruitment*

Seasonal recruitment curves of the Saudi Arabia/

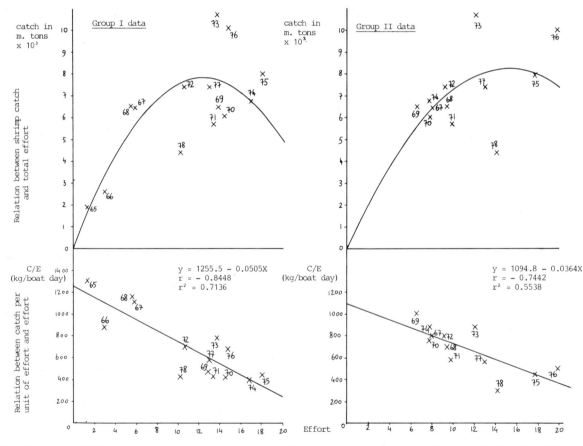

Fig 2 Saudi Arabia/Bahrain. Production models for the combined industrial and artisanal fisheries. Partly adapted from FAO/UNDP, 1982

Bahrain and of the Kuwait stocks appeared to be highly variable over the years, exhibiting a decreasing trend since the early years of the fishery (*ie* 1965), although much less so since 1969. Fishing did not seem to be the cause, as the agreement between the two curves was good, even though patterns in effort expenditure differed widely between Saudi Arabia/Bahrain and Kuwait.

Morgan and Garcia (1982) related estimates of the spawning stock abundance and the strength of the subsequent recruitment in Kuwait and found an apparent quasi linear relationship, which—if true—suggests that enlargement of the spawning stock would improve recruitment. However they pointed out that the relationship may largely be an artefact, because the trends in recruitment appeared to be unrelated to fishing. And if this is the case only an amelioration of the environmental situation could lead to a recovery of the fishery, provided it is reversible, which may not be so if land reclamation, decreased Shatt Al Arab river outflow through dam construction, *etc* were the underlying cause(s).

5 Other species

5.1 *By-catch and discards*

5.1.1 *Industrial fishery* Unlike the industrial fishery for shrimp, a similar one for fish has never been of much importance and contributes less than a quarter to the toal fish production in the Gulf at present (FAO/UNDP, 1981a).

The State Organization for Fisheries in Iraq employs 8 vessels that are trawling for fish in the Gulf, and claims to produce annually about 6 t of shrimp as by-catch.

Bromiley (1972) writes that the economic enticement of placing fish on the local markets by industrial shrimp fishing companies holds little attraction compared to that of exporting shrimp. Given the freedom of choice, the policy is not to use valuable ice and storage space early in a trip, running the risk of a shortage when shrimp are still to be caught. In general, therefore, fish landings of these companies are insignificant, although the bulk of the catch consists of fish.

Fibiger and Frederiksen (FAO, 1957) report 10% shrimp and 90% fish in 42 hauls from the Bandar Abbas area; Drobysheva (1971) 23% shrimp and 77% fish, of which 11% marketable, for Northern Iran in 1965–67, and FAO (1970) 9% shrimp and 91% fish, of which 10% marketable, in 51 hauls from Kuwait waters.

Grantham (1980) lists 33 important by-catch species or groups of species for the west coast fisheries, of which five exceed 10% of the catch weight: mojarras (*Gerres* sp.), ponyfishes (Leiognathidae), tripod fishes (*Pseudotriacanthus* sp.), goat fishes (Mullidae) and sharks and rays, all of low commercial value in the Gulf area.

The predominant species in the marketable fish category for the Iran coast were threadfin breams (*Nemipterus*), lizard fish (*Saurida*), catfish (*Tachysurus*), croakers (*Otolithes, Johnius*), scads (*Selar, Decapterus*) and scavengers (*Lethrinidae*) (Drobysheva, 1971).

5.1.2. *Artisanal fishery* Van Zalinge *et al* (1979a) report that the artisanal shrimp fisheries discard small shrimp, mainly consisting of such non-commercial species as *Metapenaeopsis* sp. and *Trachypenaeus* sp., as well as most of their fish catch for the same reasons as the industrial trawlers, being even more pressed for space, as shrimp is usually kept in ice-boxes on deck. All discarded shrimp and fish are returned dead to the sea, at least in summer.

Of the artisanal fish catch less than 10% is retained in summer, but this percentage may be higher in winter, while it also depends on the species composition of the catch. Retained species in Kuwait Bay (July, 1978) were generally croakers (*Otolithes*), tonguesoles (*Cynoglossus*), flounders (*Pseudorhombus*), flatheads (*Platycephalus*), threadfin breams (*Nemipterus*), crabs and squid. Discarded were the small individuals of these species as well as lizardfish (*Saurida*), catfish (*Arius*), sardines, anchovies and shads (*Thryssa, Sardinella, Rhonciscus, Nematolosa* and *Ilisha*).

5.2 *Predators*

Epinephelus tauvina, the most common grouper (Serranidae) in the Gulf (FAO/UNDP, 1981a) is an important predator on shrimp (Basson *et al*, 1977), as well as some sharks, guitar fishes and other batoids. Drobysheva (1971) names also *Otolithes ruber*, of which 67% of the non-empty stomachs contained shrimp remains. Also *Tachysurus dussumieri, Johnius soldado* and *Saurida tumbil* stomachs were often found to contain shrimp, while for a number of other fish species no mention of a shrimp diet is made. FAO/UNDP (1981a) reports shrimp in the stomach contents of *Arius thalassinus, Lutjanus malabaricus, Nemipterus japonicus, Plectorhynchus pictus* and *Argyrops spinifer*, but not in *Epinephelus tauvina* nor in *Saurida tumbil*, contrary to the above reports.

No quantitative data are available, but if in particular Nemipteridae, Ariidae, Pomadasyidae have much shrimp in their diet, their total consumption of shrimp could be high, as their biomass was estimated to be respectively 94 000, 46 000 and 41 000 t, together more than 15% of the total demersal and pelagic fish biomass in the Gulfs (FAO/UNDP, 1981a).

6 Environmental aspects

6.1 Oceanography
Descriptions of the general hydrological regime in the Gulf are found, eg in Unesco (1976); Fisheries of Kuwait (1978); and FAO/UNDP (1981b). Distributions of substrates are given by Emery (1956) and Wagner and Van der Togt (1973).

The seasonal variation in (surface) temperature is large, possibly due to the enclosure of the Gulf by arid landmasses, and ranges between 10°–36°C inshore and 18°–34°C offshore. Variations in bottom temperature can be up to 10°C smaller due to the development of a thermocline in summer (FAO/UNDP, 1981b).

Mean annual rainfall varies from less than 10 cm (Abu Dhabi, Qatar, Bahrain) to about 25 cm (Bushire).

Evaporation of Gulf waters exceeds the sum of precipitation, river discharge (Shatt Al Arab and some rivers on the Iran coast) and underseas upwelling of freshwater along the west coast, and is balanced by an inflow of Indian Ocean waters. As a result Gulf waters have a relatively high salinity of 37–42‰.

The freshwater outflow of the Shatt Al Arab river is largest in spring and is detectable in Kuwait waters as a result of the general counter-clockwise current movement in the northern Gulf (Enomoto, 1971; Drobysheva and Pomazanova, 1974; Mathews et al, 1981).

6.2 Pollution and land reclamation
With the rapid industrialization and population growth experienced by the Gulf countries pollution must have increased throughout the history of the shrimp fisheries, although as yet little factual information exists.

Oil slicks are regularly reported in the Gulf area and although an investigation into tarball concentrations—an indicator of the oil pollution in adjacent waters—on Kuwait beaches in 1977 and 1979 did not show an increase, the recorded levels of around 388 g/m belong to the highest in the world and are approximately 8–80 times higher than values quoted for the U S A coasts (Al-Harmi and Anderlini, 1979).

However, trace metal levels in Kuwait sediments are comparable with other relatively clean areas of the world, except for higher levels of Ag and Ni in Kuwait (Samhan et al, 1979).

Land reclamation, filling and dredging works are carried out in most Gulf countries (eg Bahrain, Tarut Bay in Saudi Arabia, Kuwait Bay), often affecting postlarval and juvenile habitats. Unfortunately precise information on the yearly extent of the nursery area destroyed has not become available. Mohamed et al (1979b) estimated that along the Kuwait coast 10 km^2 were reclaimed over the years till 1979. This affected approximately 15% of the entire Kuwait coastline, and more significantly 35% of the Kuwait Bay coast, which is though to harbour Kuwait's main nursery area for P. semisulcatus.

7 Management

7.1 Framework and background
Although governments are responsible for fisheries management, the Committee of the Indian Ocean Fishery Commission for the development and management of the fishery resources of the Gulf is the main body for the evaluation and subsequent recommendation to the governments of management options, that thus far have been documented, formulated and presented by FAO (cf. IOFC, 1979), occasionally accompanied or preceded by a working party or workshop on stock assessment (cf. IOFC, 1977 and FAO/UNDP, 1982). In a similar way the Kuwait Institute for Scientific Research has recently (cf. UFK/KISR, 1981) become involved on the Kuwait scene.

The increasing economic prosperity of the Arab countries on the west coast of the Gulf has led to a diminished vocational interest in fisheries. Governments and also private people are concerned about this trend, as fishing and seafaring in general are considered a national heritage, that with an eye on the future should be kept alive along familiar traditional lines. This has given rise to the peculiar situation in Kuwait, where a substantial artisanal fleet is manned exclusively by expatriates (mainly Iranians), but is owned by Kuwaitis who often take a considerable interest in the management of the vessels without actually going out to sea. In Saudi Arabia, Bahrain and Abu Dhabi this process has apparently not advanced so far and captains are often still local people, although the crew may be mainly expatriates from Iran, Oman, Pakistan and India. A consequence of this situation made itself recently felt particularly in Kuwait and Saudi Arabia, where the artisanal fleet size underwent a reduction as a result of a government policy limiting entry of expatriate crews.

A sign of the importance attached to the artisanal fisheries by some governments was the early reservation of certain fishing grounds for their exclusive use, eg Kuwait Bay since 1968 (FAO, 1970) and the Ras Bahrgan area in Iran (IOFC, 1973).

A very important aspect is also that the artisanal fisheries have the function of supplying the local markets with fresh shrimp and fish, which eg plays an important rôle in the Kuwait government policy of providing cheap basic food to the populace through price stabilization measures, contrary to the export oriented industrial shrimp fishery, whose export earnings are anyhow trivial compared to those from oil. However it may be significant that the recently established (government supported) Saudi Arabian company now caters primarily for the local markets.

7.2 Management measures

The cause of the failure of the shrimp fishery in 1979–80 has mainly been attributed to overfishing (IOFC, 1979) inducing all the west coast countries to adopt a closed season—probably because of the simplicity for execution of such a measure—for both the industrial and artisanal sectors in 1980 and 1981.

Saudi Arabia In Saudi Arabia moreover the fishing rights of foreign companies were withdrawn, while also a reduction of the artisanal fishery occurred. As a result the overfishing situation was reversed so completely that there even seems to be scope for a limited increase in effort provided the catch levels mentioned earlier (section 4.2) and the apparent decline in recruitment (section 4.3) are taken into consideration.

Kuwait The fishery was closed for three months (April–June) in 1980 and for five months (February–June) in 1981. As shrimp catches improved the revival of a defunct shrimp company aggravated the still existing excess in industrial effort. In the absence of other effort curbing measures the fishery was closed again for five months (February–June) in 1982. However this is not sufficient to limit effort to acceptable levels (section 4.2).

As was already pointed out (section 4.2) these closed seasons probably affect the artisanal sector proportionally more than the industrial sector.

Bahrain Measures similar to those taken in Kuwait were applied, although apparently with less effect in respect to the artisanal fishery. It was decided to impose again a closed season (March–June) in 1982.

Qatar and the United Arab Emirates The same policy as followed by Bahrain and Kuwait was adhered to. The shrimp stocks are small and vulnerable to fishing and in particular do not seem to tolerate any industrial fishing.

Iran Already in 1977 the Southern Fisheries Company in Iran formulated temporary regulations for fishing, prohibiting fishing in shrimp spawning and nursery areas and during a $2\frac{1}{2}$ months period considered to be the spawning season for shrimp.

8 Future Work

Emphasis has still to be put on the collection of landings and effort data, in particular for the artisanal fishery. Industrial data collection should be refined to specify fishing areas more precisely, while both for the industrial and artisanal landings species composition should be determined through special sampling programmes.

Biological studies should embrace in particular stock identification (through tagging experiments), assessments of natural mortality and growth rates, monitoring of recruitment levels and catchability patterns in relation to long- and short-term fluctuations of environmental conditions, and the determination of the effects of landreclamation and destruction of nursery areas.

In addition it is envisioned that information will be needed for monitoring the socio-economic effects of the management measures on the fishery sectors (in particular the artisanal one).

9 References

ABBES, R et FARRUGIO, H. Analyse des activité de la pêche
1977   traditionelle sur les côtes iraniennes entre Busher et Bandar Abbas. La Société des Pêcheries de Sud de l'Iran. Bulletin no. 13, 43 pages.

ABDUL QADER, I. The impact of the closed season on the shrimp
1982   fishery in Bahrain. In: FAO/UNDP 1982.

AL ATTAR, M H. Shrimp spawning and nursery areas in Kuwait
1981   Bay. In: UFK/KISR 1981.

AL HARMI, L and ANDERLINI, V. Survey of tar pollution on
1979   beaches of Kuwait. Kuwait Institute for Scientific Research. Annual Research Report for 1979. ISSN 0250-4065: 82–85.

AL HOSSAINI, M. Maturation and spawning season in Kuwait's
1981   shrimp fishery. In: UFK/KISR 1981.

BASSON, P W, BURCHARD, J E, HARDY, J T and PRICE, A R G.
1977   Biotopes of the western Arabian Gulf: marine life and environments of Saudi Arabia. Aramco, Bahrain, Saudi Arabia. 284 p.

BAZIGOS, G P. On the efficiency of fisheries statistical systems
1982   for shrimp fishery with special reference to the Gulf area. In: FAO/UNDP 1982.

BOEREMA, L K. The shrimp resources in the Gulf between Iran
1969   and the Arabian peninsula. FAO Fish. Circ., (310): 29 p.

BOEREMA, L K and JOB, T J. The state of shrimp and fish
1968   resources in the Gulf between Iran and the Arabian peninsula. Meeting doc. IOFC/68/Inf. 11:18 p.

BOEREMA, L K. Report on a consultancy to the shrimp resources
1979 evaluation and management project. FAO/KISR report. UTFN/KUW/006/KUW/R9: 17 p.

BROMILEY, P S. An economic feasibility study of a trawl fishery
1972 in the Gulf lying between Iran and the Arabian peninsula. FAO/UNDP Rome Indian Ocean Programme. IOFC/DEV/72/23: 64 p.

DAVIDSON, THOMAS. A study of fishing gear and methods in the
1978 artisanal fisheries of Bahrain. Consultant report to the FAO Regional Fishery Survey and Development Project. RAB/71/278: 14 p.

DROBYSHEVA, S S. The crustacean fishery. In: Commercial
1971 description of the Persian Gulf. USSR Ministry of Fisheries, Institute of Fisheries and Oceanography of the Azov Sea and Black Sea (Acherniro) (in Russian).

DROBYSHEVA, S S. The crustacean fishery. In: Commercial
1971 description of the Persian Gulf. USSR Ministry of Fisheries, Institute of Fisheries and Oceanography of the Azov Sea and Black Sea (Acherniro) (in Russian).

DROBYSHEVA, S S and POMAZANOVA, H P. The distribution of
1974 shrimp (Penaeidae) in the Persian Gulf with reference to environmental conditions. Tr. VNIRO (99): 7–17 (in Russian).

EGENAES, W H. Kuwait—10 vessel Norway-built fleet develop-
1965 ing shrimp resources. Ocean Fisheries, April 1965: 31–32.

EMERY, K O. Sediments and water of the Persian Gulf. Bull.
1956 Am. Assoc. Petrol. Geol., 40(10): 2354–86.

ENOMOTO, Y. Oceanographic survey and biological study of
1971 shrimps in the waters adjacent to the eastern coast of the State of Kuwait. Bull. Tokai. Reg. Fish. Res. Lab., (66): 1–73.

FAO, Report to the Government of Iran on the exploratory
1957 fishing survey in the Bandar Abbas region (1955/56). Based on the work of H. Fibiger and K Frederiksen, FAO masterfishermen. Rep. FAO/ETAP (676): 36 p.

FAO, Report to the Government of Kuwait on the fisheries of
1970 Kuwait. Based on the work of A Meschkat, FAO Fishery Consultant. FAO/KUW/TF 27: 36 p.

FAO, Report to the Government of Bahrain. Shrimp Project
1978 findings and recommendations. Based on the work of K H Mohamed, FAO fisheries biologist. FI: DP/BAH/74/017: 58 p.

FAO, Report to the Government of Kuwait. Shrimp resources
1980 evaluation and management. Based on the work of K H Mohamed and N P van Zalinge, FAO fisheries biologists. Kuwait Funds-in-Trust, FAO/UTFN/KUW/006/KUW/R 12: 10 p.

FAO/UNDP, Report to the Government of Iran on the South-
1968 ern Fisheries Company. Based on the work of B Andersskog, FAO/TA Fishery Adviser. Rep. FAO/UNDP/TA 2523: 20 p.

FAO/UNDP, Demersal resources of the Gulf and the Gulf of
1981a Oman. Regional Fishery Survey and Development Project. Rome, FAO. FI: DP/RAB/71/278/10: 122 p.

FAO/UNDP, Environmental conditions in the Gulf and the
1981b Gulf of Oman and their influence on the propagation of sound. Regional Fishery Survey and Development Project. Rome, FAO. FI: DP/RAB/71/278/12: 62 p.

FAO/UNDP, Report on the workshop on Assessment of the
1982 Shrimp Stocks in the West Coast of the Gulf between Iran and the Arabian peninsula, Kuwait 17–22 October 1981, FI: DP/RAB/80/015: 163 p.

FARMER, A S D and UKAWA, M. Provisional atlas for the
1980 commercially important penaeid shrimps of the Arabian Gulf. Kuwait Institute for Scientific Research Internal Report.

FARMER, A S D and AL ATTAR, M H. Results of shrimp marking
1981 programmes in Kuwait. In: Proc. int. shrimp releasing, marking and recruitment workshop, Kuwait 25–29 Nov. 1978. Kuwait Bull. Mar. Sc., (2): 53–82.

FISHERIES OF KUWAIT, Ministry of Public Works, Agricultural
1978 Department, Fisheries Division, Kuwait: 256 p. (in Arabic).

GARCIA, S and VAN ZALINGE, N P. Shrimp fishing in Kuwait:
1982 methodology for a joint analysis of the artisanal and industrial fisheries. In: FAO/UNDP 1982.

GRANTHAM, G J. Prospects for by-catch utilization in the Gulf
1980 area. Regional Fishery Survey and Development Project. Rome, FAO. FI: DP/RAB/71/278/14: 43 p.

HAMDAN, F, MATHEWS, C P and FARMER, A S D. The incidence
1981 of exceptionally large catches of shrimp. In: UFK/KISR 1981.

HULL, L E. Summary of the present state of knowledge con-
1978 cerning the Saudi Arabian Gulf shrimp fishery with particular reference to the artisanal sector. Ministry of Agriculture and Water/White Fish Authority Fisheries Development Project. Field report 24: 13 p.

IOFC, First Session of the Indian Ocean Fishery Commission
1973 special working party on stock assessment of shrimp in the Indian Ocean area, Manama, Bahrain, 29 November–2 December 1971. FAO Fish. Rep., (138): 40 p.

IOFC, Ad hoc group of the IOFC special working party on
1977 stock assessment of shrimp in the Indian Ocean area to consider the stocks in the area covered by the UNDP/FAO Regional Fishery Survey and Development Project. REM/71/278 (Bahrain, Iran, Iraq, Kuwait, Oman, Qatar, Saudi Arabia, United Arab Emirates). Doha, Qatar, 26–29 April 1976. FAO Fish. Rep. (193): 23 p.

IOFC, Indian Ocean Fishery Commission. Committee for the
1979 Development and Management of the Fishery Resources of the Gulfs. Doha, Qatar, 18–20 September 1979. FAO Fish. Rep. (223): 15 p.

JONES, D A and AL ATTAR, M H. Observations on the post
1981 larval and juvenile habitats of Penaeus semisulcatus in Kuwait Bay and adjacent waters. In: UFK/KISR 1981.

JONES, R and VAN ZALINGE, N P. Estimates of mortality rates
1981 and population size for shrimp in Kuwait waters. In: Proc. int. shrimp releasing, marking and recruitment workshop, Kuwait 25–29 Nov. 1978. Kuwait Bull. Mar. Sc., (2): 273–288.

KEYVAN, A, KAMRAN, M and NIKOUYAN, A. Iran, present
1978 fisheries situation. In: Report on the FAO/Norway workshop on the fishery resources of the north Arabian sea. Vol. 1 IOFC/DEV/78/43.1 p. 8–10.

KRISTJONSSON, H. Techniques of finding and catching shrimp in
1968 commercial fishing. In: FAO Fish. Rep., (57), Vol. 2, p. 125–192.

KRISTJONSSON, H. Consultant report to the FAO/UNDP Reg-
1978 ional Fishery Survey and Development Project. RAB/71/278: 36 p.

LEWIS, A H, JONES, D A, GHAMRAWI, M and KHOSHAIN, S. An
1973 analysis of the Arabian Gulf shrimp resources landed in Saudi Arabia, 1965–1971. Bull. Mar. Res. Centre, Saudi Arabia (4): 8 p.

LONGHURST, A R. Crustacean resources, area reviews on living
1969 resources of the world's ocean. FAO Fish. Circ. 109, 13: 27–30.

MATHEWS, C P. The history of Kuwait's shrimp fisheries. In:
1981a UFK/KISR 1981.

MATHEWS, C P. Mortality, growth and the management of
1981b Kuwait's shrimp fishery. In: UFK/KISR 1981.

MATHEWS, C P, AL ATTAR, M H and SAMUEL, M. The formation
1981 of thermal, salinity and nutrient fronts in Kuwait Bay. In: UFK/KISR 1981.

MOHAMED, K H, VAN ZALINGE, N P, JONES, R., EL MUSA, M,
1979a EL HUSSAINI, M and EL GHAFFAR, A R. Mark recapture experiments on the Gulf shrimp, Penaeus semisulcatus, de Haan in Kuwait waters. FAO/Kuwait Institute for Scientific Research. Shrimp stock evaluation and management project UTFN/KUW/006/KUW/R10: 61 p.

MOHAMED, K H, EL MUSA, M, EL HUSSAINI, M and EL
1979b GHAFFAR, A R. Observations on the biology of the exploited species of shrimp of Kuwait. FAO/Kuwait Institute for Scientific Research. Shrimp stock evaluation and management project UTFN/KUW/006/KUW/R11: 55 p.

MOHAMED, K H, EL MUSA, M and EL GHAFFAR, A R. Observa-
1981   tions on the biology of an exploited species of shrimp, *Penaeus semisulcatus* De Haan, in Kuwait. *In:* Proc. int. shrimp releasing, marking and recruitment workshop, Kuwait 25–29 Nov. 1978. *Kuwait Bull. Mar. Sc.* (2): 33–52.

MORGAN, G R and GARCIA, S. The relationship between stock
1982   and recruitment in the shrimp stocks of Kuwait and Saudi Arabia. Paper presented at the workshop on the scientific basis for the management of penaeid shrimps. Key West, 18–24 Nov. 1981.

PRICE, A R G. Temporal variations in abundance of penaeid
1979   shrimp larvae and oceanographic conditions off Ras Tanura, western Arabian Gulf. *Estuarine and Coastal Marine Science*, (9): 451–465.

PRICE, A R G and JONES, D A. Commercial and biological
1975   aspects of the Saudi Arabian Gulf shrimp fishery. *Bull. Mar. Res. Centre, Saudi Arabia*, (6): 48 p.

SAMHAN, O, ANDERLINI, V and ZARBA, M. Preliminary investi-
1979   gation of the trace metal levels in the sediments of Kuwait. Kuwait Institute for Scientific Research. Annual Research Report for 1979. ISSN 0250-4065: 93–96.

SELIM, H. Detailed study on fishing methods and gear used in
1968   Kuwait. Proceedings of the Convention of the Arab League on water resources and oceanography. Cairo, October 1968: 22 p.

SHIMURA, T. General report on fishery statistics, covering
1978   period April–July 1978. Regional Fishery Survey and Development Project, RAB/71/278. Industry Development Group Report No. 9: 49 p.

SOUTHERN FISHERIES COMPANY OF IRAN (PERSIAN GULF
1977   FISHERIES CO.). Temporary regulations for catching fish in the Persian Gulf, Oman Sea and all rivers of southern parts of Iran. Scientific and Technical Marine Fisheries Research Institute. Bulletin No. 10, 2nd edition: 11 p.

UNESCO. Study of the feasibility of a marine resources research
1976   centre in the United Arab Emirates, by K Grasshoff and S B Saila. FMR/SC/OCE/75/247 (FIT 9349): 89 p.

UFK/KISR. Proceedings shrimp fisheries management work-
1981   shop, United Fisheries of Kuwait and Kuwait Institute for Scientific Research, Kuwait, 17 Jan. 1981.

VAN ZALINGE, N P. Kuwait's artisanal fisheries. A sample
1979   programme for shrimp landings. FAO/Kuwait Institute for Scientific Research. Shrimp stock evaluation and management project UTFN/KUW/006/KUW/R5: 23 p.

VAN ZALINGE, N P. Report on the shrimp resources of the west
1980   coast of the Gulf (Bahrain, Kuwait, Saudi Arabia). Third Session of the IOFC Committee for the development and management of the fishery resources of the Gulfs. Doha, Qatar, 28–30 September 1980: 22 p.

VAN ZALINGE, N P, EL MUSA, M and EL GHAFFAR, A R. Mesh
1979a selectivity and discarding practises in Kuwait's shrimp fishery. FAO/Kuwait Institute for Scientific Research. Shrimp stock evaluation and management project. UTFN/KUW/006/KUW/R6: 17 p.

VAN ZALINGE, N P, EL MUSA, M, EL HUSSAINI, M and EL
1979b GHAFFER, A R. The Kuwait shrimp fishery and the shrimp resources in Kuwait waters. FAO/Kuwait Institute for Scientific Research. Shrimp stock evaluation and management project. UTFN/KUW/006/KUW/R7: 59 p.

VAN ZALINGE, N P, EL MUSA, M and EL GHAFFAR, A R. The
1981   development of the Kuwait shrimp fishery and a preliminary analysis of its present status. *In:* Proc. int. shrimp releasing, marking and recruitment workshop, Kuwait 25–29 Nov. 1978. *Kuwait Bull. Mar. Sc.* (2): 11–32.

WAGNER, C W and VAN DER TOGT, C. Regional distribution of
1973   the sediment types in the Persian Gulf. *In:* The Persian Gulf. Ed. B H Purser. Springer Verlag: 123–155.

A review of the shrimp fisheries of India: a scientific basis for the management of the resources

E G Silas,
M J George
and T Jacob

Abstract

Shrimp are caught along most of the 6000 km long coastline of India. Total catches expanded rapidly between 1965 and 1973, but since then catches have fluctuated, and most recent catches have been below the peak of 220 000 tons in 1975. A number of different species are caught, with a variety of gears, including trawlers in the offshore grounds, and several types of traditional gear in the inshore and lagoon areas. Examination of catch and effort statistics from different areas suggest that many stocks are fully exploited. In most fisheries shrimp make up only a small part of the total catch, but except for some large trawlers on the east coast, this by-catch is brought ashore. The geographical and seasonal variations in the by-catch are discussed.

   The varying objectives of management – biological, economic and social (including reducing con-
flicts between user groups) – are discussed. The problems include protection of the nursery areas in brackish waters, and control of fishing on the adult stocks. The various techniques that could be applied – seasonal and area closures, mesh regulation, limitation of fishing effort, and catch restrictions – are discussed.

1 Statement of the problem

The shrimp resources along the 6 100 km long coastline of India are being increasingly exploited both by the artisanal as well as industrial sectors. The pressure of fishing on existing stocks within the 75 m depth zone along the different regions of the coast is increasing with additional inputs of effort brought in by various programmes of mechanisation of the country craft and boats being implemented by the different maritime States as well as the entry of large business houses into shrimp fishing. These activities have resulted in placing India as the top ranking nation for shrimp production in the world ever since 1973.

However, a study of the trend in total production of shrimp over the past several years would indicate that the steady increase in production was maintained till 1973, the catch almost doubling, by this time. Although showing a decrease in 1974, the maximum of 220 thousand tonnes was recorded in 1975. But thereafter there is a downward trend, showing wide fluctuations below 200 thousand tonnes. These fluctuations in total shrimp production in recent years has resulted in the decline in catches in some areas causing apprehensions of depletion of the resource. This concern is genuine calling for evolving urgent measures for the proper management and conservation of this resource. The problem is of a complex nature on account of the multi-species nature of the fishery in which a wide variety of craft and gear are used.

In order to keep track of the stock position of the resources of different regions, in addition to the data collected on the total shrimp landings by various gears and crafts, regular monitoring of the biological and population characteristics is maintained at important shrimp landing centres in the different maritime States. In order to determine whether there are any indications of biological or economic overfishing in any of the areas under exploitation, macro and micro analysis of shrimp catch and effort data of the areas concerned is necessary. Scientific input applying resource assessment models to the available data pertaining to different regions is required in order to resolve the problems concerning conservation and management of the exploited shrimp stocks and for the continued sustenance of the fishery. The Central Marine Fisheries Research Institute is conducting this analysis to a certain extent. However, further effort on these studies is necessary.

2 Shrimp fisheries

2.1 The biology of major species

Shrimp fisheries of the different regions and the biological aspects of economically important species of India are fairly well documented. Full bibliographies and reviews of the main features of shrimp biology are available in species synopsis papers and other publications by George (1970a, 1970b, 1970c, 1970d, 1972, 1978, 1979), Kunju (1970), Mohamed (1970a, 1970b, 1973), Rao (1970, 1973) and Kurien and Sebastian (1975).

About 62 species of prawns and shrimps of the family Penaeidae, of which some are either commercially exploited at present or have great commercial potentialities, occur in the Indian waters. The others belong to families Sergestidae, Palaemonidae, Oplophoridae, Hippolytidae, Pandalidae and Atyidae. Based on the natural habitat of the adults, these species can be broadly grouped under three categories of penaeids and non-penaeids, namely deep-water species, littoral species and fresh water species. Of these, the important species contributing to the fishery are: *Penaeus indicus* H. Milne Edwards, *P. monodon* Fabricius, *P. semisulcatus* de Haan, *P. merguiensis* de Man, *Metapenaeus dobsoni* (Miers), *M. monoceros* (Fabricius), *M. affinis* (H. Milne Edwards), *Parapenaeopsis stylifera* (H. Milne Edwards), *P. sculptilis* (Heller), *P. hardwickii* (Miers), *Solenocera crassicornis* (H. Milne Edwards), *Hippolysmata ensirostris* Kemp, *Exopalaemon styliferus* (H. Milne Edwards), *Nematopalaemon tenuipes* (Henderson), *Acetes indicus* (H. Milne Edwards), *Macrobrachium rosenbergii* (de Man) and *M. malcolmsoni* (H. Milne Edwards). Various aspects like distribution of adults as well as different stages of life history, reproduction, spawning, larval history, adult history, population and exploitation concerning the different species are dealt with in these reviews.

Recently studies on the movement and migration of some of the commercial species of shrimps in the fishery by mark recovery experiments have been initiated and these have given some interesting results. Tagging of different species in CMFRI by using the Petersen disc tagging method commenced in 1972. A total of 3 189 tagged shrimps, mostly *Penaeus indicus*, *Metapenaeus dobsoni*, *M. affinis*, and *M. monoceros* were released from Goa (424 shrimp), Cochin (1 564 shrimp) and Madras (1 201 shrimp) between 1972 and 1974. A recovery of 2·1% was obtained, indicating localised movements, ranging to a maximum of 19 km from the place of release, except a specimen of *M. dobsoni* recovered 60 km away from the release position and another 25 km away after periods of 10 and 8 days from release respectively. In all these places none of the shrimps released in the backwaters were recovered from the sea.

During the years 1976 to 1980 more concentrated efforts were made in tagging of shrimps at Cochin using the loop tag and releasing them in the sea as well as in the backwaters. Out of a total of 15 830 *P. indicus* and *M. dobsoni* released in the sea off Cochin, 1·6% were recovered, all of them within a period of a fortnight after release and up to 10 km from the site of release, indicating only extremely limited movement. From 38 233 juvenile shrimps released in the backwaters of Cochin during these years only 0·8% was recovered. Among these recoveries only 6 specimens of *P. indicus* were obtained from the sea.

Although these results may probably be taken as pointing to the fact that the emigrant shrimps from the backwaters of Cochin are not the sole support of the shrimp fishery in the sea and that part of the brood produced in the sea remains there itself, only part of it migrating into the inside waters, further confirmatory evidence is necessary. A few recent recoveries of tagged shrimps using the loop tags released on the south west coast and captured from the south east coast may probably have far reaching implications concerning our approach to the assessment of the shrimp stocks of the different areas, their exploitation and management.

2.2 Description of the fishery

A section of the shrimp fishery of the country continues with the traditional crafts and gears, while mechanisation is slowly replacing the indigenous sector in several areas. The major development in mechanisation of shrimp fishery took place in the fifties with the introduction of shrimp trawling and at present trawling is being increasingly practised in most of the areas. In addition some of the indigenous gears like the 'dhol nets' of Maharashtra and gill nets of other areas are operated by mechanised boats. *Table 1* indicates the extent of shrimps landed in the country by the mechanised and non-mechanised sectors (average for the past 6 years).

Table 1

STATEWISE AVERAGE CATCHES (1974–79) OF SHRIMPS BY MECHANISED AND NON-MECHANISED SECTORS

State	Shrimp catch mechanised	Shrimp catch non-mechanised
Orissa	2 544	458
Andhra Pradesh	5 357	6 300
Tamil Nadu	9 109	4 000
Kerala	26 814	2 800
Karnataka	4 633	2 200
Maharashtra	100 000	9 600
Gujarat	4 600	6 400
Total	153 057	31 758

2.2.1 *Fishing crafts* A variety of indigenous crafts is used in shrimp fishing, from the simple catamarans of the east coast to the well-built canoes of Maharashtra on the west coast. Motorised pablo boats and small and large sized trawlers are engaged in shrimp trawling. Ramamurthy and

Muthu (1969) gave a detailed review of the fishing crafts and gears employed in the shrimp fishery of the country. Although the process of mechanisation of crafts has been in progress for the past several years, indigenous crafts like catamarans, canoes and plank-built boats are still operating in the small scale sector. According to 1973–77 census there were 106 480 non-mechanised crafts.

2.2.1.1 *Catamarans* The catamarans are primitive type of crafts used on the surf beaten coast, consisting of 3 to 5 logs tied together in a raft fashion. In different areas the size and number of the logs used vary slightly. Usually 2 to 4 men operate the craft.

2.2.1.2 *Canoes* Dug-out canoes are most common along the west coast, made by hollowing out a single log of wood and of varying sizes from 6·10 to 12·5 m length. Boat seines, shore seines, gill nets and cast nets are operated from those canoes often with a crew of 4 to 8 men. Plank-built canoes, out-rigger canoes and flat-bottom canoes are also in use in different areas.

2.2.1.3 *Plank-built boats* These are sturdy boats used in the northern part of both east and west coasts, used for bag net fishing. Manned by 7 to 12 men, these are considered most suited for mechanisation and quite a number of them have been mechanised. The length of the boat ranges from 6·5 to 13·0 m. The various types of plank-built boats have been indigenously evolved on the basis of their suitability for operation in the respective local conditions.

2.2.1.4 *Mechanised crafts* Motorisation of the indigenous crafts was the first step in the mechanisation of shrimp fishing. In due course many designs of small and medium sized mechanised boats to be operated from harbours and sheltered bays were introduced. The number of mechanised crafts currently in operation is 12 000. Shrimp trawling is mostly carried out by the Dan boats (6·6 × 2·2 × 1·0 m), Pablo boats (7·4 × 2·1 × 1·05 m) and shrimp trawlers (9·6 × 3·0 × 1·2 m and above). The horse power of the smaller boats ranged from 10 to 60. The larger of these boats are partly or fully decked and with trawling winches. Larger steel trawlers fitted with 90–300 HP engines and refrigerated fish holds are operated by some of the big firms as well as the Exploratory Fisheries Projects of the Government. The number of larger trawlers amounts to 75–100.

2.2.2 *Fishing gear* As in the case of fishing crafts, a variety of indigenous gears are operated for capturing shrimps in addition to the trawl nets. Nearly 0·7 million gears of assorted types are operated in the country according to the 1973–77 census. Ramamurthy and Muthu (1969) reviewed the different types of gears in operation in shrimp fishing. According to the mode of operation the gears can be grouped under the following categories.

2.2.2.1 *Fixed or stationary nets*   These include the various types and sizes of bag nets and stake nets operated against the flow of the tide in both inshore waters and brackish water areas. The bag nets constitute the most important gears for shrimp fishing in Bombay and Gunarat coasts, where they are locally known as 'Dol nets'. Depending on the manner in which these nets are operated there are two types, namely *Khunt fishing* and *sus fishing*. The nets are conical in shape, with a wide rectangular mouth. The size varies considerably, from 12 to 200 m in length with cod end mesh size of 10 mm. There are different types of bag nets operated in West Bengal and Andhra Pradesh also, locally known as '*Behundijal*' and '*Thoka vala*' respectively in these two areas. The fixed nets known as stake nets are in operation in the backwaters of west coast as well as east coast.

2.2.2.2 *Seine nets*   The seine nets include the seines with or without bags (and wings). They are known as boat-seines or shore seines depending upon whether they are hauled from a boat or from the beach. One of the important gears operated by the indigenous craft along Kerala coast is the boat seine known as *Thangu vala* of various dimensions, usually operated by two dug-out canoes with 6 to 10 men. Boat seines of different types and dimensions are in operation for catching shrimps in other areas also.

Although the shore seines are mostly used for catching inshore pelagic fishes, prawns are also caught in these nets. Shore seines of varying sizes are in use in all the areas of the coastline.

2.2.2.3 *Cast nets or falling nets*   These are very common and primitive gears used all along the coast and limited in their efficiency. They are operated by a single person very near the shore in the open sea as well as in the creeks and estuaries. The size of the net varies from 2·5 to 6·0 m in radius with webbing of mesh size 10 to 20 mm. The net is cast fully spread and as it closes traps the fishes and prawns in the water column below the net.

2.2.2.4 *Scoop nets or skimming nets*   These are employed exclusively in the creeks and backwaters and comprise of the hand net, push net and lift net.

The Chinese dip nets of Kerala backwaters is a type of lift net.

2.2.2.5 *Drift nets*   The drift nets, also called gill nets, are passive wall nets of selective nature made of cotton, hemp or synthetic fibre. The gill nets are at present increasingly used in fishing larger sized shrimps from the sea in certain regions.

2.2.2.6 *Trawl nets*   With the increase in demand for shrimps for processing and export, and the spread of mechanisation, stern trawling, particularly for shrimps, was attempted even with small mechanised boats and met with unprecedented success. Consequent to the expansion of the shrimp industry in a big way this new fishing method has come to stay, although indigenous crafts and gears are also being operated for catching shrimps to a certain extent.

Otter trawls are the most effective gears operated for shrimp fishing, the sizes of the trawl nets varying with the sizes of the crafts from which they are operated. Generally two or four seam trawl nets, overhang or non-overhang type with headline length of 7 to 27 m between the upper wing ends are used. Depending on the dimensions of the net and the towing power required the size and weight of the otter boats vary. The Indian Standards Institution has brought out requisite standards for the stern trawling gears for the different class of vessels.

Several new designs of trawling gear were introduced during the last few years. Design of a 15·25 m four-seam trawl for operation from a 9·45 m trawler is very popular. In addition to these trawls, bulged belly trawls are also in use. A 15 m bulged belly trawl suitable for 10·97 m trawler is being increasingly used. Some of the larger trawlers are resorting to out-rigger trawling.

2.2.3 *Historical review of catch trends*

2.2.3.1 *Total shrimp production*   A look at the trends in the total production (*Table 2* and *Fig. 1*) over the past 20 years shows that from 1962 through 1968 the catch, although increasing, remained below 100 thousand tonnes. Banerji (1969) has indicated this by statistical analysis of the catches. Between 1969 and 1973 there was a steep increase, the catch almost doubling. After 1973 there were fluctuations from year to year, the production decreasing in 1974 to 170 thousand tonnes, reaching the maximum of 220 thousand tonnes in 1975 and again going down to less than 200 thousand tonnes in subsequent years. The shrimp production in 1979 was 177 582 tonnes, that for 1980 170 737 tonnes and 144 969 tonnes in 1981. The trend in triennial average catch shows an overall increase

Table 2
MARINE PRAWN LANDINGS OF INDIA (IN TONNES) FROM 1962 TO 1981

Year	Penaeid prawns	Non-penaeid prawns	Total
1962	48 251	34 984	83 235
1963	41 071	40 522	81 593
1964	63 389	31 506	94 895
1965	38 085	41 415	79 500
1966	56 146	34 768	90 914
1967	63 310	31 112	94 422
1968	69 514	31 922	101 436
1969	72 133	33 965	106 098
1970	89 857	31 834	121 691
1971	72 109	76 734	148 843
1972	78 361	85 488	163 849
1973	136 514	66 955	203 469
1974	114 934	55 244	170 178
1975	141 713	79 038	220 751
1976	114 640	76 787	191 427
1977	96 472	73 992	170 464
1978	129 204	50 652	179 856
1979	113 665	63 917	177 582
1980	112 037	58 700	170 737
1981	83 539	61 430	144 969

of more than 125% in the landings from 1962 to 64 and 1972 to 76, falling to less than 100% in 1977–79 and slightly less in later years.

2.2.3.2 Shrimp landings from the west and east coasts of India

It is well known that the west coast of India accounts for more than 85% of the total marine prawn landings. As a result, the trend in the catches of this coast determines the trend in the total landings. This is clearly seen in the trend of catches of west coast which remained at a steady

level up to 1968, thereafter showing a steep increase up to 1973 and then fluctuating. The percentage of increase in the triennial averages over the years is the same as that of the total landings.

The picture of the trend in catches along the east coast is quite different. Forming less than 15% of the total landings, the catches remained below 12 thousand tonnes up to 1966. In 1967, a sharp increase to above 24 thousand tonnes is noticed and this is kept up in the subsequent year also. Then, there is a steep decline through 1972 to about the landing figures of 1966. Once again, the catches rise and reach the maximum of above 28 thousand tonnes in 1975 with slight reductions in subsequent years. Although there is a sharp decline in the catches during 1970–72 the overall increase in percentage in the triennial average is about 150%.

2.3 Statewise production of shrimp

The major contribution of the fishery being from the west coast, the general trend in the total production is set by the landings of this coast, in which Maharashtra and Kerala States account for the bulk of the catch, 48% and 31% respectively. In the northern states of Gujarat, Maharashtra and Goa the maximum catches were in 1976, and were slightly lower in subsequent years (*Table 3*). Along the Kerala and Karnataka coasts the highest catches were recorded in 1973 and 1974 respectively with subsequent decrease. In Kerala slight improvement was noticed from 1976 onwards, but during 1979 there was a decline, reaching very low

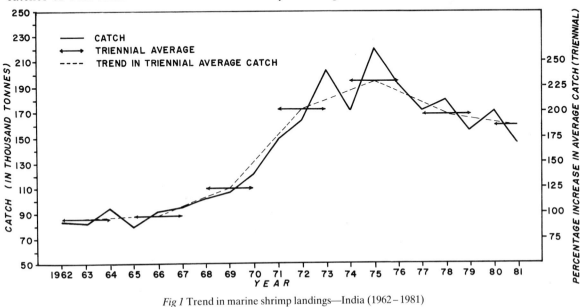

Fig 1 Trend in marine shrimp landings—India (1962–1981)

Table 3
Total prawn landings in 13 years (1969–1981)

Maritime States	1969	1970	1971	1972	1973	1974	1975	1976	1977	1978	1979	1980	1981
West Bengal and Orissa	5638	3016	1500	1471	3051	3487	5707	5635	1690	3879	3588	1304	1495
Andhra Pradesh	6064	6890	9205	5582	8839	12699	10675	11108	11375	9563	11814	10006	8335
Tamil Nadu	5814	5264	3699	5033	5789	8106	12033	9033	8356	13912	11119	10028	14252
Pondicherry	614	447	290	182	41	29	64	93	105	316	604	527	389
Kerala	34368	36954	32813	36577	85751	60829	77962	34533	40324	45428	29597	54375	22428
Karnataka	3980	7539	4420	8075	8236	12696	3074	2594	3335	8440	4660	3226	4126
Goa	559	627	279	561	785	1448	1762	4643	1460	1673	1594	1853	2237
Maharashtra	45780	57345	93611	104125	80349	64737	93665	104474	93653	85346	101846	70742	74571
Gujarat	3273	3599	3014	2231	10620	6119	15781	19275	10121	11034	11953	18590	15727
Andamans	8	10	12	12	8	28	28	39	45	265	64	54	26
Larger private trawlers	—	—	—	—	—	—	—	—	—	—	743	32	1383
All India total	106098	121691	148843	163849	203469	170178	220751	191427	170464	179856	177582	170737	144969

production and showing improvement again in 1980. In 1981 an all time low catch is recorded. In Karnataka there was considerable improvement in landings in 1978, but a decrease in 1979 to 1981. On the east coast both Andhra Pradesh and Tamil Nadu show significant improvement in shrimp landings in recent years.

2.3.1 *Data base*   The Central Marine Fisheries Research Institute (CMFRI) is the nodal organisation in India for the collection of marine fish catch statistics and data on biological and related oceanographic and ecological characteristics on a nation-wide basis. The data are collected on a sample system throughout the year.

The Institute has played a pioneering role in developing a suitable sampling design for the collection for catch data from a large number of landing centres spread over the entire coastline. The procedures have been undergoing modifications to accommodate the innovations introduced in the fishing industry from time to time.

At present the Institute is following a stratified multistage probability sampling design for estimation of marine fish landings in the country. The design involves a space-time stratification. Each maritime state is divided into zones based on criteria such as intensity of fishing, type of fishing and geographical conditions. A zone consists of about 20 to 30 landing centres. A ten-day period in a month forms the time-stratum. From the first five days of the month a day is selected randomly which together with the next 5 consecutive days form the first cluster. The next 6 days from the other two groups of ten days are so selected that a ten-day gap falls between the starting day of two consecutive clusters. Three centres are randomly selected for observations over 6 days and each selected centre is observed for two days, first day in the afternoon and second day in the morning for a six-hour duration each day. On the day of observation, based on the landing of a sample number of boats (units) selected in a systematic way, detailed recordings are made on items such as species-wise composition of catch, type of crafts and gears used and effort. The total number of boats landed during the observation period is also recorded. A sub-sample of commercially important fishes is collected for biological observation. Landings at night which are generally of a much smaller magnitude are recorded through careful enquiry.

In zones where considerable variation is observed in the landing pattern, sub-stratification is made based mostly on the intensity of landings and sampling is done from within the substratum.

In fact the stratification procedure often undergoes continuous change depending on the intensity of landings. Work programmes are prepared according to the random procedure every month afresh for implementation at the field level.

From the landings of selected boats (units) the landings for all the boats (units) landed during the observation period are estimated first. By adding the estimated quantities landed during the two six-hour periods and during the night (12 hours) the quantity landed for one day (24 hrs) at a centre is calculated. By using appropriate raising factors the monthly zonal landings are estimated. By pooling the zonal estimates for all the months the figures of annual landing are obtained. The standard errors of the estimates are also computed for the annual estimates of catch.

The Institute maintains a well-trained field staff in 42 research/field centres located along the coastline to monitor the catch. They are specially trained to identify the various species and to collect the needed biological statistics. The scientific and senior technical personnel posted at headquarters and different research centres to implement the research programmes of the Institute carry out supervision of the work of the collection of statistics at the field level.

The data collected for a month are sent within the first ten days of the succeeding month to the Data Centre maintained at the headquarters of the Institute. Scrutiny and processing of data are done by a team of qualified computing staff using partly calculators and partly programmable computers. The processed results are examined and interpreted and the information is disseminated periodically through the Institute's publications.

Some of the states like Maharashtra, Gujarat and Tamil Nadu are also collecting catch statistics from the landing centres in these states employing random sampling procedures. Frequent dialogues are arranged between the scientists of the Institute and the officials of the State Fisheries Departments to examine the figures obtained by the two agencies. Following the recommendation of the National Commission on Agriculture, an integrated methodology is being evolved so that the CMFRI and the State Departments may be able to combine their efforts to arrive at more precise estimates. In case of states where no system of collection of catch statistics exists, the CMFRI is giving the necessary technical support.

The Institute publishes state-wise and species-wise estimates of fish catch and supplies the details to national and international agencies. The species are combined to form 27 groups among which shrimps are categorised as penaeid and non-penaeid prawns.

The types of crafts used, both mechanised and non-mechanised, with further details are recorded during the observation period. Information on total manhours of fishing is also collected. However, gearwise estimates of effort for any particular species poses a major problem as the fishery is one of multiple species operated by multigears. Standardisation of the effort for selected commercially important species with reference to the most important gear prevalent in an area is being attempted.

2.4 Assessment of shrimp stocks

2.4.1 *Models for stock assessment* The study of the effect of different levels of fishing on the fish stock is essential for arriving at suitable management policies. The analytical method usually employed aims at estimating the yield per recruit under a particular set of fishing conditions. The Beverton and Holt model is commonly used for the purpose and is expressed as

$$Y/R = f(F, M, K, W, T_0, t_r)$$

where F, M are fishing and natural mortalities, K, W and t_0 are parameters of growth and t_r the age of recruitment to the fishery. For changes in the parametric values, it is possible to predict the corresponding yield per recruit. Reliable estimates of the various parameters are, however, required for the application of the model. Mortality estimates for shrimps have been studied by several workers (Banerji and George, 1967; George *et al*, 1968; Kurup and Rao, 1974; and others).

Another model which is based on the law of diminishing returns can be derived under equilibrium conditions, as

$$Y/f = a - bf$$

where Y is the catch and f is the effort. In a heavily exploited stock the catch per unit effort (CPUE) generally decreases as the effort is increased. Using data on Y/f and f for several years the constants a and b can be estimated by least square procedure. The corresponding yield curve is given by

$$Y = af - bf^2$$

The equation shows that the catch increases with initial stages of increase in effort, reaches a maximum at a particular level of effort and then decreases with further increase in effort. The curve has a maximum at $f = a/2b$ and the maximum sustainable yield (MSY) will be $a^2/4b$.

2.4.2 *Yield curve at particular centres* Using the data on CPUE and effort in respect of shrimp fishery for about ten years in nine important centres in the Indian coast, the MSY and the corresponding optimum effort have been worked out. The results are summarised in *Table 4*.

dock, Karwar, Mangalore, Calicut, Cochin, Neendakara, Mandapam and Madras the indication is that increasing the effort beyond the optimum value is not likely to increase the yield. In Kakinada on the other hand, the slope of the regression line of CPUE on effort was positive indicating that the fishing effort does not have any detectable effect on the CPUE.

Table 4

ESTIMATED MSY AND CORRESPONDING EFFORT AT MAJOR PRAWN LANDING AREAS IN INDIA. VALUES FOR 1980 ARE SHOWN FOR COMPARISON

| Centre | Values at MSY | | 1980 values | |
	Catch (tons)	Effort ('000 boat days)	Catch	Effort
Sasson Dock (Bombay)	2 980	39·8	3 914	18·0
Karwar	538	10·7	557	5·2
Mangalore	1 715	30·5	980	31·5
Calicut	760	18·7	355	8·7
Cochin	4 426	48·7	3 516	44·0
Neendakara	53 487	465·9	36 568	150·0
Mandapam	363	23·8	(b)	
Madras	920	94·0	(b)	
Kakinada	(a)	(a)	2 580	40·0

(a) c.p.u.e. increases with effort; so no fit to production model possible
(b) not available.

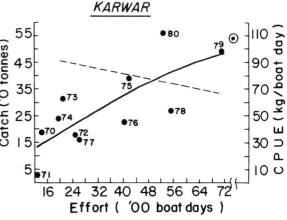

Fig 3 Catch and catch per unit effort related to effort in shrimp catches at Karwar (Karnataka) (1970–80)

Figures 2–10 show the graphs of catch plotted against effort and the fitted equation relating the two. The relationship between CPUE and effort is also shown in the same graph as a dotted line. As seen from *Table 4*, for the fisheries around Sasson

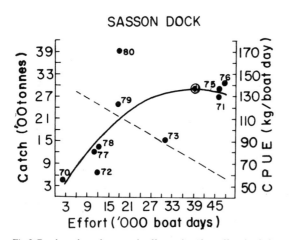

Fig 2 Catch and catch per unit effort related to effort in shrimp catches at Sasson Dock, Bombay (Maharashtra) (1979–80). ⊙ Indicates level of maximum sustainable yield (MSY) in figures 4–12

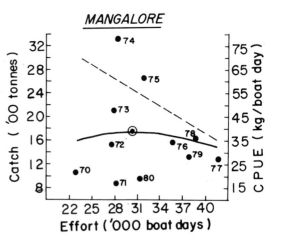

Fig 4 Catch and catch per unit effort related to effort in shrimp catches at Mangalore (Karnataka) (1970–80)

90

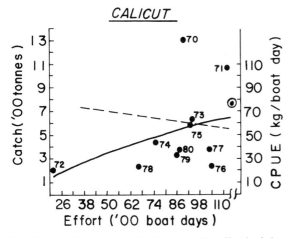

Fig 5 Catch and catch per unit effort related to effort in shrimp catches at Calicut (Kerala) (1970–80)

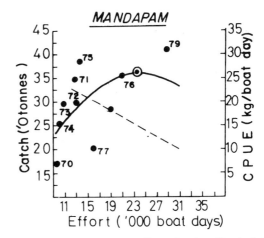

Fig 8 Catch and catch per unit effort related to effort in shrimp catches at Mandapam (Tamil Nadu) (1970–79)

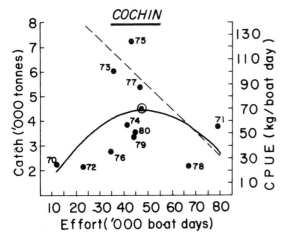

Fig. 6 Catch and catch per unit effort related to effort in shrimp catches at Cochin (Kerala) (1970–80)

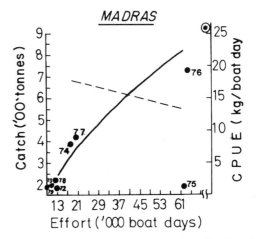

Fig 9 Catch and catch per unit effort related to effort in shrimp catches at Madras (Tamil Nadu) (1972–79)

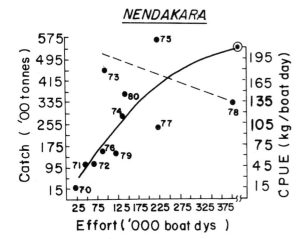

Fig 7 Catch and catch per unit effort related to effort in shrimp catches at Neendakara (Kerala) (1970–80)

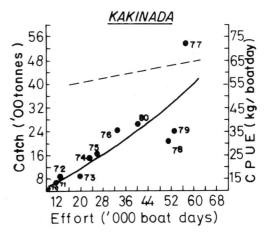

Fig 10 Catch and catch per unit effort related to effort in shrimp catches at Kakinada (Andhra Pradesh) (1970–80)

2.4.3. *Estimation of growth parameters* There are conventional methods for estimation of growth and mortality parameters. However, these need to be critically examined, especially when we consider their application to tropical fisheries with their special features like multiple spawning in one year and short life span. The Institute is examining these problems both from the theoretical and applied aspects. The results of a case study for estimation of growth parameter of one of the species of commercial shrimp are included here. Monthwise length frequency data (total length in cm) for 1980 pertaining to *Metapenaeus dobsoni* from Cochin area of Kerala State have been analysed following the procedure given by Pauly and David (1981) for extraction of growth parameters from length frequency data which involves tracing through a series of length frequency samples sequentially arranged in time, using a range of alternative growth curves and selecting the single curve that passes through a maximum number of peaks. Using a suitable computer programme the best fitting curve was selected from the series of combinations generated. The estimates of asymptotic length and the growth coefficient that best explains the peaks were $L_\infty = 12.5$ cm and $K = 0.13$ respectively. These are very close to the estimates of these parameters for the same species in the fishery of Cochin in earlier years by Banerji and George (1967).

3 Other species in shrimp fishery

3.1 *Estimates of catches*

A considerable quantity of fishes by way of by-catch from shrimp trawling as well as indigenous shrimp fishery, consisting of both trash fishes of cheaper varieties and quality table fishes is landed in India. Thus a bottom fishery or demersal fishery of very high magnitude exists in the country. These landings are estimated in the same way as the catches as described elsewhere in this review. In a total marine landing of 1 388 380 tonnes in 1979, 640 027 tonnes were contributed by demersal catches, inclusive of indigenous fishery. The statewise details are given in *Table 5*. In the total landings of 398 945 tonnes in 1979 by smaller shrimp trawlers the fish and other miscellaneous by-catches apart from shrimp amounted to 315 902 tonnes forming 79·18% of the total (*Table 6*).

The details of landings (provisional) of commercial shrimp trawlers at some selected centres in the different maritime states during 1980 is given in *Table 7*. It is seen that among all the centres Sakthikulangara (Neendakara) in Kerala State shows the maximum number of units operated as well as the greatest landings of both fish by-catches and shrimps. It is interesting to note that the percentage of by-catch during the year is also at the minimum of 54·98 in this Centre. At Cochin, the other centre of observation in Kerala, the percentage of by-catch is also comparatively low. Sassoon Dock in Bombay comes next in the quantity of by-catch and shrimps landed by the trawlers, as can be seen in the table.

From the total by-catch including various groups of fishes and miscellaneous items consisting of crustaceans other than shrimps, cephalopods *etc* only a negligible quantity is discarded. In addition to the landed by-catch of 315 902 tonnes in 1979, only an insignificant quantity consisting of Squilla and miscellaneous items such as young ones of

Table 5
STATEWISE DISTRIBUTION OF BOTTOM FISHERY DURING 1979 (IN TONNES)

| State | Trawler catch | | Total demersal catch including indigenous | Total marine catch |
	Shrimps	Fish		
West Bengal	—	—	4 325	10 744
Orissa	2 160	7 275	28 675	51 808
Andhra Pradesh	5 373	23 312	49 377	91 426
Tamil Nadu	8 216	83 496	122 085	235 008
Pondicherry	492	3 158	4 273	10 068
Kerala	24 512	54 952	102 237	330 509
Karnataka	3 857	18 157	28 495	126 384
Goa	1 559	6 493	9 558	25 388
Maharashtra	31 242	48 788	186 102	293 326
Gujarat	5 632	70 271	86 836	191 312
Andamans	—	—	576	1 721
Lakshadwip	—	—	648	3 846
Private trawlers (Large)	743	16 097	16 840	16 840
Total	83 786	331 999	640 027	1 388 380

Table 6

LANDINGS OF SHRIMPS AND BY-CATCHES OF COMMERCIAL TRAWLERS IN DIFFERENT MARITIME STATES DURING 1979 TO 1981 (IN TONNES)

	1979				1980				1981			
States	Shrimp	By-catch	Total	% of by-catch	Shrimp	By-catch	Total	% of by-catch	Shrimp	By-catch	Total	% of by-catch
Gujarat	5 632	70 271	75 903	92·6	10 315	73 718	84 033	87·7	8 805	89 994	98 799	91·1
Maharashtra	31 242	48 788	80 030	60·9	15 107	41 394	56 501	73·3	14 908	35 858	50 766	70·6
Goa	1 559	6 493	8 052	80·6	1 707	10 095	11 802	85·5	2 025	12 051	14 076	85·6
Karnataka	3 857	18 157	22 014	82·5	2 445	17 034	19 479	87·4	3 235	29 235	32 470	90·1
Kerala	24 512	54 952	79 464	69·2	46 161	59 900	106 061	56·5	16 305	33 008	49 313	67·0
Tamil Nadu	8 216	83 496	91 712	91·1	6 114	76 333	82 447	92·6	11 684	92 761	104 445	88·8
Pondicherry	492	3 158	3 650	86·5	443	2 556	2 999	85·2	300	3 227	3 527	91·5
Andhra Pradesh	5 373	23 312	28 685	81·3	3 891	15 768	19 659	80·2	5 045	21 128	26 173	80·7
Orissa	2 160	7 275	9 435	77·1	843	7 678	8 521	90·1	1 241	4 781	6 022	79·4
All India	83 043	315 902	398 945	79·2	87 026	304 476	391 502	77·8	63 548	322 043	385 591	83·5

Table 7

LANDINGS OF SHRIMP AND BY-CATCHES (IN TONNES) OF COMMERCIAL SHRIMP TRAWLERS AT SELECTED CENTRES DURING 1980 (PROVISIONAL)

					By-catch			
Centres	Number of units operated	Total landings	Prawn catch	Other Crustaceans	Fish	Miscellaneous items	Total	Percentage of by-catch in total landings
Bombay (Sassoon Dock)	21 469	18 144	5 138	4	12 924	78	13 006	71·68
Mangalore (Tadri)	7 922	2 417	353	1	1 779	284	2 064	85·39
Cochin	46 096	7 912	3 514	704	3 416	278	4 398	55·58
Sakthikulangara (Neendakara)	172 732	81 213	36 559	4 167	36 607	3 880	44 654	54·98
Tuticorin	31 517	6 417	534	12	5 871	—	5 883	91·67
Mandapam	25 143	2 533	217	151	2 047	118	2 316	91·43
Rameswaram	78 758	14 378	1 367	602	11 692	717	13 011	90·49
Nagapatnam	9 307	2 007	125	26	1 729	127	1 882	93·77
Cuddalore	16 012	1 969	121	31	1 642	175	1 848	93·85
Pudumanikuppam	13 154	1 416	165	62	919	270	1 251	88·34
Kakinada	41 174	9 025	2 698	352	5 557	418	6 327	70·10
Visakhapatnam	35 406	8 051	784	400	6 325	542	7 267	90·26

fishes and shrimps and crabs were normally discarded. In addition in the case of the smaller trawlers when the shrimp catches are unusually heavy occasionally the fish by-catches are discarded over board due to lack of space for storage and the quantity thus thrown out is not known. From the larger trawlers operated, most of the smaller fish by-catch is discarded at sea, about which data is not available.

3.1.1 *Species composition of fish in by-catch* The by-catches landed by shrimp trawlers include a wide variety of demersal fishes and a few species of cephalopods and crustaceans other than prawns. Among fishes, the most common items represented in the landings of the different maritime states are: Elasmobranchs, eels, catfishes, dorabs, lizard fish, perches, polynemids, sciaenids, ribbonfishes, carangids, silver bellies, white fish, barracudas and soles.

Elasmobranchs: This group, represented by sharks, skates and rays, is one of the common items of fish caught in shrimp trawls all along the west and east coasts and constitute nearly 5% of the annual landings of by-catches in the country. Out of 15 336 tonnes of elasmobranchs landed during 1979 more than 68% was recorded from Maharashtra and Tamil Nadu alone, forming 10·3% and 6·6% of the by-catches of these states. They occur in the trawl nets almost throughout the year, with peak landings during November–March along the coasts of Gujarat, Maharashtra and Kerala and during August–November in Tamil Nadu and Andhra Pradesh. The sharks caught are generally smaller in size than the huge fish captured very often in hooks and lines and other indigenous gears. Rays of all sizes up to about 2 m across the disc are encountered and they form the major component of the elasmobranch catch.

Eels: Although not so very abundant as other groups, two species *Muraenesox talabonoides* and *M. cinereus* are met with very often in the trawl catches of northwest coast and to a limited extent in the east coast. The highest catches are recorded in Maharashtra where they accounted for 5·4% of the total by-catches during 1979. *M. talabonoides* (wam) is the more common and measures a maximum of about 2 m in total length.

Catfishes: The catfishes form one of the common elements of the by-catches all along the Indian coasts and contribute to about 3% of the total landings. Maximum catches are recorded from Maharashtra and Kerala where they are caught throughout the year with peak abundance during

the summer period (March–May). The catfishes obtained in shrimp trawls include a large number of species mostly of the genus *Tachysurus* and are generally in the size range 15–75 cm.

Dorabs: The dorabs or wolf herrings (*Chirocentrus dorab* and *C. nudus*) occur rarely in the trawl catches. In Gujarat they are encountered almost regularly from November onwards till the onset of monsoon.

Lizard fish: Contributing to about 3% of the total by-catches the lizard fish *Saurida tumbil* and allied forms constitute an important item of the catch landed in Kerala with peak abundance during the monsoon period. In other states like Maharashtra, Tamil Nadu and Andhra Pradesh also fair quantities are recorded occasionally during the non monsoon periods. The common size range is about 15–40 cm.

Perches: The occurrence of several varieties of small and medium sized perches in shrimp trawls is a regular feature throughout the west and east coasts. *Nemipterus japonicus* (Kilimeen in Malayalam) is the most common species caught in Kerala where this group forms the largest component of the by-catches (31·1%) landed. Its peak abundance is observed during the southwest monsoon period. Other common perches are species of *Pomadasys*, *Lutjanus*, *Gerres*, *Kurtus*, *Sillago*, *Drepane* and *Therapon*. *P. hasta* locally known as 'Karkara' is a highly sought-after species occurring more frequently in Bombay–Saurashtra waters.

Polynemids: The thread fins contribute to the by-catches in minor quantities in Maharashtra, Tamil Nadu, Andhra Pradesh and Orissa. In Bombay waters they form a sizeable portion of the catch and are chiefly represented by two species namely *Polynemus heptadactylus* (Shende) and *P. indicus* (Dara), the former growing up to about 30 cm and the latter to 140 cm in total length.

Sciaenids: Of all the by-catch categories sciaenids, popularly known as 'Jewfishes', are the most common and are represented by various sizes up to about 120 cm. In 1979 they accounted for nearly 21% of the total by-catches of the country ranking first among the categories. The bulk of the landings was contributed by Gujarat and Maharashtra where the catch consists of two large growing species namely *Pseudosciaena diacanthus* (Ghol) and *Otolithoides brunneus* (Koth) and a number of smaller species collectively known as 'Dhoma' belonging to the genera *Johnius*, *Otolithus* and *Sciaena*. Substantial quantities of sciaenids are also landed in Tamil Nadu and other neighbouring areas of the east coast.

Ribbon fishes: They occur in moderate quantities all along the Indian coast contributing to nearly 5% of the total by-catches. Maximum landings are recorded in Gujarat and Maharashtra.

Carangids: Several species of *Caranx* and allied forms are often encountered in shrimp trawls as minor catches all along the west and east coast. The carangids thus caught are generally smaller in size rarely exceeding 30 cm. They are relatively more common on the east coast where peak landings are recorded during the first quarter of the year.

Silver bellies: This is the second dominant item of the by-catches and contributes to about 14% of the annual landings. Out of 43 728 tonnes landed during 1979 nearly 90% was caught from Tamil Nadu coast alone. A number of species of the genus *Leiognathus* and a single species of *Gazza* (*G. minuta*) comprise the silver bellies catch, the former group being dominant. In Tamil Nadu the silver-bellies are caught in trawl nets throughout the year, the maximum catches being recorded from February to May.

White fish: The white fish *Lactarius lactarius* is one of the quality fishes caught in travels occasionally. It is more common in the northwest coast and Tamil Nadu where the catch is generally represented by small and medium sized fish measuring 5 to 15 cm length.

Pomfrets: Like the white fish, pomfrets are also quality fishes occurring all along the Indian coasts but generally as stray numbers in the by-catches. In Bombay–Saurashtra coast, however, they are caught quite often in fair quantities and are represented by three types of which the brown-pomfret (*Parastromateus niger*) and silver-pomfret (*Pampus argenteus*) are dominant. The former species grows to fairly large size, with the sizes ranging from 10 to 30 cm.

Barracudas: One of the less common groups of by-catches, the barracudas are represented by a few species of *Sphyraena*.

Soles: The soles and other flat fishes form a regular component of the trawler landings throughout the Indian coasts and contribute to about 3% of the annual by-catch production. Maximum quantity is landed in Kerala. Except for a few large growing species like the Indian halibut *Psettodes erumei* (Aayirampalli in Malayalam) and the large 'tongue soles' *Cynoglossus dubius* and others caught occasionally the bulk of the catch is constituted by the smaller species *C. semifaciatus* popularly known as Malabar sole. The usual size is about 8–15 cm. At Neendakara in Kerala they are caught almost throughout the year with peak landings during the monsoon period.

Other crustaceans: Besides shrimps, the trawlers land considerable quantities of other crustaceans as by-catches, amounting to about 5%. Crabs and stomatopods are the most common and they occur in more or less equal proportions in the total catch. In 1979 nearly 50% of these items were landed in Kerala, followed by substantial quantities in Karnataka and Tamil Nadu. The crabs are predominantly represented by *Portunus pelagicus* and *P. sanguinolentus* and the stomatopods by a single species namely *Oratosquilla nepa*. Peak landings are observed during November–February along the west coast and May–October on the east coast. The spiny lobster *Panulirus polyphagus* is another important crustacean by-catch landed on Maharashtra and Gujarat coasts. Similarly, the deep-sea spiny lobster *Puerulus sewelli* has been trawled in considerable quantities along with shrimps from 250–400 m depth off Kerala coast.

Cephalopods: Squids and cuttlefishes contribute nearly 4% of the total by-catches of the country. Out of 10 229 tonnes landed in 1979 over 73% was obtained in Maharashtra and Gujarat. November–May is the period when the maximum catch is landed. In Kerala also substantial quantities are landed particularly during August to November. *Loligo duvaucelli, Sepia pharaonis, S. aculeata* and *Sepiella inermis* are the common species, the former three being mostly represented by the size group 10–25 cm and the other by 5–10 cm in mantle length.

Miscellaneous: In addition, several species of trash fishes, both demersal and mid-water forms, are landed regularly at all centres of the coast. In the overall by-catch landings they collectively account for about 25%.

The distribution and occurrence of these varieties of fishes vary considerably in the different shrimp fishing grounds along the coast. Their exploitation, distribution and resources in general have been studied by Rao (1973) and Silas *et al* (1976) and different varieties in particular by others. However, no assessment of the direct effect of the fish stocks and their fisheries on shrimp fishing has been made.

3.2 *Biological interaction between fish and shrimp*

The relative abundance of different fishes in the trawl grounds in relation to shrimp catches in various regions along the north west coast was studied by Rao *et al* (1966). Similar studies in other areas along the coast of India were undertaken by Rao (1973) and others. The catch rates were given and these are reliable measures for determining the difference in abundance between regions.

George *et al* (1968) studying the fishery in the trawl grounds off Cochin from 1956−57 to 1962−63 recorded considerable variation in the percentage of fish in the shrimp grounds. The variations in the composition of fishes in these trawling grounds could be seen in *Table 8*. The wide monthly variations in the percentage composition of the shrimps and fishes may be brought about by the biological interaction between these organisms in these grounds. Such variations in percentage composition of shrimps and fishes in the catches from trawling grounds have been noticed in other areas also (George *et al*, 1980),

3.3 *Shrimp catch in non-shrimp fishery*

A few shrimp are taken in the mechanised fisheries using gears other than trawls, but the quantities are very small (see *Table 9*).

4 Environmental aspects

4.1 *Natural variations*

In a detailed study of the various aspects of the shrimp fishery of Cochin area based on the catches of the trawling vessels of the Exploratory Fisheries Project and the Integrated Fisheries Project through the seasons 1956−57 to 1962−63, George *et al* (1968) explained the decrease in the total catch and the catch rate by natural fluctuations in the fishery. George *et al* (1968) also indicated that the fluctuations in the fishery are due to natural causes. Banerji (1969) studying the shrimp production in India through the years 1959 to 1968 found a rising trend with minor fluctuations and according to him these fluctuations round the trend were random brought about by natural causes. Mohamed (1973) analysing the trend in fishery of penaeid shrimps of the country also found only natural fluctuations. Natural variations in recruit-

ment of different species into the fishery South of Cochin were studied by Kurup and Rao (1974).

Table 9
DETAILS OF SHRIMP CATCH (IN TONNES) IN OTHER FISHERIES DURING 1979

Maritime States	Drift/Gill net (mechanised)		Purse seine	
	Shrimp	Total catch	Shrimp	Total catch
Gujarat	6	20 929	—	—
Goa	1	2 199	1	8 196
Karnataka	—	—	1	66 198
Tamil Nadu	3	262	—	—
All India total	10	23 390	2	75 104

4.1.1 *Relation to environmental factors*   Changes brought about by the physico-chemical disturbances in the environment are known to influence shrimp landings in the areas of 'mud bank' formations during the south west monsoon period along the south west coast of India (Banse, 1959; George, 1961; and George *et al*, 1968). In the 'mud bank' area George (1961) recorded in the monsoon months of active fishery higher shrimp landings along with decreased oxygen and temperature and higher salinity and pH in comparison with areas of lesser fishery, indicating an apparent influence of these environmental factors on the fishery.

Factors influencing shrimp catches in the backwater fishery of Cochin were studied by Menon and Raman (1961). They particularly observed the influence of rainfall on the catches, as there is a regular belief among fishermen and others connected with shrimp fishing that rainfall exerts a direct influence on the fishery. Their study indicates that there was direct relationship between

Table 8
PERCENTAGE COMPOSITION OF BY-CATCH OF FISHES IN THE FISHERY OFF COCHIN DURING 1956−57 TO 1962−63

Months	1956−7	1957−8	1958−9	1959−60	1960−1	1961−2	1962−3
August	—	54·6	—	—	—	—	—
September	—	—	99·7	—	—	96·3	78·5
October	—	94·0	99·8	98·2	98·7	82·4	68·8
November	—	83·7	72·8	52·6	96·5	77·3	71·9
December	—	39·5	63·8	84·3	63·6	47·6	55·3
January	88·1	41·6	48·2	74·5	80·8	53·8	69·3
February	46·8	78·6	64·7	67·3	85·2	54·8	51·0
March	58·3	78·7	47·0	86·8	79·1	64·0	78·2
April	57·6	49·6	23·9	84·5	69·1	53·4	88·5
May	35·0	44·7	38·3	72·8	81·0	47·6	76·6
June	26·0	18·8	—	18·0	69·9	40·0	29·7
Average	46·5	52·9	52·8	87·3	80·9	57·8	70·6

the catches and the annual rainfall. They also indicated the possibility of the monthly catches being directly affected by rainfall in one year, probably the stock abundance or recruitment being influenced by rainfall.

Relationship of the backwater shrimp catches with the phases of the moon also was investigated by Menon and Raman (1961). During the two year period of their study the highest catches were recorded on the day of new or full moon or a day or two later, indicating a relationship between the availability of shrimps and lunar periodicity. They were of opinion that the stronger tidal currents on these days forcing a larger volume of water through the nets as well as the lunar phase causing active movements of these shrimps, together resulted in improved catches. The effect of lunar periodicity on the catches of shrimps and their movement in and out of the Godavari estuarine system was studied by Subramanyam (1965).

4.1.2 *Possibilities of prediction* The possibility of using postlarval abundance as an index of fishing success in shrimps was indicated by George (1963). He tried to correlate the recruitment of the postlarvae into the backwaters of Cochin and the subsequent population contributing to both the backwater and marine shrimp fishery and expressed the opinion that this factor could possibly be used for forecasting the magnitude of the fishery.

4.2 *Impact of other human activities on the fishery* Man's activities involving steady incursion into the brackish-water areas such as large scale reclamation for agricultural purposes, destruction of mangroves, diversions of water flows in and out for industrial and agricultural needs, usage of waters for disposal of industrial wastes, exploitation for extraction of underlying mineral deposits *etc* pose an ever-increasing threat to the fishery resources dependent on these areas. George (1973) discussed the influence of backwaters and estuaries on shrimp resources of the country and established that in view of the inseparable link between shrimps and brackishwater environments, the least possible disturbance of the brackishwater habitat is essential for continued high productivity of their resources.

The changes brought about by the activities of man are those caused by (1) changes in total area of brackishwater habitat resulting from large-scale reclamation for agricultural purposes as is happening in the southern half of the Vembanad Lake, (2) protective works such as stream diversion spillways, salt-water barriers like the Thanneermuk-kom bund being constructed in Vembanad Lake, and tide control structures, (3) pollution of the waters by domestic, industrial and agricultural wastes such as dumping of solid wastes from mining operation of iron ore, industrial sludge from steel mills *etc* and (4) development of mineral resources by dredging of fossil shell deposits. These changes in the environments affect the estuarine phase of the life of the commercial shrimp in any one or more of the following ways, namely, general reduction in acreage of the habitat, change in circulation and thus affecting distribution of salinity, temperature, *etc*, lessening of average depth, impeded exchange of fresh and salt-water, loss of tidal exchange benefits, restricted influx of salt-water, change in water chemistry due to presence of toxic compounds, increased silt load, *etc*. In general the estuarine habitats are affected in two major ways, either by a net loss of total acreage available or by a change in mean salt content and chemical composition. Among others Menon (1967) expressed grave concern about the possible detrimental effects of environmental changes in Cochin backwaters to the prawn resources of the area.

As a result of the rapidly advancing civilisation deterioration in the estuarine habitat is undoubtedly taking place in several areas. The question is what steps could be taken to prevent serious damages to the important biological resources of the estuaries and consequently on the dependent marine resources. Supposing, for instance, that the erection of a particular spill-way or salt water barrier would create a situation of chronically high or low salinity in an area and that over the years this condition in turn would result in decline in production from the coastal fisheries, the amount of projected monetary loss and the benefit-cost ratio will have to be taken into consideration before finalising the project. The Thanneermuk-kam barrier constructed across the southern half of the Vembanad lake in Kerala State is a salt water extrusion project and its effect on the fishery resources of the area has been under investigation. George and Suseelan (1980) reported that the distribution of the juveniles of penaeid shrimps into the upper reaches of the lake has been affected by the construction of the bund.

5 Management

In view of the fast tempo of development in increasing exploitation and its effect on total and regional shrimp production from the coastal fishery, there is great need for increased research into the resources and devising of proper approach to management by determining several factors like the permissible yield, suitable gear which will not destroy the resource or damage the habitat and restrictions to be effected in the fishery, if any, in order to get the maximum economic return. There are several factors which make the problem of proper management of the coastal shrimp resources, as in the case of any other resource, highly complex. The facts are very often difficult to ascertain with certainty or quite often when some management measure is indicated based on proven facts the implementation poses problems.

The penaeid shrimp resources of the country which constitutes the coastal shrimp fishery occupy different eco-systems such as estuaries, inshore and offshore waters and in most of the areas along the coastline commercial fishing is well established in these different environments. The constituent species and fishery in each of these environments being the same and interdependent, any approach to coastal shrimp fishery management would involve proper study of the various biological and other aspects of the fishery of all these environments.

Secondly the coastal prawn fishery is a multi-species fishery, with each of the species having individual distribution patterns, growing to different sizes and different in breeding activities. This situation calls for monitoring of the population characteristics of each species separately in a continuing programme in order to keep track of the effects of exploitation on the stocks, which is necessary for proper management.

Thirdly various types of gears with different meshes are in use in the exploitation of the resource. In the traditional indigenous sector different kinds of seine nets, filtering type fixed bagnets, gill nets and various other gears are in use. In the rapid mechanisation taking place at various centres the trawl nets are being increasingly introduced into the mechanised sector. Standardised effort data is, therefore, essential for working out the different population parameters and also to study the trend of catches in relation to effort.

With the introduction of increasing number of mechanised fishing vessels a certain clash of interest between the traditional sector operating in the near shore areas and the mechanised sector is natural and this has been showing up in several areas in recent years. This has even resulted in violence leading to the closure of fishing operations in some places. For management of the fishery free of such conflicts it may become necessary to delineate areas of operation of inshore and offshore fishermen. For a proper delineation, detailed data on the seasonal, geographic and bathymetric distribution of the main species of prawns are essential and this has to be made available on a continuing basis.

All these inherent problems render the management of the coastal shrimp fishery difficult. Continuous monitoring of the catch and effort data both regionwise and gearwise and biological as well as population characteristics of the constituent species are the essential prerequisites for attempting this management.

5.1 *Management objectives*
Taking into consideration all these inherent problems of the shrimp fishery, management objectives have to be defined, based on which suitable management techniques have to be adopted. There could be different views about the specific objectives in management. One view is that it is difficult to manage shrimp fisheries in the biological sense but purely economic measures should motivate management so that the fishery could be based on orderly exploitation for maximum economic return. Another viewpoint is that the aim should be maximum development of the resources and increase of production. Yet another management motive is concerned with socio-economic aspect in which management aim should be to create maximum employment potential in the fishing sector. Shrimps are annual stocks with high natural mortality and because of this fact it would be advisable to fish hard except for giving protection to the nursery areas, partly to safeguard the young prawns and partly to protect their habitat. This would also indicate that the socio-economic factors were of far more importance than biological factors.

A close look at all these viewpoints might only lead to the conclusion that a combination of all these should be the ideal objective in proper management approach in shrimp fishery. But it has to be admitted that it would be extremely difficult to evolve suitable management techniques with all these objectives put together. The management objectives with which the problems are approached in this country has been oriented towards higher production and maximum development of the resources.

The problems connected with shrimp resource management revolves around (1) management of the resource at the earlier phases of each species in relation to its recruitment in and out of the brackish water systems and (2) proper management of the adult population in the fishing grounds in order to limit the exploitation at the optimum level of sustainable yield. As far as the earlier phase of the shrimp is concerned, a proper understanding of the breeding season, the migratory pattern particularly bathymetrical, recruitment to the brackish water region and growth pattern of juveniles are necessary. As regards the resources in the offshore fishing grounds, a close monitoring of the total catch, catch per unit effort, sizes of the component species and their recruitment and mortality rates will be necessary. Depletion in shrimp stocks are probably indicated by (1) catches decreasing regularly over the years, (2) reduction in the average sizes, (3) imbalances in the sex ratios, and so on. In a capital intensive fishery the fall in catch rate may result in economic over-fishing. At the same time the resources may be still biologically viable capable of speedy recovery or even attaining stability of production over a number of years. In the case of shrimp resources in certain areas of the coastline of India there has been records of such decreasing trends in catch rate in the '60s and '70s and subsequent revival of the fishery. In other words conservation in the biological sense was of minor importance, but exploitation for maximum economic return should be the consideration, so much so the problem of management is a socio-economic one.

5.1.1 *Fishery economics in relation to management* Studies on economics of fishing operations, the socio-economic impact of mechanisation on traditional fisheries and the behaviour of market mechanisms are important prerequisites for the formulation of suitable management policies. Such studies will also help in identifying bottlenecks and in arriving at modifications in the management strategies to suit changing conditions. Similar studies dealing with the mechanised sector are in progress. The year 1980–81 also witnessed a revolution in the mechanisation of canoes both with outboard motor and inboard diesel engines in the Quilon – Cochin sector of Kerala coast was well as other areas. The impact of this in production as well as socio-economic conditions is under study.

Investigations on the economics of production of shrimps at least in some representative areas indicate a considerable increase in costs for the effort expended for catching shrimps in recent years, brought about by various factors like the rise in fuel prices. According to the Mechanised Boat Owner's Association the break even point for operation of a small mechanised boat taking into account all expenses including fuel cost, maintenance, depreciation *etc* at Neendakara in Kerala State towards the close of 1979 worked out to

Rs 990/- per day of operation and this was expected to cross Rs 1000/- in 1980. This break even point had more than doubled in the course of the previous four years. According to the same source the average return from a day's catch amounted to only Rs 750/-, resulting in a net loss to the boat operator. Although this may be a little biased, the management objective in the fishery here would be to increase the economic return per boat so as to enable the small boat operator to get a reasonable profit and the management technique has to be evolved in order to achieve this goal. This type of economic analysis is needed from other areas too.

With the large scale introduction of trawlers and purse-seiners in the inshore waters, total shrimp and fish production no doubt has increased. However, the process of modernisation often affects the socio-economic conditions of the fisherfolk engaged in traditional fisheries. There is a need to undertake socio-economic surveys to study the impact, if any, and to arrive at timely remedial measures. The CMFRI has undertaken for instance, an intensive study of the impact of introduction of commercial purse-seining on the traditional rampan fishery for pelagic fish in the Karnataka coast in India. Several measures such as delimiting the area of operation of purse-seiners and preference for employment for rampani operators in purse-seine boats were suggested and are being implemented. Similar impact studies involving trawlers have been taken up.

Neendakara area in Kerala has been the biggest single centre for fishing where over 15 000 tonnes of shrimps are landed every year by mechanised boats. However, the economic condition of the large number of fishermen engaged in fishing and allied operations continues to be poor and many are in debt. The CMFRI conducted an investigation there to study the extent of rural indebtedness of fishermen families and it was observed that majority of the community is in debt to private money lenders. Steps such as liberalisation of the conditions for credit facilities by nationalised banks and promoting saving habits have been suggested to gradually bring them into a more organised form of social life.

Another important aspect which needs attention is the techno-economics of fishing operations involving input – output analysis. A case study on the various factors of expenditure and the profitability of operations in small scale fisheries has been undertaken and the results have been published.

5.2 *Management techniques*
The techniques for the management of a fishery

would largely depend on the objective. Thus there are several methods which could be adopted for proper management of the shrimp fishery.

5.2.1 *Closure of fishery* Enforcement of a temporary closure of the fishery is a method adopted in conservation. This can be either by closed seasons or closed areas. Off the south west coast of India the south-west monsoon season used to act almost as a closed season for the mechanised fishery, so that another closed season was not necessary. The stoppage of the fishery due to the south west monsoon takes place at a time when the mean sizes of all the species in the fishery are at their lowest. The absence of fishing operations for two or three months at that time acted as a natural conservation measure. But in recent years fishing operations are being carried out during these monsoon months in several centres. For example at Neendakara the maximum fishing operations and peak landings now occur during the south-west monsoon period. In such places a closed season to conserve the resources could be thought of. In locational closures the approach is to close one particular locality or fishing ground for a certain period based on the data available.

5.2.2 *Mesh regulation* Regulation of mesh sizes of the gears used in the fishery is another important method. However, in the case of the shrimp fishery, being a multispecies fishery constituted by species growing to different sizes, catches of small or medium sized shrimps include adult specimens of the species growing to smaller sizes as well as the juveniles of species growing to medium or larger sizes. Hence enforcement of a larger mesh size with a view to catch only the larger sizes would prevent the capturing of adults of the smaller species which would thus be lost to the fishery. Thus mesh regulation at larger sizes will not be helpful in Indian shrimp fishery of the inshore regions. But in the estuarine fishery involving smaller sizes of shrimps only, limiting the mesh size at larger dimensions may help in sparing the very small shrimps being subjected to the fishery.

5.2.3 *Limitation of fishing effort* When the available shrimp resources in the fishing grounds is subjected to increasing exploitation by introduction of more and more effort, a stage would be reached when further input of effort results in uneconomic returns. In such cases it would be helpful if some limitation is enforced on additional input of effort, which may be achieved by licence limitation or any other governmental control, aiming at optimum efficiency for individual fishing unit. Its implementation would require biological, sociological, economic and political value judgements.

5.2.4 *Catch restrictions* Limiting the catch per boat from a particular area or enforcing a quota system for catch per boat may also be used as a conservation measure. Enforcing catch quota or restricted hours of fishing for the boats in operation aims at restricting the total catch to an optimum level.

5.2.5 *Diversification* Where the existing fishery is predominantly trawling for shrimp, diversified fishery during certain seasons may be suggested for tapping other resources. In such cases adequate incentives and subsidies may be provided to help the diversification process.

5.2.6 *Protection of brackishwater nursery grounds* In view of the inseparable link between the marine shrimp of commercial importance and brackish water nursery grounds, in any conservation measure concerning the marine shrimp resources the protection of these nursery areas would have to be considered. Complete prohibition of shrimp fishing in these environments as practised in some countries of the world would be ideal. This being almost impracticable under the present circumstances, alternative methods of protection of the young ones in these environments have to be thought of. Closing the areas for shrimp fishing during certain seasons, mesh size regulation of the various gears in use in the fishery, limiting the unrestricted operation of fixed gears and such other small mesh nets or even phasing out the operation of these gears, total ban of export of count sizes of shrimps below a fixed minimum level *etc* are some of the methods for consideration in the protection of the nursery.

5.3 *Management advice and implementation*
In most of the resources, implementation is the primary problem to be faced on account of the social angle. However, with the objective to have a constructive management and conservation programme, a Committee appointed by the Government of India to study the shrimp resources and its conservation, has suggested the following draft recommendations.

5.3.1 *Inshore fishery*
    5.3.1.1 *Reliable estimates of mechanised boats*
It is estimated that over 12000 mechanised boats operate in the coastal fishery. Under the Plan

Programmes each of the maritime State/Union Territory is introducing additional boats to the existing fleet, while several of the boats are decommissioned, rendering it difficult to know the reliable number of boats in effective operation. The Committee therefore recommends that:

Each of the maritime state/Union Territory may arrange for obtaining statistics of the mechanised boats in operation.

5.3.1.2 *Monitoring of inshore fisheries* The inshore fishery both at the national and regional levels shows wide seasonal and annual fluctuations. In certain centres along the coast, fishing pressure, due to the concentration of mechanised boats and the infrastructure facilities available, has considerably increased in recent years causing apprehension about the over-exploitation of resources. For a pragmatic appraisal of the situation and for framing suitable management policies to obtain maximum sustainable yield it is essential that the production and effort trends and biological and environmental characteristics of the exploited populations are clearly understood. Keeping this in view the Committee recommends that:

Intensive monitoring of the inshore fisheries on both fisheries independent and dependent factors be undertaken.

5.3.1.3 *Fisheries Commission to monitor northeast coast shrimp fishery* Following the location of lucrative fishing grounds for shrimps off the Orissa and West Bengal coasts the fishing vessels based in these states as well as from neighbouring states, have stepped up their operation in this region. Consequent on this there are indications that the shrimp resources in these grounds are heavily exploited. Taking note of these developments and the imperative need to safeguard the resource the Committee recommends that:

Appropriate Fisheries Commission comprising of the representatives of the government and the industry in these states be constituted to make periodic review of the situation and to prescribe the total allowable catch and adequate arrangements be made to implement the recommendations made by the Commission and to prohibit illicit catching, with the assistance of coast guards.

5.3.1.4 *Conflicts and delimitation of fishing zones* Several maritime state governments have stipulated fishing zones for the operation of non-mechanised fishing vessels, small mechanised fishing vessels and larger vessels. However, conflicts between the non-mechanised fishing vessels and mechanised vessels are being reported. Viewing this with profound anxiousness the committee recommends that:

The operation of different types and categories of vessels in the prescribed fishing zones be strictly adhered to and implemented for the benefit of all.

5.3.2 *Brackishwater and estuarine areas*

5.3.2.1 *Nursery grounds as reserve areas* It is well-known that estuaries and mangrove swamp areas serve as nursery grounds for several species of fin fishes and shell fishes of both marine and fresh water origin. Besides, these ecosystems also play a vital role in maintaining the balance of nature. In view of this the Committee strongly recommends that:

Certain areas of such ecosystems be demarcated as reserved areas and for this purpose necessary provision be included in the Wildlife Act and Indian Fisheries Act.

5.3.2.2 *Regulation of fixed gears* Large number of fixed gears (stake nets) are operated in several of the estuaries of the country. Unrestricted operation of these nets and the smaller mesh sizes bring in appreciable quantities of small sized juveniles of commercial species. The Committee discussed in depth the destructive nature of fishing by these gears, its effect on the fisheries of the adjoining marine and fresh water regions, the pros and cons of total banning of these gears and the consequent effect on the socio-economic conditions of the fishermen of the area. Observing the social, economical, political and ethnological problems involved in total banning of these gears the Committee recommends that:

Those gears having cod end mesh size of less than 20 mm be phased out and that such of the fishermen affected by the implementation of this proposal be provided with alternate vocation through aquaculture programme.

5.3.2.3 *Implementation of regulatory orders* Several states have enacted legislation and regulatory orders under the Fisheries Act in respect of closed seasons or mesh regulation for the conservation of the fisheries resources in the inland as well as territorial waters. However, the Committee notes that they are not fully implemented and hence recommends that:

The maritime States/Union Territories should take necessary steps for effective implementation of these regulations.

5.3.2.4 *Prevention of pollution* With the advancement of industrialisation several agro-industrial complexes have been established in the vicinities of the estuaries. Large-scale domestic and industrial discharges into the estuaries cause considerable damage to the living resources often resulting in heavy fish mortalities. Noting the grave hazards of this pollution the Committee strongly recommends that:

Adequate precautionary measures be taken to prevent the effluents reaching the estuaries.

5.4 *Implementation*
Based on these recommendations and earlier studies and consultations, the enactment of a Marine Fishing Regulation Bill is under active consideration of the Government of India in consultation with the different states. This would in turn be fully utilised by the States for formulating appropriate State Fisheries Acts. Some of the States, like for instance Kerala State have already passed Acts delimiting the fishing zones for the operation of indigenous fishing vessels and mechanised fishing vessels and also banning of operation of mechanised fishing vessel activities in certain areas along the coast during certain seasons and these are being implemented.

6 Future work

6.1 *Constraints in data collection and change necessary*
In order to formulate suitable policies for efficient management of the shrimp fishery, data acquisition, processing and dissemination need strengthening. A number of fishing harbours are being constructed in different parts of the country. Many of these would require exclusive coverage for the estimation of landings at these places. This combined with the changing patterns of fishing necessitates more intensive monitoring. The sampling fraction is at present about 2% and it is desirable to increase this fraction in order to improve the precision of the estimate. Programmes involving increase of frequency of collection of data from landing centres, round the clock collection of data from fishing harbours and other centres of concentration of boats and evolution of special designs for collection of data from operation of gears such as purse-seines are necessary. Retrieval of data from the operation of the larger type of boats belonging to the big firms needs streamlining.

Another area which requires immediate and urgent attention is the data collection from the estuarine and brackishwater systems. Development of a proper catch monitoring system involving a suitable statistical design is quite essential.

6.2 *Biological studies needing priority attention*
For a proper understanding of certain phenomena in shrimp fishery, like for instance the heavy landings at Neendakara in Kerala State during the monsoon season and the shrimp fishery related to the mud bank formation in adjacent region, each having its own species composition, it would be desirable to collect intensive data on oceanographic, both biological and physico-chemical, parameters from in and around the fishing grounds on a continuous monitoring basis along with the data on production. Each prawn fishing ground has its own particular problems which need to be considered individually. Detailed studies on biological parameters for the determination of various population characteristics and delineation of stocks are necessary.

6.3 *Technological, economic and other research needs*
Fishery economic studies at micro and macro levels play an important role in the formulation of management policies. The input of fishing effort is very much linked not only with the resources but also with the state as well as national economic activities. Studies on topics such as cost economics of production, input–output relationship, variation in price structure dependent on various factors, demand supply and other market mechanisms, impact of mechanisation on the socio-economic conditions and repercussions of legislative measures on fishing activities are quite essential in order to have a proper insight into the problems facing the industry. The creation of the new Economic Division in CMFRI would go a long way in initiation of such studies. The process of modernisation of the traditional fisheries has its own impact on the socio-economic conditions of the fisher-folk. Therefore, there is a need for making socio-economic surveys at a number of centres so as to assess any adverse impact and arrive at timely remedial measures. Studies are also required to be undertaken to get an insight into the economic structure governing various activities directly or indirectly related to fisheries.

There is also need for strengthening of research

basis on all aspects including handling, preservation and utilization of products.

6.4 *Need for new models and methods of analysis*

Factors such as recruitment, growth and mortality, both natural and fishing, are all interlinked within a population and between populations. Most of the models now used treat a population by itself without taking into consideration the interacting factors. Analytical models for the interacting situations can be built up; but explicit solutions may be quite difficult to obtain. A step-by-step procedure will have to be followed to study the effort of one factor over the other and consequently the overall picture. Considering the inherent complexities of the fisheries and the lack of suitable data, an alternative procedure is to resort to simulation techniques involving use of mathematical and statistical models. Such studies could be initiated, especially now that electronic computing facilities are becoming increasingly available.

References

BANERJI, S K. Crustacean production in India. In Prawn
1969 Fisheries of India. *CMFRI Bulletin No. 14:* 259–272.

BANERJI, S K and GEORGE, M J. Size distribution and growth of
1967 *Metapenaeus dobsoni* (Miers) and their effect on the trawler catches off Kerala. *Proc. Symp. Crustacea. Mar. biol. Ass. India*, Part II, 634–648.

BANSE, K. On the upwelling and bottom trawling off the
1959 southwest coast of India. *J. Mar. biol. Ass. India*, (1): 33–49.

GEORGE, M J. Studies on the prawn fishery of Cochin and
1961 Alleppy coast. *Indian J. Fish.*, 8(1): 75–95.

GEORGE, M J. Post larval abundance as a possible index of
1963 fishing success in the prawn *Metapenaeus dobsoni* (Miers). *Indian J. Fish.*, 10A(1): 135–39.

GEORGE, M J. Synopsis of biological data on penaeid prawn
1970a *Metapenaeus dobsoni* (Miers), 1878. *FAO Fish. Rep.*, (57), Vol. 4, 1335–1357.

GEORGE, M J. Synopsis of biological data on penaeid prawn
1970b *Metapenaeus affinis* (H. Milne Edwards) 1837. *FAO Fish Rep.*, (57), Vol. 4: 1359–1375.

GEORGE, M J. Synopsis of biological data on penaeid prawn
1970c *Metapenaeus monoceros* (Fabricius) 1798. *FAO Fish. Rep.*, (57), Vol. 4, 1539–1557.

GEORGE, M J. Synopsis of biological data on penaeid prawns
1970d *Metapenaeus brevicornis* (H. Milne Edwards) 1837. *FAO Fish Rep.*, (57), Vol. 4, 1559–1573.

GEORGE, M J. On the zoogeographic distribution of Indian
1972 Penaeidae. *Indian J. Mar. Sci.*, 1(1): 89–92.

GEORGE, M J. The influence of backwaters and estuaries on
1973 marine prawn resources. *Proc. Symp. Living Resources of the Seas around India. Spl. Publ. CMFRI:* 563–569.

GEORGE, M J. Trends in marine prawn production in India.
1978 *Indian Seafoods*, 13(4) and 14(1): 19–24.

GEORGE, M J, RAMAN, K and NAIR, P K. Observations on the off
1968 shore prawn fishery of Cochin. *Indian J. Fish.*, 10(2): 460–499.

GEORGE, M J, BANERJI, S K and MOHAMED, K H. Size distribu-
1968 tion and movement of the commercial prawns of the south west coast of India. *FAO Fish. Rep.*, (57), Vol. 2: 265–284.

GEORGE, M J and SUSEELAN, C. Distribution of species of
1980 prawns in the back-waters and estuaries of India with reference to coastal aquaculture. *Symp. Coastal Aquaculture Mar. biol. Ass. India*, January 1980.

GEORGE, M J, SUSEELAN, C, THOMAS, M M and KURUP, N S. A
1980 case of overfishing: Depletion of shrimp resources along Neendakara coast, Kerala. *Mar. Fish. Infor. Serv. T and E Ser.*, 18: 1–8.

KUNJU, M M. Synopsis of biological data on the penaeid prawn
1970 *Solenocera indica* Nataraj (1945). *FAO Fish Rep.*, (57), Vol. 4, 1317–33.

KURIAN, C V and SEBASTIAN, V O. Prawns and prawn fisheries
1975 of India Hindustan Publishing Corporation (India): 1–280.

KURUP, N S and RAO, P V. Population characteristics and
1974 exploitation of important marine prawns of Ambalapuzha, Kerala. *Indian J. Fish.*, 21(1): 183–210.

MENON, M K and RAMAN, K. Observations on the prawn fishery
1961 of the Cochin backwaters with special reference to the stake-net catches. *Indian J. Fish.*, 8(1): 1–23.

MENON, M D. 'Carpe Diem' *Seafood Trade Journal*, 2(1):
1967 99–106.

MOHAMED, K H. Synopsis of biological data on the jumbo tiger
1970a prawn *Penaeus monodon* Fabricius, 1798. *FAO Fish. Rep.*, (57): Vol. 4: 1251–1256.

MOHAMED, K H. Synopsis of biological data on the Indian
1970b prawn *Penaeus indicus* H. Milne Edwards, 1837. *FAO Fish Rep.*, (57): Vol. 4: 1267–1288.

MOHAMED, K H. Penaeid prawn resources of India. *Proc.*
1973 *Symp. Living Resources of the Seas around India. Spl. Publ. CMFRI:* 548–556.

PAULY, D and DAVID, N. ELEFAN I, a BASIC program for the
1981 objective extraction of growth parameters from length frequency data. *Meeresforsch*, 28, 205–211.

RAMAMURTHY, S and MUTHU, M S. Prawn fishing methods. In
1969 prawn fisheries of India. *CMFRI Bulletin 18:* 235–257.

RAO, P VEDAVYASA. A synopsis of biological data on penaeid
1970 prawn *Parapenaeopsis stylifera* (H. Milne Edwards) 1837. *FAO Fish Rep.*, (57), Vol. 4, 1575–1606.

RAO, K VIRABHADRA. Distribution pattern of the major
1973 exploited marine fishery resources of India. *Proc. Symp. Living Resources of the Seas around India. Spl. Publ. CMFRI:* 18–101.

RAO, K VIRABHADRA, MEENAKSHISUNDARAM, P T and
1966 DORAIRAJ, K. Relative abundance of trawl fishes in the Bombay and Saurashtra waters. *J. Mar. biol. Ass. India*, 8: 205–212.

SILAS, E G, DHARMARAJA, S K and RENGARAJAN, K. Exploited
1976 marine fishery resources of India—A synoptic survey with comments on potential resources. *CMFRI Bulletin* 27: 1–25.

SUBRAMANYAM, C B. A note on the annual reproductive cycle of
1963 the prawns *Penaeus indicus* (M. Edwards) of Madras coast. *Curr. Sci.* 32(4): 165–166.

SUBRAMANYAM, M. Lunar, diurnal and tidal periodicity in rela-
1965 tion to the prawn abundance and migration in the Godavari estuarine systems. *Fish. Tech.* 2(1): 26–33.

A review of the Indonesian shrimp fisheries and their management

M Unar and
N Naamin

Abstract

Total catches of shrimp in Indonesian waters have totalled over 100 000 tons annually, with peak export earnings of over U S $200 million in 1980. Shrimp are caught both by a variety of traditional gears, and by relatively modern trawlers. This has led to conflicts between the two groups, and therefore since January 1981 trawling has been banned through Indonesia except in the eastern waters. Several species of shrimp are caught, and the biology of the major species is described. Recruitment to the fishery seems to be related to conditions in the nursery grounds, and there is a linear relation between the size of the stocks in different parts of Indonesia, and the extent of mangrove areas.

Management of the shrimp fisheries is carried out within the general objectives of the Indonesian national plan, which are to increase fish production, to improve the livelihood of fishermen; to increase employment opportunities, and to maintain the biological yield of the resources.

1 Introduction

The shrimp fishery in Indonesian waters has developed rapidly since commercial trawling started around 1966. As an export commodity it attained its peak in 1979 when the export value reached U S $200 483 000. In 1980 the value dropped to U S $185 100 000, associated with the latest trawl regulation. The total number of trawlers in the entire waters of Indonesia has to be reduced from 3 500 trawlers down to 1 000 trawlers. The vessels most concerned are the so-called 'baby-trawlers'. Starting in October 1980, trawling has been totally banned in Java, Bali and Sumatra.

Trawling has some major positive aspects, such as: increasing shrimp and fish production, foreign exchange earnings and employment. It also has negative aspects. Due to the high value of shrimp, ever-increasing demand, and the limited size of the stocks, overexploitation is occurring in a number of coastal areas where fishing gear and fishermen are overcrowded, particularly in Java and Sumatra. More seriously, such conditions give rise to frictions or conflicts between traditional inshore fishermen and the trawlers. It is to be regretted that several regulations introduced before the banning could not be implemented successfully. On the other hand, shrimp trawling in the less-populated areas off Kalimantan area and in the Arafura sea (west of Irian Jaya) can still be maintained and managed along the lines of the Government policy.

The problems of the Indonesian fisheries are therefore complex. The Government has several objectives – improving the conditions of the small-scale fisherman, increasing shrimp and fish production, increasing foreign exchange earning, *etc* – and these conflict. Resolution of these conflicts, the establishment of priorities and the choice and implementation of appropriate policies will require, among other matters, detailed biological advice.

2 The shrimp fisheries

2.1 *Historical review of the fisheries*

A major group of shrimp that back up the Indonesian shrimp fishery is the Penaeid shrimp. Penaeid shrimp are fished in practically all the coastal areas of Indonesia, especially in the shallow water near estuarine and mangrove areas. More than 42 species are utilized with the most important species being: banana (*Penaeus merguiensis, P. indicus, P. chinensis*), tiger (*P. monodon, P. semisulcatus*), king (*P. latisulcatus*), endeavour (*Metapenaeus monoceros, M. ensis, M. elegans*), rainbow or cat (*Parapenaeopsis sculptilis, P. coromandelica, P. gracillima*), and pink shrimp (*Solenocera crassicornis*), *P. merguiensis, M. ensis* and *P. coromandelica* constitute the bulk of the catch.

Another group of shrimp which occupies an important role in the production and diet on a national scale is the Sergested and Mysid, the so-called rebon, jambret or blacan. A mixture of these small shrimp is made into shrimp paste which is widely used in the daily food.

The major shrimping grounds are on the west coast of Sumatra (Meulaboh, Sibolga, Air Bangis, Painan), the Malacca Strait (Indonesian side), the east coast of Sumatra (Riau, Jambi, south Sumatra, Lampung, Sunda Strait), the south coast of Java (Pangandaran, Cilacap, Pacitan, Nusa Barung, Grajagan), the north coast of Java (from Banten to Madura Strait), off Kalimantan (west, central, south and east) south Sulawesi (Bone Gulf and west coast), Woworanda Bay, south of east Timor, Aru Island in Moluccas and the Arafura Sea west coast of Irian Jaya.

There are two groups of gears employed to catch shrimp, *ie*, the traditional group such as tidal trap, lift net, gillnet, can-trang (Danish seine), beach seine, push net, and the modern group, such as trawl nets.

These traditional gears are practised in Sumatra, Kalimantan and Java, where tidal traps play the most important role in shrimp fishing along the east coast of Sumatra, followed by gillnets. Since 1 January 1981, trawling is limited to Kalimantan waters and the eastern part of Indonesia. Small sampan-type vessels ranging between 7 and 35 tons dominate the trawling fleet. A one-day trip is common for the smallest vessels, while the rather bigger ones have an average of five days operation for each trip. Approximately 800 of such 'baby trawlers' are reported from Kalimantan and South-Sulawesi. In addition, 200 big trawlers ranging between 90 and 350 tons are licensed to operate in Kalimantan and Irian Jaya.

Penaeid shrimp catches in 1972–79 fluctuated from 49 000 to 133 000 t (*Table 1*). Since trawling was banned, in Java and Sumatra shrimp landings and activities relating to shrimp processing and marketing have declined. In certain shrimp centres such as at Cilacap on the south coast of Java, where formerly trawling was the predominant gear, shrimp catch dropped by about 80 percent. In Kalimantan and eastern Indonesia, shrimp catches are still increasing.

2.2 The biology of major species
Data available on the biological aspects are mainly

for *P. merguiensis*, *P. monodon*, *P. semisulcatus*, *M. monoceros*, *M. ensis* and *Parapenaeopsis coromandelica*.

2.2.1 *Migration and distribution of life history stages*
Based on catch studies and the experiences of the fishermen, there are two types of shrimp migration, ie, horizontal and vertical migration. Horizontal migration can be observed in all life stages. The larval and post-larval stages move from offshore spawning ground, estuarine and lagoon nursery areas. Post-larva entering the east channel of the Segera Anakan lagoon through the current of ebb tide and the juvenile and adolescent stages are caught by bamboo screen of 3–4 kg/unit/night during the current of low tide (Van Zalinge and Naamin, 1976). Juveniles migrate back to the sea, where the adults make horizontal movements connected with feeding and spawning. Vertical migration is mainly a daily cycle, making shrimp more or less available to bottom fishing gear at certain times of day.

Even though spawning of the penaeid shrimp in Indonesia occurs year round, it is likely that there are two or three peaks. In Arafura Sea it was noted that for *P. merguiensis*, *P. semisulcatus* and *M. ensis*, several peaks of spawning might have occurred in a year (Naamin and Yamamoto, 1977). Van Zalinge and Naamin (1975) based on the occurrence of maturing or mature females of most-important species (*viz*, *P. merguiensis*, *M. ensis* and *M.*

Table 1

MARINE SHRIMP CATCHES IN INDONESIA 1972–79 (IN '000 T)
(DIRECTORATE GENERAL OF FISHERIES STATISTICS 1972–79)

Coastal area	1972	1973	1974	1975	1976	1977	1978	1979
Sumatra	26·1	20·8	22·9	24·1	66·5	73·6	74·1	74·8
– west coast				1·4	1·2	1·1	1·0	1·3
– Malacca Strait				20·6	62·8	66·5	70·0	68·2
– east coast				2·0	2·5	5·0	3·1	5·3
Java	8·9	5·9	4·7	9·8	13·3	22·3	17·9	17·5
– north coast				6·7	10·3	17·4	12·7	11·3
– south coast				3·1	3·0	4·9	5·2	6·2
Kalimantan	11·9	11·6	7·9	11·7	13·1	16·6	19·7	24·9
– south west				7·5	8·2	10·9	11·9	14·1
– east coast				4·3	4·9	5·7	7·8	10·8
Sulawesi	5·3	5·1	3·5	3·2	3·4	4·1	5·2	5·2
– south Sulawesi				2·5	3·3	4·0	5·2	5·1
– north Sulawesi				0·2	0·1	0·1	—	0·1
Bali and Nusa Tenggara	0·3	0·3	0·5	0·3	0·1	0·1	0·1	0·6
Molluca and West Irian	7·6	9·4	10·3	9·3	8·9	8·8	9·0	10·1
Total	60·1	53·0	49·8	58·5	106·7	124·5	125·2	133·1

Notes: 1. Until 1975 landings were not specified by coastal area
2. Figures are of whole shrimp and include small shrimp (Acetes, Sergestes)

elegens) reported that spawning apparently takes place all year round, although at fluctuating rates. Size/frequency samples taken monthly show that recruitment to the fishery occurs throughout the year. There appears to be a period of five months spawning and recruitment to the fishery.

2.2.2 *Growth* Naamin and Yamamoto (1977) studied the growth rate of three important species of shrimp in the Arafura Sea, and found that the growth rates of adult *P. merguiensis, P. semisulcatus* and *M. ensis* are 1.8 g, 2.7 g and 1.5 g per month, respectively.

By manipulating the Von Bertalanffy growth equation, for *P. merguiensis* caught in Tanjung Karawang, Martosubroto (1977) calculated the theoretical longevity and L_∞ to be more than three years and 246 mm total length, respectively. By using the same method as Martosubroto above, Badrudin (1977) studied age, growth, sex ratio and gonadal maturity of *P. monodon* in Meulaboh – Aceh. The L_∞ of 350 mm total length was estimated, whilst the largest shrimp caught was 289 mm.

Poerwanto (1975) reported that the average size at first maturity of *P. merguiensis* in Tanjung Karawang – north coast of west Java – was 131 mm total length. Martosubroto (1977), using the regression line method, found that the average sizes at first maturity of *P. merguiensis* and *M. ensis* were 26 mm and 20 mm, carapace length, or 125 mm and 95 mm total length, respectively.

2.3 *Sources of data*

The objective of the statistical data collecting programme is to obtain the basic statistics pertaining to catch, effort and gear. It is necessary to meet a minimum level of accuracy. At present some landings are mis-classified and most figures based upon the information from local fishery staff are guesses. This will be different at district, regency and provincial offices. It has caused a lot of trouble and is confusing to the user.

The data on the quantity and value of landings, species, size, and trip can be obtained from records of shrimp dealers (first buyers) or auction (if any, it is available mostly in Java). Data on area (grounds), trawling depth, and the days fished can be obtained from a random sample of the trips by interviewing the captain or members of the crew. The data on area, depths and days fished are expanded in a normal manner to totals for the entire operating fleet. This programme has been carried out in Cilacap (southern Java) since mid-1974.

Another programme is the logbook programme, which has been carried out in the joint venture shrimp fishery in the Arafura Sea since 1974. The primary objective of the logbook programme is to obtain continuing indices of cpue by fishing area. These indices, expressed in kg/h of trawling, are used to monitor index of abundance (stock abundance). This logbook programme uses a revised reporting system which consists of two parts. In the first (Form A), asks for records of catch and fishing effort by fishing areas in terms of time of capture, and also catch landed in terms of time of landing. The second (Form B), asks for records of the size composition of shrimp catch in terms of internationally standardized size categories of frozen shrimps. Both Forms A and B were designed for processing by an electronic computer.

Most of the data available are: catch by species by fishing areas and by gears. Gears are specified by type of gears, and by boats which are also specified by ton and hp classes. For the Arafura Sea it seems that the logbook system is the more effective because the issue of fishing licences is limited to the receipt of an adequate report. There is good cooperation between the fishing company and the Directorate General of Fisheries and the Research Institute for Marine Fisheries. For other areas it seems that records from the shrimp dealer and auction, combined with an interview with the captain or crew, are the more effective.

2.4 *The effect of fishing*

Since the most available data are catch and effort statistics, the production model has been the only model used to evaluate the stock. From the operational viewpoint, general production models possess the advantage of requiring only catch and effort data, which are usually available at relatively little expense. These models are particularly useful when age determination and tagging are difficult. Available assessment models are valid to control the number of effort, *ie*, the number of boats, days fished.

Based on the analysis using the production models above (linear and exponential), it has been found that in some areas, such as the Malacca Strait, south and east Kalimantan, Cilacap and the Arafura Sea, the shrimp stocks have been fully or overexploited (see *Table 2*). In these areas the size of the shrimp has decreased tremendously. This provides an independent confirmation of the severe effect of fishing on the stocks. One thing that caused this fact is that the extent of the effort could not be controlled, due to the fishery having developed so fast that the data was available too

Table 2
NUMBER OF TRAWLERS, NUMBER OF BOAT-MONTHS, TOTAL CATCH (WHOLE SHRIMP) AND CATCH
PER BOAT-MONTH IN CILACAP – SOUTH COAST OF JAVA

Year	No. of trawlers	No. of boat-months	Total catch (t)	Catch per boat-month (t)	Catch/ trawler
1971	13	50	179·0	3·58	
1972	122	356	3 798·0	10·67	31·13
1973	237	1 395	2 084·7	1·78	8·80
1974	186	1 414	2 910·5	2·06	15·65
1975	166	1 421	3 005·0	2·11	18·10
1976	200	1 959	3 054·3	1·56	15·27
1977	234	2 225	4 899·6	2·20	20·94
1978	209	2 094	5 204·7	2·49	24·90
1979	167	2 124	5 241·5	2·46	31·39

Source: Naamin (1980)

late to predict the fishery and to advise the fishery administrator.

3 Other species

3.1 *Catches of fish in shrimp fisheries*

More than sixty species of fish are caught during shrimp trawling or by other gear that catch shrimp. The main species or group of species are: bombay duck (*Harpodon nehereus*), pomfrets (*Stromateus* spp.), croakers (*Scianidae*), Spanish mackerels (*Scomberomerus*), rays (*Batodei*), sharks (*Selachii*), catfishes (*Arius*), Hairtails (*Trichiurus*), seaperches (*Lutjannidae*), snappers (*Serranidae*), groupers, squid (*Loligo* spp.), and swimming crab. The estimate of those kind of fish caught are as follows:

Arafura Sea; varies from 3·6 to 8·5 t/trawler/day (see *Table 3*)

South and east Kalimantan; varies from 0.97 to 2.5 t/trawler/day

Cilacap and south coast of Java; varies from 0·89 to 1·79 t/trawler/day

Table 3
NUMBER OF TRAWLERS, WEIGHT OF SHRIMP AND FISH CAUGHT IN
THE ARAFURA SEA, 1970–78

Year	No. of trawlers	No. of days fishing	Catch (t) Shrimp	Catch (t) Fish	Catch/day/ trawler
1970	17	2 202	812·4	15 436	7·0
1971	42	6 684	2 493·3	47 367	7·0
1972	65	12 418	4 358·6	82 650	6·7
1973	79	16 019	6 891·5	131 100	8·2
1974	95	21 552	6 532·0	123 500	5·7
1975	104	24 570	4 737·1	90 060	3·7
1976	120	29 441	5 567·4	105 830	3·6
1977	120	28 575	5 687·7	107 920	3·8
1978	125	30 172	5 984·0	113 620	3·8

Source: Naamin (1979)

Malacca Strait; varies from 1·1 to 1·8 t/trawler/day

East coast of Sumatra; varies from 1·5 to 2·5 t/trawler/day

The size of those fishes ranged from small types of a few centimetres, like anchovies, to the larger sizes of more than one metre, like sharks and rays or groupers.

The ratio of shrimp to fish caught also depends upon the distance of the fishing ground. The closer the ground to the shore, estuarine and lagoon, the more shrimp were caught; the ratio could be 1:1 or 1:0·3, and the further the ground from the shore, estuarine and lagoon, the fewer shrimp were caught; the ratio falling to 1:20 or 1:30.

Most of the fish and other species caught by shrimp fishing are landed, except the fish and other species that are caught by the industrial trawlers in the Arafura Sea, south and east Kalimantan. On these only table fish for food on board are kept by the fishermen, others are discarded.

There is no research on the assessment of the direct effect on the fish stock and their fisheries of shrimp fishing.

3.2 *Biological interaction between fish and shrimp*

Some species of fish can be used as shrimp indicators. The presence of many of these species in an area is usually an indication of a good shrimping ground. The species are: bombay duck (*Harpodon nehereus*), croaker (*Scianidae*), hairtail (*Trichiurus* spp.). It is not known whether this association involves any biological interaction with shrimp.

3.3 *Shrimp caught as by catch in non-shrimp fisheries*

There is no shrimp caught as by catch in non-shrimp fisheries.

107

4 Environmental aspects

4.1 Natural variations in recruitment

Van Zalinge and Naamin (1975) studied the correlation between the rainy season and the spawning period in the Cilacap shrimp fishery. Generally a low level of spawning is found in the period November to February, which could correspond with the reduced catch rates in April to June, while high catch rates in September and October seem to originate from high spawning rates around April and May. The March to September period of relative high spawning coincides with the time of the year that rainfall decreases and the dry season, possibly favouring the development of the spawn.

At sea the catch rates are generally highest during the wettest part of the year (September to December), in which river run-off is highest as well, affecting the environmental conditions in the shallow brackish water lagoon and mangrove areas (Segara Anakan). According to Djajadiredja and Sachlan (1956) the maximum catches are taken in the Segara Anakan during the dry season in the months from June to August, while in the rainy season shrimp is caught in small numbers only.

In the Arafura Sea and Kalimantan areas, the peak season of good catch at sea also occurs during the wettest or rainy season, August to September and November to December respectively.

4.2 Human activities other than fishing

There is no record on the impact of coastal development such as land reclamation and destruction of mangroves on recruitment of shrimp. Nearly all the mangrove areas in the north coast of Java have been for many years converted into brackishwater fish or shrimp ponds or reclaimed for human settlement. Although there is no direct data on the effect of this on the shrimp fishery, it is noticeable that the shrimp catches along the north coast of Java are far less than its adjacent areas where mangroves are still available, such as the east coast of Sumatra, off Kalimantan and the south coast of Java.

An attempt has been made by Martosubroto and Naamin (1977) to relate the surface areas of tidal forest (mangroves) to the commercial shrimp catch in those and other areas in Indonesia. A significant linear relationship was obtained between these two variables as seen in their Figure 2. This relation indicates that the shrimp production increases with the size of tidal forest areas, implying that any reduction of tidal forest is likely to reduce shrimp production. According to Garcia (*pers. comm.*, 1980), to ensure that this relationship is a cause – effect one, we still need some more evidence that the destruction of mangroves has decreased the production in some areas. This linear relationship has been criticized by Marr (1976), who suggests that the relationship is more likely to be exponential. It would also be better to use the shrimp potential in each area, instead of shrimp production, because the latter is also related to fishing effort, although since most stocks are heavily fished the difference is probably not important.

5 Management

Along the lines of the general development programme for Indonesia in the present National Five-Year-Plan, fisheries management objectives should include the following aspects:

(*1*) to increase fish production in order to supply the needs of the domestic and export market;

(*2*) to increase the livelihood and welfare of the fishermen and fish-farmers by increasing their incomes;

(*3*) to extend employment opportunities;

(*4*) to maintain a maximum biological yield of the resources.

Tropical coastal waters of the shelf area have no doubt a very high productivity, especially in areas where many rivers or a big river flows in, and have most diverse fishery resources. Over 42 species of shrimp occur in the catch of the entire country and this diversity of species results in complex management problems. The distribution of the small type of penaeid shrimp such as *Metapenaeus* spp., and *Parapenaeopsis* spp., is limited to the estuarine areas, and the areas close to the shore where they mature and spawn. The large species, such as *Penaeus* spp. also spend their juvenile and adolescent stages in those areas, but later move offshore.

About 90 percent of the catch is taken by small-scale fishermen, whose distribution is directly related to the population density, being concentrated particularly in Java and the north-east coast of Sumatra. Because of the large numbers of fishermen, and their scattered distribution, statistics of catch and effort and other data necessary for stock assessment are difficult to collect, although the statistics are improving. During the course of the shrimp fishery development, overcapitalization and excess fleet capacity can be observed in some of the more populated areas along the east coast of Sumatra, and on both coasts of Java and Kalimantan. The fleet expanded considerably, but this did not lead to an increase of the total catch. Catch per vessel decreased sharply and after a certain period there are signs that the total catch is also decreasing. This overexploitation of the shrimp stock calls for management. In managing shrimp fisheries several regulations for trawling were introduced before the banning. Chronologically these regulations are as follows:

(*1*) 1973 and 1974 – two regulations by the Minister of Agriculture were introduced primarily to encourage the trawlers to utilize the large by-catches, especially in new fishing grounds, such as Irian Jaya. For the western part of Indonesia there were not too many problems arising from the regulation, since most of these by-catches are accepted by local people.

(*2*) 1975 – in this year the Government issued a warning that, taking into account of the demand of conservation and the programme of full utilization, conservation measures could be applied to certain stocks, including shrimp. Several methods were mentioned: closed areas and season, control of efforts, mesh size and other fishing gear limitations.

In the same year shrimp regulations were enforced in Irian Jaya for the first time. These measures included:

— the prohibition of trawling in depths less than 10 m;
— prohibition of the use of pair trawls and of a cod-end mesh size less than 3·0 cm;
— the number of trawlers to be licensed to operate in that area will be defined on 1 April each year.

These measures were designed to maintain the catch at a level corresponding to the maximum sustainable yield of the stock.

(*3*) 1976 – in the western part of Indonesia the inshore small-scale fishermen had been playing an increasingly important role in shrimp fishing. The number of 'baby trawlers' also increased considerably. Total reported landings of shrimp in 1976 almost doubled compared with the previous year, but overcrowding could not be avoided, usually causing frictions followed by sporadic incidents between the two groups. Regulations were introduced which emphasized the protection of resources harvested by inshore fishermen, and found directly off their villages, and which lessened the competition between these fishermen and the trawlers. The coastal waters was divided into three fishing zones, and trawling was prohibited in the first zone – an area of three miles wide, taken from the low water level onshore.

In most regions the densest shrimp school are located in the first zone. For small trawlers profitable operations depend on good shrimp catches, *ie*, fishing in the near-shore zone, and it is difficult to ensure that these trawlers fish only in the deeper zones further offshore.

(*4*) 1978 – an interdepartmental regulation to control and slow down the construction of trawlers was established in 1978 involving more than 40 boatyards along the east coast of north Sumatra and the Riouw provinces.

In the same year another shrimp trawling regulation based on the capacity of some shrimp centres, such as Cilacap on the south coast of central Java, was introduced. On 1 June 1978 the number of trawlers based in Cilacap was reduced from 160 to 89 vessels. This figure was based upon the optimum number of effort suggested by production model calculation. The excess trawlers were directed to transfer their operations to neighbouring, less overcrowded, provinces. These were usually less favourable for shrimp fishing. Besides low catches, landing facilities were inadequate, so the trawlers made great efforts to re-enter their former fishing ground.

(*5*) 1980 – finally in this year all trawling activities were banned in Java, Bali and Sumatra. Vessels registered or licensed in these areas were not permitted to move their trawling operations or bases to other regions. Before the banning, trawlers were encouraged to transfer their operations to other fishing gear, such as purse seining, gill-netting or even skipjack pole and line fishing and soft loans were provided to those who needed them. After the banning was implemented, soft loans were also provided to small-scale inshore fishermen to be able to equip their small vessels with engines and new fishing gear. This was a crash programme to modernize their fishing operations to take advantage of the reduction in total effort. Landings can be expected to increase accordingly.

In managing trawl fisheries for shrimp, where there is no artisanal fishery, experience suggests that control of effort by limiting entries, as it has been practised in Irian Jaya waters, is the most effective method. The optimum number of vessels is estimated from production model analysis. Here we have to be careful in defining fishing effort and to really control the fishing mortality, since fishing mortality applied by a given unit of effort depends on different parameters, such as type of vesssel, horse power, size and type of gear and the distribution of effort on ages, space and time. In several Indonesian waters there are tendencies toward increasing horse powers of existing vessels, employing larger nets and conducting prolonged trips by fishing almost 24 hours a day.

Another possible measure is mesh-size regulation, which faces problems in the implementation and enforcement. This measure is commonly applied to trawls, but could be considered for similar gears, such as the tidal trap. These traps, which depend on tidal currents in their operation, are trawl-like nets with each wing tied to wooden stakes implanted in the ground. They commonly have cod-end mesh sizes less than one centimetre. They are found along the east coast of Sumatra sometimes arranged in batteries close to estuaries. Traditionally their major target was sergestid shrimp, but juveniles of bigger shrimp, such as *Penaeus merguiensis* are common in their catches and sometimes in big quantities. Consideration has been given to introduce a levy on shrimp export, which would ideally also control fishing effort and support the management of shrimp generally, but so far no such steps have been experimented.

6 Future work

Collection of shrimp catch and effort data in Irian Jaya is carried out through cooperation of the trawlers' captains and their companies. Logbooks and forms for this purpose are provided by the Director General of Fisheries and the Research Institute for Marine Fisheries. Starting a few years ago, length composition of shrimp caught in this area is also collected in the same way and would be an important base for long-term studies supporting the management of Irian Jaya shrimp fisheries. In other areas, depending on the progress of their development in shrimp fisheries, experience shows that trained personnel recruited from local fishery services can be assigned to do data collection on catch, effort and local shrimp marketing structure.

In the areas where trawling has been banned, surveys are carried out to observe new developments in fishing gear, shrimp landings and socio-economic aspects of the shrimp industry. Bottom gillnetting for shrimp is developing in most banned areas, but regretfully, on the east coast of Sumatra the number of tidal traps is also increasing. Experimental fishing conducted in the south coast of Java about ten months after the banning was implemented gave remarkably high catches of shrimp, showing that the stock is recovering. The highest percentage of shrimp workers suffering from this banning are those who were involved in post harvest activities, such as processing and mar-keting, especially in areas where trawling was concentrated, such as in the north Sumatra province. The need of designing new management alternatives in shrimp fisheries for such areas is urgent, including solving the common difficulties of enforcing the management programmes.

In environmental studies a cooperative research project is being carried out to conduct research on the role of mangrove forests in the distribution of shrimp, particularly of the nursery grounds. Also shrimp tagging is planned to be included in the national shrimp research programme.

7 References

BADRUDIN, M. Preliminary study on age, growth, sex ratio and
1977 ovary maturity of *Penaeus monodon* in Meulaboh waters, West Aceh. *In* Second National Seminar on Shrimp, 15–18 March 1977, Jakarta, pp. 150–7 (in Indonesian)

DJAJADIREDJA, R and SACHLAN, M. Shrimp and prawn fisheries
1956 in Indonesia with special reference to the Kroya district. *Proc. IPFC*, 6(2–3), 366–77

INDONESIA, DIRECTORATE GENERAL OF FISHERIES, 1972–1979.
1972– Fisheries statistics for the years 1972–1979. Kuala
1979 Lumpur, Directorate General of Fisheries

MARR, J C. The Citanduy River Development Project. Segara
1976 Anakan fishery resources aspect. ECI, Denver, Banjar – Indonesia

MARTOSUBROTO, P. The spawning season and growth of *P.*
1977 *merguiensis* and *M. ensis* in the waters of Tanjung Karawang. *In* Second National Seminar on shrimp, 15–18 March 1977, Jakarta, pp. 145–9 (in Indonesia)

MARTOSUBROTO, P and NAAMIN, N. Relationship between tidal
1977 forest (mangroves) and commercial shrimp production in Indonesia. *Mar. Res. Indonesia*, No. (18), 81–6.

NAAMIN, N. An evaluation to the joint venture shrimp fishing
1979 company fishing in Arafura sea. Bogor, Graduate School, Bogor Agricultural University, 17 p. (in Indonesian) (mimeo)

NAAMIN, N. The present status of the shrimp fishery in the
1980 Cilacap area and some problems of its management. *In* Report of the Workshop on the biology and resources of penaeid shrimp in the South China Sea area. Part I. Manila, South China Sea Fisheries Programme, SCS/Gen/80/26: 49–54

NAAMIN, N and YAMAMOTO, T. Some thoughts on the conserva-
1977 tion of shrimp resources in West Irian waters (Arafura Sea). Paper presented at the Second National Seminar on shrimp, 15–18 March 1977, Jakarta, 48 p. (mimeo)

POERWANTO, D. The maturity state, spawning and fecundity of
1975 *P. merguiensis* in Tg. Karawang. M.S. Thesis. Bogor Agricultural University, 42 p. (in Indonesian)

THE PRESIDENT OF THE REPUBLIC OF INDONESIA, Decree No.
1980 39/1980, Concerning the trawl fishery banned in Indonesia, 5 p. (in Indonesian)

VAN ZALINGE, N P and NAAMIN, N. The Cilacap based trawl
1975 fishery shrimp along the south coast of Java. *Liporan Penelitian Perikanan Laut/Mar. Fish Res. Rep.*, 1(1975), 1–44

VAN ZALINGE, N P and NAAMIN N. Trip report to Cilacap.
1976 (mimeo)

Biologie et exploitation de la crevette pénaeide au Sénégal

*F Lhomme
et S Garcia*

Abstract

This paper gives a digest of the main results obtained during many years of research in Senegal and concerns all aspects of shrimp biology, dynamics and exploitation, such as growth, mortality, migration, reproduction, selectivity, recruitment life cycles, seasonal and long-term variability, artisanal and industrial fisheries, potential productivity and management. It provides for an assessment of priorities for research and management.

1 Introduction

La pêcherie crevettière du Sénégal est intéressante dans la mesure où son historique, ainsi que la structure des ressources exploitées et les interactions entre les divers modes d'exploitation sont une bonne illustration des difficultés que l'on peut rencontrer dans l'aménagement d'une ressource multispécifique commune à deux pays, exploitée, à la fois par une pêche industrielle et artisanale, et dont la stratégie d'exploitation a nettement évolué avec le temps, sous les contraintes biologiques et économiques.

Ce travail résume les résultats obtenus après plusieurs années de recherche et de collecte des données sur la pêche crevettière du Sénégal et plus généralement sur la pêche chalutière. Pour garder au document un format raisonnable, seuls les résultats les plus pertinents pour la compréhension et l'aménagement du système d'exploitation ont été conservés. La quantité de résultats est cependant importante et les lecteurs voudront bien excuser les difficultés de compréhension qui peuvent en découler, et se référer aux références d'origine pour de plus amples détails.

Dans un premier temps, le document résume les informations biologiques disponibles en matière de croissance, mortalité, migrations, reproduction, sélectivité, recrutement, chronologie des cycles vitaux et variations des caractéristiques biologiques des stocks, insistant sur les différences entre les deux stocks du Sénégal (St-Louis au nord et Roxo-Bissagos au sud) et les éventuelles variations interannuelles.

Dans un deuxième temps, on procède à la description des pêcheries, artisanale et industrielle (en insistant sur l'aspect multispécifique de cette dernière), de la collecte et du traitement des statistiques de pêche, des caractéristiques des engins de pêche et des captures, des prises accessoires et rejets, *etc.*

L'utilisation de ces statistiques permet ensuite d'évaluer les potentiels de capture en mer, dans les conditions actuelles d'exploitation des juvéniles en estuaire.

Les variations naturelles interannuelles de production étant une caractéristique importante des stocks à vie courte, un paragraphe a été consacré à leur analyse en mer et en estuaire.

En ce qui concerne l'aménagement, les mesures actuellement en vigueur ont été résumées et le point sur les informations pertinentes actuellement disponibles a été fait: potentiels et niveaux d'exploitation, interactions entre types d'exploitation, interactions entre les exploitations des divers groupes-cible de la pêche démersale sénégalaise, partage des ressources entre le Sénégal et la Guinée–Bissau.

Le travail se termine par une série de recommandations en matière de recherche et d'aménagement.

2 Résumé des informations biologiques disponibles

2.1 *Environnement et distribution géographique*

La zone concernée par cette étude se situe entre 11°30 N et 17° N (*fig 1A* et *1B*). Bien que située au-dessous du tropique du Cancer, cette région est le siège de variations saisonnières de température très importantes sur le plateau continental (environ 12°C d'amplitude) qui lui donnent un caractère tempéré. Les fleuves sont, en revanche, de type nettement tropical avec une saison de crue très marquée et très courte, et des apports d'eaux douces plutôt limités (les variations de salinité sont indiquées sur la *figure 2*).

Si le fleuve Sénégal mérite son nom, les autres affluents continentaux situés plus au sud (Sine Saloum et Casamance) sont en réalité des rias submergées, avec un bassin versant réduit, des apports d'eau douce faibles et des périodes de sursalure saisonnières.

Il existe deux stocks de crevettes, l'un situé au nord du Cap Vert (stock de St-Louis) et l'autre au sud (stock de Roxo-Bissagos). La différence essentielle entre les deux stocks résidant dans une importance bien moindre des apports d'eau douce,

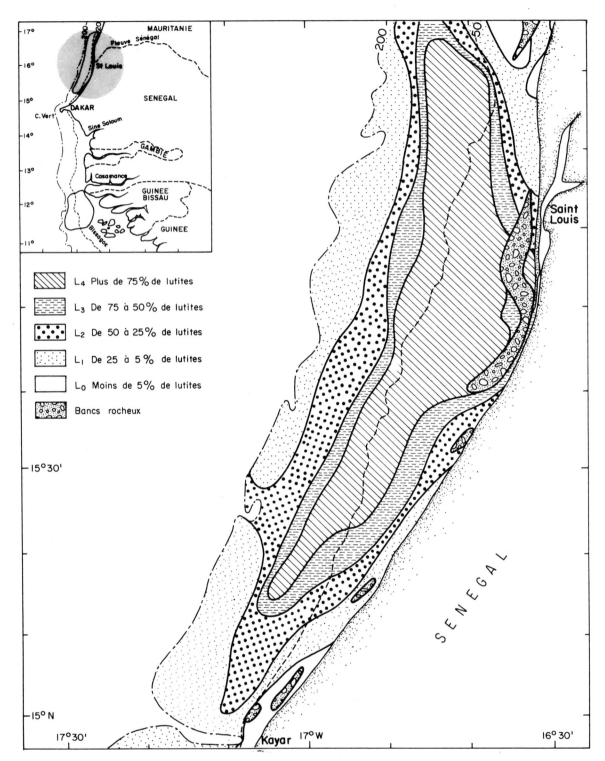

Fig 1 Position et nature sédimentologique des fonds de pêche à la crevette du Sénégal (d'après Domain, 1977)

Légende:

▨	L₄	Plus de 75% de lutites
▦	L₃	De 75 à 50% de lutites
▪	L₂	De 50 à 25% de lutites
∵	L₁	De 25 à 5% de lutites
☐	L₀	Moins de 5% de lutites
⬡		Bancs rocheux

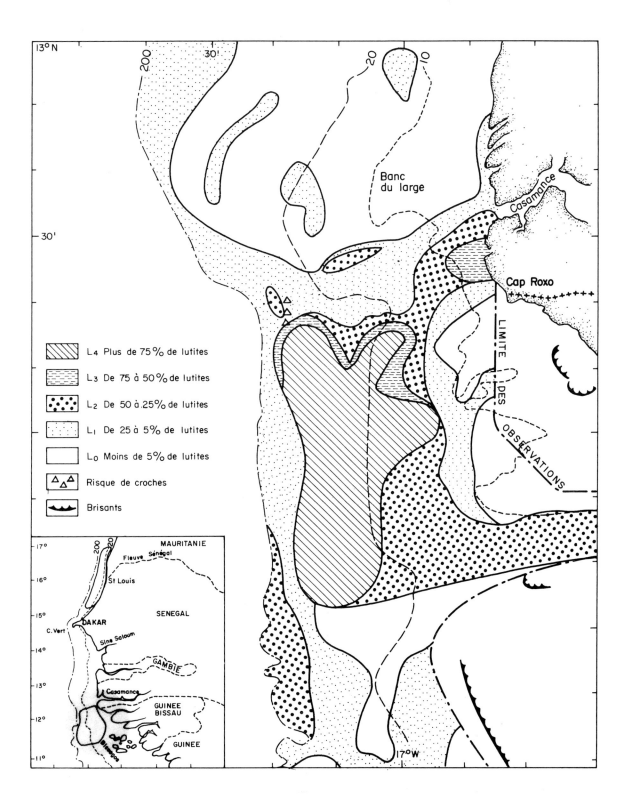

13°N

200

30'

30'

20

10

Banc
du large

Casamance

Cap Roxo

LIMITE DES OBSERVATIONS

L₄ Plus de 75% de lutites

L₃ De 75 à 50% de lutites

L₂ De 50 à 25% de lutites

L₁ De 25 à 5% de lutites

L₀ Moins de 5% de lutites

△△△ Risque de croches

Brisants

17°

200

20

MAURITANIE

Fleuve Sénégal

16°

St Louis

SENEGAL

15°

DAKAR

C. Vert

Sine Saloum

14°

GAMBIE

13°

Casamance

12°

GUINEE
BISSAU

Bissagos

GUINEE

11°

17°W

113

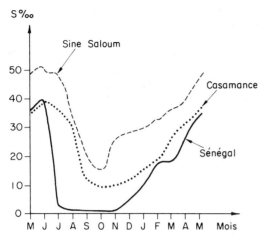

S‰

Sine Saloum

Casamance

Sénégal

M J J A S O N D J F M A M Mois

Fig 2 Evolution saisonnière des salinités dans le fleuve Sénégal (Rochette, 1964), Foundiougne (dans le Sine Saloum, De Bondy, 1968) et Ziguinchor (Casamance, d'après Brunet-Moret, 1970)

une pêche artisanale plus faible et des nurseries plus limitées, dans le Nord.

La carte sédimentologique simplifiée des fonds situés au large du Sénégal est donnée sur les *figures 1A* et *1B*. Elle montre clairement l'existence de deux grandes zones vaseuses au nord du Sénégal (fonds de St-Louis) et entre le Sénégal et la Guinée Bissau (fonds de Roxo Bissagos), encadrées de fonds plus durs, plus ou moins rocheux juste au sud du cap Vert ou constitués de sables purs, au large de la Casamance. Ces deux zones sont le siège d'une intense exploitation crevettière depuis 1966, date d'une profonde reconversion de la pêche chalutière sénégalaise (voir paragraphe 3).

La pêche artisanale des crevettes est une activité traditionnelle au Sénégal. Elle se déroule dans les principaux milieux d'estuaire: l'embouchure du fleuve Sénégal, le complexe Sine – Saloum – Gambie ainsi que dans la Casamance.

2.2 Croissance

2.2.1 *Croissance des juvéniles* Elle a été étudiée par la méthode des progressions modales sur des échantillons collectés sur le terrain (dans le cas du Sine – Saloum) ou au marché de St-Louis (dans le cas du fleuve Sénégal). Dans ce dernier cas seulement, les paramètres de l'équations de von Bertalanffy ont été calculés. Les accroissements observés sont les suivants:

âge relatif* (semaines)	0	1	2	3	4	5
taille (LC mm)[1/]	16·0	17·6	19·2	20·8	22·0	23·4
accroissements (mmLC)	1·6	1·6	1·6	1·2	1·4	1·4
	6	7	8	9	10	
	24·8	26·0	27·2	28·2	29·4	
	1·2	1·2	1·0	1·2	—	

*a partir d'une borne arbitraire t = 0 à 16 mm de longeur céphalothoracique

[1]LC indique la longueur céphalothoracique et LT la longueur totale

Les paramètres de l'équation de von Bertalanffy sont $LC_\infty = 52·8$ mm, $LT_\infty = 25·4$ cm et K = 0·045 si t est exprimé en semaines ou K = 0·19 si t est exprimé en mois. Les accroissements théoriques exprimés en longueur totale varient de 4·6 à 2·9 cm LT/mois pour des juvéniles de 1 à 10 cm de longueur totale.

Dans le Sine Saloum, seule une croissance moyenne a pu être calculée sur une période relativement courte (5 semaines) et on a obtenu 5 mm LC/mois (entre 11 et 16 mm LC) ou 2·4 cm LT/mois (de 4·8 à 7·2 cm LT).

La comparaison des accroissements moyens à une taille de 6·0 cm LT dans ces 2 nurseries indiquerait une croissance plus rapide (3·6 cm LT/mois) dans le fleuve Sénégal (contre 2·4 cm LT/mois dans le Sine Saloum qui est un milieu sursalé en permanence).

2.2.2 *Croissance des adultes en mer* Son étude a été rendue difficile par l'existence de variations saisonnières de croissances importantes liées aux variations hydroclimatiques. De bons résultats ont été obtenus par marquage et la comparaison de ces derniers avec les progressions modales a permis, d'une part, de confirmer l'existence des variations saisonnières, et d'autre part de calculer des courbes de croissance différentes pour les diverses saisons. Les résultats obtenus pour les deux sexes sont donnés ci-après:

	Femelles		Mâles	
Période	K/mois	LC_∞ (mm)	K/mois	LC_∞ (mm)
1 – 11/1 – 2	0·17	39·0	0·22	29·0
1 – 2/15 – 4		Croissance nulle		
15 – 4/1 – 8	0·38	43·2	0·50	31·3
1 – 8/1 – 11	0·27	42·8	0·32	28·6

114

On a pu aussi montrer qu'il existait une relation entre K et la température. Par exemple, les valeurs trouvées pour les femelles sont données ci-après:

K	0	0·17	0·27	0·38
t°C	14·3	15·6	18·3	22·1

On verra ultérieurement que le recrutement est pratiquement continu et il est clair que chaque cohorte aura, en fait, un schéma de croissance propre, dépendant de la date de naissance.

Pour schématiser ce phénomène, nous avons représenté sur la *figure 3* les croissances théoriques des 12 cohortes recrutées chaque mois à 25 mm LC sur le stock de St-Louis et qui disparaissent de la population après 18 mois d'existence en mer. Dans cette région où le recrutement le plus important se produit de décembre à mai (période où la croissance est faible ou nulle), il existe donc, en fin d'année, un groupe dominant de taille homogène mais composé d'individus d'âges très différents.

Il est important de noter qu'il n'existe donc pas au Sénégal de loi de croissance ni de relation âge-longueur unique. Cet aspect de la croissance entrainant une certaine variabilité dans l'âge à la maturation, l'âge à la première capture, *etc* . . . il nous a paru utile de donner à titre d'exemple la courbe de croissance et la chronologie des évènements principaux pour les cohortes nées sur le fond de St. Louis en septembre, en saison chaude (il s'agit de l'une des plus importantes) (*fig* 4) et sur le

fond de Roxo-Bissagos en janvier, en saison froide (*fig* 5).

Il est clair cependant que ces courbes ne sont pas généralisables à toutes les cohortes et permettent seulement de souligner quelques différences dans la chronologie des cycles vitaux pour les cohortes nées en saison chaude ou froide, à St-Louis ou à Roxo-Bissagos.

2.3 *Mortalités*
Elles n'ont pu être étudiées que pour les adultes en utilisant les techniques du marquage à l'aide de disques de Petersen (Neal, 1969).

2.3.1 *Mortalité en fonction de l'âge* L'analyse des taux de recapture en fonction de la taille (ou de l'âge) permet d'approcher ce phénomène. Deux cas fréquents sont donnés sur la *figure 6*. L'exemple le plus classique correspond à la *figure 6A*: le taux de recapture augmente jusqu'à 25–30 mm LC puis décroit régulièrement. La partie gauche de la courbe est liée à une mortalité additionnelle due à la marque et au phénomène de recrutement partiel. Le maximum observé correspond le plus souvent à la classe de taille ou au groupe le plus important. La partie droite de la courbe peut être due à une diminution de la mortalité totale et en particulier à une diminution de la mortalité par pêche avec la taille ou avec l'âge.

La *figure 6B* donne un autre exemple de courbe, plus rare. Ce type de courbe a également été noté par Garcia (1978) en Côte d'Ivoire où il est

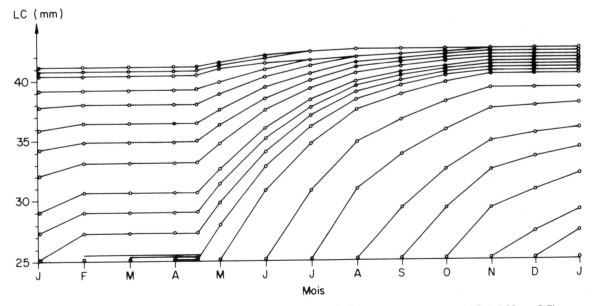

Fig 3 Croissance théorique des femelles en mer sur le fond de St-Louis (taille moyenne au recrutement fixée à 25 mm LC)

115

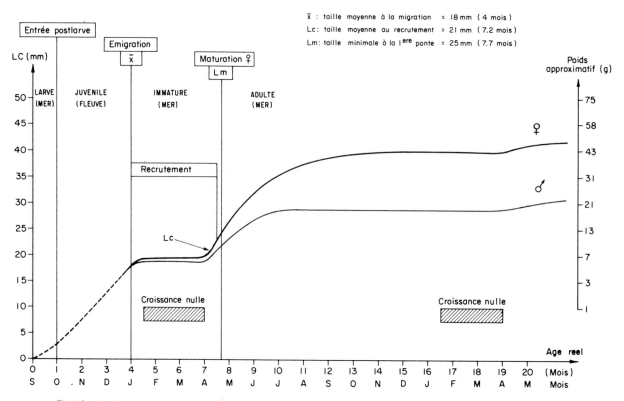

Fig 4 Courbe de croissance théorique des individus nés en septembre (saison chaude) sur le fond de pêche de St-Louis

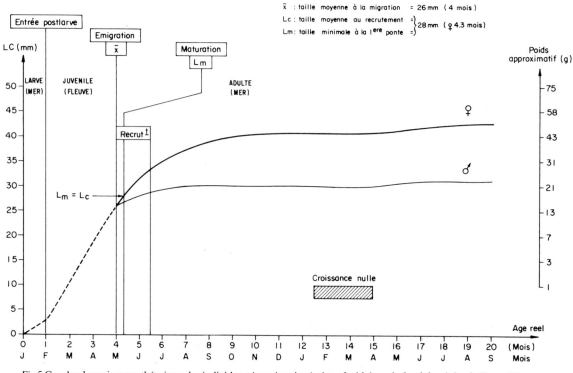

Fig 5 Courbe de croissance théorique des individus nés en janvier (saison froide) sur le fond de pêche de Roxo-Bissagos

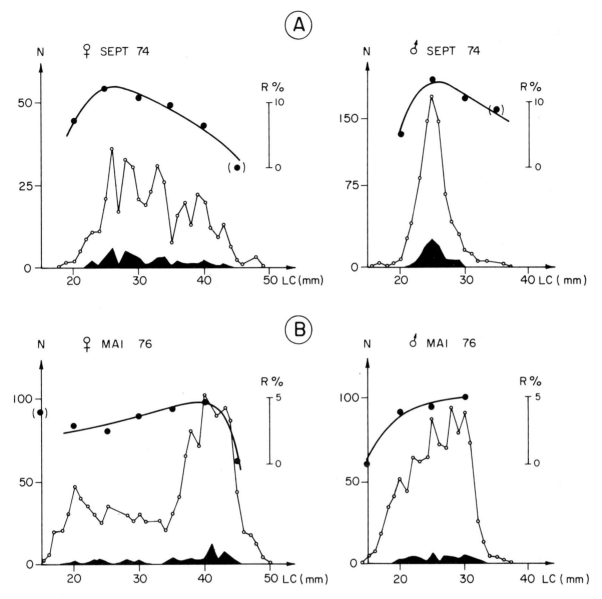

Fig 6 Variations du taux de recapture en fonction de la taille et distribution de fréquence des tailles des individus marqués (en blanc) et recapturés (en noir) sur le fond de pêche de Roxo-Bissagos

observé en même temps que se produisent d'importants changements saisonniers de capturabilité. Il est intéressant de noter que le maximum correspond là encore aux groupes les plus abondants. Cette observation tend à confirmer l'analyse de Garcia (1978) suivant laquelle chez les pénaeides, où la pêche est très dirigée, la mortalité par pêche est liée à l'abondance.

2.3.2 *Mortalité naturelle apparente et mortalité par pêche* Une série de marquages a été entreprise sur les fonds de St-Louis et Roxo-Bissagos entre 1975 et 1976. Les résultats suivants, obtenus par les méthodes préconisées par Gulland (1969)[1] ont été retenus.

	q	$Z/mois$	$F/mois$	$X/mois$
St-Louis	0·0007	1·19	0·22	0·97
Roxo-Bissagos	0·0005	1·11	0·16	0·95

[1]Etude de l'evolution dans le temps des recaptures par unité d'effort de pêche

117

Les valeurs de q peuvent être comparées en les ramenant à une surface standard de 100 milles carrés et on a obtenu les valeurs de 0·0030 pour St-Louis et 0·0040 pour Roxo-Bissagos. Il est intéressant de noter:

- la similitude des valeurs de X (la mortalité naturelle apparente), qui paraissent en outre donner une estimation très surestimée de M si on se réfère aux données de la littérature.
- la similitude des valeurs de F et de q compte tenu de la précision (non calculée) que l'on peut espérer obtenir par l'utilisation de telles méthodes d'estimation.

Dans la mesure où il est bien connu que les marquages donnent souvent une valeur biaisée de M et si l'on admet avec Gulland (1971) que F est proche de M quand le stock est exploité à un niveau proche du MSY, il découle des résultats précédents et de l'analyse des modèles de production (voir plus loin) que M devrait être proche de 0·22 et supérieure à 0·16. Cette conclusion tend à corroborer de manière indirecte les travaux de Garcia (1978) qui donne M = 0·20 – 0·25, pour la même espèce en Côte d'Ivoire.

2.4 Migrations des adultes en mer

Elles ont été étudiées par marquage. Les déplacements observés sont faibles et lents. Ils sont parallèles à la côte pour le fond de St-Louis qui est très allongé et où les courants sont également parallèles à la côte. Ils n'ont pas de direction privilégiée sur le fond de Roxo beaucoup plus vaste et où le système des courants est plus complexe. Une liaison très nette entre les changements saisonniers des courants sur le fond de St-Louis et les changements de direction des migrations a été mise en évidence.

2.5 Reproduction

2.5.1 *Taille à la première ponte* Le critère de maturation choisi pour cette étude (observation visuelle externe, échelle de De Vries et Lefèvre, 1969, stades préponte IV et ponte V) conduit à déterminer une taille à la première ponte. L'analyse de l'évolution du pourcentage d'individus au stade IV + V en fonction de la taille, dans les débarquements des crevettiers de Dakar par un échantillonnage intensif et régulier de 1973 à 1977 indique que la taille à la première ponte (déterminée par l'intersection de la courbe lissée avec l'axe des x) est de 25 mm LC pour St-Louis et 28 mm LC pour Roxo-Bissagos. A titre tout à fait indicatif cela correspond par exemple à un âge de 7·7 mois et 4·3 mois (respectivement) pour les cohortes principales définies précédemment

(paragraphe 2.1, *figures 4* et *5*).

2.5.2 *Saisons de ponte* Un échantillonnage régulier au port a permis d'obtenir le pourcentage de femelles au stade (IV + V), mois par mois, de 1972 à 1977. L'importance de la reproduction dépendant non seulement de ce pourcentage mais aussi de l'abondance réelle du stock, un *indice de fécondité potentielle* a été calculé en utilisant:

- la cpue mensuelle en nombre (indice d'abondance)
- le sex-ratio mensuel
- le % de femelles mûres
- la fécondité individuelle moyenne des femelles mûres, fonction de la taille moyenne des femelles capturées et de la relation de fécondité. Cette dernière étant inconnue pour *P. notialis* nous avons utilisé celle de Martosubroto (1974) pour *P. duorarum duorarum* espèce très voisine et longtemps confondue avec l'espèce considérée ici.

Le produit de ces 4 paramètres permet d'obtenir un indice de fécondité de la population qui tient compte des variations saisonnières de l'abondance des femelles, de leur état de maturation et de leur âge, comme l'ont recommandé Le Reste (1977) et Garcia (1978).

Les variations de l'indice de fécondité potentielle sur une année moyenne sont indiquées sur les *figures 7A* et *8A*.

Pour le stock de St-Louis on distingue une saison de ponte bien marquée d'août à décembre, avec un maximum très bien défini de septembre à novembre. Certaines années un schéma bimodal est observé, avec un maximum en début et un autre en fin de saison chaude.

Pour le stock de Roxo-Bissagos, la reproduction s'étend sur toute l'année et se présente sous l'aspect de pics irréguliers en chronologie et en intensité. Des maximums importants sont observés en saison chaude (août) en période de transition (novembre) mais *également en pleine saison hydrologique froide* (janvier). On peut considérer qu'il existe en moyenne un cycle saisonnier peu marqué avec une période de pont relativement plus intense de août à janvier.

2.5.3 *Variations interannuelles de reproduction* Il a déjà été indiqué qu'il existait quelques différences dans les variations saisonnières de reproduction, d'une année sur l'autre. Bien que la reproduction n'ait pas été analysée en détail il a paru important d'attirer l'attention sur un autre type de variations interannuelle. La *figure 9* montre l'évolution du pourcentage de complètes femelles

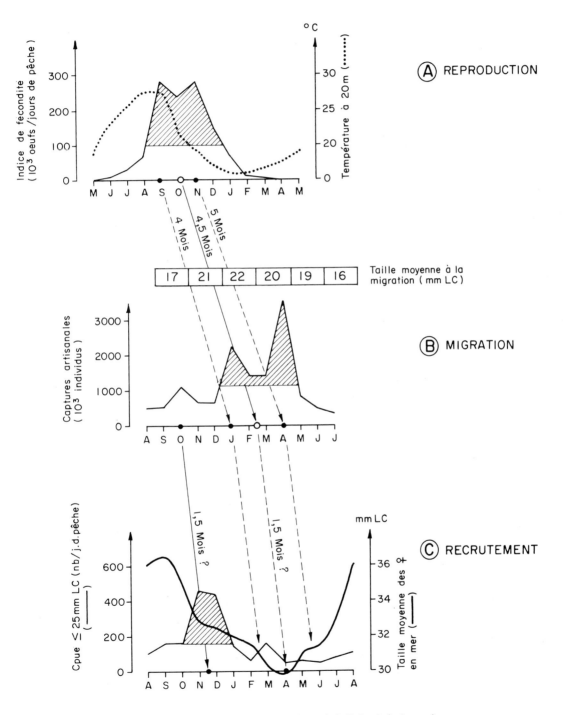

 Valeurs supérieures à la moyenne

A REPRODUCTION

B MIGRATION

C RECRUTEMENT

Fig 7 Chronologie du cycle vital pour le stock de St-Louis (voir texte)

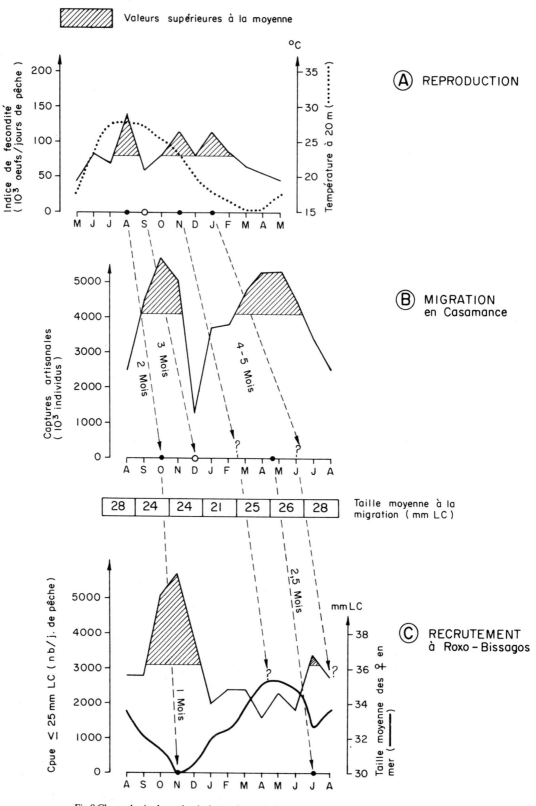

Fig 8 Chronologie du cycle vital pour le stock de Roxo-Bissagos (voir texte)

120

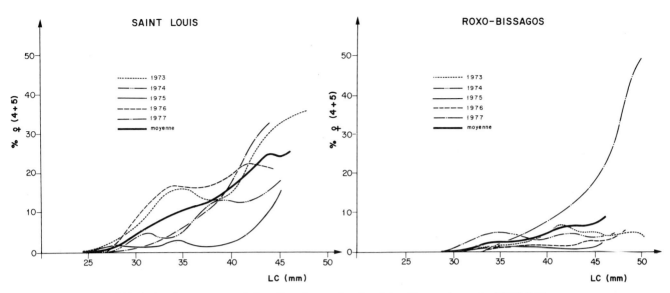

Fig 9 Evolution du pourcentage de femelles mûres en fonction de la taille pour les années 1973 à 1977

mûres avec la taille de 1973 à 1977 chaque courbe est basée sur l'échantillonnage d'une année ainsi que l'évolution moyenne pour toute la periode considerée. Ces courbes très globales sont difficiles à analyser mais elles sont une manifestation résumée de l'intensité de la reproduction à l'échelle de l'année et elles montrent qu'il existe des variations très importantes d'une année sur l'autre qui ne sont pas liées, comme on pourrait le penser, à des différences dans la distribution saisonnière de l'effort de pêche annuel d'une année sur l'autre.

Les pourcentages observés sont excessivement faibles en 1975 pour les 2 zones de pêche considérées alors qu'ils sont anormalement élevés en 1977 à Roxo-Bissagos.

La relation précise entre ces observations et la reproduction des stocks n'est pas établie mais il est certain que des variations interannuelles de reproduction existent.

2.5.4 *Variations géographiques* Bien que les températures et leurs variations saisonnières dans les 2 fonds de pêche soient comparables, il existe des différences nettes entre les deux fonds

(*a*) les pourcentages de femelles mûres observées sont toujours beaucoup plus faibles à Roxo-Bissagos quelles que soient les tailles ou les années considérées;

(*b*) la ponte se produit en moyenne à une taille plus faible (25 mm LC) à St-Louis qu'à Roxo-Bissagos (28 mm LC);

(*c*) les variations saisonnières de reproduction ont des amplitudes très différentes.

2.5.5 *Ponte et conditions de milieu* La plupart des travaux sur la reproduction des pénaeides relient la ponte à la température de l'eau et à ses variations. L'amplitude importante des variations thermiques au Sénégal (12°C) devrait conduire à l'existence de saisons de ponte très marquées (type tempéré). C'est bien le cas à St-Louis où la saison de ponte correspond à la transition entre saison chaude et saison froide. Lorsque deux maximums existent, certaines années, ils correspondent aux deux transitions entre la saison chaude et la saison froide (*Fig 7A*). Ce phénomène est d'ailleurs généralement observé dans le Golfe de Guinée (Isra-Orstom, 1979).

Le schéma observé à Roxo-Bissagos diffère légèrement dans la mesure où les individus aux stades IV et V existent toute l'année. Les valeurs les plus élevées sont cependant observées pendant la pleine saison chaude et pendant la transition entre la période chaude et la période froide et ce jusqu'en janvier où la température est déjà très basse (environ 17° en moyenne).

2.6 *Migration des subadultes vers la mer*
La signification de l'effort de pêche dans une pêcherie artisanale de ce type a été discutée très en détail par Garcia (1978) ainsi que la validité des divers indices d'abondance que l'on peut obtenir comme indicateurs de l'intensité de la migration. Il a en particulier indiqué que l'effort et l'abondance variaient saisonnièrement dans le même sens, et que la capture totale était le plus souvent un indice de migration utilisable, dont les variations saisonnières étaient cependant vraisemblablement

121

exagérées. C'est cet indice (captures totales en nombre) que nous avons retenu, en l'absence de données fiables d'effort. Dans le fleuve Sénégal, le cycle saisonnier est net. La migration est continue et présente un pic secondaire en octobre (correspondant à la crue annuelle) et une saison principale de décembre – janvier à avril – mai (*Fig. 7B*). Dans la Casamance, la migration est également continue et présente deux pics d'égale importance: le premier en octobre, correspondant à la crue annuelle, le second en avril – mai (*Fig. 8B*). Malgré leur différence d'allure, les deux cycles de migration concordent relativement bien, le pic d'octobre étant mieux marqué en Casamance que dans le fleuve Sénégal.

2.7 *Sélectivité*

L'étude a été réalisée par la méthode de la double poche avec des mailles de 20, 25, 30, 35, 40 et 50 mm de côté correspondant à des ouvertures de maille[1] de 37, 46, 54, 62, 73 et 94 mm. Le maillage de la double poche était de 10 mm de côté. Les courbes obtenues sont données sur la *figure 10*. On remarque que les courbes correspondent aux mailles de 37 à 62 mm d'ouverture sont très voisines tandis que les courbes correspondant aux mailles de 73 et 94 mm, identiques entre elles sont très largement décalée et de pente plus forte (avec un intervalle de sélection beaucoup plus étendu.

On remarque également que les courbes ne sont pas exactement symétriques (la longueur à 50%, l_{50}, est différente de la longueur moyenne de sélectivité, l_s). Les valeurs moyennes des coefficients de sélection sont respectivement 0·450 et 2·08, lorsque les longueurs céphalothoraciques ou les longueurs totales sont utilisées.

----
[1]mesurées à la jauge CIEM

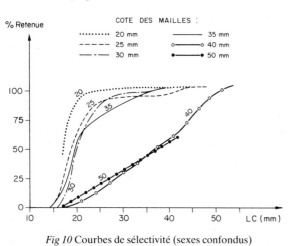

Fig 10 Courbes de sélectivité (sexes confondus)

Des différences liées au sexe ont été observées: au-dessous de 20 mm de longueur céphalothoracique, le pourcentage de rétention des mâles semble inférieur à celui des femelles. La signification de cette différence n'a pas été testée mais elle avait déjà été suggérée pour la même espèce en Côte d'Ivoire (Garcia, 1974). Le phénomène s'inverse au-dessus de 20 mm.

L'analyse des conséquences d'un changement de maillage, utilisant la méthode de Cadima (1977), indique que le passage de la maille actuelle de 37 mm à une maille de 62 mm, entrainerait une perte immédiate, en poids, négligeable (dans les quelques semaines suivant l'application du nouveau maillage) de l'ordre de 3% et un gain à long terme équivalent si le taux d'exploitation atteint 0·5. Ce résultat signifie que la maille des crevettiers pourrait être portée à 62 mm sans affecter de manière appréciable la rentabilité de la pêcherie, tout en améliorant sensiblement la survie des juvéniles des espèces associées.

2.8 *Recrutement*

2.8.1 *Périodes de recrutement* Elles sont indiquées de manière indirecte par les périodes de migration hors des lagunes (cf. paragraphe 2.5). On a également essayé de les suivre par l'abondance de juvéniles de petite taille (LC ≤ 25 mm) dans les captures pour déceler un éventuel temps de latence entre migration et arrivée dans les captures (*Fig. 7C* et *8C*).

Les courbes obtenues sont très similaires pour les deux stocks considérés avec un maximum très net en novembre ou novembre – décembre sur un cycle saisonnier bien marqué à recrutement continu.

Dans les deux cas, il semble que les recrues quittant l'estuaire en octobre soient parfaitement reconnaissables dans les captures dès novembre après un temps de latence similaire à celui que l'on observe, par exemple, en Côte d'Ivoire (Garcia, 1978). En revanche, dans les deux cas, la forte migration de janvier à mai n'est pas représentée au niveau des captures de la pêche industrielle (*Fig 7* et *8C*).

En ce qui concerne le stock de St-Louis, l'évolution saisonnière de la taille moyenne dans la population, qui passe par un minimum en mars – avril – mai, confirme cependant que le recrutement principal se produit de janvier à juin avec un maximum en avril. Il se produirait donc à une taille supérieure à 25 mm LC et on doit noter en effet que c'est à cette période que la taille à la migration est la plus élevée (*Fig 7B*). Le temps de latence serait d'environ 1·5 mois.

En ce qui concerne le stock de Roxo-Bissagos, la situation paraît plus complexe. En effet, l'évolution de la proportion de juvéniles de petite taille dans les captures et celle de la taille moyenne dans la population suggèrent toutes deux un recrutement important en octobre – novembre – décembre, vraisemblablement lié à la période des crues alors que l'importante migration de janvier à juin n'apparaît pas au niveau du recrutement. Notons cependant que cette migration se produit également à une taille très élevée, plus élevée que celle que l'on observe à St-Louis et pour la plupart des stocks de pénaeides côtiers du monde (Ceci doit vraisemblablement être mis en liaison avec le biotope d'antiestuaire très particulier de la Casamance dont la salinité est supérieure à celle de la mer pendant une bonne partie de l'année). Les jeunes recrues de grande taille seraient donc confondues avec les crevettes plus âgées, recrutées auparavant d'autant que le stock d'adultes étant situé loin dans le sud (*Fig. 1*) le temps de latence pourrait être plus long (2·5 mois) et susceptible de permettre une croissance importante en cette période chaude de l'année. Le résultat serait que le recrutement de janvier à juin ne peut être détecté directement à partir des captures si ce n'est par le petit pic de juillet, auquel correspond d'ailleurs une petite chute de la taille moyenne (*Fig. 8C*) ce qui donnerait un temps de latence très approximatif de 2·5 mois.

Dans la suite de cette analyse nous admettrons faute de données contraires, que les évènements qui se déroulent dans la Casamance sont un bon exemple de ce qui se passe dans les autres très nombreuses nurseries qui alimentent le fond de pêche de Roxo Bissagos. Ceci devrait cependant être vérifié, en particulier en Guinée Bissau, avant que les conclusions de ce travail pour ce stock soient considérées comme définitives.

2.8.2 *Taille et âge à la première capture* La combinaison de la courbe de sélectivité et de la courbe de recrutement permet de calculer la courbe réelle d'entrée dans la pêcherie. La courbe de sélectivité pour la maille utilisée au Sénégal (20 mm de côté) est donnée sur la *figure 10*. La courbe de recrutement indiquant pour chaque taille le pourcentage de la population totale (mer + estuaire) qui se trouve sur le fond de pêche en mer est difficile à obtenir. Elle a été approchée en considérant qu'en deçà d'une certaine taille ($\bar{X} - 2$ écarts type) tous les individus sont encore en estuaire (recrutement = 0%) et qu'au delà d'une taille ($\bar{X} + 2$ écarts type) tous les individus sont en mer (recrutement 100%). La taille $\bar{X}$ est la taille moyenne à la

migration et l'écart type est celui de la distribution des crevettes en migration. Entre les points 0 et 100% une simple droite a été tracée.

La *figure 11* montre les résultats.[1] On voit en particulier, qu'à cause d'une taille à la migration plus grande, la taille moyenne à la première capture est plus élevée à Roxo Bissagos (28 mm LC) qu'à St-Louis (21 mm LC).

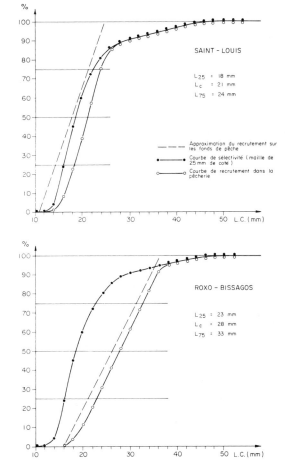

Fig 11 Détermination de la taille au recrutement pour les fonds de pêche de St-Louis et Roxo-Bissagos

A titre indicatif et toujours pour souligner la variabilité des phénomènes la courbe de recrutement en fonction de la taille a été transformée en courbe de recrutement en fonction de l'âge (*Fig. 12*) pour les 2 cohortes types définies (St-Louis/septembre et Roxo Bissagos/janvier). On voit que l'âge à la première capture varie, pour des

[1] la courbe de sélectivité utilisée ici est celle correspondant au maillage de 25 mm de côté, très proche du maillage le plus couramment utilisé au Sénégal et pour laquelle de meilleures données étaient disponibles pour les petites tailles.

raisons naturelles de 4·3 mois dans le premier cas à 7·4 mois dans le deuxième, la différence étant essentiellement due aux différences dans la taille à la migration et à la date du recrutement en mer par rapport à la période où la croissance est nulle.

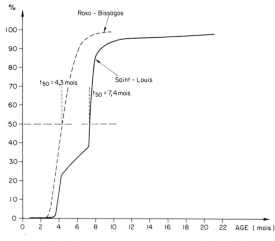

Fig 12 Courbes de recrutement en mer en fonction de l'âge (femelles) pour deux cohortes particulières (St-Louis, nès en septembre et Roxo-Bissagos, nès en janvier)

2.9 Chronologie des cycles vitaux

La comparaison des cycles saisonniers de reproduction, migration et recrutement (*Fig. 7 et 8*) permet de proposer une chronologie moyenne pour les cycles vitaux des 2 populations considérées.

En ce qui concerne St-Louis, la *figure 7* montre que le décalage entre la ponte et la migration est de 4 à 5 mois approximativement. Si la période larvaire est d'environ 1 mois comme on l'admet généralement la durée du séjour en lagune serait de 3 à 4 mois. Le recrutement dans les captures se ferait 1·5 mois plus tard à âge de 5·5 – 6·5 mois.

Sur le fond de Roxo-Bissagos, la *figure 8*, plus difficile à interpréter suggère un âge à la migration très variable, de 2 – 3 mois au moment de la crue à 4 – 5 mois en pleine saison sèche (lorsque la salinité de la Casamance est plus élevée). Le recrutement dans les captures se produirait 1 mois plus tard en période de crue et environ 2·5 mois plus tard en saison sèche.

Ces valeurs sont, bien entendu données à titre indicatif car d'une saison à l'autre, la durée du séjour en estuaire varie de même que les vitesses de croissance entraînant l'existence de chronologies différentes d'une cohorte à l'autre.

Il existe également des variations interannuelles dans la taille à la migration (cf. paragraphe 3.2.2), susceptibles de modifier ce schéma moyen.

2.10 Variations des caractéristiques biologiques
2.10.1 Variations saisonnières de la densité
Les variations de la prise, de l'effort et de la cpue, considérées ici comme un indice de densité, sont indiquées sur la *figure 13*. Ces dernières sont bien marquées à St-Louis et à peine perceptibles à Roxo-Bissagos. Les deux cycles ne sont pas exactement synchrones mais dans les deux cas la densité est plus élevée pendant la deuxième moitié de l'année.

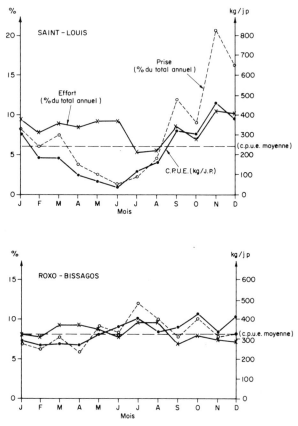

Fig 13 Variations saisonnières moyennes (1973 – 1976) de la prise, de l'effort de pêche, de la cpue

2.10.2 Variations de la taille moyenne

a) Variations saisonnières: elles sont représentées sur la *figure 14*. Il existe un cycle annuel très net. Les variations sont plus nettes pour les femelles que pour les mâles. Il est intéressant de noter que bien que les deux stocks de St-Louis et Roxo-Bissagos aient des hydroclimats très semblables (tout au moins en ce qui concerne la température), les variations saisonnières de taille moyenne sont pratiquement opposées.

b) Variations interannuelles. La *figure 15* représente les variations de la taille moyenne annuelle

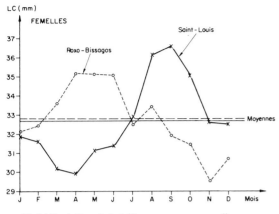

Fig 14 Evolution de la taille moyenne mensuelle en mer

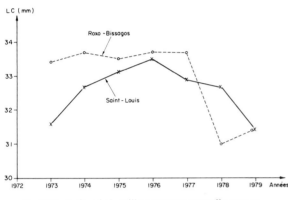

Fig 15 Evolution de la taille moyenne annuelle en mer

dans les captures de 1973 à 1978. On note une diminution progressive sur St-Louis depuis 1976, et une chute plus rapide sur Roxo depuis 1977.

2.10.3 Variations du sex-ratio

a) En fonction de la taille (Fig. 16). La courbe a une forme très caractéristique semblable à celle donnée par Garcia (1974) pour la même espèce en Côte d'Ivoire, mais largement décalée vers les petites tailles. En revanche les courbes de St-Louis et Roxo sont très proches. Ceci indique que ce type de courbe n'est pas caractéristique d'une espèce mais plus vraisemblablement d'un stock précis sous un certain régime d'exploitation (Garcia et Le Reste, 1981).

b) Variations saisonnières. Des variations ont été observées mais elles sont difficiles à interpréter. Le pourcentage de femelles est élevé en février, mai–juin et novembre–décembre à St-Louis. Il est élevé de février à juin et en novembre–décembre à Roxo-Bissagos.

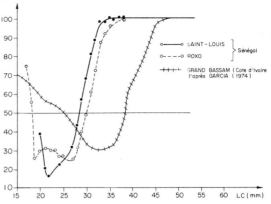

Fig 16 Variations du sex-ratio en fonction de la taille, au Sénégal et en Côte d'Ivoire

3 Descriptions des pêcheries

3.1 *La pêche industrielle en mer*

3.1.1 *Les espèces exploitées* La pêche chalutière a commencé vers 1950 au Sénégal. Jusqu'en 1965, elle a exploité le large plateau sénégalais au sud du Cap Vert, jusqu'à la Casamance. Les espèces recherchées sont des poissons nobles de fonds durs: dorades roses (*Pagrus ehrenbergi*), pageots (*Pagellus coupei*), dorades grises (*Diagramma mediterraneum*) et mérous (*Ephinephelus aeneus*). Les rougets (*Pseudupeneus prayensis*) sont fréquents et les soles (*Cynoglossus spp*) plutôt rares. Les débarquements totaux sont donnés sur *fig. 17*.

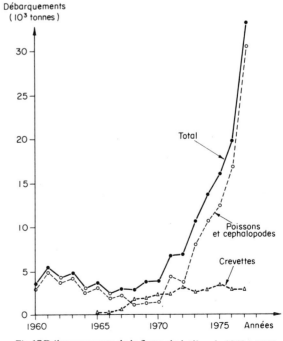

Fig 17 Débarquements de la flotte chalutière de 1960 à 1977

Dès 1965 la découverte et la mise en exploitation des fonds à crevettes au nord et au sud du Sénégal entraînent une mutation radicale de la pêcherie. Comme le montre la *figure 18*, le nombre de bateaux consacrés à la crevette augmente très rapidement de 1965 à 1970 au détriment des chalutiers pêchant le rouget ou les autres espèces. La composition spécifique des débarquements à Dakar change également profondément comme le montre la *figure 19*. Les espèces 'grises' caractéristiques des fonds vaseux dominent dès 1968 avec les crevettes. Ce sont surtout des soles (*Cynoglossus spp*) des capitaines (*Pseudotolithus*), des 'thieckems' (*Galeoïdes*).

A partir de 1970 une évolution inverse se produit. Les stocks de crevettes étant pleinement exploités, la flottille se diversifie à nouveau et plusieurs activités spécifiques nouvelles se dégagent: pêche spécialisée des rougets, des céphalopodes, des brotules, des soles, *etc.* Les diverses redistributions de l'effort de pêche de la flotte chalutière du Sénégal sur une ressource multispécifique compliquent l'analyse des stocks de crevettes et de leurs réactions à l'exploitation.

3.1.2 *La flotte chalutière* L'évolution des effectifs de la flottille chalutière dakaroise et de sa composition sont données sur la *figure 18*. Composée d'unités de 100 à 300 ch. avant 1965, la flotte s'enrichit de nombreuses unités de 300 à 450 ch. entre 1965 et 1972, destinées surtout à la pêche à la crevette.

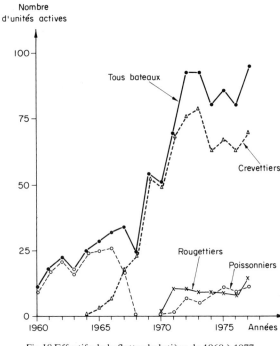

Fig 18 Effectifs de la flotte chalutière de 1960 à 1977

3.1.3 *Les fonds de pêche à la crevette* La répartition géographique détaillée de l'effort dirigé sur les crevettes est indiquée sur les *figures 20* et *21*. La comparaison de ces cartes avec la *figure 1* montre la relation entre les concentrations de crevettes et la répartition des sédiments fins. Pour le fond de St-Louis la concordance et excellente et la pêche

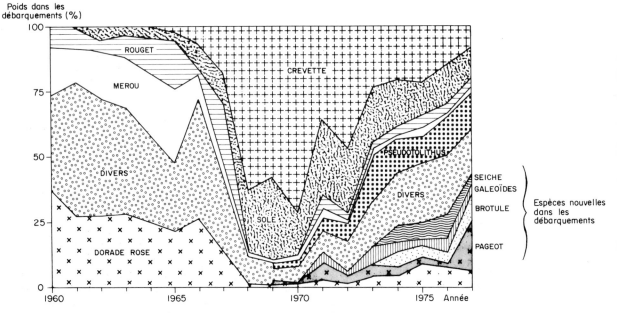

Fig 19 Evolution de la composition spécifique des débarquements de la flotte chalutière de 1960 à 1977

126

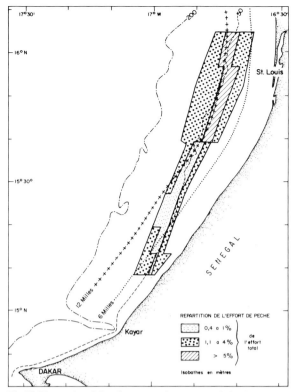

Fig 20 Zones de pêche des crevettiers au Nord du Cap Vert (fonds de St-Louis)

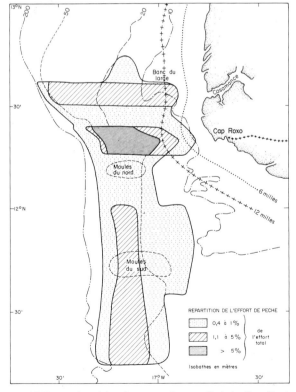

Fig 21 Zones de pêche des crevettiers au Sud du Cap Vert (fonds de Roxo-Bissagos)

s'effectue sur les zones où le pourcentage de lutites (particules de taille inférieure à 60 microns) dépasse 50%. La relation est beaucoup moins nette pour le fond de Roxo-Bissagos où l'espèce est exploitée sur des fonds sablo-vaseux ou même sableux.

La répartition bathymétrique de l'effort de pêche sur les deux fonds est donnée sur la *figure 22*.

La surface des fonds de pêche exploités a été estimée à 420 milles carrés (St-Louis) et 800 milles carrés (Roxo-Bissagos), à partir de la distribution de l'effort de pêche.

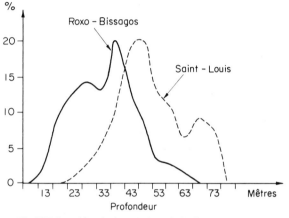

Fig 22 Répartition bathymétrique de l'effort total annuel

3.1.4 *Collecte des données statistiques* Trois sources ont été utilisées:
  – le pointage quotidien des unités présentes au port, permettant de calculer le temps d'absence au port;
  – les enquêtes à bord des navires au moment du débarquement, permettant d'obtenir la durée de la marée, la zone de pêche, une estimation du poids total débarqué et une ventilation approximative par espèce;
  – les fiches de débarquement en usine représentant la source principale d'informations, recoupent les précédentes et donnent la distribution précise des captures par espèce ou groupe d'espèces. La collecte presque exhaustive de ces données permet de disposer de données de qualité.

3.1.5 *Traitement des données* Le principal problème rencontré pour le traitement des données historiques est celui de l'évaluation de l'effort de pêche dirigé sur la crevette dans une pêcherie démersale multispécifique de ce type où le biotope à exploiter constitue une mosaïque de peuplements différents que le chalutier peut exploiter

127

successivement d'une marée à l'autre et parfois pendant la même marée. Il est en effet primordial de bien identifier cet effort pour pouvoir calculer des indices d'abondance non biaisés par des modifications progressives de stratégie de la flottille. Les chalutiers dakarois exploitent des fonds durs (les cibles sont le rouget, les sparidés, les céphalopodes) ou divers types de fonds meubles très côtiers (pour les soles et les Sciaenidés), du plateau (pour les crevettes et les *Galeoïdes*), ou du sommet du talus (pour la brotule).

La présence des crevettes dans les captures, en pourcentage très dominant, accompagnées de soles et autres poissons 'gris' (Sciaenidés, *Galeoïdes*, etc . . .) indique que la pêche a eu lieu sur les fonds meubles sablo-vaseux ou vaseux caractéristiques des fonds à crevettes. Le mélange d'espèces incompatibles (rougets ou brotules par exemple) permet d'éliminer aisément un certain nombre de marées mixtes non utilisées pour calculer la cpue. A l'intérieur du groupe crevette/sole/poisson 'gris', au sein duquel s'est effectuée la plus importante redistribution de l'effort des crevettiers après 1970, les marées orientées plutôt vers l'un de ces éléments sont définies à l'aide d'un 'seuil critique' (pourcentage en poids) de présence de l'espèce-cible dans la capture. Ces seuils ont été calculés après examen de la composition spécifique des marées de tous les bateaux (une centaine) pendant un an et en utilisant les connaissances empiriques disponibles sur certaines unités de la flotte de pêche et leurs habitudes. Le traitement final a été effectué sur ordinateur à l'aide d'un programme de tri séquentiel.

Les marées 'pures', dirigées vers les crevettes, ainsi définies ont servi à calculer la cpue. Le rapport des prises totales et des cpue permet de calculer un effort total appliqué à la crevette.

Le problème de la définition d'un effort spécifique dans une pêcherie multispécifique est toujours très ardu. Il est difficile de trouver une solution entièrement satisfaisante et en l'occurence il est clair que la procédure utilisée ici a pour résultat d'éliminer les marées où faute d'une abondance suffisante des crevettes le chalutier a été contraint de changer de stratégie en cours de marée. En conséquence il est possible que la diminution de l'abondance sous l'action de l'effort de pêche ait été quelque peu sous-estimée.

Les modifications progressives de puissance motrice des unités de pêche dans la flottille ont également été prises en compte en effectuant une standardisation des puissances de pêche relatives à partir d'une relation entre la cpue moyenne annuelle des bateaux et leur puissance motrices.

3.1.6 *Résultats*

a) Evolution de la stratégie La *figure 23* montre que le pourcentage de la capture totale de crevettes qui a été obtenu au cours de marées considérées comme des 'marées à crevettes' a très sensiblement baissé depuis 1969, ce qui traduit bien le changement progressif de stratégie de pêche des patrons qui, de plus en plus, effectuent des marées mixtes au gré des abondances rencontrées, et recherchent de plus en plus souvent directement le poisson, pour des raisons économiques liées à une diminution progressive de l'abondance des crevettes et à une hausse des coûts de production, consécutive à la hausse du prix du carburant.

Une étude récente non publiée de la FAO montre que ces changements de stratégie ont affecté différemment les bateaux de types différents. A titre d'exemple, en 1978, les chalutiers congélateurs et les glaciers[1] de 50 tonneaux de jauge brute d'un armement de Dakar ont débarqué les espèces suivantes:

	Crevette	Sole	Capitaine	autres
Congélateurs	66%	19%	2%	13%
Glaciers	7%	32%	38%	23%

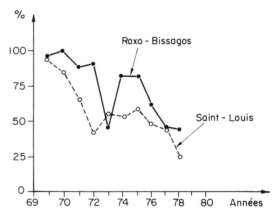

Fig 23 Pourcentage de la capture totale annuelle obtenue au cours de marées dirigées exclusivement sur les crevettes

On voit bien que les 'crevettiers' glaciers ont nettement changé de stratégie et exploitent maintenant en réalité le poisson, ce qui n'est pas le cas des congélateurs. La *figure 25* donne l'évolution des rendements par groupes d'espèces depuis 1971. Il est clair que les congélateurs se sont davantage spécialisés sur la crevette après 1973 alors que les glaciers ont diversifié leur activité, diminuant leurs rejets mais dirigeant surtout une part de plus en

[1] Bateaux conservant leur capture dans la glace.

128

plus importante de leur effort sur d'autres espèces que la crevette.

Cette même étude montre également des différences importantes dans la distribution géographique de l'effort de pêche des différents types de chalutiers (entre St-Louis et Roxo-Bissagos) et dans l'évolution au cours du temps de cette distribution. Par exemple lés congélateurs de 50 TJB ont pêché au Nord de Dakar jusqu'en 1973, puis uniquement au Sud dès 1974. Les congélateurs et glaciers de 120 TJB fréquentent exclusivement le fond sud (situé plus loin mais plus riche où l'abondance est plus stable). En revanche, les glaciers de 50 TJB ont pêché sur le fond nord jusqu'en 1975 et partagent leur activité entre le Nord et le Sud depuis 1976. Ils sont donc les seuls à exploiter St-Louis actuellement.

Il n'est pas douteux que ces ajustements sont la conséquence des modifications des conditions économiques d'exploitation (coût du carburant) et qu'ils sont gouvernés étroitement par le coût de fonctionnement, l'autonomie en glace, la capacité de stockage, le coût du stockage, *etc*.

b) Standardisation de l'effort de pêche des crevettiers 'purs' (fig. 24) Les relations calculées sont les suivantes:

  – espèce-cible seule: cpue = 0·023 P + 10·80
  – toutes espèces cumulées:
  cpue = 0·024P + 18·70

La cpue est exprimée en kg/heure et la puissance motrice P en chevaux. La classe de bateaux dont la puissance moyenne est 425 ch a été prise comme standard. C'est la classe la mieux représentée dans la pêcherie. Sa puissance de pêche relative (PPR) a été fixée à 1 et celle correspondant aux autres classes a été calculée simplement par le rapport des cpue correspondantes. On a obtenu les coefficients suivants.

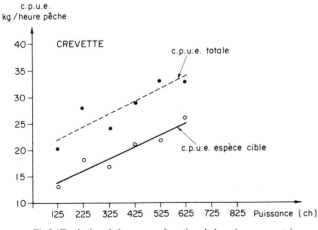

Fig 24 Evolution de la cpue en fonction de la puissance motrice

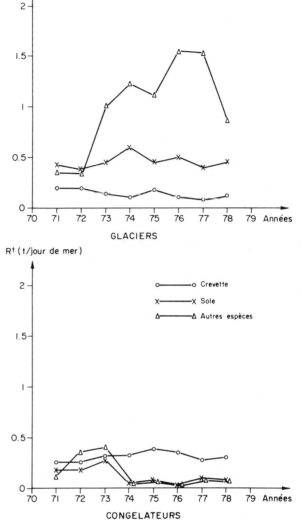

Fig 25 Evolution des rendements journaliers moyens annuels pour les crevettiers glaciers et congélateurs (50 TJB) de 1971 à 1978 (dernière année incomplète)

Puissance motrice (ch)	125	225	325	425	525	625
Puissance de pêche relative	0·66	0·77	0·89	1·00	1·11	1·22

c) Capture accessoire La *figure 24* montre également l'évolution des cpue, toutes espèces comprises, en fonction de la puissance motrice. L'écart entre les 2 droites représente l'évolution du volume de la capture accessoire (conservée à bord) avec la puissance motrice. Les droites sont parallèles et indiquent que cette dernière est relativement constante en valeur absolue. Elle décroit cependant en valeur relative.

P(ch)	125	225	325	425	525	625
Capture accessoire %	37	34	31	29	27	25

La capture des crevettiers 'purs' se compose (en 1978) de 9% de *Sciaenidés*, 7% de soles, 4% de crabes, 2% de *Galéoides*, 1% de pageots et 5% de divers pour 72% de crevettes.

3.1.7 *Les rejets* Ils sont très importants dans la pêche crevettière et ont sensiblement évolué au cours du temps. Malheureusement peu d'informations quantitatives sur cet aspect sont disponibles.

Les espèces régulièrement capturées par les crevettiers peuvent être classées en 3 groupes:
- espèces totalement conservées: *Penaeus notialis*, *Cynoglossus* spp, *Pseudotolithus* et crabes *Neptunes validus* (seules les pinces sont conservées)
- espèces totalement rejetées: *Ilisha africana*, *Brachydeuterus auritus* et *Balistes capriscus*
- espèces partiellement rejetées: *Galeoïdes decadactylus*, *Pomadasys* spp., *Pagellus coupeï*, *Arius* spp. Dans ce cas, la proportion rejetée dépend du taux de remplissage des cales du bateau et de l'abondance relative des différentes espèces pondérée par leur valeur marchande. Des données non publiées disponible au CRODT[1] ont montré une baisse récente de la taille moyenne des pageots débarqués, liée à une diminution des rejets par suite d'un relèvement de la valeur marchande des pageots de petite taille.

Ces rejets sont composés d'espèces non commercialisables et de jeunes individuels d'espèces commercialisables.

Les crevettiers congélateurs effectuent des rejets plus importants que les glaciers. Une enquête entreprise en 1978 auprès de quelques patrons de pêche a permis d'estimer les captures réelles de poissons divers par les crevettiers à environ 3·5 tonnes par jour (pour 400 ch.). Cette valeur est valable pour les 2 types de crevettiers. En revanche les débarquements journaliers de ces espèces atteignent 0·5 t par jour pour un congélateur et 2·5 t par jour pour un glacier. Les rejets peuvent donc être très grossièrement estimés à 3 t par jour pour les congélateurs et 1·2 t par jour pour les glaciers soit respectivement 86% et 34% de la capture totale.

Si l'on extrapole à toute la flotte chalutière du Sénégal (on dispose de quelques données fragmen-taires pour les chalutiers non crevettiers) on arrive à 50 000 t de rejets environ par an alors que les débarquements totaux en poissons sont inférieurs à 40 000 tonnes.

Il est urgent d'intensifier l'étude de ce phénomène et en particulier de situer la taille moyenne des poissons rejetés par rapport à leur taille de première maturité sexuelle.

3.2 *La pêche artisanale en estuaire*
Elle existe dans 3 systèmes estuariens: le fleuve Sénégal dont le régime est de type tropical pur, le Sine Saloum qui est une ria envahie par la mer dont la salinité est toujours égale ou nettement supérieure à 35‰ et la Casamance, petit fleuve côtier dont le débit et la pente sont très faibles, qui est alimenté directement par la nappe phréatique, et qui présente un noyau sursalé à certaines périodes de l'année.

La pêcherie est très importante dans la Casamance (1 000–1 550 t par an), limitée dans le fleuve Sénégal (200–300 t par an) et très irrégulière dans le Sine Saloum (0 à 250 t par an).[1] Deux engins sont utilisés: le kili qui est une poche de filet tractée par deux hommes (ouverture 4 m × 1 m et ouverture de maille 16 mm) et le filet fixe (ou chalut à l'étalage) dont l'ouverture atteint 8 m × 1 m et dont la maille fait 22 mm d'ouverture. Cette poche est maintenue ouverte par deux tangons de bois et utilisée à partir d'une pirogue mouillée dans le lit du fleuve. La pirogue porte généralement 2 filets soutenus à leur autre extrémité par 2 futs métalliques servant de flotteurs.

Le kili est la seule méthode utilisée dans le Sine Saloum où les courants sont faibles, le chalut à l'étalage est le seul engin autorisé en Casamance.

3.2.1 *Prises, efforts, saisons et zones de pêche*
a) Fleuve Sénégal Les débarquements totaux annuels de 1960 à 1977 sont disponibles ainsi que les débarquements mensuels de 1975 à 1977 (*Annexe I*). La capture totale annuelle n'a jamais dépassé 270 tonnes.

La pêche n'est jamais nulle et les meilleures prises sont obtenues en moyenne de janvier à avril (*Fig. 26*). Elle s'effectue tout près de l'embouchure du fleuve ou à St-Louis même.

b) Sine Saloum Les données disponibles sont regroupées dans *l'Annexe II*. Aucune donnée, même approximative, sur l'effort n'est connue. Les meilleures captures sont obtenues en sep-

---

[1]Centre de Recherches Océanographiques de Dakar Thiaroye – B.P. 2241 – Dakar – Sénégal.

[1]on peut noter qu'il existe également une petite pêche artisanale mal connue dans la Gambie, produisant 200–400 t/an.

tembre –octobre (*Fig. 26*). Les captures annuelles n'ont pas dépassé les 310 tonnes et les principaux centres de débarquement sont Fatik, Kaolack et Foundiougne.

c) Casamance   C'est une exploitation traditionnelle qui remonte au moins à 1959. Le nombre de pirogues participant à la pêche a parfois été répertorié par l'usine de regroupement et traitement des captures de Ziguinchor et il est disponible pour 1963 (valeur moyenne uniquement), 1964 à 1966 et 1970–1971 (*Annexe III*). Des données de captures, plus ou moins complètes, sont disponibles depuis 1960 (*Annexes IV* et *V*).

Les captures, très importantes, ne sont jamais nulles. Le schéma saisonnier est nettement bimodal avec un maximum en septembre – octobre et un autre en avril – mai – juin (*Fig. 26*).

La zone de pêche autorisée actuellement est comprise entre Ziguinchor et Goudomp situés respectivement à 63 et 115 km de l'embouchure.

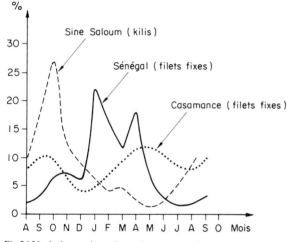

Fig 26 Variations saisonnières des captures (en pourcentage du total annuel) dans les 3 pêcheries artisanales du Sénégal

3.2.2 *Taille des crevettes capturées*

a) *Fleuve Sénégal*   L'évolution saisonnière des tailles moyennes est donnée sur la *figure 27*. Il existe un cycle saisonnier simple avec des valeurs faibles pendant la période où la baisse de salinité est la plus brusque (cf. *fig 2*). L'histogramme total annuel est donné sur la *figure 28*, pour 1977–78 et la taille moyenne à la migration est de 18·9 mm (LC) soit environ 9 cm de longueur totale ($\sigma^2 = 14\cdot12$, n = 7 349).

b) *Sine Saloum*   Les données disponibles sont incomplètes. Elles indiquent une taille minimale en septembre correspondant à la période de dessalure. L'histogramme total obtenu pendant la période d'observation (de juillet à novembre 1973)

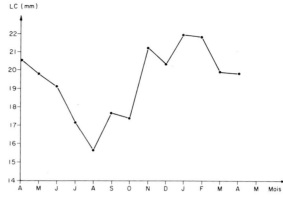

Fig 27 Evolution de la taille moyenne mensuelle des échantillons provenant du fleuve Sénégal (avril 1977 à avril 1978)

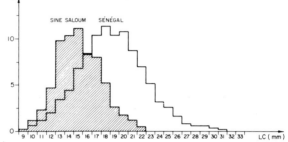

Fig 28 Histogramme total des tailles des juvéniles capturées dans le fleuve Sénégal (en 1977–1978), et dans le Sine Saloum (juillet–novembre 1973)

est donné sur la *figure 28* et la taille moyenne correspondante est 15·2 mmLC ou environ 7 cm de longueur totale ($\sigma^2 = 5\cdot82$ et n = 2011).

c) *Casamance*   L'évolution saisonnière paraît plus compliquée que dans le fleuve Sénégal. Les tailles capturées sont élevées en saison sèche et diminuent pendant la dessalure. Cependant, un second minimum plus difficile à expliquer par l'hydrologie existe en janvier (*Figure 29*).

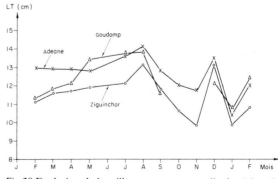

Fig 29 Evolution de la taille moyenne mensuelle des échantillons provenant de trois zones de pêche de la Casamance (1976–1977)

L'analyse de la taille moyenne pose un problème qui semble particulier à la Casamance. En effet, les données anciennes collectées par De Bondy (1968) indiquent pour 1966 une taille moyenne de 9·5 cm(LT) proche de celle observée dans le fleuve Sénégal, et des valeurs généralement rencontrées en Afrique de l'ouest (Garcia, 1978) avec une variance de 1·72 (pour 17 598 individus mesurés). En 1976, 47 825 individus ont été mesurés et la taille moyenne est 12·2 cm(LT) ($\sigma^2 = 4·91$). Toutes les données disponibles pour la période récente indiquent que les crevettes migrent à une taille moyenne beaucoup plus forte, que par le passé et anormale pour les penaeides côtiers de ce type (*Fig 30*). Ce phénomène a été relié par Le Reste (1980) aux variations interannuelles de salinité de la Casamance en liaison avec la grande sécheresse sahélienne. Selon l'auteur, les crevettes resteraient plus longtemps dans un milieu de salinité plus élevée, permettant une croissance jusqu'à une taille à la migration plus élevée. Garcia (1978) à partir de données sur plusieurs stocks ouest-africains avait également conclu que la durée du séjour en estuaire dépendait des conditions ambiantes et en particulier de la salinité.

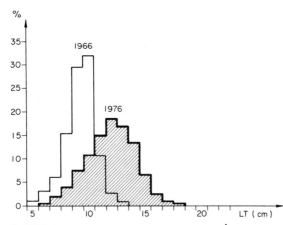

Fig 30 Histogramme total annuel des tailles dans les échantillons de captures récoltés en 1966 (d'après De Bondy, 1968), en 1976 dans la région de Ziguinchor

Une autre particularité de la Casamance réside dans la distribution des tailles en fonction de la distance à l'embouchure. Contrairement à tous les schémas généralement admis, les crevettes sont d'autant plus petites que l'on se rapproche de la mer (*Fig 31*). Leur taille est maximale vers Tambacoumba et tendrait peut-être à diminuer ensuite. Ce phénomène très particulier dont l'importance pour la pêcherie est considérable, pourrait être lié au fait qu'il existe un noyau sursalé saisonnier situé en amont de Ziguinchor.

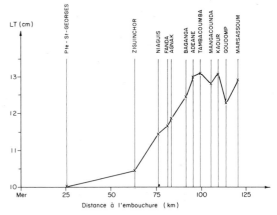

Fig 31 Variation de la taille moyenne dans les échantillons de captures, en fonction de la distance à l'embouchure (Casamance, février 1976–février 1977)

4 Evaluation des potentiels en mer

Les données disponibles depuis 1965 ont été utilisées pour calculer les potentiels à l'aide d'un modèle de Fox (1970) en utilisant le programme PRODFIT (Fox, 1975) (*Figs 32* et *33*).

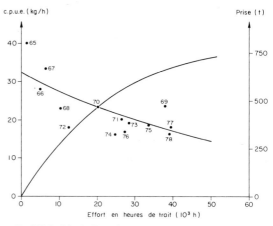

Fig 32 Modèle de Fox obtenu pour le stock de St-Louis

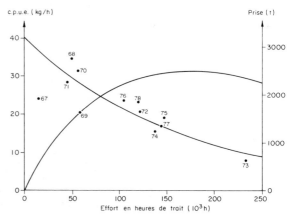

Fig 33 Modèle de Fox obtenu pour le stock de Roxo-Bissagos

	Saint-Louis			Roxo-Bissagos		
Année	Prise (t)	Effort (10^3h)	c.p.u.e. (kg/h)	Prise (t)	Effort (10^3h)	c.p.u.e. (kg/h)
1965	55	1·4	40·1	—	—	—
1966	143	5·1	27·9	6	1·6*	3·7*
1967	218	6·5	33·5	358	14·9	24·0
1968	227	9·9	23·0	1 700	49·5	34·3
1969	896	38·1	23·5	1 205	58·8	20·5
1970	486	20·5	23·7	1 826	58·3	31·3
1971	532	26·7	19·9	1 322	46·2	28·6
1972	224	12·6	17·8	2 529	122·8	20·6
1973	540	28·3	19·1	1 768	235·7	7·5
1974	405	25·2	16·1	2 096	137·0	15·3
1975	617	33·7	18·3	2 814	148·1	19·0
1976	464	27·6	16·8	2 438	104·6	23·3
1977	707	39·7	17·8	2 424	144·3	16·8
1978	637	38·9	16·4	2 716	119·1	22·8

*valeurs non utilisées

La prise maximale moyenne (ou MSY) n'étant pas un objectif à atteindre par la flottille, pour des raisons économiques, nous avons calculé également la prise optimale moyenne (Y_{opt}) et l'effort correspondant (f_{opt}) en utilisant le critère de rendement marginal de Gulland et Boerema (1973). Ce niveau correspond à celui où l'augmentation des captures par unité d'effort devient inférieure à 10% de la cpue correspondant à une très faible taux d'exploitation. Les résultats obtenus sont:

	St-Louis	Roxo-Bissagos
M.S.Y.	750 tonnes	2 500 tonnes
f_{MSY}	$63·10^3$ heures	$168·10^3$ heures
$cpue_{MSY}$	11·19 kg/h	14·8 kg/h
Y_{opt}	730 tonnes	2 400 tonnes
f_{opt}	$49·10^3$ heures	131·10 heures
$cpue_{opt}$	14·9 kg/h	18·4 kg/h
f (1975−78)	$35·10^3$ heures	$129·10^3$ heures
f/f_{opt}	0·72	0·98

Le stock de Roxo-Bissagos semble donc exploité au mieux de ses possibilités alors que celui de St-Louis pourrait supporter un accroissement d'effort de 28%. Ces résultats sont valables pour le régime d'exploitation actuel. Toute modification du vecteur de F en fonction de l'âge (par changement de maille ou modification du taux d'exploitation en estuaire) conduira à une révision de ces évaluations, Garcia et Lhomme (1979) ayant procédé à l'estimation des potentiels de divers stocks d'Afrique de l'Ouest, on peut comparer les productions optimales par unité de surface.

Etant donné que la productivité d'un fond de pêche en mer dépend aussi du prélèvement effectué sur le recrutement par la pêche artisanale, nous avons tenté d'en tenir compte en ajoutant la capture de la pêche artisanale au potentiel en mer calculé dans les conditions actuelles d'exploitation. Cette approche, seule possible pour le moment, ne devrait pas introduire d'erreur importante car

		Production optimale		Productivité (tonnes/mille^2/an)	
Fonds de Pêche	Surface milles^2	mer	mer + estuaire	mer	mer + estuaire
St-Louis	430	730	880	1·7	2·0
Roxo-Bissagos	800	2 400	3 900[2]	3·0	4·9
Sherbro[1]	400	1 800	2 140	4·5	5·3
Côte d'Ivoire	390	570	1 220	1·4	4·1
Bénin − Nigéria Cameroun[3]	1 500	≈3 500	4 000	2·3	2·7
Gabon	875	1 500[4]	1 500	1·7	1·7

[1]Sierra Leone + Liberia
[2]Production artisanale considerée: 200 t (Saloum) 200 t (Gambie) et 1 100 t Casamance, valeur moyenne des dernieres ann1ees)
[3]Selon Garcia et Lhomme (1979)
[4]Capture actuelle, proche du potentiel maximum

selon Garcia (1978), à des niveaux d'exploitation en mer proches du MSY, les variations simulées du taux d'exploitation des juvéniles en estuaire influent très peu sur la prise totale (mer + lagune).

5 Les variations naturelles de production

5.1 *En Casamance*

Le Reste (1980) à analysé les variations interannuelles de capture dans la pêche artisanale de Casamance. L'analyse repose sur les éléments suivants:

- la taille moyenne à la migration a changé entre 1966 et 1976 (cf para. 3.2.2 *c*)
- la taille moyenne (et très vraisemblablement l'âge) à la migration varie d'une année à l'autre avec la salinité ambiante.
- la salinité ambiante de la Casamance est liée aux pluies des années antérieures par le jeu du lessivage des sols sursalés et de la percolation de la nappe phréatique directement dans la Casamance qui est un système hydraulique très particulier (Le Reste, 1980)

Si on représente par Y_i les captures et par P_i les pluies de l'année i, il indique que la corrélation est absente ou mauvaise entre:

- Y_i et P_i
- Y_i et $(P_i + P_{i-1})$ ou $(P_i + P_{i-1} + P_{i-2})$
- Y_i et P_{i-1}

En revanche, il y a une très bonne corrélation entre:

$$Y_i \text{ et } \frac{(P_{i-1} + P_{i-2})}{2} \text{où } R = 0.91 \ (n = 11)$$

La corrélation est également très bonne entre

$$Y_i \text{et} \left(\frac{1}{2} P_{i-1} + \frac{1}{4} P_{i-2} + \frac{1}{8} P_{i-3} + \frac{1}{16} P_{i-4} + \frac{1}{32} P_{i-5} \right)$$

où $\check{R} = 0.90 \ (n = 11)$. La grandeur entre parenthèses est appelée 'volume déterminant' et permet de prendre en considération un certain nombre d'années antérieures avec une pondération décroissante, reflétant ainsi la 'mémoire' des évènements passés conservés par ce système hydraulique particulier grâce à l'effet tampon de la nappe phréatique.

Cette dernière corrélation a été conservée par l'auteur pour servir de modèle prédictif (*Fig 34*) car elle rendait mieux compte des phénomènes passés (*Fig 35*). L'utilité de cette relation pour prédire réellement les captures doit encore être testée dans l'avenir.

Au vu de ces résultats très intéressants on peut proposer, pour expliquer le phénomène, l'hypothèse suivante: la salinité ambiante agit sur la durée

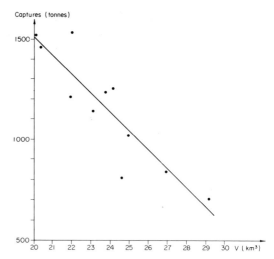

Fig 34 Captures de l'année et 'volume déterminant' pour le fleuve Casamance (d'après Le Reste, 1980), voir texte

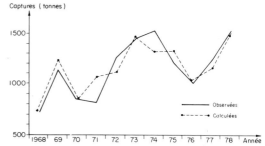

Fig 35 Captures observées et calculées pour le fleuve Casamance (d'après Le Reste, 1980)

du séjour en lagune, donc sur la taille à la migration et par là même sur la taille à la première capture dans la pêche artisanale au filet fixe et, par la suite, dans la pêche industrielle en mer. *La taille à la première capture serait donc soumise à des variations naturelles d'une année à l'autre.* Ces variations aurient apparemment pour conséquence, en estuaire, l'augmentation de biomasse exploitable, ce qui indiquerait d'ailleurs que la mortalité naturelle à ce stade est inférieure à la croissance en poids.

Une augmentation de biomasse associée à une augmentation de la taille à la première capture, se traduirait par une augmentation de capture dans la pêcherie artisanale même avec un effort de pêche stable et inversement.

5.2 *En mer*

Une relation entre les pluies et la production des stocks en mer ayant souvent été indiquée, elle a été recherchée entre les pluviométries sur le bassin versant du Sénégal et de la Casamance et la production des fonds de St-Louis et Roxo-Bissagos.

Nous avons choisi la cpue standardisée annuelle comme indice d'abondance moyenne (supposant que d'éventuelles petites variations de la distribution saisonnière de l'effort de pêche n'avaient sur cet indice qu'un impact négligeable). Cette cpue varie également avec le niveau de l'effort de pêche qui a beaucoup varié durant les dix dernières années. Pour supprimer cette cause de variation on a considéré que les modèles de Fox, calculés pour les deux stocks considérés, représentaient la réalité et que les écarts des points par rapport à la courbe étaient des anomalies positives ou négatives (l'exploitation portant sur une seule classe d'âge les captures annuelles peuvent être considérées comme des valeurs à l'équilibre).

Ces anomalies de l'abondance ont été reliées respectivement aux anomalies de pluviométrie (écart annuel par rapport à la moyenne sur la période considérée) sur le bassin versant de la Casamance (pour Roxo-Bissagos) et aux anomalies de débit annuel moyen pour le fleuve Sénégal (dans le cas de St-Louis).

Les résultats obtenus sont résumés sur la *figure 36*

– St-Louis:

$$Y = 0·013X - 0·574$$
$$N = 14 \text{ couples}$$
$$r = 0·72 \text{ (hautement significatif)}$$

– Roxo-Bissagos:

$$Y = -0·62X - 0·75$$
$$n = 12$$
$$r = 0·52 \text{ (significatif)}$$

Les anomalies sont donc très bien corrélées mais les corrélations sont de signe opposé pour les deux fonds de pêche. A St-Louis, l'abondance en mer paraît d'autant plus élevée que les apports d'eau douce sont importants alors que l'inverse est observé à Roxo-Bissagos.

L'interprétation d'une différence aussi radicale entre les deux fonds de pêche est difficile avec les éléments dont nous disposons. Il est certain qu'elle ne provient pas de l'utilisation de données climatiques différentes (pluviométrie dans un cas et débit du fleuve dans l'autre) car ces données sont très fortement corrélées dans ce cas particulier. Il semble, bien que l'ajustement des modeles de Fox ne soit pas toujours satisfaisant (cf. *fig 32*) qu'il ne s'agisse pas d'un artéfact car on observe effectivement que les années situées dans la partie supérieure du nuage de points pour St-Louis sont situées dans la partie inférieure pour Roxo-Bissagos.

On doit également noter que ce phénomène d'inversion apparente de la relation a été observé dans le Golfe du Mexique: Gunter et Hildebrandt (1954) ont montré une relation positive, pour *P.*

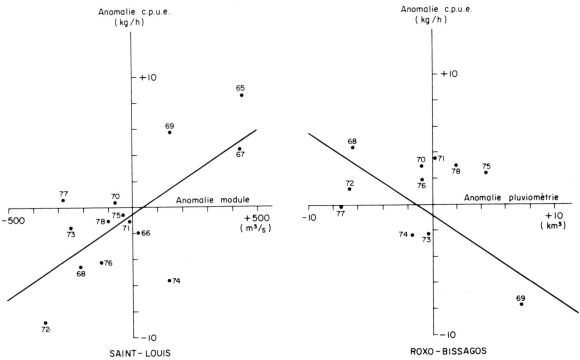

Fig 36 Relation entre les anomalies de la capture en mer et celles de la pluviométrie (stock de Roxo-Bissagos) ou des apports fluviaux (stock de St-Louis)

setiferus au Texas, entre les pluies et l'abondance en mer alors que Barret et Gillespie (1975) notent une relation négative entre les apports fluviaux et l'abondance pour la même espèce en Louisiane. Il est intéressant de noter, à ce sujet, que le Texas et la région de St-Louis, au nord du Sénégal, sont deux régions sèches avec des apports continentaux faibles (la relation y est positive) alors que la Louisiane, comme la région des Bissagos, sont des zones où les apports d'eau douce sont plus importants (la relation y est négative). Ces observations pourraient indiquer, comme le soulignent Garcia et Le Reste (1981) que la relation entre les apports d'eau douce et la production de crevettes ne serait pas linéaire mais présenterait un seuil critique au-delà duquel elle s'inverserait.

Diverses théories ont été proposées pour expliquer les relations observées: Gunter et Edwards (1969) suggèrent une action sur la surface des nurseries disponibles par l'intermédiaire de la salinité. Subramanyam (1966) envisage une action directe de la pluviométrie sur le recrutement des postlarves: par une modification des gradients de salinité entre les nurseries et la mer. Ruello (1973) pense que les relations sont plus complexes et que les apports d'eaux douces affectent à la fois la reproduction en mer, la survie et le recrutement des postlarves, la survie et la croissance des immatures ainsi que leur recrutement en mer. Les résultats obtenus en Afrique de l'Ouest permettent, en outre, de mettre en évidence l'action des apports d'eau douce sur l'âge à la migration et donc sur l'âge à la première capture dans les pêcheries. Ces modifications de l'âge à la première capture se traduisent par des variations de rendement par recrue et auraient donc une influence sur les captures totales annuelles.

6 Amenagement

6.1 *Mesures actuellement en vigueur*
6.1.1 *En mer* La maille actuellement couramment utilisée au Sénégal est une maille de 20 mm de côté (et 37 mm d'ouverture en moyenne). Les zones de pêche sont également règlementées. Le chalutage est interdit dans la bande côtière des 6 milles. Les chalutiers glaciers peuvent opérer à partir des 6 milles et les congélateurs à partir des 12 milles (ces limites sont indiquées sur les *figures 20* et *21*). Il est nécessaire d'obtenir un permis de pêche et, dans le cas de la pêche crevettière, la jauge autoriséé est limitée.

6.1.2 *En estuaire* On se limite au cas de la Casamance où s'est développée l'exploitation la plus importante. Une maille minimale de 12 mm de côté est imposée et les arts trainants sont interdits. La pêche n'est autorisée qu'entre Ziguinchor et Goudomp (où la taille des crevettes est élevée). Ele est en outre autorisée dans le chenal central et interdite dans les chenaux secondaires qui servent de nurseries.

6.2 *Considérations sur la situation actuelle*
La mise en place de schémas d'aménagement doit reposer sur une connaissance approfondie de la pêcherie, de ses caractéristiques et de ses contraintes. Dans cet ordre d'idées, on résumera dans les paragraphes suivants les données pertinentes dont le schéma d'aménagement définitif devra tenir compte.

6.2.1 *Les potentiels et les niveaux d'exploitation*
Ils ont été estimés au paragraphe 4. Le niveau 'optimum' se situerait autour de 730 tonnes pour St-Louis et 2400 tonnes pour Roxo-Bissagos. Le premier stock est exploité à 72% de ses possibilités et le second à 98%. Il n'y a donc pas surexploitation biologique et le stock de St-Louis pourrait même supporter un accroissement de l'effort de pêche (les rendements y sont cependant déjà plus faibles qu'à Roxo-Bissagos).

6.2.2 *Les interactions entre la pêche industrielle et artisanale* Les évaluations ci-dessus ne sont valables que dans l'état actuel d'exploitation des nurseries par la pêche artisanale et seront modifiées par tout changement dans la mortalité par pêche appliquée aux juvéniles.

Les interactions entre pêche artisanale et pêche industrielle peuvent théoriquement se produire dans les deux sens, la pêche artisanale réduisant le recrutement des juvéniles dans la pêche industrielle et cette dernière réduisant le potentiel reproducteur et, par conséquent, le nombre de larves disponibles pour repeupler les estuaires. En réalité, on tend à négliger ce dernier phénomène car jusqu'à présent on n'a pas pu mettre clairement en évidence de relation entre l'abondance des reproducteurs et celle des recrues pour les crevettes penaeïdes (Garcia et Le Reste, 1981).

En ce qui concerne le fond de St-Louis (*figure 37*) les relations sont simples, la nurserie est limitée à l'embouchure du fleuve Sénégal, la pêche artisanale est réduite et capture 200 tonnes/an en moyenne.

En ce qui concerne le fond de Roxo-Bissagos ces relations sont plus complexes car des nurseries existent sur la côte sud du Sénégal (Siné Saloum et Casamance), en Gambie (dans le fleuve Gambie) et en Guinée–Bissau (Rio Cacheu, Rio Geba, etc . . .). La pêcherie artisanale de Casamance produit plus de 1000 tonnes par an en moyenne dans la seule partie médiane du fleuve entre Ziguinchor et Goudomp (et il n'est pas sûr que ce secteur, fortement exploité, fournisse encore un nombre significatif de recrues au fond de pêche de Roxo-Bissagos). Il n'est pas possible actuellement de connaître, même approximativement, la proportion de recrutement potentiel total du stock de Roxo-Bissagos qui est capturée en Casamance. Si l'on suppose que cette fraction est du même ordre de grandeur que le rapport entre la superficie totale des nurseries existantes et celle de la zone exploitée de Casamance, le prélèvement réalisé par la pêche artisanale ne devrait pas être très important.

En réalité l'ignorance des relations exactes entre la nurserie de Casamance et le fond de pêche de Roxo-Bissagos rend extrêmement difficile l'élaboration d'un schéma global d'aménagement. La *figure 37* schématise la position des stocks en mer et des nurseries, les flèches indiquent la migration et donc le sens de l'interaction.

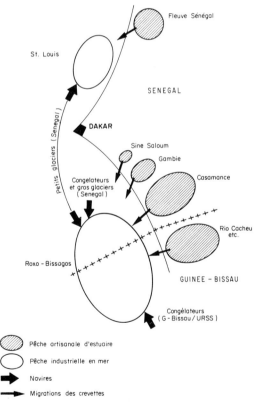

Fig 37 Schéma théorique de distribution des crevettes adultes et juvéniles, des types de pêche, et interactions éventuelles (le sens de la flèche indique le sens de l'action) sur les fonds de pêche à la crevette du Sénégal. La frontière maritime est totalement arbitraire

6.2.3 *Interactions entre les deux fonds de pêche en mer* En théorie il serait intéressant d'extraire de chaque stock la capture optimale (définie au paragraphe 4) en régulant la distribution de l'effort de pêche. La situation est, en fait, très complexe car les deux fonds sont situés à des distances différentes de Dakar (le coût du trajet, en carburant, est donc différent) et leur richesse est inégale (les crevettes sont moins abondantes à St-Louis). Les paramètres économiques doivent donc être pris en compte, une étude détaillée à l'aide de techniques de simulation, analogue à celle que Clark et Kirkwood (1979) ont réalisée en Australie, permettrait de définir la distribution optimale de l'effort sur les deux fonds.

En fait les conséquences de l'alternative qui est offerte aux pêcheurs sont nettement perceptibles à l'heure actuelle. Le fond de St-Louis, moins riche mais plus proche de Dakar, n'est exploité qu'à 72% de ses possibilités par des chalutiers glaciers de moins de 50 TJB (exploitant non seulement les crevettes mais aussi, et de plus en plus, les espèces de poissons associées) alors que le fond de Roxo-Bissagos, plus lointain mais plus riche, est exploité à 98% de ses possibilités par de gros congélateurs et des glaciers de plus gros tonnage (les petits glaciers sont les seuls à se déplacer saisonnièrement de St-Louis à Roxo-Bissagos). Il est permis de penser que ceci découle empiriquement de la recherche d'un certain optimum économique fonction de la richesse des fonds et des coûts respectifs d'exploitation des divers types de bateaux. Nous verrons plus loin que le fond de Roxo-Bissagos pose également le problème d'une exploitation coordonnée entre le Sénégal et la Guinée – Bissau.

La position des deux fonds de pêche par rapport à Dakar, ainsi que les types de bateaux qui les exploitent et les transferts d'effort d'un fond à l'autre, sont indiqués sur la *figure 37*.

6.2.4 *Interactions entre les exploitations des diverses espèces cibles* L'une des grandes difficultés de l'aménagement des pêches en zone tropicale est le caractère multispécifique de l'exploitation. C'est particulièrement vrai au Sénégal où les chalutiers exploitent une mosaïque de peuplements sur lesquels l'effort est distribué en fonction de leur abondance, de leur valeur, des possibilités du marché, *etc.* . . Ces considérations ont gouverné l'évolution brutale de la pêcherie pendant les 15 dernières années (cf. paragraphe 3.1.1), évolution qui apparaît nettement si on examine la structure spécifique des débarquements (*Fig 19*).

On a porté sur la *figure 38* les interactions existant actuellement entre l'exploitation de la crevette

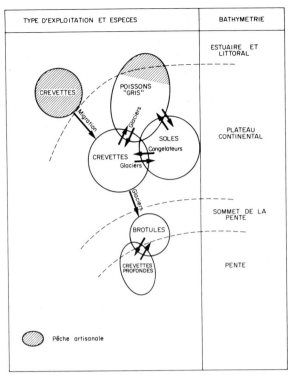

TYPE D'EXPLOITATION ET ESPECES	BATHYMETRIE

ESTUAIRE ET LITTORAL

CREVETTES

POISSONS "GRIS"

Migration

Glaciers

SOLES Congelateurs

PLATEAU CONTINENTAL

CREVETTES Glaciers

Glaciers

SOMMET DE LA PENTE

BROTULES

CREVETTES PROFONDES

PENTE

Pêche artisanale

Fig 38 Interaction entre l'exploitation des principaux groupes cible exploités par la pêche crevettière artisanale et industrielle au Sénégal

et celle des autres espèces de fond meuble du plateau, ou des espèces du sommet et de la pente continentale. Les flèches indiquent le sens des transferts d'effort de pêche et le type de bateau concerné est précisé. On distingue trois cibles principales: crevettes, soles (*Cynoglossus* spp) et poissons 'gris' (*Pseudotolithus*, *Galeoïdes*), ces deux dernières exploitations spécialisées étant entreprises par des crevettiers reconvertis. Selon les marées, les mêmes bateaux exploitent l'une ou l'autre de ces cibles. Les congélateurs exploitent seulement les espèces de plus grande valeur telles-que les soles et les crevettes.

Il est clair que dans ces conditions l'aménagement de chaque type de ressource, ou groupe d'espèces, séparément, poserait quelques problèmes.[1] En ce qui concerne les maillages optimaux, ils sont vraisemblablement différents pour les crevettes et les poissons (on a vu au paragraphe 2.6 qu'une maille de 62 mm pourrait être utilisée sans dommage à la place de la maille actuelle de 37 mm), surtout si l'on considère les gros Sciaenidés et Sparidés. Cependant, dans la situation

---

[1] Nous admettons que l'optimisation de l'exploitation de chaque espèce est impossible et ne considérons ici que la possibilité d'aménager les biotopes principaux, étant entendu que certaines espèces appartiennent à plus d'un biotope.

actuelle, une maille unique de 60 mm telle qu'elle a été recommandée par le Comité des Pêches de l'Atlantique Centre-Est (COPACE) pour toutes les espèces démersales du plateau (FAO-COPACE, 1980) faciliterait grandement la mise en oeuvre et le contrôle d'une telle règlementation, en attendant que le développement des services de contrôle permette la mise en place effective d'une règlementation plus fine. Dans tous les cas, cette dernière devrait être assortie de normes précises concernant les pourcentages d'espèces secondaires compatibles avec chaque type de maillage. Cette règlementation serait complexe et poserait de très sérieux problèmes de contrôle, à terre et surtout en mer.

En ce qui concerne la limitation des taux d'exploitation des différentes espèces cibles, ou groupes d'espèces, la mise en oeuvre d'une limitation par cible serait également très difficile pour les mêmes raisons.

Il est intéressant de rappeler les conclusions de Garrod (1973, p. 1984) concernant le problème de l'exploitation d'une ressource multiple. Cet auteur indique 'That multiple-stock fishery resources form a robust system; within limits they can tolerate wide variations in fishing mortality on individual stocks and, in themselves, these do not have an adverse long-term effect on the stock or its yield. The important feature is the overall mean level of fishing effort applied to the complex. . . From an administrative point of view of fisheries management this also means that there is no advantage in trying to secure a precise control of fishing mortality'. Il ajoute que les risques de destruction de stocks élémentaires diminuent si la distribution de la mortalité, sur ces stocks, peut fluctuer à court terme en fonction de leur abondance, cette stratégie fournissant un 'tampon' permettant de dévier les excès d'efforts de pêche sur d'autres éléments de la ressource.

Cette réflexion est essentielle dans le cas particulier traité dans ce travail car elle impliquerait qu'un niveau global 'optimal' d'effort soit défini, au moins pour le peuplement de fonds meubles (poissons gris, crevettes et soles) dans son ensemble, et que la distribution de cet effort soit laissée à l'appréciation des prédateurs que sont les pêcheurs. Une difficulté majeure réside dans le fait que le pêcheur n'est pas guidé par l'abondance de la ressource mais par la valeur de la capture qu'il obtient. Dans ces conditions, lorsque des différences importantes existent dans la valeur des captures provenant des divers éléments d'une ressource multiple, on peut craindre une surexploitation grave ou une extinction de l'élément dont la valeur est la plus élevée, si aucun contrôle visant à répartir l'effort de pêche n'est assuré. La distribution fine de l'effort sur une telle mosaïque d'espèces étant difficile à contrôler directement, elle peut l'être indirectement, au moins en partie, par l'intermédiaire de taxes au débarquement. En effet, la perception d'une taxe, différente selon les espèces, modulée en fonction de leur valeur marchande et des résultats que l'on veut obtenir, devrait permettre d'égaliser à long terme les revenus procurés par l'exploitation des divers éléments de la ressource. Il est même permis d'imaginer que l'exploitation de certains éléments peu rémunérateurs (balistes, par exemple) puisse être favorisée à l'aide d'une prime au débarquement prélevée sur les taxes provenant des espèces les plus chères.

Il faut enfin noter que le mode d'exploitation d'une ressource multiple proposé par Garrod (1973) implique que les unités de pêche soient polyvalentes et puissent passer d'un élément à l'autre de la ressource. Ceci n'est pas toujours possible et la pêcherie démersale du Sénégal montre que, pour des raisons technologiques et économiques, certains types de bateaux se spécialisent parfois sur un groupe d'espèces où ils sont rentables et que leur transfert sur d'autres éléments de la ressource est difficilement réalisable.

6.2.5 *Partage de la ressource et aménagement coordonné entre le Sénégal et la Guinée – Bissau* Le stock de Roxo-Bissagos est commun à ces deux pays, quelle que soit la frontière maritime qui sera définie ultérieurement. Il est exploité à l'heure actuelle par des flottes basées à Dakar et à Bissau, et il n'existe pas d'accord bilatéral formel entre les deux pays permettant de fixer d'un commun accord le niveau d'exploitation du stock ou la maille autorisée pour les chaluts. Cependant, les pays côtiers de la région ont institué une Conférence Interministérielle sous-régionale sur la préservation, la conservation et l'exploitation des ressources halieutiques au sein de laquelle ce type de problème doit être négocié. Pour sa part, le Comité des Pêches de l'Atlantique Centre-Est (COPACE), organisé sous l'égide de la FAO, a déjà abordé à plusieurs reprises le thème de l'aménagement des stocks communs pour essayer de dégager au niveau régional un consensus sur les méthodes de répartition des ressources communes. Dans le cas d'un stock comme celui-ci, le problème ne devrait pas être trop difficile à résoudre en ce qui concerne les adultes car la répartition est bien connue et les migrations semblent être surtout des mouvements irréguliers pouv-

ant conduire à un brassage de la population. Une étude détaillée de la distribution des crevettes sur l'ensemble du fond, et une estimation du taux réel de brassage devraient permettre de définir dans quelle mesure il est nécessaire de coordonner les aménagements nationaux.

Le problème est plus complexe si l'on considère également l'exploitation artisanale et s'il apparaît qu'une intensification de cette pêche dans l'un des pays peut affecter l'abondance des adultes dans l'autre pays. La pêche artisanale des juvéniles est actuellement plus développée au Sénégal qu'en Guinée – Bissau et les possibilités d'expansion de cette dernière ne sont pas connues. Des campagnes de marquages effectuées sur les juvéniles en migration, au Sénégal et en Guinée – Bissau, permettraient de définir les relations réelles entre les diverses nurseries potentielles et le stock d'adultes.[1]

7 Conclusions: priorités en matière de recherche et d'aménagement

7.1 *En matière de recherche*

7.1.1 *Etude des maillages et essais de chaluts sélectifs.* Les conséquences biologiques et économiques (en ce qui concerne les poissons) de l'utilisation d'une maille unique de 62 mm pour tout le complexe crevette – sole – poissons gris, doivent être évaluées car le groupe d'espèces doit être étudié dans son ensemble.

Des essais de chaluts sélectifs, permettant d'éviter la destruction des juvéniles de poissons présents sur les fonds à crevettes, devraient être entrepris (dans le cadre régional) pour essayer de résoudre également le grave problème des rejets.

7.1.2 *Etude des liaisons entre nurseries et fonds de pêche* Un programme de marquages des juvéniles serait de nature à permettre une meilleure analyse de l'effet de la pêche artisanale dans le fleuve Casamance, sur le recrutement du stock marin de Roxo-Bissagos.

7.1.3 *Etude détaillée de la distribution des crevettes à Roxo-Bissagos à l'aide de campagnes de prospection par chalutages (échantillonnage stratifié):* d'une part pour obtenir une meilleure connaissance de la distribution des crevettes, et d'autre part pour préciser la nécessité éventuelle d'une coordination de l'aménagement entre le Sénégal et la Guinée – Bissau.

---

[1]Ces marquages pourraient être réalisés à l'aide de rubans de vinyl, qui permettent de marquer, avec succès, des crevettes de 50 – 60 mm de longueur totale (Marullo, 1976; Klima et Parrack, 1980).

7.1.4 *Etude économique détaillée de l'exploitation artisanale et industrielle:* pour donner aux autorités chargées de la gestion les éléments nécessaires à leurs prises de décisions en matière d'aménagement et de développement respectifs de ces deux types de pêcheries dont l'impact sur le pays côtier (tant en ce qui concerne la capture et sa valeur que la distribution des revenus) est totalement différent.

7.1.5 *Application d'un modèle de simulation pour définir l'effet de la pêche artisanale sur la pêche industrielle*

7.1.6 *Détermination du niveau d'effort globalement admissible pour l'ensemble des espèces de fonds meubles sablo-vaseux du plateau*

7.2 *En matière d'aménagement*

7.2.1 *Réglementation de l'effort en mer* Compte tenu des évaluations réalisées et du comportement des unités de pêche (cf. paragraphe 3.1.6a), on peut émettre les recommandations suivantes:

a) ne pas accroître le nombre de crevettiers congélateurs car ils seraient contraints d'exploiter Roxo-Bissagos déjà pleinement exploité.

b) Tenir compte des captures de crevettes par les chalutiers non spécialisé dans le calcul du nombre total de crevettiers autorisés. Sur St-Louis, l'effort correspondant à l'optimum est de 49 000 heures de chalutage, d'un crevettier glacier type de 400 ch et 50 TJB exploitant exclusivement la crevette, ce qui correspondrait à environ 17 bateaux (pêchant 180 jours ou 2 900 heures par an, performances observées en 1978). Cependant, l'examen des données détaillées pour 1978 indique que les 'crevettiers' ne consacrent en fait que 60% de leur temps de pêche à la recherche de la crevette, le reste de l'effort étant orienté vers les poissons gris et les soles. Dans ces conditions, 24 glaciers de 50 TJB et 400 ch peuvent être autorisés. Leur nombre pourrait encore être augmenté si une part encore plus importante de leur effort était dirigée vers les poissons et si le potentiel de ces derniers le permet.

Notons cependant qu'une étude récente (FAO, non publié) a montré que l'achat de nouvelles unités ne pourrait être rentabilisé qu'avec des crevettiers glaciers de 120 TJB et 400 ch. Ces bateaux étant apparemment moins rentables, à l'heure actuelle, sur le stock de St-Louis, il pourrait s'avérer nécessaire de diminuer encore davantage l'effort de pêche appliqué aux crevettes avant qu'un renouvellement ou un accroissement de cette flottille soit possible.

En ce qui concerne le fond de Roxo-Bissagos, l'effort moyen appliqué a été de 280 jours de mer

par chalutier congélateurs de 400 ch (ou 4600 heures de chalutage). L'effort optimum étant de 131 000 heures de trait, le stock peut supporter 28 crevettiers congélateurs. Cependant, les données indiquant que 20% des captures de crevettes sont obtenues actuellement par les chalutiers classiques recherchent le poisson, 23 crevettiers congélateurs peuvent donc être autorisés et ce nombre devra êtra revu si la capture de crevettes par les chalutiers classiques venait à augmenter.

c) Augmenter la polyvalence des crevettiers congélateurs actuels pour permettre une exploitation plus large du complexe d'espèces démersales de fonds meubles sablo-vaseux.

d) Limiter la flotte de pêche démersale à son niveau actuel (car des transferts d'effort sont encore possibles, des peuplements de fonds durs vers ceux de fonds meubles, des rougets, pageots, céphalopodes, vers les crevettes, soles et poissons gris), ou ne permettre que des accroissements très progressifs en surveillant la réaction de l'ensemble de la ressource et surtout les changements d'espèces cibles éventuels.

7.2.2 *Réglementation du maillage en mer* Il est nécessaire de prendre sérieusement en considération la recommandation du COPACE pour l'utilisation d'une maille unique de 60 mm pour les espèces démersales. Dans le cas du complexe d'espèces des fonds sablo-vaseux, les données disponibles permettent de justifier cette mesure qui n'étant pas néfaste à l'exploitation des crevettes, améliorerait celle des poissons associés et faciliterait grandement le contrôle au port et en mer.

7.2.3 *Recherche des bases d'un aménagement coordonné avec la Guinée–Bissau*

7.2.4 *Aménagement de la pêche artisanale* Dans l'état actuel des connaissances, la capture en lagune paraît surtout dépendre des fluctuations climatiques (cf. 5.1) tout au moins au dans les conditions présentes d'exploitation.

Il paraît cependant judicieux de conserver les règlementations déjà vigueur (maillage de 12 mm de côté, chalutage et filets trainants interdits, pêche interdite dans les nurseries).

Bibliographie

BARRET, B B et GILLESPIE, M C. Environmental conditions
1975 relative to shrimp production in Louisiana. *NMFS/
      NOAA, Louisiana Wildl. Fish. Comm., Tech. Bull.,*
      15: 22 p.
BONDY, E T (DE). Observations sur la biologie de *Penaeus
1968 duorarum* au Sénégal. *Doc. Sci. Centre Rech.
      Oceanogr. Dakar–Thiaroye,* 16: 50 p.

BRUNET-MORET, Y. Etudes hydrologiques en Casamance—
1970   Rapport définitif. *ORSTOM, Paris.*

CADIMA, E L. Effet sur la production d'un changement de l'âge
1977   de première capture. *FAO circ. des pêches*, 701: 45 – 51.

CLARK, W C et KIRKWOOD, G P. Bioeconomic model of the Gulf
1979   of Carpentaria prawn fishery. *J. Fish. Res. Board Can.*,
       36(11): 1304 – 12.

DE VRIES, J et LEFEVERE, S. A maturity key for *Penaeus*
1979   *duorarum* of both sexes. *In:* Actes Symposium
       oceanogr. Ress. Halieut. Atlant. trop. *UNESCO Abid-
       jan*, (1966): 419 – 424.

DOMAIN, F. Les fonds de pêche du plateau continental ouest
1976   africain entre 17°N et 12°N. *Doc. Sci. Centre rech.
       Oceanogr. Dakar – Thiaroye*, 61, 2 cartes.

DOMAIN, F. Carte sédimentologique du plateau continental
1977   sénégambien. Extension à une partie du plateau conti-
       nental de la Mauritanie et de la Guinée – Bissau.
       ORSTOM, notice explicative, 68: 17 p., 3 cartes
       couleur.

FAO/COPACE. Rapport du groupe de travail ad hoc sur l'exploita-
1977   tion de la crevette (*Penaeus duorarum notialis*) du
       secteur Mauritanie–Libéria. *COPACE/PACE, séries
       77/5*: 85 p.

FAO/COPACE. Rapport de la sixième session du Comité des
1980   Pêches de l'Atlantique Centre-Est (COPACE), Agadir
       11 – 14, décembre 1979. *FAO fish. Rep.*, 229: 67 p.

FOX, W W Jr. Fitting the generalized stock production model
1975   by least squares and equilibrium approximation. *Fish.
       Bull NOAA/NMFS*, 73(1): 23 – 37.

GARCIA, S. Biologie de *Penaeus duorarum notialis* en Côte
1974   d'Ivoire. IV. Relation entre la répartition et les condi-
       tions du milieu. Etude des variations du sex ratio.
       Doc. Sci. Centre rech. Océanogr. Abidjan, 5(3–4):
       1 – 39.

GARCIA, S. Biologie et dynamique des populations de crevette
1977   rose *Penaeus duorarum notialis* Perez Farfante 1967, en
       Côte d'Ivoire. *Trav. Doc. ORSTOM*, 79: 221 p.

GARCIA, S et LE RESTE, L. Cycles vitaux, dynamique, exploita-
1981   tion et aménagement des stocks de crevettes penaeïdes
       côtières. *FAO Doc. Tech. Pêches*, T 203 (en prépara-
       tion).

GARCIA, S et LHOMME, F. L'exploitation de la crevette blanche
1977   (*Penaeus duorarum notialis*) au Sénégal. Historique
       des pêcheries en mer et en fleuve, évaluation des poten-
       tiels de capture. *In:* Rapport du groupe de travail ad hoc
       sur l'exploitation de la crevette *Penaeus duorarum
       notialis* du secteur Mauritanie – Libéria. *COPACE/
       PACE, séries 77/5: 17 – 40.*

GARCIA, S et LHOMME, F. Les ressources de crevettes roses
1979   (*Penaeus duorarum notialis. In:* Les ressources
       halieutiques de l'Atlantique Centre Est; première par-
       tie: les ressources du Golfe de Guinée de l'Angola à la
       Mauritanie. Sous la direction de J P Troadec et S
       Garcia. *FAO doc. Tech. sur les pêches*, 186: 123 – 148.

GARROD, D J. Management of multiple resources. *J. Fish. Res.
1973   Bd. Canada*, 30(12), pt. 2: 1977 – 85.

GULLAND, J A. The fish resources of the ocean. *Fishing News
1971   (books) Ltd.*, West Byfleet: 225 p. Rev. Ed. of *FAO
       fish. Tech. Pap.*, 97: 425 p. (1971).

GULLAND, J A et BOEREMA, L K. Scientific advise on catch
1973   levels *Fish. Bull. NOAA/NMFS:* 325 – 335.

GUNTER, G et EDWARDS, J C. The relation of rainfall and
1969   freshwater drainage to the production of the penaeïd
       shrimps (*Penaeus fluviatilis* Say and *Penaeus aztecus*
       Ives) in Texas and Louisiana waters. *FAO, fish. Rep.*,
       53(3): 875 – 92.

GUNTER, G et HILDEBRANDT, H H. The relation of total rainfall
1954   of the state and catch of the marine shrimp *Penaeus
       setiferus* in Texas waters. *Bull. Mar. Sci.*, 4(2): 95 – 103.

ISRA-ORSTOM. La reproduction des espèces exploitées dans le
1979   golfe de Guinée. Rapport du groupe de travail ISRA-
       ORSTOM, Dakar, 7 – 12 décembre 1977. *Doc. Scient.
       Centre Rech. Océanogr. Dakar – Thiaroye*, 68: 213 p.

KLIMA, E D et PARRACK, M L. A technique to obtain informa-
1980   tion on shrimp management in the western central
       Atlantic ocean. *In:* Proceedings of the W.P. on shrimp
       fisheries of northeastern south America, Panama
       city, Panama, 23 – 27 April 1979, report of the meeting
       contributions. FAO/UNDP interregional project for
       the development of fisheries in the western central
       Atlantic. *WECAF rep.*, 28: 104 – 22.

LE RESTE, L. Biologie et dynamique des populations de la
1977   crevette *Penaeus indicus* H. Milne Edwards 1837, au
       Nord Ouest de Madagascar. *Trav. Doc. ORSTOM*, 99,
       291 p.

LE RESTE, L. The relation of rainfall to the production of the
1980   penaeïd shrimp (*Penaeus duorarum*) in the Casamance
       estuary (Senegal). Paper presented at the Symposium
       of tropical ecology, Kuala Lumpur.

LHOMME, F. Biologie et dynamique de *Penaeus (Farfante-
1981   penaeus) notialis* Perex-Farfante, 1967) au Sénégal.
       Thèse de doctorat d'Etat. *Univ. Paris VI*, 225 p.

MARULLO, F. A vinyl streamer tag for shrimp (*Penaeus spp*).
1976   *Trans. Amer. Fish. Soc.*, 105(6): 658 – 63.

NEAL, R A. Methods of marking shrimp. *FAO Fish. Rep.*,
1969   57(3): 1149 – 1165.

ROCHETTE, C. Remontée des eaux marines dans le Sénégal.
1964   Rapport de la mission d'aménagement du fleuve
       Sénégal, *ORSTOM:* 81 p.

RUELLO, N V. The influence of rainfall on the distribution and
1973   abundance of the school prawn *Metapenaeus macleayi*
       in the Hunter River region (Australia). *Mar. Biol.*,
       23(3): 221 – 228.

SCET INTERNATIONAL. La pêche maritime au Gabon; diagnostic
1980   et perspectives. *In:* Etude régionale sur la pêche
       maritime dans le golfe de Guinée. *CCE/FED, annexe
       III:* 109 p.

SUBRAMANYAM, M. Fluctuations in prawn landings in the
1966   Goodavari estuarine system. *Proc. IPFC*, 11: 44 – 51.

Annexe I

ÉVOLUTION DE LA PRISE TOTALE ANNUELLE (TONNES) DANS LE FLEUVE SÉNÉGAL DE 1960 À 1978 (DONNÉES D.O.P.M.)

Année	1960	1961	1962	1963	1964	1965	1966	1967	1968	1969	1970	1971	1972	1973	1974	1975	1976	1977	1978
Prise totale (tonnes)	7	21	35	21	21	7	10	28	8	13	10	55	260	127	267	122	99	141	113

ÉVOLUTION DE LA PRISE TOTALE MENSUELLE (KG) DANS LE FLEUVE SÉNÉGAL DE 1975 À 1977 (DONNÉES D.O.P.M.)

	J	F	M	A	M	J	J	A	S	O	N	D
1975	25 500	18 900	14 100	24 000	1 800	600	300	600	3 450	7 200	13 500	12 000
1976	20 000	18 500	15 800	18 900	2 500	1 600	3 500	2 700	4 000	6 000	3 000	3 000
1977	18 300	1 800	2 000	7 300	12 800	7 500	1 400	2 200	1 200	4 200	2 700	1 600
Pourcentage de la prise annuelle (moyenne)	22·4	13·8	11·2	17·7	6·0	3·4	1·8	1·9	3·0	6·1	6·8	5·9

Annexe II

ÉVOLUTION DE LA PRISE TOTALE ANNUELLE (TONNES) DANS LE SINE SALOUM DE 1959 À 1978 (DONNÉES D.O.P.M.)

	1959	1960	1961	1962	1963	1964	1965	1966	1967	1968	1969	1970	1971	1972	1973	1974	1975	1976	1977	1978
Prise totale (tonnes)	213	0	50	100	61	61	22	177	196	192	93	67	124	104	36	83	307	0	0	63

ÉVOLUTION DE LA PRISE MENSUELLE (KG) DANS LES PORTS DE FATICK FOUNDIOUGNE ET KAOLACK EN 1959 ET DE 1967 À 1971 (DONNÉES DIR. OCEANOGR. PÊCHES MARIT. SÉNÉGAL)

	J	F	M	A	M	J	J	A	S	O	N	D
1959	1 000	1 000	4 000	18 000	18 000	36 000	45 000	45 000	31 000	18 000	4 500	2 000
1967	3 550	2 900	9 970	6 200	2 500	8 490	21 830	32 019	49 262	24 920	6 000	8 624
1968	11 600	14 060	17 157	10 240	11 550	9 900	16 974	16 055	31 517	20 240	20 750	16 050
1969	4 950	1 700	0	0	0	0	0	6 470	17 120	37 695	15 450	6 360
1970	9 780	5 180	3 460	580	0	40	1 050	2 570	7 805	36 795	16 410	6 015
1971	100	0	0	50	300	50	1 800	2 720	29 525	56 610	38 355	21 985
Pourc. de la prise annuelle (moy. 1967–1970)	6·1	4·1	4·6	2·3	1·6	2·5	5·5	9·1	17·9	26·9	12·4	6·7

Annexe III

EFFORTS MENSUELS EXPRIMÉS EN NOMBRE DE PIROGUES POUR LA CASAMANCE DE 1963 À 1966 ET EN 1970–1971 (DONNÉES AMERGER)

	J	F	M	A	M	J	J	A	S	O	N	D	Moyenne
1963	Données mensuelles non retrouvées												196
1964	(238)	238	275	275	282	326	266	330	266	369	410	407	307
1965	387	346	412	349	(456)	456	563	418	427	388	(414)	(414)	419
1966	(414)	(414)	440	540	491	668	688	708	709	(619)	(619)	(619)	577
1970	360	458	457	867	827	827	827	827	600	500	500	500	629
1971	0	600	(700)	800	763	727	930	(930)	940	940	943	1036	846

Les chiffres entre parenthèses sont interpolés

Annexe IV

PRISES MENSUELLES DE CREVETTES TRAITÉES PAR LES USINES DE ZIGUINCHOR DE 1960 À 1968 (EN KG)

	J	F	M	A	M	J	J	A	S	O	N	D	TOTAL
1960	2600	1190	1857	8395	21918	24356	17096	8726	11677	11288	10640	4398	124391
1961	4900	2240	3500	15820	41300	45610	33740	16440	22000	21200	20050	8200	235000
1962	2956	3167	4763	7115	39562	44720	65955	31839	29016	25670	15487	6956	277755
1963	5356	6182	8165	16801	30700	101166	43094	56705	40546	27642	19430	16760	372547
1964	18613	18400	25373	37166	51643	44869	45685	77916	67663	55031	29405	32298	504122
1965	13421	28037	45510	55430	?	?	174851	65694	68549	32883	?	?	(748880)
1966	?	?	22345	35456	68089	78406	66838	73994	92705	73121	?	?	(611241)
1967	Données mensuelles non retrouvées												562000
1968													713000

Sources: 1960 à 1962: MONOD (1966)
1963 à 1966: statistiques relevées aux usines par le C.R.O.D.T.
Données incomplètes

143

Annexe V

PRISES MENSUELLES DE CREVETTES TRAITÉES PAR LES USINES DE ZIGUINCHOR DE 1969 À 1977 (EN KG). STATISTIQUES RELEVÉES AUX USINES PAR LE C.R.O.D.T.

	J	F	M	A	M	J	J	A	S	O	N	D	TOTAL
1969	37 454 incomplet	98 295	130 818	187 116	124 546	110 326	105 725	130 638	117 022	52 854	12 546	14 023	1 128 563
1970	26 726	33 287	49 669	74 931	94 687	136 600	112 583	129 104	78 334	39 554	35 479	29 888	840 842
1971	34 771	42 553	56 361	109 154	101 028	84 611	90 062	57 227	61 101	72 965	52 032	49 994	811 859
1972	44 552	64 676	108 345	131 655	107 251	76 286	65 438	53 155	176 644	217 275	128 655	76 010	1 249 942
1973	97 856	152 872	173 806	126 572	136 982	102 912	96 627	106 685	178 736	109 123	117 904	58 658	1 458 733
1974	65 470	137 444	136 622	158 474	213 387	155 180	110 866	118 128	138 447	122 233	100 035	79 892	1 536 178
1975	81 675	70 515	110 741	133 384	167 282	97 204	107 565	103 637	125 801	127 633	50 472	40 302	1 216 211
1976	23 382	64 994	104 771	140 650	181 549	174 292	73 904	51 420	44 090	52 631	64 576	35 179	1 022 438
1977	48 770	62 517	87 899	126 211	116 403	147 127	110 526	67 883	111 077	192 953	113 168	49 702	1 234 236
Moy.	54 275	80 795	107 359	132 016	138 124	120 504	97 033	90 875	114 584	109 691	74 985	47 993	1 116 556
Moyen%	4·7	6·9	9·2	11·3	11·8	10·3	8·3	7·8	9·8	9·4	6·4	4·1	
Ecart-type	25 172	40 895	39 589	30 938	40 734	34 027	17 261	33 239	47 202	62 989	41 084	22 758	247 513

The dynamics and management of shrimp in the Northern Gulf of Mexico

Brian J Rothschild
and
Susan L Brunenmeister

Abstract

An improved understanding of stock structure, stock productivity, effects of fishing on recruitment, effects of harvest size on biological production, effects of environment on reproductive success, and interactions among species will facilitate management of the U S Gulf of Mexico shrimp fishery. The fishery takes primarily brown shrimp (*Penaeus aztecus* Ives), white shrimp [*Penaeus setiferous* (L.)] and pink shrimp (*Penaeus duorarum* Burkenroad). Annual landings are presently about 77 000 mt.

Technological improvements in fishing vessels and gear, and increases in the number of craft have led to increased fishing pressure on the stocks. Current high levels of fishing effort which are at the transition point between parabolic and non-parabolic production functions estimated for each shrimp species suggest careful monitoring of the stocks is necessary and that reductions in effort would almost certainly lead to overall economic benefits, given the potential of limited returns with increasing effort and risk of sharp declines in the stock. While there is some support for the hypothesis that environmental variables control recruitment and not stock size, a correlation between stock and recruitment is demonstrable for brown shrimp. Such a correlation may also obtain for white shrimp suggesting again that increases in fishing effort should be viewed with caution. While catch and some effort statistics are available for the offshore fishery, little data exists on the important recreational and commercial inshore fisheries. This situation and the lack of current statistics creates problems in developing timely management strategies.

Each shrimp species has a similar life history pattern with respect to offshore spawning and juvenile development in estuaries. However, temporal and spatial recruitment patterns of the species differ with regard to inshore and offshore fisheries. Regulations imposing minimum size limits on landed shrimp vary from state to state and along with market conditions have contributed to sizeable discards of small shrimp (especially brown shrimp and pink shrimp) whose magnitudes are not known. This situation and concern for optimizing yield-per-recruit have led to present management practices of closing fishing in certain areas and times. Available estimates of growth and mortality rates of each shrimp species, however, vary and in many aspects are not completely known. Such gaps along with incompletely understood recruitment phenomena affect management advice and the success of intended management measures.

1 Statement of the problem

In recent years the landings in the U S Gulf of Mexico shrimp fishery have amounted to about 77 000 metric tons, (ex-vessel value about U S $322 million). Management of these shrimp stocks has been the subject of considerable concern. Prior to the U S extension of fishery jurisdiction, only the U S States bordering the Gulf of Mexico actively participated in shrimp management. However, in recent years, the States' management activities have been joined by the U S Federal Government, which through the passage of the Fisheries Conservation and Management Act of 1976, extended fishery management jurisdiction to 200 miles from the coast. The Act created *inter alia* eight Regional Fisheries Management Councils. One of these Councils, the Gulf of Mexico Regional Fisheries Management Council (referred to here as the 'Council') is responsible for developing fishery management plans for the northern Gulf of Mexico. The Council has recently prepared the document 'Fisheries Management Plan and Proposed Regulations for the Shrimp Fishery of the Gulf of Mexico, United States Waters'* (referred to here as the 'Management Plan').

For the shrimp fishery, the Council identified a multiplicity of management problems. These are specified in the Management Plan:

'(*1*) Conflict among user groups as to area and size at which shrimp are to be harvested.

(*2*) Discard of shrimp through the wasteful practice of culling.

(*3*) The continuing decline in the quality and quantity of estuarine and associated inland habitats.

---

*Shrimp Fishery of the Gulf of Mexico, Plan Approval and Proposed Regulations, Gulf of Mexico Fishery Management Council, November, 1980, Federal Register 45 (218): 74178–74308.

(4) Lack of comprehensive, coordinated and easily ascertainable management authorities over shrimp resources through their range.

(5) Conflicts with other fisheries such as the stone crab fishery in southern Florida, the groundfish fishery of the north central Gulf, and the Gulf's reef fish fishery.

(6) Incidental capture of sea turtles.

(7) Loss of gear at trawling grounds due to man-made underwater obstructions.

(8) Partial lack of basic data needed for management.' (p. 74192)

While some of these problems are political or institutional in nature, others are scientific and relate, for example, to the status of the stocks. Resolution of political or institutional fishery problems is, however, often facilitated through an improved understanding of scientific problems such as determining the status of the stocks. Understanding the status of the stocks includes topics such as stock structure, stock productivity, the effect of fishing on recruitment, the effect of harvest size on biological production, the effect of the environment on reproductive success, and interactions among stocks in the fishery and associated stocks.

Views on these scientific and technical matters are continuing to evolve. For example, the Council has visualized a particular uniqueness in the 'annual' nature of the shrimp stocks. The Council has concluded that 1) there is no demonstrable relation between stock size and recruitment levels for Gulf of Mexico shrimp stocks, 2) it is impossible to 'recruitment overfish' shrimp stocks, 3) there should be no constraint on the quantity of shrimp taken each year, 4) the environment, especially temperature and salinity and not stock size, controls the success of recruitment, and 5) surplus-production models are inadequate for providing guidance on the relation between shrimp production and the amount of fishing. It is easy to see that these interpretations are critical to the management of shrimp stocks in the Gulf. They, in fact, underlie interpretations of some of the political and institutional problems articulated by the Council. In this paper, we discuss these issues and specifically consider the shrimp fisheries, environmental aspects and management.

2 Shrimp fisheries

Description of the fishery

The Gulf of Mexico shrimp fishery is characterized by heterogeneity. The species composition of the catch varies temporally and geographically; both commercial and recreational fishermen use a vari-ety of fishing craft and gear types.

The catch includes primarily seven taxa: brown shrimp (*Penaeus aztecus* Ives), white shrimp (*Penaeus setiferus* L.), pink shrimp (*Penaeus duorarum* Burkenroad), royal red shrimp (*Hymenopenaeus robustus* Smith), seabob (*Xiphopenaeus kroyeri* Heller), rock shrimp (*Sicyonia brevirostris* Stimpson), and broken-necked or sugar shrimp (*Trachypenaeus* spp.). Brown, white and pink shrimp constitute the bulk of the landings while seabob and sugar shrimp are caught incidentally with brown and white shrimp, and rock shrimp are caught incidentally with pink shrimp.

Trends in landings of the six commercially important species from 1960 to 1981 from U S waters of the Gulf of Mexico are shown in *Figure 1*. Brown shrimp landings have generally increased from 1960 to 1981. Peak years during this period were 1960, 1967, 1972, 1977 and 1981. Inshore and offshore landing trends were similar, with inshore landings accounting for around one-fourth of the total catch. White shrimp landings also increased from 1960 through 1981 with peaks in 1963, 1970, 1978 and 1981. Inshore and offshore landing trends of white shrimp were similar, with inshore landings constituting about one-third of the total. Landings of pink shrimp also increased slowly from 1960 to 1981. Inshore and offshore landings of pink shrimp showed similar trends, however, the inshore landings of this species constitute a much lower average percentage (about 4%) of the total catch than is the case for brown or white shrimp. Minimal landings of royal red shrimp were first recorded in 1962 and although landings have fluctuated considerably from year to year, they have increased overall through 1981. Recorded landings of seabob and rock shrimp also show overall increases through 1981, despite considerable annual variation. Landings of seabob in 1980 increased dramatically compared to former years, with inshore landings comprising little of the total catch in any year.

The abundance and relative importance of each species to the fishery varies geographically and temporally. Pink shrimp and rock shrimp are caught primarily in eastern Gulf waters off west Florida and the Florida Keys, usually in depths of 60–70 fathoms (110–128 m). White and brown shrimp and associated seabob and sugar shrimp are caught primarily off the coasts of Louisiana and Texas. The bulk of the catch of white shrimp is taken in depths less than 15 fathoms (27·4 m), but brown shrimp are caught in deeper waters, up to 30–40 fathoms (55–73 m). While white shrimp catches are greatest along the Louisiana and north-

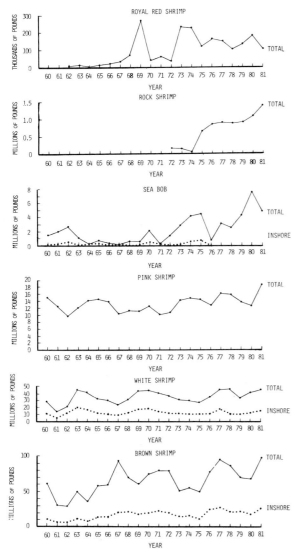

Fig 1 Trends in inshore and offshore landings (weights are 'heads-off') of shrimp in the northern Gulf of Mexico. Note differing scales

ern coast of Texas, brown shrimp catches from offshore waters in these areas are greater in magnitude. Royal red shrimp are caught only in deep offshore waters (200 fms; 366 m).

As shown in Brunenmeister (1983, Figures 7, 8, and 9), fishing effort for brown, white and pink shrimp varies seasonally. In general, effort peaks for brown shrimp during the summer, for white shrimp in the late spring and the fall, and for pink shrimp in late winter, early spring. Paralleling patterns in recruitment, peak landings of brown shrimp occur during July and August, those of white shrimp during September to November. Those of pink shrimp are more temporally prot-

racted, occurring during the period December to May.

Shrimp fishing was first recorded to have taken place using haul seines in inshore waters. In 1913, otter trawls were first used to catch shrimp off the northeast coast of Florida. The success of these early ventures resulted in the replacement of haul seines by otter trawls by 1917. Anderson (1948) noted that the annual haul seine catch amounted to 20 million pounds (heads-on) and that the fishery rapidly expanded with the adoption of the otter trawl. By 1923, around 47 million pounds of shrimp heads-on (or about 28 million pounds heads-off) were landed. During the 1920's, the typical fishing craft changed from small open boats (15–25' in length) powered by gasoline engines to the Florida-type and the Biloxi-type boat designs which are still used. The most common 'Florida-type' design has a forward wheelhouse and engine; the less common 'Biloxi-type' design which is utilized primarily in inshore waters has an aft wheelhouse and engine room. During the 1930's, propulsion systems changed from gasoline to diesel engines (Captiva, 1967). Anderson (1948) noted that the early engines used in the shrimp fishery boats were modified automobile engines for which no parts or replacement models existed during World War II. He states that the conversion to marine diesel engines at the end of the war resulted in increased fishing power since diesel powered boats could pull larger nets at a greater speed.

White shrimp constituted the bulk of the catch up to the 1950's, with Louisiana waters producing two-thirds of the entire landings along the southeast Atlantic and Gulf of Mexico coasts. Although brown shrimp were caught during this period, their exploitation was retarded by their low market values (Lyles, 1950). The offshore fishery underwent a rapid expansion during the 1940's and through the 1950's as shrimp fishermen searched for alternative stocks because of low white shrimp availability, increased fleet size, and economic difficulties. Exploitation of new grounds off Louisiana and Texas, and in the Tortugas off Florida occurred with directed harvesting of brown and pink shrimp. Lyles (1950) states the first catch consisting entirely of brown shrimp was landed at Port Isabel, Texas, in 1940. Exploitation of brown shrimp off the Texas coast was aided by depth locators and was made feasible through the success of governmental education campaigns aimed at public acceptance of more pigmented shrimp.

It was during this period of expansion that double-rig trawls were adopted, primarily by vessels fishing brown and pink shrimp. Shrimp fishermen

147

at Rockport, Texas, in late 1955 were the first to employ this technology. Subsequent improvements and demonstrated results led to major conversions by Texas shrimpers in 1957 and subsequent adoption by Florida shrimpers (Knake, Murdock and Cating, 1958). Replacement of the single large otter trawl net, which ranged from 80 to 100 feet in width, by two small trawls, typically 40 to 50' in width, resulted in increased efficiency and decreased costs.

The second major innovation was the double-rig twin-trawl. In this arrangement, two small 25-foot trawls are towed on each side of the vessel. First experimentally fished off the Texas coast in 1971 (Bullis and Floyd, 1972), these four-trawl rigs are used primarily off the Texas coast where minimal turning while fishing is required.

Average vessel size and horsepower has increased from the period of offshore expansion to the present. In the construction of larger vessels, steel and aluminum began to replace wood as hull materials. The first all-steel trawlers began fishing in the 1940's and by the early 1960's, 50% of the boats constructed were all steel. Captiva (1967) noted that steel hulls have greater fuel and water capacity and strength than wooden hulls. The first all-aluminum shrimp trawler fished in the Gulf in 1969 (Anon., 1969a).

Increases in the length and range of fishing trips have been accompanied by the installation of catch-handling and freezing equipment, and electronic navigation aids. Although the first freezer trawlers appeared in the 1940's (Captiva, 1967), dependable freezer units for small trawlers were developed in 1966 and by the following summer, these units were used by over 50 vessels (Anon., 1968b). Ironically, the improvement in gear and vessel technology led to a manpower shortage during a period of the fishery. Mehos (1969) stated that the industry's greatest problem (then) was a shortage of skilled personnel.

Characteristics of the shrimp fleet operating in offshore waters in 1977 (as reflected by vessels whose captains were interviewed by U S port agents during that year) indicated that vessels averaged 76 gross tons and 61' in length, were powered by 285 horsepower engines, fished 2 nets 52' in width and were approximately 12 years old. Comparisons between wooden and steel-hulled vessels in this sample showed that most wooden vessels were smaller (61 versus 105 gross tons) less powerful per unit size (242 hp versus 366 hp), pulled 2 rather than 4 nets, and were older (14·5 years versus 6·7 years).

In inshore waters, in addition to the commercial fishery discussed above, significant quantities of shrimp are taken by recreational fishermen and by commercial live-bait fishermen.

Data base

Prior to 1956, landing statistics were based on state tax records and a limited number of shrimpdealer records. The statistics included annual landings by state and type of gear, the number of vessels or boats fishing commercially, and the number of commercial fishermen fishing for shrimp.

The system used from 1956 to the present encompasses the entire northern Gulf of Mexico (see Snow, 1969). This system involves three kinds of commercial fishery data: 1) landings, 2) fishing effort and location, and 3) vessel characteristics. Copies of sales receipts are collected from commercial shrimp-house dealers. These data include a vessel identification number and date of sale, statistical area of catch (*Figure 2*), the weight and species of shrimp by market category, and the prices. Other information includes the number of trips actually made if the sales record was a composite of landings of several trips, whether the shrimp were landed heads-off or with the heads-on, and whether the shrimp were sorted into the various market categories by hand or by machine.

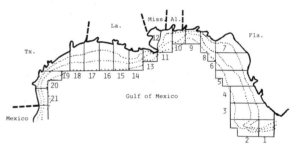

Fig 2 Statistical areas used for recording statistics on the U S shrimp fishery in the northern Gulf of Mexico

Fishing effort and location data are obtained from a sample of trips by interview methods. The depth and the approximate location fished, whether fishing occurred during the day and/or night and the amount of time spent in actual fishing are recorded. Amounts of shrimp, fish or other bycatch which are discarded, however, are not systematically recorded.

Vessel characteristics are available for each vessel (craft 5 gross tons and larger). These records specify engine horsepower, vessel length, gross tonnage, year of construction, number and size of trawls used, crew size, and hull material. Unfortunately, some attributes that can significantly affect fishing power, such as the years of experience of the captain or kind of navigational instruments aboard, are not recorded.

However, the available data does not represent the entire fishery. Catches and characteristics of the bait fishery and recreational fishery are not systematically recorded or sampled, although some data has been collected in selected areas. Various other kinds of data pertaining to the shrimp fishery have been collected by state and/or federal agencies. These include meteorological and hydrographic data, shrimp abundance and size data from research cruises or site sampling, and growth and mortality data from mark – release – recapture studies. A general summary of shrimp monitoring programs and research activities of Gulf States is provided by Christmas and Etzold (1977). Data collection activities of the NMFS in the Gulf and southeast Atlantic region are outlined in Prytherch (1980) and Southeast Fisheries Center (1982). Poffenberger (1982) provides a review of the economic data available for the shrimp fisheries in the southeastern United States.

Biology of the major species
Life-history synopses have been compiled for each major species (brown shrimp, Cook and Lindner, 1970; pink shrimp, Costello and Allen, 1970; and white shrimp, Lindner and Cook, 1970). These species of shrimp prawn offshore and the plank- tonic larvae drift shoreward. After passing through several larval stages, the very young shrimp move into estuaries where they become benthic. After a few months, the young shrimp emigrate from the estuaries to the sea. Spawning and recruitment of penaeid shrimp in the Gulf of Mexico occur virtu- ally throughout the year, but there are well-defined seasonal peaks which are described below. In addi- tion, movements of very small shrimp into estuaries and those of juveniles out of the estuaries into the sea have been observed to peak during periods of maximal tidal currents in conjunction with dark and full phases of the moon.

The distribution and stock structure of brown shrimp, white shrimp and pink shrimp may be considered in terms of the distribution of landings, catch-per-unit-of-effort, and results from tagging studies. With respect to landings, the Management Plan (Figure 3.2-4) shows the relative landings of brown shrimp are less in Areas 15, 16, and 17 (*v Figure 2*), than to the east and west of these Areas. The relative landings of white shrimp are less in area 16 than in areas to the east and west. The relative landings of pink shrimp are by far the greatest in Area 2.

Landings, however, reflect the distribution of fishing and may not always reflect the actual distri- bution of the particular species that is being studied. The distribution of catch-per-unit-of- effort indices is often a better index of distribution. Brunenmeister (1983) studied the correlations in annual standardized CPUE among statistical reporting areas in the Gulf of Mexico. Her analysis reflects heterogeneities in geographical distribu- tions of CPUE. For example, for brown shrimp, Area 8 appeared fairly distinct from other statisti- cal areas and one interpretation of the data showed discontinuity between Areas 13, 14, 15, 11, 16 and Areas 17–21. For white shrimp, Area 17 appeared to be distinct from the other areas, as did Areas 8 and 20 (there is little likelihood of a biological relation between these two areas). For pink shrimp, three groupings were inferred: Areas 2–4, Areas 5–8, and Area 11.

Tagging data also provide a basis for inference regarding the distribution and stock structure of shrimp. Large numbers of brown, white, and pink shrimp were tagged during the years 1977 to 1980. Locations of releases and recaptures are shown in *Table 1*. As might be expected from the very high *apparent* mortality rates of tagged shrimp, most tagged shrimp that were recovered were recovered in the area where they were tagged. Most of the movement of shrimp among statistical areas occur- red between the statistical area of tagging and adjacent areas. However, it does appear, particu- larly for brown shrimp and particularly for those in the eastern Gulf, that a net westward movement occurs.

Thus, for the northern Gulf of Mexico, there are geographical heterogeneities in shrimp densities as evidenced by catch-per-unit-effort data. These heterogeneities are not particularly evident from tagging data since time at liberty for tagged shrimp is relatively short. On the other hand, since shrimp move between adjoining statistical areas, separate stocks may not exist. If there are stock separations, brown shrimp may consist of two stocks separated at Areas 16, 17; white shrimp may consist of one or many stocks; and pink shrimp possibly consist of northern and southern stocks along the west coast of Florida, and an additional stock in the western Gulf.

Recruitment patterns
Recruitment patterns may be deduced from aver- age catch patterns. The Management Plan (Figures 3.2-6, 3.2-7, 3.2-8) gives the average distribution of catches according to shrimp size, depth and season. For the purpose of this discussion, we are

Table 1
RELEASE AND RECAPTURE OF TAGGED SHRIMP BY STATISTICAL AREA

Brown Shrimp

1977–1980 REL AREA	Recapture Area															
	8	9	10	11	12	13	14	15	16	17	18	19	20	21	MEX	TOT
8	0	0	0	0	0	0	0	0	0	0	0	0	0	0	0	0
9	0	0	0	0	0	0	0	0	0	0	0	0	0	0	0	0
10	0	0	0	0	0	0	0	0	0	0	0	0	0	0	0	0
11	0	1	0	366	17	3	0	0	0	0	0	0	0	0	0	387
12	0	0	0	0	7	0	0	0	0	0	0	0	0	0	0	7
13	0	0	0	1	1	405	98	5	2	0	0	0	0	0	0	512
14	0	0	0	0	0	0	2 846	1 115	19	4	2	2	1	0	0	3 989
15	0	0	0	1	0	2	11	68	15	4	2	1	1	0	0	105
16	0	0	0	0	0	0	0	0	1	2	0	0	0	0	0	3
17	0	0	0	0	0	0	1	7	39	238	27	8	1	0	0	321
18	0	0	0	0	0	0	0	1	0	0	51	39	3	0	0	94
19	0	0	0	0	0	0	1	3	6	12	193	1 793	208	91	1	2 308
20	0	0	0	0	0	0	1	0	0	1	2	42	617	248	10	921
21	0	0	0	0	0	0	0	0	0	0	2	4	40	552	32	630
MEXICO	0	0	0	0	0	0	0	0	0	0	1	3	2	46	3 666	3 718
TOTAL	0	1	0	368	25	410	2 956	1 199	82	261	280	1 892	873	937	3 709	12 995

White Shrimp

1977–1980 REL AREA	Recapture Area															
	8	9	10	11	12	13	14	15	16	17	18	19	20	21	MEX	TOT
8	0	0	0	0	0	0	0	0	0	0	0	0	0	0	0	0
9	0	0	0	0	0	0	0	0	0	0	0	0	0	0	0	0
10	0	0	0	0	0	0	0	0	0	0	0	0	0	0	0	0
11	1	0	0	583	4	2	0	0	0	0	0	0	0	0	0	570
12	0	0	0	1	18	0	0	0	0	0	0	0	0	0	0	19
13	0	0	0	0	1	42	76	8	3	0	0	0	0	0	0	130
14	0	0	0	0	2	7	3 744	1 269	142	19	5	1	1	0	2	5 192
15	0	0	0	0	0	1	135	263	39	3	2	0	0	0	0	443
16	0	0	0	0	0	0	7	41	49	13	1	0	0	0	0	111
17	0	0	0	0	0	0	0	1	12	799	25	2	0	0	0	839
18	0	0	0	1	0	0	0	2	3	57	807	20	1	0	0	891
19	0	0	0	0	0	0	0	1	1	2	119	499	22	4	0	648
20	0	0	0	0	0	0	0	0	1	0	2	53	126	6	0	188
21	0	0	0	0	0	0	0	0	0	0	0	2	2	10	0	14
MEXICO	0	0	0	0	0	0	0	0	0	0	8	53	2	3	12	78
TOTAL	1	0	0	565	25	52	3 962	1 585	250	893	969	630	154	23	14	9 123

Pink Shrimp

1977–1980 REL AREA	Recapture Area															
	8	9	10	11	12	13	14	15	16	17	18	19	20	21	MEX	TOT
8	0	0	0	0	0	0	0	0	0	0	0	0	0	0	0	0
9	0	0	0	0	0	0	0	0	0	0	0	0	0	0	0	0
10	0	0	0	0	0	0	0	0	0	0	0	0	0	0	0	0
11	0	0	0	0	0	0	0	0	0	0	0	0	0	0	0	0
12	0	0	0	0	0	0	0	0	0	0	0	0	0	0	0	0
13	0	0	0	0	0	0	0	0	0	0	0	0	0	0	0	0
14	0	0	0	0	0	0	0	0	0	0	0	0	0	0	0	0
15	0	0	0	0	0	0	0	0	0	0	0	0	0	0	0	0
16	0	0	0	0	0	0	0	0	0	0	0	0	0	0	0	0
17	0	0	0	0	0	0	0	0	0	0	0	0	0	0	0	0
18	0	0	0	0	0	0	0	0	0	0	1	0	0	0	0	1
19	0	0	0	0	0	0	0	0	0	0	4	7	1	0	0	12
20	0	0	0	0	0	0	0	0	0	0	1	5	27	31	1	65
21	0	0	0	0	0	0	0	0	0	0	4	31	231	3 589	125	3 980
MEXICO	0	0	0	0	0	0	0	0	0	0	0	1	7	68	880	956
TOTAL	0	0	0	0	0	0	0	0	0	0	10	44	266	3 688	1 008	5 014

interested primarily in the distribution of the smallest shrimp (*ie* the recruits) which are identified as Size-Class 8. Shrimp of this size class are those with a 'heads off' weight of 1/68 pound (·0067 kg; about 110 mm total length) or less. Using catch as an index of abundance, we can see that brown shrimp tend to be recruited in inland waters (there is only a small amount of recruitment in water 0–5 fathoms deep). Landings of small brown shrimp from inshore waters average approximately 3–4 times greater in magnitude than those from offshore waters, although the peak months of landings are similar (*Figure 3*). Peak months of recruitment to both inshore and offshore fisheries occurs during May, June and July with minimal recruitment occurring during the rest of the year. The ratio of catch magnitudes for the peak months is about 2:3:1, respectively.

White shrimp are recruited not only in inland waters but also in offshore waters 0–5 fathoms deep. The recruitment in inland waters takes place August through December with peaks in October and November. The ratio among months in this period is roughly 1:1:3:3:1. As the season progresses, recruitment begins to take place in the 0–5 fathom zone, with the peak recruitment in this zone occurring during November, December, and January. Recruitment patterns to the inshore and offshore fisheries are shown in *Figure 4*.

Pink shrimp recruitment appears to take place mostly in water 11–15 fathoms deep. Recruitment appears to occur year-round, with peaks in February, March, April, and May. *Figure 5* shows the monthly landing patterns of catches of small pink shrimp from inshore and offshore waters.

The relatively *large* magnitude of catches of very small shrimp is also of interest. These catches average perhaps, 12 million pounds (heads-off) of brown shrimp and 5 or 6 million pounds (heads-off) of white shrimp. The catch of these very small shrimp amounts to perhaps ten percent of the total catch.

Growth patterns

As there are presently no techniques for directly ageing shrimp, growth patterns have been inferred from mark–recapture data or other data from which temporal changes in size can be estimated. Various functional forms, metrics (*eg* total length, carapace length, and weight, *etc*) and independent variables (*eg* temperature and length) have been employed in the estimation of shrimp growth. Growth functions for pink, white or brown shrimp populations in the Gulf of Mexico (*Table 2*; *Figures 6a* and *6b*) indicate sexually dimorphic growth

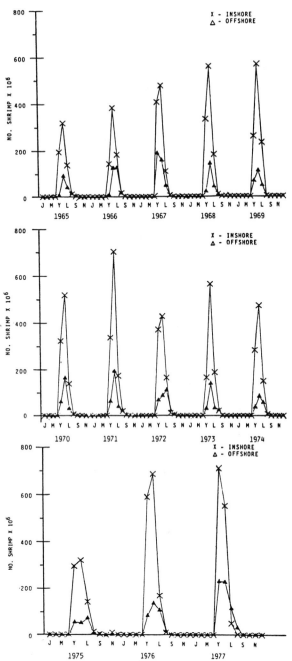

Fig 3 Estimated number of 'recruitment size' (weighing less than 1/68 pound, heads-off) brown shrimp landed from inshore and offshore waters by month and year. Note that May is represented by 'Y' and July by 'L'

patterns. Comparison among studies shows that growth rates estimated for male or female pink shrimp by Iverson and Jones (1961) differ dramatically from those estimated by Berry (1967). Growth patterns of brown shrimp estimated by Parrack (1979) and Chavez (1973) for males are

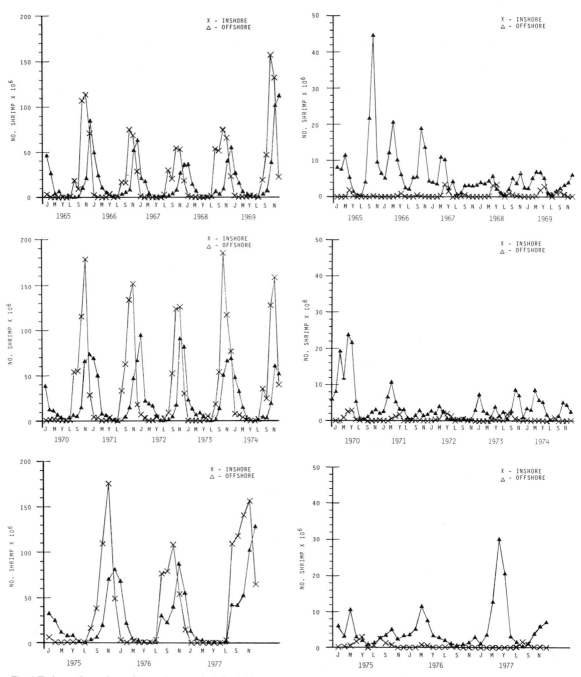

Fig 4 Estimated number of 'recruitment size' (weighing less than 1/68 pound, heads-off) white shrimp landed from inshore and offshore waters by month and year. Note that May is represented by 'Y' and July by 'L'

Fig 5 Estimated number of 'recruitment size' (weighing less than 1/68 pound, heads-off) pink shrimp landed from inshore and offshore waters by month and year. Note that May is represented by 'Y' and July by 'L'

quite similar while their estimates for females are somewhat disparate. Comparison between species shows that growth rates of male and female white shrimp estimated by Lindner and Anderson (1956) are similar to those estimated for male and female pink shrimp, respectively, in winter months by Iverson and Jones (1961). However, growth rates of pink shrimp estimated by Berry (1967) are similar to those estimated for brown shrimp by Parrack (1979) and Chavez (1973). Thus, pub-

Table 2
VARIOUS REPORTED GROWTH EQUATIONS FOR GULF OF MEXICO SHRIMP SPECIES[*]

Species	Sex	Growth Model	Source
Brown	Male	$L = 168·7(1 - 0·9979e^{-0·3357T})$	Parrack, 1979
	Female	$L = 193·6(1 - 0·9982e^{-0·3363T})$	Parrack, 1979
	Male	$L = 178·1(1 - e^{-0·2567(T+0·2388)})$	Chavez, 1973
	Female	$L = 236·0(1 - e^{-0·162(T+0·759)})$	Chavez, 1973
Pink	Male	$L = 168(1 - e^{-0·046(t+5·68)})$	Berry, 1967
	Female	$L = 199(1 - e^{-0·055(t-0·06)})$	Berry, 1967
	Male, (s)	$C = 46·44(1 - e^{-0·934T})$	Iverson and Jones, 1961
	Male, (w)	$C = 35·99(1 - e^{-0·876T})$	Iverson and Jones, 1961
	Female, (s)	$C = 45·68(1 - e^{-0·912T})$	Iverson and Jones, 1961
	Female, (w)	$C = 51·58(1 - e^{-0·919T})$	Iverson and Jones, 1961
White	Male	$L = 170(1 - e^{-0·761T})$	Lindner et al, 1956
	Female	$L = 190(1 - e^{-0·817T})$	Lindner et al, 1956
	All	$L = 214(1 - e^{-0·09(t+0·2)})$	Klima, 1974
	All	$W = 87(1 - e^{-0·12(t+0·57)})^3$	Klima, 1964

[*] L = total length in mm;
W = total weight in grams;
C = carapace length in mm;
T = age in months;
t = age in weeks;
(s) = growth computed for spring, summer, and fall
(w) = growth computed for winter

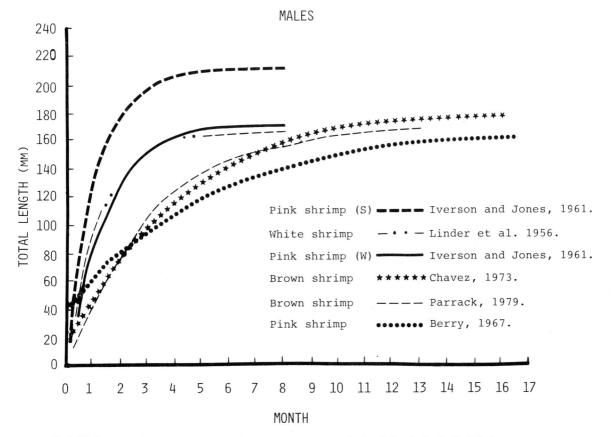

Fig 6a Various growth equations estimated for male brown, white or pink shrimp in the Gulf of Mexico

153

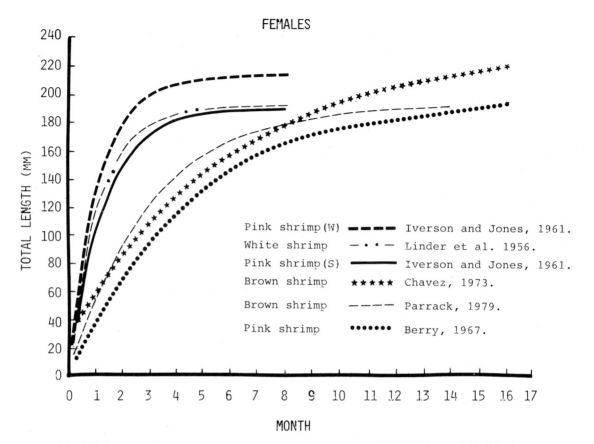

FEMALES

Fig 6b Various growth equations estimated for female brown, white or pink shrimp in the Gulf of Mexico

lished estimates of growth rates based on mark–recapture data show relatively wide variations.

Seasonal growth has been documented for pink shrimp (Berry, 1967) and white shrimp (Lindner and Anderson, 1956). The effect of temperature on growth has been studied with varied results. Phares (1980) found that including temperature as an independent variable in growth models of white shrimp explained only 4 to 8 percent additional variation in size changes with time. Nichols (1981a) related the instantaneous rate of growth, $\Delta l/\Delta t$, to length and temperature. The difficulties in estimating growth of white shrimp are illustrated by this latter analysis which shows that for any fixed temperature, the rate of growth tended to increase with increasing shrimp size up to 90–100 mm in tail length and thereafter tended to decrease.

Mortality rates

Most mortality rates estimated for shrimp have been derived from mark–recapture data (*Table 3*). These estimates vary considerably and, in some cases, seem unusually high. For example, estimated annual natural mortality rates range from $M = 11$ for brown shrimp, $M = 2$ to $M = 6$ for white shrimp, and $M = 1$ to $M = 28$ for pink shrimp. The higher mortality rates are most likely in error. This deduction is evident from the expression derived by Holt (1965) for the average age (T) of individuals in a population:

$$\overline{T} = \frac{t_1 - t_2 e^{-Z(t_2-t_1)}}{1 - e^{-Z(t_2-t_1)}} + \frac{1}{Z}$$

where t_1 is recruitment age and t_2 is cohort age. If a typical 'high' value of Z, say $Z = 10$, is substituted in the above expression, the increase in the average age of the population at each month past recruitment is ·04, ·06, ·08, ·09, ·10, ·10 years, respectively. In other words, if $z = 10·0$, the average age in the population converges to only $t_1 + 0·1$ years ($t_1 + 1·2$ months) or about 5 weeks beyond recruitment; hence, such a population would be composed of quite small shrimp indeed. Therefore, most of the mortality rates which have been estimated appear to be too high. One conclusion is that shrimp tagged in the Gulf of Mexico have higher mortality rates than untagged shrimp.

It has been difficult to partition mortality into

Table 3

ESTIMATES OF MORTALITY RATES FOR GULF OF MEXICO SHRIMP. THE ORIGINAL RATES
WERE CONVERTED INTO INSTANTANEOUS ANNUAL RATES. M = NATURAL MORTALITY;
F = FISHING MORTALITY; $Z = M + F$

Species	Z	F	M	Source
Brown	14·04	3·12	10·92	Klima, 1964
		1·04 – 16·38		Neal, 1968
White	23·92			Klima, 1964
	7·28 – 14·04	3·12 – 9·88	4·16	Klima and Benigo, 1965
	8·52 – 11·75	5·41 – 6·81	2·13 – 6·29	Klima, 1974
Pink	39·52	10·92	28·60	Kutkuhn, 1966
	11·44 – 14·04	8·32 – 11·80	1·25 – 3·17	Berry, 1967
	5·72 – 9·36	1·56 – 3·64	4·16 – 5·72	Costello and Allen, 1968
	5·72	5·68	1·04	Berry, 1970
	13·00		3·90 – 6·50	Lindner, 1966

immediate and longer-term losses although it should be noted that mortality rates of tagged shrimp may be very dependent upon the size of the shrimp tagged. This may be observed, for example, in Lindner and Anderson (1956), whose results indicated that white shrimp tagged at less than 14 cm in length were less likely to be recovered than shrimp tagged at larger sizes.

The above observations suggest that tagging studies on shrimp for the purpose of investigating mortality or growth rates should include careful attention to tagging procedures, and that analyses should examine the effect of shrimp size on survival and take this into consideration concerning whatever parameters are being estimated.

Recently, the following approaches have provided estimates of mortality rates which seem more reasonable.

(1) Utilizing data from a tagging experiment on brown shrimp conducted off of Freeport, Texas in November 1969, Parrack (1981) correlated recaptures over a 14-month period with unstandardized brown shrimp effort and obtained average estimates of $M = 2·76$ and $F = 0·36$. The catchability was $·1642 \times 10^{-4}$. This analysis suggests an exploitation rate of ten percent.

(2) As is well known, catch-at-age data may be used to estimate total mortality rates which, when regressed on fishing effort, provide estimates of average natural mortality and fishing mortality rates and the catchability coefficient. Using this procedure on estimated average catch-at-age of brown shrimp. Parrack (1981) estimated $M = 1·8$ to 2·5 at catchabilities $·1661$ to $·1763 \times 10^{-4}$, respectively. Analogously, declines in catch-per-unit effort of separate cohorts from one period to the next can be used to estimate total mortalities employing catch curve analysis. Using this approach, Parrack (1981) estimated $M = 1·8$ with

a catchability coefficient of $·2819 \times 10^{-4}$ for brown shrimp.

(3) A third possibility is to analyze the decline in numbers caught per unit effort over time in cases where significant recruitment to the fishery does not occur after a definable point in time, or to employ DeLury-type methods. We used the former approach to estimate brown and white shrimp mortality rates pertaining to the offshore fishery. For each year from 1965 to 1980, the log of the catch in numbers per unit of monthly standardized regional effort, $CPUE_n$ (v Brunenmeister, 1983, for effort computation) was regressed on month to estimate Z for each 'biological' year.

For brown shrimp, a biological year was defined from May through April since these months demarcate best the observed monthly cycles in the average size of shrimp landed and in $CPUE_n$ (Figure 7). Average size of landed brown shrimp was highest from January to April and lowest from May to July. The pattern in $CPUE_n$ indicates that peak catch rates occurred during June or July; thereafter, $CPUE_n$ declined through the months until the following May, when the cycle began again. As noted above, recruitment of brown shrimp to both inshore and offshore fisheries was usually confined to a period of 2–3 months with a peak in May or, more often, in June or July. Thus, peak months in brown shrimp recruitment to the offshore fishery corresponded to the peak months of observed $CPUE_n$ in most years. Regressions of log $CPUE_n$ on month for each biological year were computed from the peak month in $CPUE_n$ through December. December was chosen as the final month since the period from May through December accounted for ca. 90% of the total effort expended on brown shrimp during the biological year and change in slope was often marked in the curves after December.

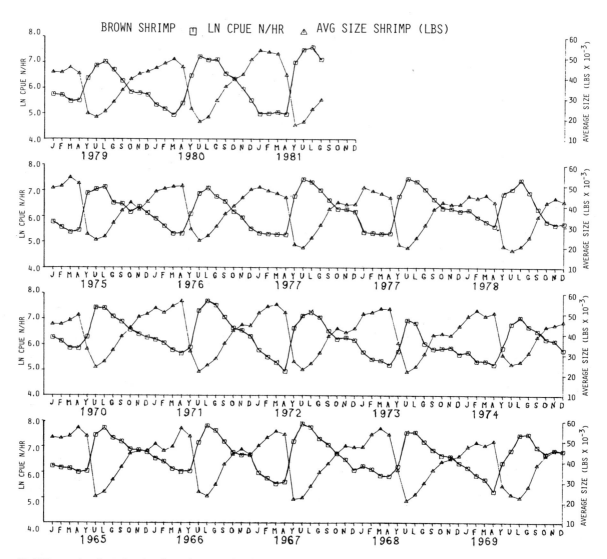

Fig 7 Time series of relative abundance (expressed as the natural log of the catch in number per unit of annual standardized regional effort) and average weight (in pounds, heads-off) of brown shrimp in landings

For white shrimp, the biological year was represented by the period from September to August of the subsequent year since cycles in $CPUE_n$ and effort fell within these bounds (*Figure 8*). The average size of white shrimp landed per month was also strongly seasonal. The largest average sizes were landed during July; a second peak in size of smaller shrimp occurred in October. $CPUE_n$ varied inversely with the average size of shrimp landed. $CPUE_n$ peaked during the winter, between November and January, and declined through July or August. As with brown shrimp, peaks in the landings of small shrimp correspond well with those months in which highest $CPUE_n$ was observed. Estimates of monthly Z for white shrimp were obtained by regressing the log of $CPUE_n$ on

month over the period from the peak in $CPUE_n$ to its nadir during a biological year.

For brown shrimp, annualized estimates of monthly Z ranged from 1·716 to 4·380 corresponding to survival rates of 18·0% and 1·3%, respectively. Correlations ranged from −0·757 to −0·995; hence, declines in $CPUE_n$ with month were relatively well defined. For white shrimp, the annualized Z estimate obtained for each year ranged from 2·580 to 4·968, corresponding to survivorship rates from 7·6% to 0·7%. Correlations ranged from −0·816 to −0·990 indicating monthly declines in $CPUE_n$ of white shrimp were relatively constant.

The Z's for each year pertaining to each species were then regressed on the annual standardized

156

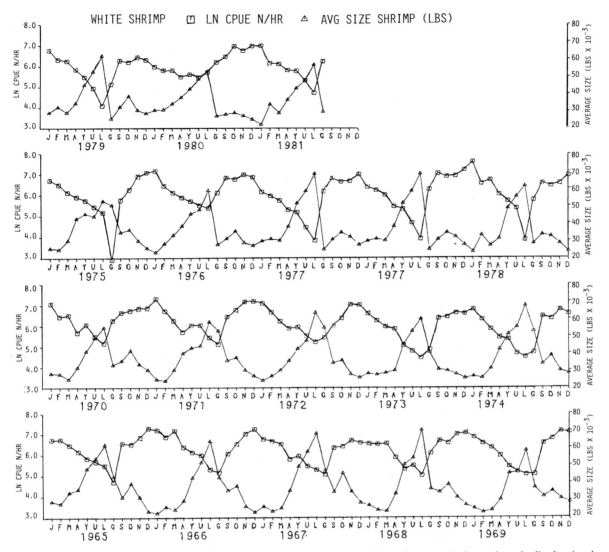

Fig 8 Time series of relative abundance (expressed as the natural log of the catch in number per unit of annual standardized regional effort) and average weight (in pounds, heads-off) of white shrimp in landings

regional fishing effort estimated for brown and white shrimp, respectively in a biological year (*v* Brunenmeister, 1983) to obtain estimates of average Z and M. This approach was not suitable for pink shrimp due to the relatively extended recruitment pattern of this species compared to brown and white shrimp and the lack of consistent definable cycles in $CPUE_n$ (*Figure 9*).

For brown shrimp, M was estimated to be 1·7; F in 1965 was ·7, and F in 1980 was 1·7. The estimated exploitation rates thus increased over the years from somewhat less than 30% to about 50%. The catchability coefficient, q, was estimated at 0·00091 in thousands of (standardized) hours fished. While a linear relationship between effort and Z was apparent (*Figure 10a*), the amount of scatter

$(r_s = 0·511, p < 0·05)$ suggests that natural mortality rates or q may not have been constant from year to year. Estimates of Z obtained from the regression indicate annual survival rates ranged between 2·9% and 9·0%, with an average of 5·4% during the period studied. These survival rates are much higher than those previously reported (*v Table 3*); however, they are similar to those estimated by Parrack (1981).

For white shrimp, M was estimated to be 2·6, with a lowest F in 1967 of ·6, and a highest F in 1977 of 1·9. The exploitation rates thus ranged from about 20% to around 40%. The catchability coefficient, q, was estimated at 0·00073 in terms of thousands of (standardized) hours fished. As in brown shrimp, a relatively large amount of scatter

157

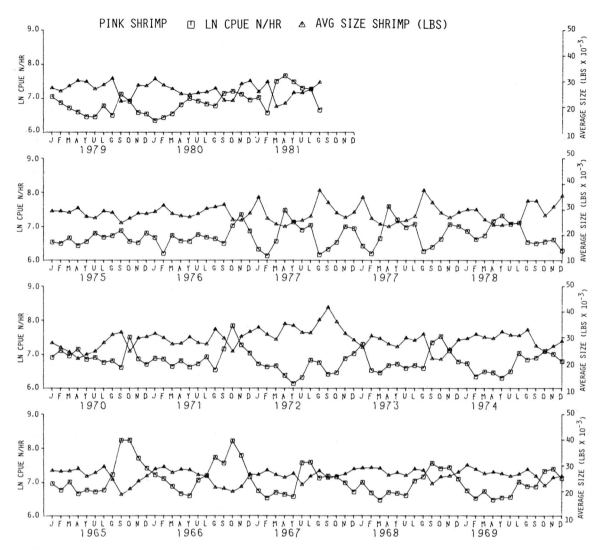

Fig 9 Time series of relative abundance (expressed as the natural log of the catch in number per unit of annual standardized regional effort) and average weight (in pounds, heads-off) of pink shrimp in landings

in Z was observed with effort levels ($r_s = 0.442$; $0.10 < p < 0.20$; *Figure 10b*). Estimates of Z obtained from the regression equation ranged from 3.141 to 4.505, indicating annual survival rates of 4.32% and 1.11%, respectively. These total mortality rates are significantly lower than any previously estimated for white shrimp in offshore waters. Klima (1964) estimated a biweekly survivorship rate of 40% from recoveries of stained white shrimp over a six-week period (September – November, 1962) that were released off the Louisiana coast in Statistical Areas 16 and 17. His estimate was derived from the rate of decline in recoveries adjusted by biweekly (unstandardized) effort levels and yields an extremely high annualized total instantaneous mortality rate of

$Z = 23.9$. Other estimates of mortality rates published for offshore waters were preliminary calculations (Klima and Benigno, 1965) or stated by the author to be unreliable (Klima, 1974).

There are a number of sources of possible error in using this approach for estimating natural mortality. The sources of these errors relate to phenomena that either increase or decrease the apparent number of shrimp early or later in the season. For example, if recruitment to the 'fishable stock' was not accounted for, then mortality would be underestimated. Another source involves changes in the catchability coefficient which might be induced by changes in fishing practices. Catchability might, for example, increase when shrimp reach larger sizes and/or are more valuable, and

158

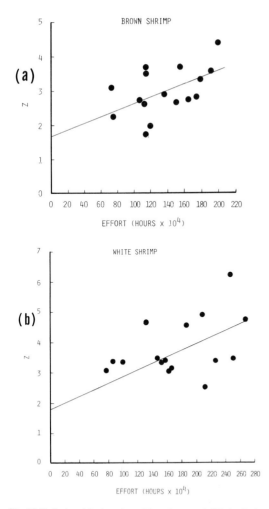

Fig 10 Relationship between Z and annual (biological year) standardized regional fishing effort for a) brown and b) white shrimp. The intercept is an estimate of the natural mortality and the slope is an estimate of the catchability coefficient

possibly subject to more concentrated, more effective fishing. On the other hand, when shrimp are at larger sizes, they are generally at lower densities and thus may be more difficult to catch unless they concentrate in detectable patches.

Another important source of error, particularly with respect to the brown shrimp analysis, is the amount of discarding of small shrimp that occurred. As shown in *Table 4*, discard rates may be high at times and they vary among years. Peak discards occurring early in the season, would bias the mortality rate downward. In addition, the variability in discard rate would contribute to the variability in the estimate of Z, which induces variation in apparent natural mortality.

(4) A fourth possibility is to use cohort analysis. The above or similar estimates could be used for M and a terminal value of F could be used to initiate the iterative solution to the catch equation. It would, of course, be necessary to estimate the age structure of the catch, by comparing the size distribution with growth patterns in order to estimate size at age. Parrack (1981) used such an approach to estimate annual fishing mortality rates of brown shrimp. There are, however, biases in his estimates because it was assumed that the sex ratio was 1:1 in each size class below the maximum length of males, which is less than that of females.

Thus, several approaches have been used to estimate shrimp mortality and each have errors in estimates associated with them. Tagging data are difficult to interpret, the mortality rates seem too high, and there is some question as to whether tagged animals behave as the rest of the population. Average-age data depends on various assumptions including the time of recruitment and whether cohorts as coherent units within a year can

Table 4

ESTIMATES OF THE PERCENTAGE OF SHRIMP DISCARDED BY WEIGHT FROM BROWN AND PINK SHRIMP CATCHES IN DIFFERENT MONTHS

	JAN	FEB	MAR	AP	MAY	JUN	JUL	AUG	SEP	OCT	NOV	DEC
Brown Shrimp												
1964[1]						8	2	<1				
1965[2]						14	4	<1				
1972[3]					32	23–45						
1973[4]						43	16	26	6	27	0	0
1974[4]						34	20	7	2	1	1	0
1974[4]					93	79						
Pink Shrimp												
1951[3]								5		39		
1963[1]										27	46	51
1964[1]	42	28	24	23	38	26	2	7	16	24	22	25
1965[1]	16	6	6	6	33	12	23	11	1	1	5	2
1966[1]	4	2	1	1	1	6						

[1]From Berry and Benton, 1969
[2]From Baxter, 1973
[3]Calculated from data given by Siebenaler, 1952
[4]Bryan and Cody, 1975

be identified for shrimp The age-free CPUE analysis produced relatively low correlations between Z's and standardized effort. Finally, cohort analysis requires numerous assumptions to construct the age distribution of the catch in order to estimate F. It is interesting to note, however, that the various estimates of natural and fishing mortality made in the recent analyses are similar. Further, they are much lower than previous estimates and are closer to intuitively expected rates than those deduced previously.

Before leaving the general subject of mortality, it is worth considering temporal changes in fishing effort and fishing mortality because monitoring and understanding such changes are critically important to understanding the dynamics of the stocks. First of all, most indices of nominal fishing effort in the Gulf of Mexico show substantial increases in effort have occurred since the early 1960s. For example, Table 5 in the Management Plan shows that from 1960 to 1975, the number of shrimp vessels increased from 2 941 to 3 780; the gross tons per vessel increased 41·3 to 64·0, and the number of shrimp boats increased from 3 089 to 5 054.

Brunenmeister (1983) estimated the change in offshore standardized effort (which should be proportional to offshore fishing mortality) for three species of shrimp and Parrack (1981) estimated changes in fishing mortality from cohort analysis. Analysis of standardized effort showed the following:

(1) From 1965 to 1980 brown shrimp standardized effort roughly doubled. This increase was not distributed uniformly. Off Louisiana, in Areas 11 – 16, effort quadrupled, while off Texas in Areas 17 – 21, it has been relatively stable since 1968 (v Brunenmeister, 1983, Figure 11).

(2) White shrimp effort has on the average at least doubled from 1965 to 1980.

(3) Pink shrimp effort has increased about 30 percent from 1965 to 1980.

Thus, considering these time series, adjusted or standardized fishing effort has in some cases increased substantially, but in other cases has been relatively stable during recent years. Stability in standardized effort and increases in nominal effort implies a decrease in catchability.

Effect of fishing on the shrimp stocks

Production models Standardized adjusted fishing effort was used by Brunenmeister (1983) to calculate production models for brown, white, and pink shrimp. These analyses were based only on the offshore fishery. Hence some variability of the observed points from this may be a result of inshore fishing. In this regard, although the majority of white and brown shrimp are caught in offshore waters, inshore landings are appreciable. This might be taken into account in future analyses by treating the inshore catch as a factor related to recruitment to the offshore fishery. The inshore catch of pink shrimp, however, is relatively small and thus probably had little effect.

Most previous calculations of production models have been based upon a calendar year rather than a biological year. Brunenmeister's calculations on the biological year basis is preferable since biological-year calculations minimize the occurrence of more than a single year-class within any single year. The production model analyses all tend to reflect similar features, *viz*

(1) In a number of instances, the biological year and calendar year estimates gave similar results, but in other instances, such as those for brown and white shrimp, the model best fit to the data differed. In general, optimum effort and MSY estimates estimated from the biological year data were slightly greater than those estimated from the calendar year data.

(2) For most cases, however, the best fit to the data occurred when $m = 0$ or was close to zero. However, the values of R^2 which were computed for 'standard values' of m (ie $m = 1·001, 2·000$) were not greatly different from one another. This means that there is not enough information within the data on hand to readily distinguish among the range of values of m.

(3) Each stock has been exposed to a several-fold increase in fishing effort.

(4) The stocks, as indicated by CPUE, are decreasing in abundance with the progression of time. More significantly, as might be expected they are decreasing with increasing fishing effort and hence, clearly show that fishing effort is affecting stock abundance.

(5) Recent levels of fishing effort are quite high. They are at a point where increases in fishing effort will yield only relatively small increases in catch. In some instances, catches have actually declined in the presence of high levels of effort, a condition usually indicative of overfishing. The high level of fishing effort and the fact that stocks are at the transition point between a parabolic and non-parabolic production function suggests that careful monitoring of the status of the stocks is necessary.

(6) The annual nature of the shrimp populations and the time lags in reporting statistics (data necessary for estimating vessel standardized effort after 1977 was not available at the time that production

model analyses were prepared) suggest that events may have occurred in the last few years which could be critical to management. It is absolutely essential with an annual stock such as shrimp to maintain real-time statistics on the fishery.

Yield per recruit   Yield-per-recruit is a function of age-specific fishing mortality, natural mortality and growth. This section will first consider a standard Beverton and Holt model and then comment upon the situation where the value of the fish increases substantially with age.

For the Beverton and Holt model, we used the von Bertanlanffy growth function to describe growth in total length. Where appropriate, we averaged the male and female values of k and L or used parameters computed for combined sexes (v Table 2).

The specific functions used were,

$L = 169[1 - \exp(-3 \cdot 9t)]$ for male brown shrimp
$L = 194[1 - \exp(-3 \cdot 9t)]$ for female brown shrimp
$L = 216[1 - \exp(-3 \cdot 9t)]$ for white shrimp
$L = 183[1 - \exp(-2 \cdot 6t)]$ for pink shrimp

where (t) refers to month.

Parameters were adjusted to annual values, then rounded, and the M/k ratios in Beverton and Holt (1966) were used. Eumetric curves were derived from these tables and are plotted in *Figure 11* along with the approximate recruitment size ranges for the various species of shrimp. The eumetric curve gives the recruitment length or age that maximizes yield-per-recruit for any fixed F. The approximate length at recruitment in the present fishery (68–100 count) is shown as horizontal bands according to species. (Conversions to total length were calculated from equations in Brunenmeister, 1980, and Fontaine and Neal, 1971). We have also plotted as vertical lines the reasonable bounds of fishing mortality, $1 < F < 2$.

Figure 11 suggests that within the range of $1 < F < 2$, the eumetric curve is particularly sensitive to the magnitude of natural mortality. If natural mortality for both brown and white shrimp is about $m = 2$, then it would appear that the minimum size of capture should be increased. Pink shrimp appear to be fished eumetrically if $M = 2 \cdot 25$, but the size at capture could be decreased if $M = 3 \cdot 75$.

In sum then, these simple analyses suggest, given our present appraisal of the magnitude of shrimp mortality, that reductions in the size of shrimp at capture should be avoided. Further, it is likely that

at higher levels of fishing mortality, increased size at capture could improve the yield-per-recruit.

Because of the value of shrimp and the seeming sensitivity of shrimp to yield-per-recruit considerations, catches of small shrimp should be carefully monitored and the costs and benefits of catching small shrimp should be evaluated. In this regard, it is important to recognize that the small shrimp catch may be partitioned into two components, one about which fairly reasonable records are kept and one about which little is known.

According to the recorded catch (see Figures 3.2-6, 3.2-7, 3.2-8 in the Management Plan), there are significant quantities of shrimp caught at sizes smaller than 68 per pound. The basically unrecorded component, much of which can be small shrimp, are the catches of recreational fishermen and of commercial bait shrimp fishermen and the shrimp that are discarded by commercial fishermen. Estimates of recreational and bait shrimp catches are given in *Table 5a* and *5b*, respectively.

Table 5a
ESTIMATES OF ANNUAL SHRIMP CATCHES TAKEN BY THE RECREA-
TIONAL FISHERY*

| | Thousands of Pounds, Headless | | |
	TOTAL	BROWN	WHITE
Alabama	257 ± 46	125	37
Mississippi	175 ± 9	94	16
Louisiana	23 600	7 329	7 662
Texas	901 ± 118	431	134

Table 5b
ESTIMATES OF ANNUAL SHRIMP CATCHES BY THE BAIT FISHERY*

| | Thousands of pounds, Headless | | | |
	TOTAL	BROWN	WHITE	PINK
Florida	1 099	—	—	1 099
Alabama	36	28	8	—
Mississippi	27	23	4	—
Louisiana	971	475	496	—
Texas	1 458	1 119	349	—

*Federal Register 45 (218), p. 74284

The quantity of shrimp that is discarded depends on state laws governing minimum landing sizes; the value and availability of small shrimp; the type of size grading used by processors; the quantity of fish caught with the shrimp; and the season, depth, and geographical area fished.

Large discards of small unmarketable shrimp stimulated the closure of fishing areas and the institution of minimum landing sizes. Minimum size limits vary. In Texas, shrimp must number 39 or less per pound, heads-on (or 65 per pound

161

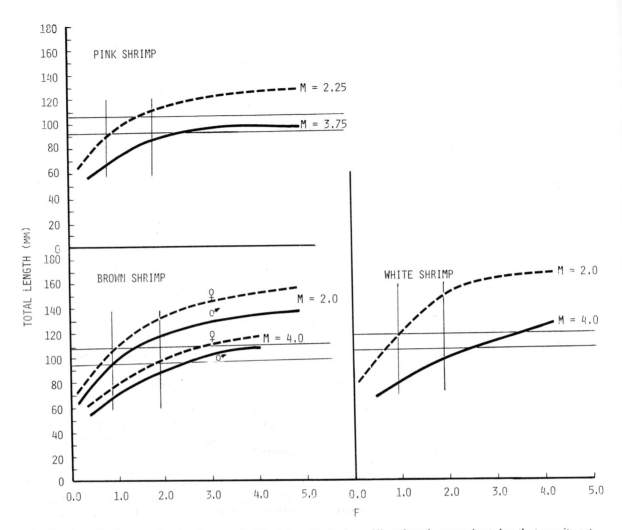

Fig 11 Eumetric fishing curves for pink, brown and white shrimp. The horizontal lines show the approximate length at recruitment. The vertical lines demarcate estimates of the ranges of likely fishing mortality rates

heads-off) except from August 15 to October 31, when a minimum count of 50 shrimp per pound, heads-on is required, or from November 1 to December 15, when no size limits are imposed. In Florida, the minimum statewide size is 47 shrimp per pound, heads-on (or 70 pound heads-off); in three counties, the size limit is 55 shrimp per pound, heads-on, inside bays and sounds. In Alabama and Mississippi, the size limit is 68 shrimp per pound, heads-on, excepting bait shrimp on which no size limit is imposed. Lastly, in Louisiana, a size limit of 68 shrimp per pound heads-on is applied except during the inshore spring season and on brown shrimp taken after November 20. In these latter cases and on shrimp caught in offshore water, no size limits are imposed. Thus, sizes of shrimp discarded by fishermen vary according to the state waters in which they are landed, since shrimp less than these minimum size restrictions cannot be landed legally.

The method used to separate or 'grade' landed shrimp into various market categories also affects discarding. Machine grading automatically sorts individual shrimp into size categories. Box-grading, however, is based on the average number of shrimp composing several five-pound samples taken from a landing. Since, in box grading, small quantities of small shrimp may be included with larger shrimp, discards at the dock tend to be somewhat reduced.

Berry and Benton (1969) found a negative relationship between discarding on the Tortugas pink-shrimp grounds and the ex-vessel price of small shrimp during the period from October 1963 to September 1966. Thus when small shrimp are plentiful, they are likely not to be marketable and are therefore discarded; if the abundance of small shrimp is low they may be retained. In this regard, general monthly declines in discard rates of brown shrimp caught off the Texas coast between June and August (*Table 4*) can be seen to be correlated with declines in catch in number per unit effort and increases in the average size of shrimp landed (*v Figure 7*).

Baxter (1973) noted that crews are unable to 'head' all the catch when catch rates are high and therefore they select only the larger valuable shrimp and discard the rest. However, Berry and Benton (1969) noted that if catches are large and contain few fish, the entire catch may be retained and culled later at the dock. With regard to these factors, Siebenaler (1952) noted that both average shrimp size and average number of fish caught per tow varied seasonally in the Tortugas pink-shrimp fishery. Thus, variations in catch composition and catch rates may be expected to differentially affect the temporal and geographical extents of discarding.

On the Texas coast, juvenile shrimp are protected by area – season closures in both offshore and inshore waters and by minimum landing size limits. In these waters, fishermen fish for large shrimp offshore and discard small ones. Also wholesale shrimp dealers in Texas are reluctant to buy small shrimp, and thus the probability of discards occurring is increased. In Louisiana, in contrast, smaller shrimp can be legally landed and more small shrimp are marketed there.

It is important to emphasize, that the simple yield-per-recruit model tends to develop *misleading* management advice. This is because the value of shrimp increases substantially as shrimp increase in size. Thus, a more appropriate 'growth' curve for *shrimp management* would reflect its increase in *value* rather than its increase in weight. If this 'growth' is rapid, it would be desirable to postpone the catching of small shrimp. It is therefore quite likely that the eumetric curves depicted in *Figure 11* suggest catching shrimp at too small sizes and hence, the minimum size of shrimp fished should be increased to a significant degree.

The extensions of the simple models depend on interpretation of the variability associated with the parameter estimates and any other special properties of shrimp populations. Several deserve mention. These include the variability of mortality estimates, growth patterns, and the timing of spawning. Nichols (1981b) discusses the effect of biases of mortality estimates and while they are probably no more biased than for other species of fish, they may be more critical for shrimp in the

Gulf of Mexico. This owes to the relatively high value of shrimp and to the fact that, since shrimp are distributed over a wide area, there may be local differences in the various mortality estimates. The variable growth pattern of white shrimp was considered by Nichols (1981a) and by us. The pattern suggests that white shrimp undergo an early period of normal growth, become dormant, then undergo rapidly accelerating growth. Owing to the relatively high mortality rate of white shrimp, the rapidly accelerating spring and summer growth might not have a particularly important yield-per-recruit effect, but it might have an important effect on the effective biomass of the spawning stock – an effect that is probably undetected in the analysis discussed herein.

Finally, it is clear that shrimp spawn over an extensive time period. Hence, a careful analysis of the effect of extended spawning on yield-per-recruit is required.

Stock and recruitment There has been a general perception that recruitment levels of shrimp in the Gulf of Mexico depend only on environmental factors and not on stock size. In order to examine this question, we have investigated the relationship between various indices of abundance in year (n) which are related to the biomass of spawners and indices of abundance in year ($n + 1$) which are related to the number of recruits.

Table 6 gives the catch-per-unit-of-effort in weight of brown shrimp in year n (an index of spawning biomass) and the catch-per-unit-of-effort in number for the month with the highest catch-per-unit-of-effort in number for year $n + 1$ (an index of recruitment). These data are plotted in *Figure 12a*. Similar data for white shrimp may also be found in *Table 6*. It should be noted that in some biological years white shrimp numerical abundance is temporally bimodal and in these instances both values are reported in the table. Only the Fall values have been plotted (*Figure 12b*).

An examination of *Figure 12a* shows that there is a good relationship between the index of stock size and the index of brown shrimp recruitment. An examination of *Figure 12b* shows that the relationship between recruitment and stock size for white shrimp is not as clear as that for brown shrimp. For white shrimp, the threshold level below which stock size must fall to obtain a sharp decline in recruitment is not known. In any event, the correlation between stock and recruitment for brown shrimp is better than that found for most fisheries and that for white shrimp is no more equivocal than that observed for many fisheries.

163

Table 6
STOCK AND RECRUITMENT RELATIONSHIPS. CATCH PER UNIT EFFORT IN WEIGHT REFERS TO THE
ANNUAL CATCH PER UNIT OF STANDARDIZED REGIONAL EFFORT FOR BIOLOGICAL YEARS; CATCH PER
UNIT EFFORT IN NUMBER REFERS TO THE MAXIMUM MONTHLY STANDARDIZED REGIONAL VALUE
OBSERVED WITHIN EACH BIOLOGICAL YEAR; MONTH OF PEAK CATCH REFERS TO THE MONTH IN WHICH
THE FOREGOING MAXIMUM WAS OBSERVED. FOR WHITE SHRIMP, IN 1965 AND IN 1974, THERE WERE
TWO WELL DEFINED PEAKS WHICH ARE INDICATED

	Biological Year (May – April)	Catch Per Unit Effort in Weight	Catch Per Unit Effort in Number	Month Of Peak Catch
Brown Shrimp	1965	58	2 242	July
	1966	59	2 437	July
	1967	65	2 658	June
	1968	46	1 889	July
	1969	39	1 731	August
	1970	48	1 664	June
	1971	49	2 135	July
	1972	37	1 361	July
	1973	24	1 015	June
	1974	30	1 130	July
	1975	33	1 237	July
	1976	32	1 182	July
	1977		1 667	June
	(Sept. – Aug.)			
White Shrimp	1965	20	1 408, 1 236	Dec., March
	1966	19	1 327	Dec.
	1967	18	765	Nov.
	1968	12	1 092	Dec.
	1969	14	1 188	Nov.
	1970	18	1 438	Jan.
	1971	18	1 273	Dec.
	1972	12	1 078	Nov.
	1973	10	849	Nov.
	1974	10	838, 819	Nov., Jan.
	1975	15	1 258	Jan.
	1976	12	1 053	Nov.

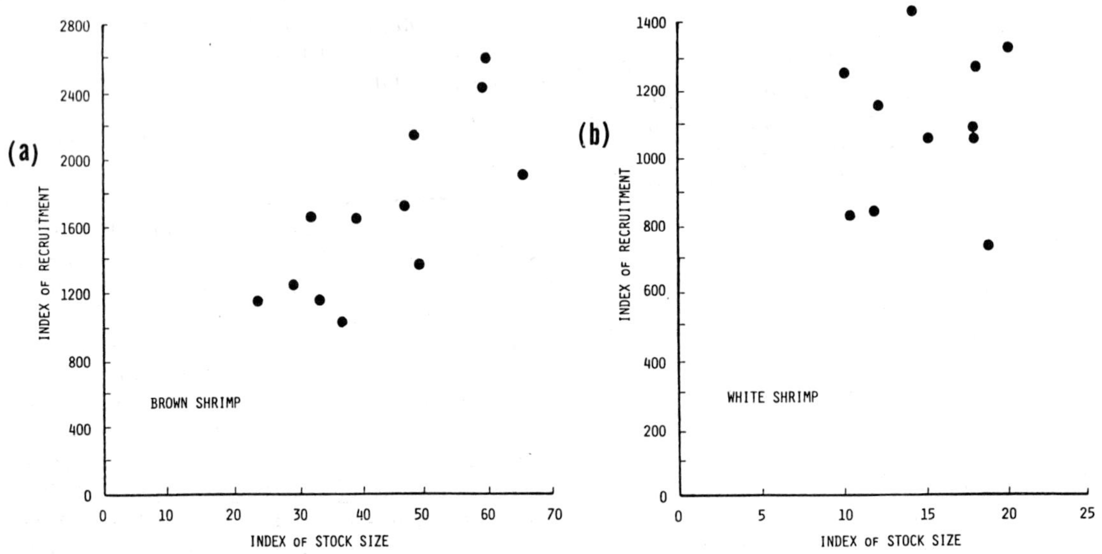

Fig 12 Relationship between index of recruitment in year (N + 1) and index of stock size in year (N) for (a) brown and (b) white shrimp from values presented in *Table 6*

164

3 Other species

Shrimp fishery by-catch

Estimates of catches of fish and other species by shrimp trawlers and the proportion of these which are landed or discarded are not well known, but available information indicates that these aspects vary according to season and location. Siebenaler (1952) studied catches of shrimp trawlers operating in the Tortugas and estimated an average catch ratio of 'trash' (fish, miscellaneous crustacea and shrimp heads) to headed shrimp at 2:1. Vincent (1950) noted on the then newly exploited south Florida grounds that fish often constituted 90% of catches.

Bryan (1980) analyzed the catches taken along the Texas coast by a research shrimp trawler which fished along side commercial shrimp trawlers and used the same types of gear. He found that fish and invertebrates other than shrimp constituted 81% (ca. 4:1) and 61% (3:2) by weight of the sample catches taken during 1973 and 1974, respectively. Compton (1961; 1962) estimated fish to shrimp catch weight ratios of 6:1 during 1960 and 7:1 during 1961 from samples taken off the central coast of Texas in Statistical Area No. 20 (see *Figure 2*) in depths of 3–20 fathoms. Blomo and Nichols (1974) estimated fish to shrimp catch ratios ranging by weight from 1:1 to 7:1, and an annual average of 4:1 in the western Gulf of Mexico.

Much higher weight ratios of fish to shrimp catches (14:1) have been cited for the Mississippi delta area, where fishery is directed to groundfish (Sheridan, Powers and Browder, 1983). Moore, Brusher and Trent (1970) found average catch weights of fish trawled during 1962 and 1963 at stations off Louisiana were significantly greater than those trawled at stations off Texas and that these differences were greatest among stations at depths less than 64 m. Hence, a wide range of bycatch estimates exists and these estimates vary according to depth, area and season in which samples were fished.

Species and sizes of fish in by-catch

The species of fish and other invertebrates that constitute bycatch vary among the shrimp fisheries. With respect to the pink shrimp fishery, Siebenaler (1952) recorded 32 species of fish and 20 species of invertebrates in the bycatch of shrimp trawlers operating on the Tortugas grounds during 1951 and 1952. He noted the dominant bycatch species were mojarras (*Eucinostomus argenteus*), swimming crabs (*Cronius ruber*), bronze grunts (*Bathystoma striatum*), rockfish (*Diplectrum formosum*) and cuban snappers (*Lutjanus* sp.).

On offshore brown shrimp grounds, Bryan (1980) counted 99 species of fish and 13 species of invertebrates in the by-catch of sample shrimp trawls taken off Texas in 1973 and 1974. The most dominant species by weight were swimming crabs (*Callinectes sapidus*), Atlantic croaker (*Micropogon undulatus*), the stromateid (*Peprilis burti*), shoal flounder (*Syacium gunteri*), stomatopods (*Squilla empusa*) and sugar shrimp (*Trachypeneaus* sp.). Hildebrand (1954) noted from his examinations of catches of large commercial shrimp trawlers that, besides species of *Penaeus*, the Gulf crab *Callinectes danae* and *Squilla empusa* were the most abundant crustaceans captured by the offshore shrimp fishery. Moore *et al* (1970), found that the longspine porgy (*Stenotomus caprinus*), Atlantic croaker, inshore lizard fish (*Synodus foetens*) and silver seatrout (*Cynoscion nothus*) were the most abundant fish in their samples which were taken off the Texas coast during 1962–1964 but they noted significant seasonal and depth differences in the dominance of different fish species in this area. Compton (1961) found that the five most abundant fish by number in shrimp trawl catches off the central Texas coast at depths from 3–20 fathoms during 1961 were the bumper (*Chloroscombrus chrysurus*), Atlantic croaker, seatrouts (*Cynoscian nothus* and *C. lanceolatus*) and menhaden (*Brevoortia patronus*) and he also noted seasonal changes in abundance of species and bycatch magnitudes. Off Louisiana, Moore *et al* (1970) observed that the most abundant species in catches were the Atlantic croaker, longspine porgy, sand seatrout (*Cynoscion arenarius*) and sea catfish (*Galeichthys felis*). And they also noted seasonal and depth changes in species dominance.

Altogether, Moore *et al* (1970) recorded 18 species of fish off Louisiana and 19 fish species of fish off Texas which constituted 1% or more of their shrimp trawl samples by weight. Bryan (1980) also recorded 19 species from shrimp trawl samples off Texas which contributed to 1% or more catches by weight; however, only 11 species were recorded in common by these authors. These differences may represent sustained changes which occurred in the composition of the fish community between the early 1960's and 1970's, or may simply represent inter-year variations or differeneces in areas sampled. Compton (1966) noted that annual changes between 1960 and 1965 in apparent abundances of common fishes in his bycatch samples were not great and that marked changes in catches of some species from year to year were generally due to unusual samples which might account for all of the marked increase. Finally, in the inshore bait shrimp fishery, Holloway (1981) noted that Atlantic croaker, pinfish (*Lagodon rhomboides*) and Gulf menhaden (*Brevoortia patronus*) comprise most of the incidental finfish catch.

Data pertaining to the sizes of fish in the by-catch of shrimp trawlers is not extensive. Roithmayr (1965) noted that fish averaging less than one-half pound (·23kg) were common and that these were usually discarded. Compton (1966) showed that the mid-point in the length range of croaker caught in 3–10 fathoms in shrimp trawls off Port Aransas, Texas, increased from approximately 4·5 cm in February to around 22·3 cm in October. Monthly size ranges of the seatrouts *Cynoscion nothus* and *C. arenarius*, however, showed no interpretable trends, although larger sizes of both species were caught in 11–20 fathoms than in 3–10 fathoms.

An indication of the species in the by-catch of shrimp trawls which are commercially valuable to shrimp fishermen is available from landings- by-gear tables in U S Fishery Statistics. These data for 1975 show these numbers of species or species groups were landed from shrimp otter trawl catches: 20, on the Florida west coast, 22 in Alabama, 16 in Mississippi, 23 in Louisiana, and 14 in Texas. While the species composition of these landings varied somewhat from state to state, Atlantic croaker, flounder (*Paralichthys* spp.), king whiting (*Menticirrhus* spp.), seatrout (*Cynoscion arenarius*, *C. nothus*) and blue crabs (*Callinectes sapidus*) were the most abundant by weight. Of these, landings from shrimp trawls amounted to a substantial proportion of the total recorded landings of flounder, king whiting, and seatrouts.

These observations indicate that the bycatch of shrimp trawlers is very diverse and varies among regions, habitats, depths and seasons. Because of this, it may be concluded that little is known about the composition of the bycatch of the U S shrimp fishery in the Gulf of Mexico and that systematic sampling will be required to improve our knowledge of this aspect of the fishery.

Other fisheries on the same species Both commercial and recreational fisheries harvest species that are taken in the bycatch of shrimp trawlers. Gutherz, Russell, Serra and Rohr (1975) indicated that the Atlantic croaker, spot, sand seatrout and silver trout contributed the major portion of the catches of the industrial bottomfish trawl fishery and croaker foodfish trawl fishery which operate between northern Florida and eastern Texas. With respect to the croaker foodfish industry, Gutherz

et al (1975) noted that approximately 75% of this catch is landed by trawlers who fish primarily for croaker, whereas the remainder is largely landed by shrimp fishermen who occasionally trawl for croaker. However, croakers are also landed by the handline fishery which operates near offshore oil platforms. These croaker average 2 lbs. in weight (18 in. length) and are larger than the trawl-caught croaker. Gutherz *et al* (1975) indicated that the sizes of croaker acceptable to the industrial and foodfish fisheries differed, with the foodfish fishery landing only croaker larger than $9\frac{1}{2}$ inches whereas those landed by the industrial fishery must be 8 inches or smaller so that they can be processed. They noted that most croaker landed by the industrial fishery weight less than one-third of a pound (·14 kg). Hence, discards of croaker by the shrimp fishery may differentially affect these commercial finfish fisheries.

Species other than croaker are utilized by the industrial fishery, except for sharks, skates, rays and small catfish (Gutherz *et al*, 1975). Roithmayer (1965) recorded 177 species which were landed by the industrial groundfish fishery operating off Louisiana, Mississippi and Alabama during the years 1959 to 1963. He noted Atlantic croaker constituted about 56% of the annual production, which along with spot, sand seatrout, silver seatrout and seasonal catches of the Atlantic cutlass fish (*Trichiurus leptorus*) and longspine porgy made up 82% of the catches.

An estimate of other fisheries which harvest species caught as bycatch by the shrimp fishery is available from landing-by-gear statistics available for the Gulf States from U S Fishery Statistics. In 1975, for example, sizeable landings of croaker and flounder in Florida were made from trammel nets and runaround gill nets as well as from fish otter trawls. In Alabama, nearly all the landings of Spanish mackerel (*Scomberomorus maculatus*) were recorded from shrimp trawls and runaround gill nets (34% and 63% of the catch, respectively) which was also generally the case in Louisiana. In Mississippi, 82% of the landings of red drum (*Sciaenops ocellata*) were accounted for by these same gears, of which 25% was attributed to shrimp trawls. In Louisiana, 98% of the sheepshead (*Archosargus probatocephalus*) sold commercially was landed by shrimp trawls, trammel nets and set nets (36%, 20% and 42%, respectively). In Texas, haul seines contributed 20% of the classified commercial landings of king whiting of which the remainder was landed from shrimp trawls. While these statistics are not exact they do show that the shrimp fishery by-catch interacts with other fisheries other than through discards. These effects may be large in the marketplace with respect to some species, like flounders or king whiting of which nearly half to all of the landings classified for the Gulf States in 1975 were attributed to shrimp trawls.

The effect of the shrimp fishery by-catch on recreational finfish fisheries is difficult to assess, where more concern has been to determine the relationship of the commercial finfish fishery to the recreational finfish fishery, especially in inshore areas (*v eg*, Juneau and Pollard, 1981). However, the effect of the shrimp fishery bycatch in inducing increased mortality of the young of many fish species of commercial or sport fishing interest or of the prey of these target species is a consideration. For example, Compton (1966) noted that quantities of juvenile red snapper, *Lutijanus blackfordi*, a popular sport fish, appeared in trawl catches taken in inshore areas in the fall off the central Texas coast. Lower and upper limits of size ranges of fish species numbering 50 individuals or more in the bycatch samples taken off the Texas coast recorded by Compton (1966) averaged 6·8 cm and 19·3 cm. Hence, quite young individuals were taken of several fish species.

Direct effects of the shrimp fishery on the fish stocks and their fisheries The primary concern of the effect of the shrimp fishery on the fish stocks and their fisheries has been the impact of the discards from the shrimp fishing bycatch. The potential magnitude of the effects of these discards can be appreciated by comparing them to the landings of the commercial fisheries. Landings of the industrial bottomfish fishery rapidly increased from 2 721 MT in 1952 to about 34 019 MT in 1958 and thereafter slowly increased to a catch level of 40–50 000 MT during the period from 1962 to 1972 (Gutherz *et al*, 1975). Landings of croaker foodfish increased from approximately 136 MT in 1967 to around 4 536 MT in 1971 (Gutherz *et al*, 1975). Bryan (1980) estimated that 45 800 MT of invertebrates other than shrimp and approximately 87 400 MT of fish were discarded by the shrimp fleet fishing on the Texas brown shrimp grounds during 1973 and 1974. Hence in the early 1970's, the estimated shrimp fleet fish discards along the Texas coast were almost double the industrial bottomfish fishery landings in the Gulf of Mexico. Also, in 1973, the estimated discards of invertebrates other than shrimp off Texas was approximately twice the tonnage of white and brown shrimp landed from offshore waters during this year.

Biological interactions
Types of biological interactions (*eg* predation; competition) between shrimp and other species were reviewed by Sheridan *et al* (1983). They indicated little study has been directed to possible competitive interactions between shrimp and fish, and concluded that the extent of predation on penaeid shrimp by fish is not well known. They cited recent quantitative studies of trawl-caught fish in offshore waters off the Gulf of Mexico in which *Penaeus* shrimp were found in less than 1% of the examined fish stomachs. They also noted that, in other work, the genus *Penaeus* was not detected in the diets of 26 abundant offshore fishes although *Penaeus* species inhabited the same waters. They did note, however, that other genera belonging to the Penaeidae have been identified as prey of fish and that studies categorizing 'penaeids' as prey have identified 42 fish species as predators.

Effects of catch and discards of fish on growth and mortality of shrimp Effects of the discards of fish on shrimp stocks were explored by Sheridan, *et al* (1983) using two modeling approaches. Results from one model suggested that the elimination of bottomfish discards would reduce shrimp production by only a small amount. Results from the other model, an energy-flow ecosystem model, estimated that reduction of the by-catch discards by 50% would reduce shrimp stocks by 25%. Reduction of by-catch magnitudes by the use of trawls 50% less efficient in catching fish than those currently in use was estimated to reduce shrimp stocks by 8%.

Shrimp caught as by-catch in non-shrimp fisheries Gutherz *et al* (1975) indicated that the foodfish croaker fishery, which operates balloon net and fish net trawls equipped with rollers, at times lands brown, white and pink shrimp of significant dollar value to the fishermen.

4 Environmental aspects

Salinity and temperature are important environmental influences on the survival and growth of shrimp. Ford and St. Amant (1971) examined factors affecting the production of brown shrimp in Louisiana waters and found that post-larval abundances were not consistently correlated with subsequent catch magnitudes. They concluded that early warming, without subsequent severe coldfront induced drops in temperature, along with salinities of 10‰ or higher, enhanced landings. However, they suggested that the importance of salinity to survival was reduced with increased spring warming over 20°C.

A positive correlation has been noted between white shrimp landings in Texas and rainfall in the same year and two previous years (Gunter and Edwards, 1969). Barrett and Gillespie (1973) observed an inverse relationship between landings of white shrimp in Louisiana and the discharges of the Mississippi and Atchafalaya Rivers. The discrepancy between these trends can be accounted for by the difference in amounts of fresh water drainage in the two areas of study and suggests that optimum conditions for white shrimp production will vary geographically according to local conditions.

Variation in shrimp growth with water temperature has also been documented. Phares (1980) improved the predictability of a model of white shrimp growth by incorporating temperature. Differences in growth rates between pink shrimp recruited in early spring and fall were observed by Berry (1967). However, Iversen and Jones (1961) reported no differences in growth rates of shrimp between the winter season (December–March) and spring, summer and fall (April–November). Nichols (1981a) found that temperature differentially affected growth rates of white shrimp according to their size.

Annual variation in temperature also affects shrimp movements. The onset of fall and the cooling of shallow bay waters elicits offshore migration of shrimp. However, infrequent severe cold fronts have decimated the shallow-water fauna along the Gulf coast. Gunter (1956) noted that a cold wave in 1940 killed practically all the shrimp on the South Atlantic coast. He noted that while the spring fishery failed, fall abundance was normal. Thus, shrimp populations may recover rapidly from natural environmental variations. While the life history patterns of these species show accommodations to environmental vagaries, they are of limited potential in compensating for adverse conditions created by man.

Human activities other than fishing

Human activities in inshore and offshore habitats of shrimp may affect recruitment and survival of stocks. Shallow water dredging for sand, gravel and oyster shell not only alters the bottom directly, but may change local current patterns leading to the erosion or siltation of productive habitats. Destruction of wetlands by development of waterfront properties results directly in loss of productive habitat acreage and in the reduction of detrital production.

Channeling or obstruction of water courses emptying into estuaries can result in loss of wetland acreage and/or changes in the salinity profile of the estuary. Lowered flow rates of drainage systems can reduce the amount of nutrients that are washed into estuaries and permanently alter the composition of shoreline communities. For example, Heald and Odum (1970) noted that accumulation of leaves in red mangrove forests could result in ecological succession to a willow and buttonwood community. Red mangrove forests are a major source of detritus in the pink shrimp food web of south Florida estuaries.

Nuclear power plants produce large quantities of heated effluent so that thermal pollution is now a consideration. As most marine animals exist just a few degrees below their critical thermal maximum during the summer months, large quantities of hot effluent could severely impoverish estuarine systems. Roessler and Zieman (1970), for example, found that the area in which all plants and animals were killed or greatly reduced in number that was adjacent to a nuclear plant outflow in Biscayne Bay, Florida, corresponded closely by the area delineated to the +4°C isotherm.

Chemical pollution of estuaries by pesticides and herbicides may have chronic as well as acute effects on estuarine organisms. Chlorinated hydrocarbons like DDT, although relatively insoluble, are adsorbed on particulate matter and hence enter the food chain. Duke (1970) cites studies showing 0·0002 parts per million of parathion and 0·0006 ppm of DDT, respectively, caused 50% mortality or loss of equilibrium in survivors of juvenile penaeid shrimp. Although not all organisms are affected to the same degree by any pollutant, the number of pollutants entering a system could affect enough of its components to significantly alter its state. Chronic effects are hardest to quantify and may lead to slow impoverishment of systems. For example, pink shrimp exposed to sublethal amounts of DDT showed a gradual decrease in blood protein (Duke, 1970).

Man's activities in the offshore environment include dumping of toxic wastes and development of mineral resources. While the eventual development of technologies to significantly exploit hard mineral resources on and below the seabed may be a decade or more in the future, offshore extraction of gas, oil and sulfur is booming. The effects of the presence of offshore oil platforms in the Gulf (exploration and extraction processes) include physical interference to fishing, and biological damages.

5 Management

Managing northern Gulf of Mexico shrimp stocks is a most difficult task. There are substantial complexities in resolving allocations among user groups, improving the biological and economic efficiency with which the stocks are utilized, improving the utility and timeliness of fishery data, and increasing the capabilities of management organizations and institutions. While many of the difficult issues appear to be political or institutional, each is typically rooted in biological and economic, technical and scientific considerations. Technical or scientific considerations may generally be partitioned into those that relate to concepts and into those that relate to application.

The conceptual basis for shrimp management in the Gulf of Mexico tends to diverge somewhat from that generally underlying management of longer-lived organisms and it relates primarily to three general areas in fisheries population dynamics theory. These are (1) the stock-recruitment relationship, (2) the interaction of environment and fishing effort in generating strong and weak year classes and (3) production models. We will treat each of these in turn.

The stock–recruitment relationship

A demonstrable stock-recruitment relationship for shrimp has not emerged in previous analyses, and this had led to the conclusion that with present fishing technology, recruitment overfishing of Gulf of Mexico shrimp stocks will not occur. Therefore, the Management Plan states that the maximum allowable yield in any year is 'all the shrimp that can be caught'.

A lack of stock-recruitment relationship implies that recruitment fluctuates about an average recruitment value. Thus, small, intermediate and large populations of shrimp all produce similar recruitment levels. A model consonant with the perceived no stock-recruitment relationship (*Figure 13*) shows that recruitment is constant over all observed stock sizes above the vertical 'threshold line'. Economic or technological factors would prevent the exertion of enough effort to reduce the stock below the threshold. If such a relationship holds then it is indeed impossible to (recruitment) overfish shrimp and the maximum allowable yield in any year is 'all the shrimp that can be caught'.

The critical issue, then, relates to the location of the threshold on the stock-size axis (T) and whether, in fact, recruitment varies for stock sizes greater than T. However, the notion of a fixed threshold needs to be approached with considera-

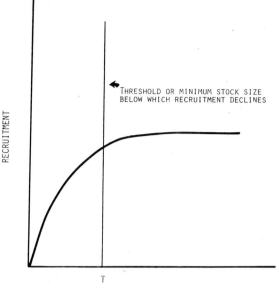

Fig 13 Hypothetical relationship between recruitment and stock showing the 'threshold' below which recruitment declines

ble caution. If, for example, a fleet is operating at the threshold level in year (n) and in year ($n + 1$) the stock declines sharply, then it will be unlikely that the fleet will be able to generate a concomitant short-term decline in nominal effort and thus, the fishing mortality on the stock will be sharply increased.

The estimation of the position of T is beyond the scope of this paper. (See Poffenberger (1983) for a review of economic questions relevant to shrimp management). However, we can see from *Figure 12* that there is a correlation between recruitment and stock for brown shrimp and a weaker relation for white shrimp. If there is a relationship for brown shrimp, then it appears that the stocks are below T. For the white shrimp either there is no relation over the range of stock sizes or the relationship is masked. Nevertheless, the practical management implications are clear that any increase in effort, particularly in brown shrimp, should be viewed with caution.

There are a number of penaeid shrimp populations around the world that have collapsed (*eg* 'The Gulfs' and India). There is no doubt that as populations decline, the risk of recruitment failure increases. The fact that shrimp are an annual stock simply intensifies the possibility of a calamitous stock collapse.

Lastly, it should be pointed out that knowledge of the stock-recruitment relationship may be useful for the development of fishing strategies. For example, if it appeared that the good correlation between recruitment and stock size for brown shrimp would maintain itself over the years, then the relationship could be used to develop an optimal multi-year fishing program.

The interrelationship of fishing effort and environment

The Management Plan asserts that estuarine conditions of temperature and salinity (or river discharge magnitudes) have a strong effect on production as indexed by landings. The Plan says, 'For example, linear regressions of catch on effort showed that effort alone explained only 38 percent of the variation in catch of Louisiana white shrimp and 57 percent of the variation of the Gulf brown shrimp catch. Multiple regressions including environmental parameters explained 89 percent and 88 percent, respectively'. (p. 74274).

The implication of the above observations is that fishing effort has very little effect on stock size, with a considerable proportion of the total variation in stock size being controlled by environmental variables. However, an alternative view is possible. This view derives from the previously discussed production model analyses. In these analyses, fishing effort accounted for 60–70 percent of the variability in catch. Hence, environmental effects would be much less important than indicated by the multiple regressions incorporating environmental variables. These above postulated relationships could be easily tested by comparing predictions from the regression equations with rainfall, temperature, catch and effort data which have been obtained subsequent to the completion of the regression analyses.

We should emphasize that we are not minimizing the effect of the environment. Environmental variables obviously play an important role. However, it seems fairly certain that the critical environmental variables are much more complex than temperature and precipitation alone, and that the effects of fishing need to be taken into account when analyzing variations in stock levels.

The usefulness of production models

The Management Plan expresses doubts about the usefulness of production models for providing management advice on shrimp since penaeid shrimp do not meet the criteria for application. The Management Plan says that surplus production models 'were designed for, and are usually applied to, species with multiple year classes' and that they '. . . assume that environmental effect are constant.' (p. 74274).

In actuality, production models are easier to interpret in single year-class fisheries like shrimp than in multiple year-class fisheries. This is because production models are not designed to deal with several year classes, each fluctuating in abundance and each exposed to different amounts of fishing effort. Inasmuch as environmental effects are concerned, production models usually incorporate an error term which takes into account the variation of environment. If, however, there are radical trends in the environment, then the form of the production model should change.

It is quite possible that improved fits to production models might be obtained by including environmental variables, but great caution would need to be exercised to be certain that improvement in the relationship was causal and not fortuitous in nature. Moreover, the increased utility of such a function for management purposes is questionable since the fitted curve would still be a function of specific past environmental conditions and prediction of future yield at a given effort level would require prediction of future environmental conditions.

The advice that the catch in any year should be the maximal amount of shrimp that can be caught could lead to drastic overfishing. As the Management Plan notes, surplus production models portray 'equilibrium conditions'. Thus, a production model suggests what will happen on the average and does not describe a trajectory of events. Such information is important, because while average conditions may not obtain in any single year, the population model at least provides an estimate of the amount of effort that should be expended on the average to obtain a particular average catch. Furthermore, production models reflect the average position of a population with respect to exploitation.

The great majority of the production models for shrimp show that relatively high levels of fishing effort are being expended. Thus, a reduction in effort would almost certainly lead to economic benefits. In contrast, an increase in effort would be of limited economic value to the fishermen and could result in an increased risk of population collapse or in sustained reduction in the production of the population. This is evident by the fact that from a statistical viewpoint it is difficult to judge the descriptiveness of the asymptotic and parabolic forms of the production model with regard to the dynamics and status of Gulf of Mexico shrimp. If the asymptotic form holds, then the population is in good condition, but if the parabolic form holds then the considered population is on the verge of collapse.

These remarks simply serve to underline the necessity of timely analyses and of careful monitoring of annual stocks. In any particular year, catches of an annual stock may be exceptionally large, but if they are large relative to the population, subsequent recruitment will likely be small. Thus, the following conclusions can be drawn as far as biological aspects of management are concerned:

(*a*) Specific management advice must be based on analyses incorporating recent statistics. Reactions to events by annual stocks are effected in a single year; hence, current statistics are required to provide sound advice.

(*b*) It would be imprudent to increase fishing effort without fully understanding the consequences. The appropriate amount of effort and the anticipated catches are listed in the section on production models. It would appear for example, that brown shrimp have been fished at effort levels which would produce declines in catch (*v* Brunenmeister, 1983, *Figure 11*).

(*c*) With respect to stock and recruitment relationships, the consequences of increases in effort should be carefully considered. Increased effort could cause the stock to decline and hence, recruitment to decline. It is possible, given the relationship in *Figure 12*, to fish brown shrimp stocks, for example, to optimize recruitment over a series of years.

(*d*) Catching small shrimp should generally be avoided. If brown and white shrimp natural mortality rates are relatively low, as suggested by our analyses, important gains in yield-per-recruit could be derived. This is, however, a tricky problem, since estimates of mortality rates need to be improved and the practicality of any size enhancement scheme needs to be explored in considerable detail. In any event, the rapid increase in value as shrimp size increases suggests that from an economic point of view, biologically determined minimum sizes are too small.

Acknowledgements

William Fox Jr. encouraged this study. Appreciation is extended to Joseph Powers and Michael Parrack of the National Marine Fisheries Service Miami Laboratory for their contributions. We also thank Harry Hornick, Philip Jones and Janice Silverstein of the Chesapeake Biological Laboratory, University of Maryland for their assistance in the production of statistics and figures.

References

ANDERSON, W W. Some problems of the shrimp industry. *Proc: Gulf Carib. Fish Inst.* 1:12–14. 1948

ANONYMOUS. First all-aluminum shrimper will fish Gulf of Mexico in 1969. *Com. Fish. Rev.*, 30(12):14. 1968a

ANONYMOUS. More freezer trawlers active in shrimp fishery. *Com. Fish. Rev.*, 30(5) 68. 1968b

BARRETT, B B and GILLESPIE, M C. Primary factors which influence commercial shrimp production in coastal Louisiana. *Louisiana Wild. Fish. Comm. Tech. Bull.*, 9, 28 pp. 1973

BAXTER, R N. Shrimp discarding by the commercial fishery in the Western Gulf of Mexico. *Mar. Fish. Rev.* 35:26. 1973

BERRY, R J. Dynamics of the Tortugas (Florida) pink shrimp population. Ph.D. dissertation, University Microfilms, Ann Arbor, Mich, 177 pp. 1967

BERRY, R J. Shrimp mortality rates derived from fishery statistics. *Proc: Gulf Carib. Fish. Inst.* 22:66–78. 1970

BERRY, R J and BENTON, R C. Discarding practices in the Gulf of Mexico shrimp fishery. *FAO Fish Rept. Dept.* 3:983–999. 1969

BEVERTON, R J H and HOLT, S J. Manual of methods for fish stock assessment. Part II. Tables of yield functions. *FAO Fish. Tech. Pap. No. 38* (Rev. 1), 67 p. 1966

BLOMO, V and NICHOLS, J P. Utilization of finfishes caught incidental to shrimp trawling in the Western Gulf of Mexico. Part I: Evaluation of markets. Texas A and M Univ., College Station, Texas. *Sea Grant Publ.* TAMU-SG-74-212, 85 pp. 1974

BRUNENMEISTER, S. Commercial brown, white and pink shrimp tail size: total size conversions. *NOAA Tech. Mem.* NMFS-SEFC-20, 7 p. 1980

BRUNENMEISTER, S. Standardization of fishing effort and production models for brown, white and pink shrimp stocks fished in U S waters of the Gulf of Mexico. In this volume. 1983

BRYAN, C E. Organisms captured by the commercial shrimp fleet on the Texas brown shrimp (*Penaeus aztecus* Ives) grounds. Thesis. Corpus Christi State University, Division of Biology, Corpus Christi, Texas, 44 pp. 1980

BRYAN, C E and CODY, T J. A study of commercial shrimp, rock shrimp and potentially commercial finfish 1973–1975. Part III. Discarding of shrimp and associated organisms on the Texas brown shrimp grounds. Coastal Fisheries Branch, Texas Parks Wild., Austin, Texas, 38 pp. 1975

BULLIS, H R and FLOYD, H. Double-rig twin shrimp-trawling gear used in Gulf of Mexico. *Mar. Fish. Rev.* 34(11–12):26–31. 1972

CAPTIVA, F J. Trends in shrimp trawler design and construction over the past five decades. *Proc: Gulf Carib. Fish. Inst.* 19:23–30. 1967

CHAVEZ, E A. A study on the growth rate of brown shrimp (*Penaeus aztecus* Ives, 1891) from the coasts of Veracruz and Tampaulipas, Mexico. *Gulf Res. Rep.* 4(2):278–299. 1973

CHRISTMAS, J Y and ETZOLD, D J. The shrimp fishery of the Gulf of Mexico United States: a regional management plan. Gulf Coast Res. Lab., Ocean Springs, Miss., *Tech. Rep. Ser. No. 2*, 128 pp. 1977

COMPTON, H. Survey of the commercial shrimp and associated organisms of Gulf Area 20. Texas Game Fish. Comm., Mar. Fish. Div., *Proj. Repts. 1959–1960*, 16 pp. 1961

COMPTON, H. Survey of the commercial shrimp and associated organisms of Gulf Area 20. Texas Game Fish. Comm., Mar. Fish. Div., *Proj. Repts. 1960–1961*, 19 pp. 1962

COMPTON, H. A survey of fish populations in the inshore Gulf of Mexico off Texas. Texas Parks Wildl. Dept., *Coastal Fish., Proj. Rept. for 1965*:55–86. 1966

COOK, H L and LINDNER, M J. Synopsis of biological data on the brown shrimp *Penaeus aztecus* Ives, 1981. *FAO Fish. Rep.* 57:1471–1497. 1970

COSTELLO, T J and ALLEN, D M. Mortality rates in populations of pink shrimp, *Penaeus duorarum*, on the Sanibel and Tortugas grounds, Florida. *U S Fish Wild. Serv., Fish. Bull.* 65(2):313–338. 1968

COSTELLO, T J and ALLEN, D M. Synopsis of biological data on
1970   the pink shrimp *Penaeus duorarum duorarum* Burkenroad, 1939. *FAO. Fish. Rep.* 57:1499–1537.

DUKE, T W. Estuarine Pesticide Research—Bureau of Commercial Fisheries. *Proc:* Gulf Carib. Fish. Inst. 22:146–153.
1970

FONTAINE, C T and NEAL, R A. Length-weight relations for
1971   three commercially important penaeid shrimp of the
        Gulf of Mexico. *Trans. Amer. Fish. Soc.* 100(3):584–586.

FORD, T B and ST. AMANT, L S. Management Guideline for
1971   predicting brown shrimp, *Penaeus aztecus*, production
        in Louisiana. *Proc:* Gulf Carib. Fish. Inst. 23:149–160.

GUNTER, G. Principles of shrimp fishery management. *Proc:*
1956   Gulf Carib. Fish. Inst. 8:99–106.

GUNTER, G and EDWARDS, J C. The relation of rainfall and fresh
1969   water drainage to the production of penaeid shrimp
        (*Penaeus fluviatlis* Say and *Penaeus aztecus* Ives) in
        Texas and Louisiana waters. *FAO Fish Rep.* 57:875–892.

GUTHERZ, E J, RUSSELL, G M, SERRA, A F and ROHR, B A.
1975   Synopsis of the northern Gulf of Mexico industrial and
        foodfish industries. *Mar. Fish. Rev.* 37(7):1–11.

HEALD, E J and ODUM, W E. The contribution of mangrove
1970   swamps to Florida fisheries. *Proc:* Gulf Carib. Fish.
        Inst. 22:130–135.

HILDEBRAND, H H. A study of the fauna of the brown shrimp
1954   (*Penaeus aztecus* Ives) grounds in the western Gulf of
        Mexico. *Publ. Inst. Mar. Sci.* 3:233–366.

HOLLOWAY, STEPHEN L. The shrimp fishery. *In:* Report of the
1981   Workshop on the Ecological Interactions between
        Shrimp and Bottomfishes, April, 1980, edited by Peter
        F. Sheridan and Sammy M. Ray. *NOAA Tech. Mem.
        NMFS-SEFC-63*:11–12.

HOLT, J S. A note on the relation between the mortality rate and
1965   the duration of life in an exploited fish population.
        ICNAF Res. Bull. 2:73–75.

IVERSEN, E S and JONES, A C. Growth and migrations of the
1961   Tortugas pink shrimp, *Penaeus duorarum*, and changes
        in the catch per unit of effort of the fishery. *Florida St.
        Bd. Conserv., Tech. Ser.* 34, 22 p.

JUNEAU, CONRAD L, JR and POLLARD JUDD, F. A survey of the
1981   recreational shrimp and finfish harvests of the Vermilion Bay area and their impact on commercial fishery
        resources. *La. Dept. Wildl. Fish. Tech. Bull. No. 33,*
        40 pp.

KLIMA, E F. Mark–recapture experiments with brown and
1964   white shrimp in the northern Gulf of Mexico. *Proc:*
        Gulf Carib. Fish Inst. 16:52–64.

KLIMA, E F. A white shrimp mark–recapture study. *Trans.
1974   Am. Fish. Soc.* 103(1): 107–113.

KLIMA, E F and BENIGNO, J A. Mark–recapture experiments.
1965   *In:* Biological Laboratory, Galveston, Texas, Fishery
        Research for the year ending June 30, 1964. *U S Fish.
        Wildl. Serv. Cir.* 230:38–40.

KNAKE, B D, MURDOCK, J F and CATING, J P. Double-rig shrimp
1958   trawling in the Gulf of Mexico. *U S Fish. Wildl. Serv.
        Fish. Leafl.* 470, 11 p.

KUTKUHN, J H. Dynamics of a penaeid shrimp population and
1966   management implications. *Fish. Bull* 65(2):313–338.

LINDNER, M J. What we know about shrimp size and the
1966   Tortugas fishery. *Proc:* Gulf Carib. Fish. Inst. 18:18–26.

LINDNER, M J and COOK, H L. Synopsis of biological data on the
1970   white shrimp *Penaeus setiferus* (Linn.), 1767. *FAO
        Fish. Res. Rep.* 57:1439–1469.

LINDNER, M T and ANDERSON, W W. Growth, migrations,
1956   spawning and size distribution of shrimp *Penaeus
        setiferus. Fish. Bull.* 106: 553–645.

LYLES, C H. The development of the brown shrimp fishery in
1951   Texas. *Proc:* Gulf. Carib. Fish. Inst. 3:50–51.

MEHOS, J. The shrimp industry's main problem: manpower.
1969   *Proc:* Gulf Carib. Fish. Inst. 21:42–44.

MOORE, D, BRUSHER, H A and TRENT, L. Relative abundance,
1970   seasonal distribution and species composition of demersal fishes off Louisiana and Texas, 1962–1964. *Contr.
        Mar. Sci.* 15:45–70.

NEAL, R A. An application of the virtual population technique
1968   to penaeid shrimp. *Proc:* S. E. Assoc. Game and Fish
        Comm. 21:264–272.

NICHOLS, S. Growth rates of white shrimp as a function of
1981a   shrimp size and water temperature. Paper presented at
        the Workshop on the Scientific Basis for the Management of Penaeid Shrimp. Key West, Florida, 18–24
        November 1981. Sponsored by Southeast Fisheries
        Center, U S NMFS, the Gulf States Marine Fisheries
        Commission, in collaboration with FAO, 9 pp.
        (mimeo).

NICHOLS, S. Impacts of variation in growth and mortality rates
1981b   on management of the white shrimp fishery in the U S
        Gulf of Mexico. Manuscript, Paper presented at the
        Workshop on the Scientific Basis for the Management
        of Penaeid Shrimp. Key West, Florida, 18–24
        November 1981. Sponsored by Southeast Fisheries
        Center, U S NMFS, the Gulf States Marine Fisheries
        Commission, in collaboration with FAO, 19 pp.
        (mimeo).

PARRACK, M L. Aspects of brown shrimp, *Penaeus aztecus*,
1979   growth in the northern Gulf of Mexico. *Fish. Bull.* 76(4)
        827–836.

PARRACK, M L. Some aspects of brown shrimp exploitation in
1981   the northern Gulf of Mexico. Paper presented at the
        Workshop on the Scientific Basis for the Management
        of Penaeid Shrimp. Key West, Florida, 18–24
        November 1981. Sponsored by Southeast Fisheries
        Center, U S NMFS, the Gulf States Marine Fisheries
        Commission, in collaboration with FAO, 50 pp.
        (mimeo).

PHARES, P L. Temperature associated growth of white shrimp in
1980   Louisiana. *NOAA Tech. Mem. NMFS-SEFC-56,*
        16 pp.

POFFENBERGER, J. A report on the available economic data for
1982   the shrimp fisheries in the southeastern United States.
        Dept. of Commerce. *NOAA Tech. Mem. NMFS-
        SEFC-100*, 21 p.

POFFENBERGER, J R. An economic perspective of problems in
1983   the management of penaeid shrimp fisheries. In this
        volume.

PRYTHERCH, H F. A directory of fishery data collection activities
1980   conducted by the Statistical Surveys Division in the
        southeast region of the United States. U S Dept. Commerce. *NOAA Tech. Mem. NMFS-SEFC-16*, 91 pp.

ROESSLER, M A and ZIEMAN, JR, J C. The effects of thermal
1970   additions of southern Biscayne Bay, Florida. *Proc:*
        Gulf Carib. Fish. Inst. 22:136–145.

ROITHMAYR, CHARLES M. Industrial bottomfish fishery of the
1965   northern Gulf of Mexico, 1959–63. U S Fish Wildl.
        Serv. Special Scientific Report-Fisheries, No. 518,
        23 pp.

SHERIDAN, P F, BROWDER, J A and POWERS, J E. 1983. Ecologi-
1983   cal interactions between penaeid shrimp and bottomfish
        assemblages. In this volume.

SIEBENALER, J B. 1952. Studies of 'Trash' caught by shrimp
1952   trawlers in Florida. *Proc:* Gulf Carib. Fish. Inst. 4:94–99.

SNOW, G W. Detailed shrimp statistical program in the Gulf
1969   states. *FAO Fish. Rep.*, (57) Vol. 3:947–956.

SOUTHEAST FISHERIES CENTER. A general overview of the Statis-
1982   tical Surveys Division's fisheries data collection
        activities in the southeastern region of the United
        States. National Marine Fisheries Service, Southeast
        Fisheries Center, Miami, Florida, 82 pp.

Biology

The behavior and catchability of some commercially exploited penaeids and their relationship to stock and recruitment

J W Penn

Abstract

The behavior of a number of commercially important Penaeid species has been reviewed in relation to its effect on trawl catchability. This review suggests that most *Penaeus* species fit into one or other of three broad behavioral categories and that these behavior types are often related to the general level of turbidity in the species preferred adult habitat.

The effects of these described behavioral patterns on trawl catchability, particularly through differences in vulnerability and level of aggregation have been noted and used to rank the three behavioral types in order of their potential for population reduction by fishing.

Using the fishing histories of a number of stocks representing each behavior/catchability type, a relationship between the occurrence of declining yields and the general catchability of the species being exploited has been shown. Of the three groups, the high catchability species which often exhibit schooling behavior, appear to be most at risk in terms of reduced yields at high levels of effort.

The possibility that recruitment overfishing has contributed to these declines in yield has been discussed within the general context of the need to understand spawner-recruit relationships in the management of Penaeid stocks.

Introduction

In the virtual absence of any published information on stock recruitment relationships for crustaceans (Hancock, 1973), it was been generally assumed that the relationship for the commercially important penaeid shrimp species, would be of the Beverton and Holt (1957) asymptotic form (Le Reste and Marcille, 1973; Neal, 1975). Neal (1975) after reviewing spawning stock-recruit relationships for the northern Gulf of Mexico penaeid fisheries, observed, that because of the high fecundity of shrimp, a relatively small population of spawners is required to maintain stock levels. Secondly he suggested those fisheries were operating at levels of spawner abundance somewhere on the flat part of the curve, so that environmentally caused fluctuations in recruitment rather than changes in spawner abundance were the primary cause of stock fluctuations; and finally that overfishing was unlikely because fishing becomes unprofitable at levels of abundance which are represented on the flat part of the curve.

This broad appraisal of the penaeid spawner recruit relationship and its implications for management, agrees generally with the observations of Boerema (1974) who had reported previously, 'that in the absence of any clear examples of declining recruitment due to high levels of fishing, the danger of such an occurrence seemed to be small.' However, Neal (1975) also noted that the long standing fishery for *Penaeus setiferus* in both Louisiana and Georgia/east Florida waters had shown a marked decline in catch while effort increased, and suggested that there may be situations where the reproductive capacity of a penaeid shrimp stock might have been reduced. More recently, as fishing pressure on penaeid stocks around the world has continued to increase and data for additional years has become available, downward trends in production from other fisheries, *eg* the Arabian Gulf (Van Zalinge, 1980), Mexico Pacific coast (Edwards, 1978) have become more common. These trends in yield suggest that the assumption that fishing is usually confined to the upper flat section of the stock recruitment curve may no longer be correct for all species.

The purpose of the paper is to investigate an observed relationship between Penaeid stocks which have declined with increasing effort and the catchability (behavior) of the species concerned, which may be related to spawning stock and recruitment. The hypothesis being considered is that level of catchability will be reflected in the degree to which spawning stocks and therefore recruitment can be reduced by fishing, ie those species whose spawning stocks are most easily reduced will be those first to show a decline in recruitment with increasing effort. The general level of catchability in such a situation may, therefore, be a means of identifying those stocks which are most at risk from recruitment overfishing, ie where fishing mortality reduces the spawning stock to the point where yield declines through reduced recruitment, which is the primary concern of fisheries management, (Gulland, 1978).

Shrimp behavior

The behavior of the commercially important Penaeus species has been reported extensively in the literature (Racek, 1959; Fuss and Ogren, 1966; Kutty and Murugapoopathy, 1968; Hughes, 1969; and others). Most of these studies, eg by Racek (1959), have been carried out in an attempt to explain the observed short term fluctuations in field catch rates, typical of these species. More recently, Hindley (1975), Moller and Jones (1975) and Wickham and Minkler (1975) have investigated the factors affecting activity and in particular behavioral rhythms. This latter comparative study of the behavior of the three major Gulf of Mexico penaeids found that each of the species exhibited different behavioral patterns, which would affect their potential for capture by trawling ie their catchability.

In reviewing these and other literature reports of the behavior of Penaeus species, particularly with reference to its effects on trawl capture, it has been noted that many of the commercial species tend to follow generally one or other of the behavior-catchability patterns described for the three Gulf of Mexico species ie

(1) Strongly nocturnal but often inactive or buried at night. Always buried during the day (P. duorarum).

(2) Generally nocturnal and continuously active at night, buried during the day but with a tendency to occasionally emerge (P. aztecus).

(3) Rarely buried and almost continuously active (P. setiferus).

Recent work by the author has shown that the two most important Penaeids from Western Australian waters, Penaeus latisulcatus and Penaeus esculentus, also follow this trend in having behavioral strategies similar to types 1 and 2 respectively. For example, aquarium experiments on P. latisulcatus using the methods described by Hindley and Penn (1975), have shown that this species is strongly nocturnal, almost never emerges during daylight, and usually burrows into the substratum when not actively foraging at night. The species was also found to have a well developed activity rhythm which continued for several days in constant darkness, whereas in constant light it rarely emerged at all. In addition, a series of experiments also showed (Figure 1) that the time spent foraging, ie above the substratum at night, varied significantly with the annual temperature cycle. These data support the field observations of Penn (1976) which suggested that catchability (probably related to activity) varied significantly with temperature. Other observations from this work, that P. latisulcatus catch rates varied with moon phase also correspond to the activity observations by Fuss (1964) for P. duorarum.

In a similar series of aquarium experiments P. esculentus was also found to be generally nocturnal (Table 1). However, it always emerged during darkness, remaining above the substratum until the start of the next light period. In addition, P. esculentus always emerged when the lights were turned off regardless of time of day or temperature and did not burrow significantly in constant darkness over a period of 5 days. During this time considerable periods were spent quiescent on the surface of the substratum. In addition, when the light portion of the light/dark cycle was reduced, some individuals did emerge during the light period unlike the more strongly nocturnal species P. latisulcatus. These behavioral attributes of P.

Table 1

RATES OF EXPOSURE (% VULNERABLE) OF P. esculentus UNDER VARYING LIGHT CYCLE AND TEMPERATURE CONDITION IN AQUARIUM EXPERIMENTS

| Exper. No. | No. of Shrimp | Experimental treatment | | Mean % Exposure | |
		Light cycle	Temp.	Dark	Light
1	21	12L : 12D	22·2°C	95·8%	0%
2	18	12L : 12D	22·2°C	97·3%	0%
3	16	12L : 12D	20·2°C	95·5%	0%
4	16	12L : 12D	20·0°C	94·0%	0%
5	21	12L : 12D	18·2°C	96·0%	0%
6	15	0L : 24D	22·2°C	94·5%	—
7	21	16L : 8D	22·2°C	93·6%	4·2%
8	21	20L : 4D	22·2°C	96·2%	5·1%

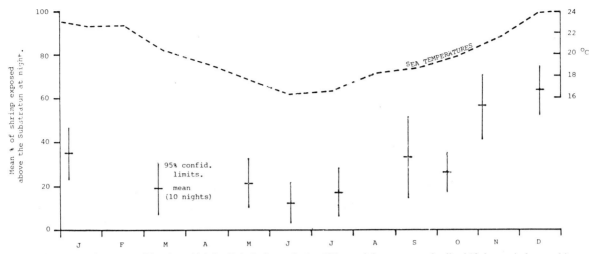

Fig 1 Mean rates of exposure (% vulnerable) for *P. latisulcatus* during 12 hour nights over standardised 10 day periods at ambient temperatures in aquarium experiments at WAMRL. (Each experiment was carried out using a freshly captured group of shrimp representing the modal size in the stock at that time in near by Cockburn Sound.)

esculentus parallel reasonably closely those reported for *P. aztecus*.

The behavior of the third major Western Australian species *P. merguiensis* appears to follow the type 3 pattern. From the literature *P. merguiensis* is not known to burrow into the substratum (Hindley, 1975) and demersal trawl catches do not usually vary significantly between day and night, Munro (1975). However, adults of the species frequently form dense semipelagic schools during daylight hours which is when most significant commercial catches are made, Lucas *et al* (1979). This species which does not apparently burrow to any extent and is typically fished in daylight hours, corresponds generally in behavioral characteristics to *P. setiferus* except that *P. setiferus* does not form schools. However, this lack of schooling appears to be a recent change in behavior for *P. setiferus*, since it has been reported that virtually all fishing of *P. setiferus* in the early 1900's (Moffett, 1967) was by using seine nets to catch 'schools' of this shrimp.

Behavior environmental relationships

These three general behavioral strategies used by the Gulf of Mexico and Western Australian species appear to be adaptations to the habitats occupied by the adults. The adults of both type 1 species are usually fished over sandy substrata, which are most often associated with relatively clear water habitats, ie *P. duorarum* (Williams, 1958), *P. latisulcatus* (Hall and Penn, 1979). In these habitats, the advantages of a well developed burrowing behavior which leaves the animal visually

exposed to predation only at night and then only for the minimum time required for feeding are obvious.

Type 2 species differ from type 1 in being commonly associated with softer silt substrata, often of terrigenous origin, ie *P. aztecus* (Williams, 1958), and *P. esculentus* (Hall and Penn, 1979). Such silty areas are generally associated with more turbid waters than the sandy substratum of type 1 species, which in turn correlate with the less rigid nocturnal burrowing habits of type 2 species. The observations of White (1975) that *P. esculentus* is sometimes trawled during daylight hours in the relatively turbid waters of Exmouth Gulf (Western Australia), supports this view. Another feature which appears to be a common characteristic of these two species is their camouflage colouration patterns, which the juveniles at least, use in association with vegetative cover to avoid predation, *eg P. aztecus* (Williams, 1955), and *P. esculentus*, (Young and Carpenter, 1977). The use of vegetative cover also extends to adults of *P. esculentus* in the clear waters of Shark Bay (author's observations) where the silty areas in the shallower end of the embayment usually have seagrass cover. For these reasons this group of species would appear to be quite versatile in the range of habitats that they are able to occupy as adults.

While type 1 and type 2 species occur in areas with clear to turbid waters, type 3 *P. setiferus* (Neal, 1975) and *P. merguiensis* (Munro, 1975) are almost exclusively found in areas where rivers discharge, which are usually characterised by soft mud bottoms and high turbidity. However, unlike *P. setiferus* in its present state of exploitation, *P.*

merguiensis typically forms dense schools which in themselves generate intense localised turbidity. Such turbid patches, known as 'mud boils', are the means by which fishermen locate (with aerial spotter plane assistance) schools of shrimp (Lucas *et al*, 1979). This schooling behavior and associated generation of intense turbidity tends to occur at times of slack water in the tidal cycle especially during neap tides (Munro, 1975) when turbidity falls to a minimum. From these observations, it appears that the schooling behavior and turbidity generation may be of survival value to this non burrowing species, in minimising predation at times of reduced turbidity, since predatory fishes are usually most actively feeding at times of slack tide when the water becomes clearer. The observations of Williams (1958) that *P. setiferus* in aquaria tended to actively generate turbidity in his substrate experiments suggests that *P. setiferus* may also exhibit this behavior given clear water conditions. The typically turbid habitat of *P. setiferus*, where light levels on the bottom can be expected to be very low, may have also been the reason for the apparently anomalous nocturnal activity in Wickham and Minkler's (1975) experiments, *ie* the unnaturally high light levels experienced by *P. setiferus* in aquaria in daylight may have inhibited its normal daytime activity as indicated by the fishery on this species.

The three broad behavior categories which have been described in the preceding text have been used to group the commercially exploited Penaeus species for which behavioral data are available, *Table 2*. In each case the source of the data used to classify the species has been given, however, where a species appears to fit a group, but detailed behavioral information is unavailable, the reference column contains the notation 'probable'.

Behavior and catchability

The three general behavioral strategies observed for Penaeus species, while obviously providing adequate protection from natural predators, do not provide similar defence against capture by trawl. Such differences in the probability of capture by trawling, are reflected in and quantified by the co-efficient of catchability as defined by Ricker (1975). In the case of shellfish fisheries, Caddy (1979) noted that variations in catchability were largely a function of changes in fishing power, fishing strategy, vulnerability, and stock aggregation. While fishing power and fishing strategy are a function of fishermen and are not likely to vary radically from species to species, vulnerability and aggregation of individuals which are directly related to the behavior of a species could result in major differences in catchability between species.

For example, type 1 species will have a low vulnerability to trawl capture due to their remaining buried during the day and frequent burrowing at night in response to environmental factors such as moonlight (eg *P. duorarum*, Fuss and Ogren, 1966) and temperature (eg *P. latisulcatus*, Penn, 1976). The data shown in *Figure 1* suggests that the vulnerability of *P. latisulcatus* from Cockburn Sound, Western Australia, to night-time trawl capture would vary between approximately 10% and 80% with temperature, and that these percentages may be further reduced by burrowing in response to lunar cycle, a factor shown (Penn, 1976) to also significantly alter catchrates. These Type 1 species are therefore considered to have relatively low catchabilities in general terms. By comparison, type 2 species such as *P. esculentus* will be much more vulnerable (*Table 1*) to night-time trawling, than group 1 species, since they appear not to burrow as readily in response to temperature changes, although they may burrow in response to lunar light cycle changes and moulting (White, 1975). The *P. aztecus* data presented by Wickham and Minkler (1975) suggests that this species would be similarly vulnerable to trawl capture. In comparison to type 1 species, these species are therefore considered to be approximately 100% vulnerable to night trawling as well as being vulnerable in daylight on some occasions. Their catchability will therefore largely be gear oriented.

Since type 3 species do not have well developed burrowing behaviors they must be considered vulnerable to trawl capture at the same level as type 2, except that they may be captured during both day and night. In practice, however, such species are mostly fished in daylight which gives them effectively the same vulnerability as type 2 species.

While the effects of burrowing behavior on vulnerability and therefore catchability are obvious, the other major behavioral factor affecting catchability, *ie* stock aggregation, has a potentially greater effect on catchability. Although the adult stocks of most penaeid species tend to aggregate to some extent, for example in areas of preferred habitat, the schooling behavior exhibited by a number of type 3 species such as *P. merguiensis* takes this behavior to the extreme. Such behavior generally separates the type 1 and 2 species from type 3. As a consequence, the schooling members of the type 3 species group must in particular be considered to have a level of catchability several orders of magnitude above that of type 1 and type 2 species.

Table 2
BEHAVIORAL/ENVIRONMENTAL GROUPINGS FOR SOME MAJOR COMMERCIALLY IMPORTANT PENAEUS SPECIES

Type No. and Characteristics	Species Name	Reference Source
1 Nocturnally active/ strongly burrowing/ clear water habitats (Night fishing)	*P. duorarum*	Wickham and Minkler (1975) Fuss and Ogren (1966)
	P. latisulcatus	Penn and Stalker (1979) Hall and Penn (1979)
	P. plebejus	Racek (1959)
	P. notalis	Probable
	P. brasiliensis	Probable
2 Nocturnally active/ turbid water or vegetated habitats (Night fishing)	*P. aztecus*	Wickham and Minkler (1975) Williams (1958)
	P. esculentus	Penn (unpublished data) Hall and Penn (1979), White (1975)
	P. semisulcatus	Kutty and Murugapoopathy (1968) Moller and Jones (1975)
	P. monodon	Moller and Jones (1975) Motoh (1981)
	P. japonicus	Hudinaga (1942)
3 Non burrowing/ Turbid water habitats (Daylight fishing)	*P. setiferus*	Wickham and Minkler (1975) Williams (1958) Gunter and Edwards (1969)
	P. merguiensis	Munro (1975), Lucas *et al* (1979) Hindley (1975)
	P. indicus	Marcille (1978) Kristjonsson (1969) Kutty and Murugapoopathy (1968)
	P. orientalis	Kristjonsson (1969)
	P. occidentalis	Probable
	P. schmitti	Probable

It is of interest to note that this separation of schooling from non schooling species has been used previously (Racek, 1959) as a distinguishing character for major ecological groupings within the Penaeidae. Racek (1959) described the Penaeids of the eastern Australian coastline as either 'consistent' or 'inconsistent' species, the latter being those which 'are always on the move and show a preference for turbid water and soft muddy grounds. They form pronounced age groups and dense schools and may occur sporadically in enormous quantities on inner littoral grounds'. Racek goes on to suggest that 'they are essentially perifluvial forms and therefore their abundance is closely linked with the occurrence of river floods'. *P. merguiensis* and *Metapenaeus macleayi* were described as typical 'inconsistent' species. 'Consistent' species, Racek suggests, 'prefer a certain habitat in a well defined area which may be, however, very extensive and may cover a wide range of depths. These species do not form pronounced age groups except during their mating and spawning period and are to be found, year by year, in only slightly varying abundance on the same grounds. Due to the absence of irregular migratory schooling habits, consistent prawns are rarely caught in considerable quantities, except when leaving the estuaries. The king prawn (*P. plebejus*) and the common tiger prawn (*P. esculentus*) belong to this group.'

In this paper Racek's 'consistent species' correspond to behavior types 1 and 2, while inconsistent species correspond to type 3.

In summary then, the various behavioral strategies of Penaeid species have been shown to cause differing levels of vulnerability (to fishing gear) and stock aggregation, which are major factors affecting catchability (Caddy, 1979). As a consequence of these differences the catchability of the three behavioral types can be considered as grading from low for type 1 species to high for type 3 and extremely high for those type 3 species which form schools. These general differences in catchability will be reflected in the degree to which the stocks of each behavioral type can be reduced before catchrates become uneconomic and fishing ceases. Such differences in the ability of a fishery to reduce a stock it is suggested, will have a direct effect on the size of the spawning stock present during and at the end of each year and hence on the numbers of spawners contributing to recruitment in the following year.

Stock stability

For the pioneer Penaeid shrimp fisheries of the Gulf of Mexico (Gunter and Edwards, 1969) catch records extend back to the 1920's but most other large scale shrimp fisheries have developed more recently during the 1950's and 60's. For the majority of the latter, records of total catch are available, but records of fishing effort have been generally insufficient to provide an accurate index of the recruited stock in each year. However, in fisheries where effort can be assumed to have been constant, as is often the case in mature fisheries, the total catch of adults each year may be a reasonable indicator of stock size. Total catch has been previously used in this way, to correlate stock size with environmental variables such as rainfall, *eg* Gunter and Edwards (1969), Ruello (1973) and Dall (1980).

In the following section both catch and effort, and total catch over time (where appropriate assumptions about effort can be made) have been used to review variability and long term trends in yield from a number of shrimp fisheries. Where significant declines in yields have occurred over long periods, they have been considered to indicate

recruitment over-fishing unless the decline could be explained by decreasing effort or could be attributed to growth overfishing, *ie* a fishing mortality caused decline in the size of fish caught which results in a reduction in yield. The issue of long term environmental effects on recruitment has been deliberately avoided, but from the following, it can be seen that this common explanation of declining yields must be challenged as a means of explaining stock failures. For the purposes of this review the stocks examined have been grouped using the behavior-catchability types of the species being exploited.

Type 1 Low catchability species

(*a*) *P. latisulcatus* – Shark Bay (W. Australia). The catch and effort data from the two species exploited in the Shark Bay fishery has been collected and treated using the methods described by Hall and Penn (1979). These data have been plotted as a historical sequence (*Figure 2*) for the period 1962–82*. It should be noted, however, that only raw effort data is given which underestimates the effective effort actually applied, more detailed effort statistics are given in Bowen and Hancock (this volume).

These data show that the annual variability in yield from the *P. latisulcatus* stock has been relatively low for a penaeid fishery possibly due to the general stability of the hypersaline nursery areas (Penn and Stalker, 1979) compared with the more usual but variable riverine nursery areas of related species. Secondly, the yield appears to have leveled off since about 1974 while effective effort has continued to increase. Since there has been no evidence of a decline in yield with the significantly increased effort in recent years, it must be concluded that the fishery has not been noticeably affected by either growth or recruitment overfishing.

(*b*) *P. plebejus* – Australian East Coast. The *P. plebejus* stock along the eastern Australian coastline is fished extensively throughout its range from Twofold Bay (NSW) to Tin Can Bay (Qld), with a major portion of the catch being taken from Moreton Bay and adjacent offshore grounds, Lucas (1974). Although no accurate catch and effort statistics for the entire stock are available, Haysom (1975) gives some statistics for the inshore Moreton Bay fishery from 1952 to 1970. On the basis of these data and a general knowledge of the east coast fisheries, Haysom (1975) observed that

---

*Data for 1981 and 1982 years has been added to this figure at the Editor's suggestion after the meeting.

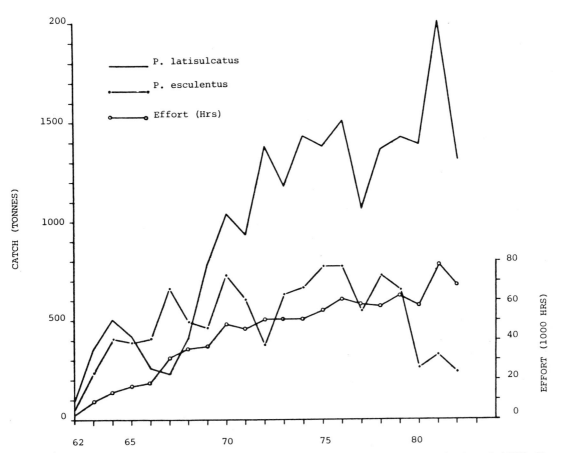

Fig 2 Catch and effort (unadjusted hours) from the Shark Bay, (Western Australia) prawn fishery for the period 1962–82

the Moreton Bay fishery generally was remarkably stable in comparison with other shrimp fisheries. Since the Moreton Bay catch represents a major portion of the annual recruitment to the offshore spawning stock and there has been no suggestion of an overall decline in catch by later workers, Lucas (1974) and Ruello (1975), it would appear that the stock has not suffered a major decline to the mid 1970's.

(*c*) *P. duorarum* – Tortugas Sanibel fishery (USA). The *P. duorarum* fishery operating on the Tortugas and Sanibel grounds off the West Florida coastline is one of the major single species fisheries of the Gulf of Mexico. Annual production from these grounds are available (Joyce and Eldred, 1966) for the period 1951–65 and 1956–1978 (Rothschild and Parrack, M. S. this meeting). However, although the two data sets conflict during the late 1950's period, the recent yields appear to be in the same order as those published by Joyce and Eldred (1966). Since fishing effort on this fishery was not restricted, and there was a general increase in size and power of vessels in the 1950's

and 1960's (Kristjonsson, 1969), which apparently continued to recent times (Blomo *et al*, 1978), it would appear that effort has almost certainly shown a general increase during the 1950–1979 period.

In this situation where yield has been relatively stable over a period while effort has apparently increased, suggests that the stock to this point in time (1978) had not been adversely affected by fishing pressure.

(*d*) *P. latisulcatus* – South Australia. Two significant stocks of *P. latisulcatus* in Spencers Gulf and Gulf St Vincent, South Australia, have been exploited since 1968. Data for the fishery have been reported, Byrne (1980), but because the catch and effort data presented were recorded for periods which were not comparable, the status of these stocks cannot be accurately assessed. In addition, the data for the Gulf St Vincent area has been complicated in later years by the development of an additional fleet in Investigator Strait, which appears to have been fishing on part of the original stock. However, Byrne (1980) suggests that the overall yield, although variable due to the influence of short term environmental factors, has remained relatively stable since the 1973–74 period. During that time 1973–9, there has been a slow, but general increase in effort, within the confines of a limited entry management regime. Since no appreciable decline has occurred over this period, it must be assumed that these stocks have not yet been adversely affected by the application of fishing effort.

In general terms, this review of type 1 stocks for which data was available to the author, suggests that although most have reached a point where further increases in effort have not increased yields appreciably, none have shown variations in yield which could not be attributed to short term environmental effects.

Type 2 Medium catchability species fisheries

(a) *P. esculentus* – Shark Bay (W. Australia). The catch and raw effort data for the *P. esculentus* stock in Shark Bay is given in *Figure 2*, and covers the period 1962–82. This data shows that the degree of variability in annual yield has been generally greater than that of the type 1 species *P. latisulcatus* exploited by the same fleet in Shark Bay. However, the yield from the *P. esculentus* stock showed a significant decline during the 1980 season, and preliminary observations suggest that a similar low yield can be expected in 1981*. This decline followed a change in the direction of effort toward *P. esculentus* during 1979 when the price of *P. esculentus* rose to approximately double that for *P. latisulcatus*. A significant portion of this increased effort came from new large replacement vessels which specifically concentrated on *P. esculentus* for economic reasons. This directing of effort by these vessels was reflected in their high fishing power on *P. esculentus*, ie up to two times that on *P. latisulcatus*, and the increase in effective effort in 1979 (Bowen and Hancock, this volume).

This additional effort on *P. esculentus* was largely achieved through the improved use of radar in both locating and fishing of small concentrations in areas of preferred habitat. It is also of interest to note that the level of effort expended on the reduced available stock during 1980 could not have occurred without the availability of the second species on the same grounds which made the combined catchrate economically viable.

Obviously more years of data at present levels of effort will be necessary to determine whether the

low recruitment in 1980 and 1981 seasons has resulted from a reduced spawning stock or an environmentally induced reduction in survival of recruits. However, considering the range of past yields from this fishery and the relatively stable (hypersaline) nursery system in Shark Bay (Penn and Stalker, 1979) it would appear that recruitment overfishing has almost certainly contributed to the current low catches.

(b) *P. esculentus* – Exmouth Gulf (Western Australia). A fishery for shrimps began in Exmouth Gulf in 1963 based mainly on *P. merguiensis*. However, by 1965 the major species in the catch was *P. esculentus* which has remained the dominant species up to 1980. Because the fleet during the 1963–1967 period operated during both day and night, depending upon the target species, the effort data in those early years was not usable as it could not logically be assigned to either species. As a consequence the effort has been restricted to the period following 1967 when the fishery was based predominantly on the *P. esculentus* stock. These catch and raw effort data during the history of the fishery are presented in *Figure 3*.

Figure 3 shows that annual yields from this stock have been more variable than those from a stock of the same species fished in Shark Bay (*Figure 2*). However, underlying these variations there appears to have been a trend towards increasing yield up to 1975. Since 1975 this trend has become less distinct suggesting that the curve may be leveling off.

Because of the high degree of variation and general upward trend in annual yield, it is concluded that present levels of effort are not appreciably affecting the size of individuals in the catch or recruitment and that the observed variations are probably due to year to year environmental effects on recruit survival*.

(c) *P. aztecus* – Northern Gulf of Mexico. Catch data for the stocks of *P. aztecus* in Texas and Louisiana waters for the period 1948–64 have been given by Gunter and Edwards (1969), in their study of the relationship between rainfall and catch. Additional data for *P. aztecus* from Louisiana up to 1970 have also been reported by Ford and St Amant (1971). Although no effort data were included in these papers, there has apparently been a general increase in effort in the northern Gulf of Mexico fisheries throughout this period (Neal, 1975; and Nichols and Griffin, 1975)

---

*Subsequent catches supporting this statement have been added to this figure after the meeting.

*Since the preparation of this paper the yield from this stock has declined to extremely low levels (see additional data *Figure 3*) in which recruitment overfishing has been implicated.

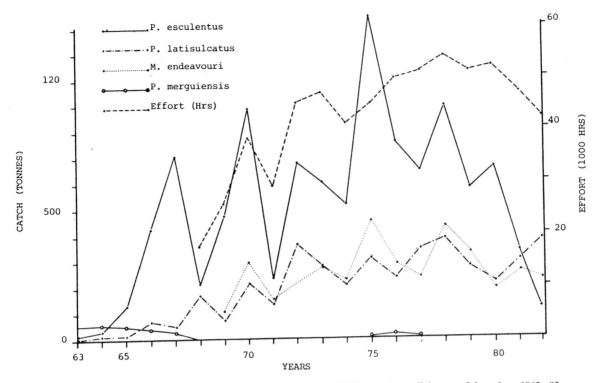

Fig 3 Catch and effort (adjusted hours) trawled in the Exmouth Gulf (Western Australia) prawn fishery from 1963–82

particularly on the *P. aztecus* stocks.

These data from Texas and Louisiana waters suggest that the yield from both areas has been highly variable on a year to year basis, particularly that from Louisiana. Secondly, the catch from Texas after reaching a peak in the early 1950's as the fishery developed, appears to have declined generally to 1964 (the end of Gunter and Edwards, 1969 data). However, more recent statistics from Blomo *et al* (1978) suggest that the catch of *P. aztecus* from the entire Gulf of Mexico may have stabilised from 1965 to 1975; while effort apparently continued to increase. This situation of a relatively stable yield over a long period of time, while effort continued to increase, lends additional support to the assessment by Neal (1975), that the reproductive capacity of the *P. aztecus* stock was unlikely to be affected by fishing within economic constraints at that time, and that differences in recruit survival were the most likely causes of variations in yield.

(*d*) *P. semisulcatus*–Western Arabian Gulf. Stocks of *P. semisulcatus* support a major fishery in the waters of the Arabian Gulf and have been the subject of a number of studies, Boerema (1969), Kristjonsson (1969), Ellis (1975) and Van Zalinge (1980). Van Zalinge (1980) presents a comprehensive stock assessment for the western Arabian Gulf

fishery based on catch and effort data from both the industrial and artisanal fishery. This assessment suggests that the yield has become increasingly variable as effort has increased and recently (1977/78 season onwards) the yield has shown a severe decline. This decline apparently began with the complete failure of one of the major stocks contributing to the fishery in 1977/78, which was followed by a severe decline in landings from all west coast grounds in the following two seasons. Van Zalinge (1980) suggests that this decline may be the result of high levels of effort causing recruitment overfishing compounded by the effects of deteriorating environmental conditions generally.

However, the use of these data as a demonstration of recruitment overfishing has been complicated by the observation of Van Zalinge (1980), that schooling 'sometimes' occurs at the time of recruitment, resulting in large catches. This observation, when compared with those of Kristjonsson (1969), who noted that catches at that time were mostly taken from dense schools, suggests that schooling behavior which was once common has now been reduced.

If such a reduction in schooling has occurred as it appears, then the consequent change in catchability makes it particularly difficult to combine catch and effort data from the schooling and non-schooling periods of the fishery's history. However, the recent trends in yield, while schooling has not been a significant feature of the fishery, suggests that the possibility of recruitment overfishing may well have become a reality.

In summary, two of these four type 2 catchability species fisheries examined appear to have suffered a reduction in yield which could well be related to recruitment overfishing.

Group 3 Stocks

(*a*) *P. merguiensis* – Exmouth Gulf (W. Australia). Shrimp fishing in Exmouth Gulf began in 1963 based on daylight fishing of schooling *P. merguiensis*. Subsequently the fishery has developed into a night fishery for *P. esculentus* (referred to previously in this paper). Yield statistics for this stock were collected from the start of fishing, however effort data although recorded (hours trawled) were considered unreliable because daylight searching time (for schools) was not available and trawling time for *P. merguiensis* could not readily be separated from trawling time aimed at the alternative species *P. esculentus* which was fished mostly at night.

The production data for both *P. merguiensis* and *P. esculentus* and hours trawled (*Figure 3*) show that the catch of *P. merguiensis* declined and the night fishing for *P. esculentus* increased. This reduction in catch of the daylight schooling *P. merguiensis* occurred while the effort (including the use of an aerial spotter plane) apparently remained at a high level, certainly sufficient to find schools had they been present. Since 1968 no catches from schools have been reported. The only catches taken since that time have been at night, while trawling for *P. esculentus*. The short recurrence of *P. merguiensis* in the fishery from 1975 to 1977, which resulted in 17 tonnes being caught in 1976 was recorded from non schooling catches taken at night.

Since this stock which occurred in Exmouth Gulf was the southernmost stock on the west coast of Australia and at the lower end of the species temperature range, the stock collapse could be explained either as an environmentally induced variation in recruit survival or as a function of reduced recruitment from a fishery induced reduction in spawning stock. While there is insufficient data to investigate the spawner-recruit alternative further, it is of interest to note that *P. merguiensis* was known to occur in the area from research vessel surveys from 1952 – 58 prior to commercial fishing (White, 1975). This consistent occurrence of *P. merguiensis* before heavy fishing pressure

was applied adds support to the view that fishing contributed to the collapse of the stock.

A similar situation to that which has occurred in Exmouth Gulf where the *P. merguiensis* stock declined as night trawling for *P. esculentus* increased, appears to be currently occurring in another W. Australian *P. merguiensis* fishery in Nickol Bay. In this fishery, the *P. merguiensis* catch, although highly variable, has registered low catches outside of the previous range in two of the three years since night trawling for the same alternative species began in 1978, despite a general increase in fishing effort.

(*b*) *P. merguiensis* – Gulf of Carpentaria (Australia). The *P. merguiensis* fishery in the Gulf of Carpentaria began in 1966 and expanded as new grounds, based on additional nursery areas, were discovered up to 1972. Since 1972 virtually all grounds within the Gulf have been fished during each season. Fishing of these stocks occurs almost enitrely during daylight on dense schools of shrimp, which are found with the assistance of both echosounders and aerial spotter planes which locate the shrimp schools by visually detecting the associated 'mud boils' described previously (Lucas *et al*, 1979).

The records of annual production catch from this stock (*Figure 4* – combined data from Lucas *et al*, 1979, and Dall, 1980) show that yields have been highly variable from year to year. This variability, which appears to be related to the preceding summer rainfall (Dall, 1980) does not, however, reflect the wider variation in catch from individual grounds, which on occasion fails completely (Lucas *et al*, 1979).

A second feature of this fishery has been the severe decrease in the length of the fishing season, *ie* from 9 months in 1968 to less than two months in 1976 (Lucas *et al*, 1979). This shortening of the fishing season which corresponds generally to the schooling period, appear to have been largely independent of the total catch in each season.

The stock assessment of this fishery up to 1976 (Lucas *et al*, 1979) suggested that the fishery had been fully exploited since 1971 and that catch variations were due to environmental influences on recruitment. However, the additional data since 1976 suggest that, while annual variation has remained high, an overall trend towards decreasing yield has occurred. This downward trend has occurred while the fleet concentrating its effort on the schooling *P. merguiensis* stock has continued to increase (Somers and Taylor, 1981). However, interpretation of these data showing a decrease in yield with increasing effort has been complicated

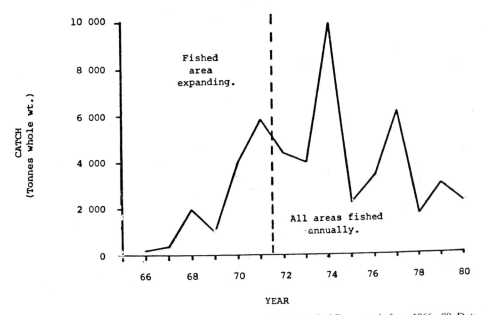

Fig 4 Annual catch (tonnes) from the *P. merguiensis* stock in the Gulf of Carpentaria from 1966–80. Data for 1966–76 and 1977–80 taken from Lucas *et al* (1979) and Dall (1980) respectively

by the reduction in mean size of shrimp caught, as well as the drastic reduction in fishing season. The reduction in average size of shrimp caught suggests that yield in numbers rather than weight has probably not shown as significant a decline (Dall, 1980), while the shortening of the season may have biased the number taken from the recruit population upward because of the reduced average time for natural mortality to occur before capture.

Obviously growth overfishing has been involved in the observed decline in yield, but the degree of decline in recent years also suggests that recruitment overfishing may have begun to have an effect. This suggestion is supported by the comments of Lucas *et al* (1979), that recruitment overfishing was a possibility, but was ruled out at that time (1976) largely because of the lack of any downward trend in catch.

(c) *P. setiferus* – Northern Gulf of Mexico. The white shrimp *P. setiferus* has been subjected to commercial fishing in the Gulf of Mexico since the early 1900's. Historical data on this fishery has been given by a number of authors including Joyce and Eldred (1966), Gunter and Edwards (1969) and Neal (1975). These data suggest that catches of *P. setiferus* have been variable from year to year and the variations have often been related to or caused by factors such as rainfall, Gunter and Edwards (1969). In addition to the short term variability, the data from a number of the States fishing this species, show a general long term decline in yield, *ie* Louisiana (Gunter and

Edwards, 1969), Georgia and Florida east coast (Neal, 1975). The stocks in Texas, although showing a major decline between the mid 1940's and 1950's apparently recovered back to the level of the mid 1940's by 1964 (the limit of the data given by Gunter and Edwards (1969). More recent data 1956–1978 (Rothschild and Parrack, MS) for the entire northern Gulf of Mexico, shows that the yield after the low period of the late 1950's has shown a slight upward (but highly variable) trend. These levels of yield, are, however, well below those reported from the same area during the early 1940's (Gunter and Edwards, 1969).

Since the *P. setiferus* stocks are the most readily accessible to fishing of the three species in the Gulf, through being closer inshore (Neal, 1975), it would appear reasonable to assume that the effort applied to these stocks would have always remained at a high level. On this basis the reduced catches from States other than Texas may well be evidence that the reproductive capacity of these stocks may have been reduced (Neal, 1975). An alternative explanation for this overall decline may be that the schooling of the species (Moffett, 1967) in earlier times decreased, as effort on these stocks escalated in the post-war period. Such a change in behavior would have significantly reduced catchability and hence cause a major change in the effectiveness of fishing effort. The apparently anomalous situation in the Texas fishery where the yield decreased significantly then recovered equally rapidly from a nil catch situation, appears to have been the result of an unusual rainfall deviation (Gunter and Edwards, 1969) possibly combined with a change in the direction of effort related to the discovery of a considerably larger stock of *P. aztecus*, in adjacent offshore waters, *ie* the catch from this area tends to show an inverse relationship to that of *P. setiferus* during the 1950–60 period.

(*d*) Other type 3 species fisheries. While there are a considerable number of type 3 species fisheries in various parts of the world, *ie* Madagascar – *P. indicus* (Marcille, 1978), India – *P. indicus*, China/Japan – *P. orientalis* (Kristjonsson, 1969), Mexico – *P. vannamei* (Edwards, 1978), the data available are generally insufficient for any specific investigation of the occurrence of recruitment overfishing. However, many of these accounts suggest that catches have not been sustained over long periods of time.

For example, a stock of midwater schooling *P. orientalis* in the Yellow Sea was 'discovered in 1959/60 and rapidly attracted a fleet of up to 200 large trawlers. However, by 1964 shrimp were no longer found in midwater, possibly due to a change in oceanographic conditions and the fishery for schooling prawns apparently ceased.' (Kristjonsson, 1979). In Madagascar, Marcille (1978) reports that a fishery based on schooling *P. indicus* began in 1967, but by 1973 had reached a critical stage where catches continued to decrease despite an increase in effort. A decrease in recruitment was implicated in this decline. A similar situation appears to have occurred along the highly productive S.W. coast of India, where, a fishery developed in the late 1950's (Kristjonsson, 1969) with schooling *P. indicus* being the most important species in the catch. Recent reports (Anon, 1980) do not mention fishing methods based on schooling *P. indicus* and show that this species now forms a small percentage of the catch. However, because of the multispecies nature of the current fishery, the trends in *P. indicus* catch cannot be evaluated.

This brief review of catchability type 3 species fisheries, suggests that they, as a group tend to exhibit a high degree of variability in annual yield, probably caused by year to year environmental influences on recruitment related to hydrological stability in their preferred habitats (discussed previously). In addition to this annual variability, the yields from those stocks examined in detail have shown an overall reduction at high levels of effort, in which recruitment overfishing appears to have been implicated.

A further common feature of most of these type 3 species stocks, has been a reduction or cessation in schooling behavior as the general level of effort has increased. Such changes in behavior, possibly related to increased turbidity, will radically alter the relationship between fishing effort and fishing mortality, making the interpretation of these statistics extremely difficult. However, a lack of schooling behavior at high levels of effort would lower catchability in general and have the secondary effect of reducing the possibility of spawning stocks being depleted to a point where recruitment could be affected.

Discussion

The failure of previous attempts to observe any spawning stock-recruitment relationship for Penaeid fisheries in the Gulf of Mexico (Neal, 1975) was attributed largely to variability in survival during the planktonic larval stages. This variability, together with the obvious variations in recruitment of some species related to environmental factors such as rainfall (Gunter and Edwards, 1969; Ruello, 1973; Glaister, 1978; Dall, 1980), has tended generally to obscure the underlying relationship between spawning stock and recruitment. However, it appears likely that the problems of producing meaningful indices of abundance for both spawning stock and recruitment using catch per effort data from these short lived species, may also have contributed significantly to the inability to demonstrate such a relationship.

For example an index of spawning stock which reflects annual variations in the reproductive capacity of a Penaeid population must take into account inter-actions between, intraseasonal changes in increasing adult catch rates (which are not necessarily related directly to absolute abundance), the timing and duration of the spawning season, and changes in average fecundity with increasing size during the spawning season. In considering such interactions for populations of *P. latisulcatus*, Penn (1980) noted that an index based on catch rate data alone will frequently bear little relationship to the reproductive output from a stock. The use of catch per effort data in estimating recruit abundance for penaeids also presents some problems because the rapid changes in catch rate (as recruits enter the fishery during each season) compounds the well known difficulties involved in estimating population size for crustaceans generally as noted by Hancock (1973). The effects of year to year changes in mean size of individuals in the catch and possibly reduced natural mortality due to the application of effort, as noted for the *P. merguiensis* fishery in the Gulf of Carpentaria, also adds to these problems. Finally studies by Ruello

(1973) and Dall (1980) have suggested that the annual catch taken from some penaeid fisheries reflects more the rate of recruits entering the adult stock, than the actual abundance of juveniles in the nursery area, ie rainfall or flooding at time of recruitment relates directly to the total catch taken.

For all of these reasons, it appears unlikely that the detailed data required to show a spawner-recruit relationship for a penaeid fishery will become available in the near future. However, since some stocks have now begun to show yield declines which may be attributable to recruitment overfishing, the inability to show directly a spawner-recruit relationship can no longer be accepted as a reason for management to disregard its potential effects.

In the preceding review of exploited *Penaeus* species stocks for which some usable data were available to the author, most type 1 (low catchability) stocks appear to be withstanding the present high levels of effort without showing substantial declines in yield. However, in constrast, yields from two of the four type 2 (medium catchability) species considered, and most of the stocks of the type 3 (high catchability) daylight fished species, particularly those which form schools, have shown a decline in yield as effort has increased.

While growth over fishing has obviously contributed to these reduced yields, particularly where an increase in artisanal fishing on juveniles has occurred, it would appear unlikely that this has been the major cause in most cases. Similarly, environmental effects on recruitment, although an acceptable explanation for short term fluctuations in yield, are unlikely to have been the major cause of long term declines in yield, as has occurred for example in the *P. setiferus* stocks (Neal, 1975) of the Gulf of Mexico. However, recruitment overfishing, although exceedingly difficult to demonstrate, does have the potential to have been the major cause of at least some of the reported decreases in yield. The observed increased likelihood of a stock showing a decline in yield with increasing effort, which corresponds to the level of catchability and hence potential reduction in numbers of spawners in the stock being exploited, is considered to provide significant support for this view.

If this circumstantial evidence that recruitment overfishing has contributed to at least some of the observed declines in yield from penaeid stocks is accepted, then the assumption that the relationship between catch and effort will be asymptotic within the normal range of effort (Boerema, 1974; Neal,

1975; Jones and Dragovich, 1977) can no longer be considered to be the general rule for Penaeid fisheries. This situation implies that control of exploitation, particularly in fisheries based on type 3 species which are more susceptible to environmental as well as fishery effects, will become increasingly necessary for conservation as well as economic reasons as technological advances continue to increase effective effort in the established fisheries.

References

ANON. A case of overfishing: depletion of shrimp resources
1980   along the Neendakara Coast, Kerala. *Mar. Fish. Infor. Serv. T and E ser. 18*, 1–8.

BEVERTON, R J H and HOLT S J. On the dynamics of exploited
1957   fish populations. *Fishery Invest., Ser. II, 19*, 533 p.

BLOMO, V, GRIFFIN, W L and NICHOLS, J P. Catch-effort and
1978   price-cost trends in the Gulf of Mexico shrimp fishery: Implications on Mexico's extended jurisdiction. *Mar. Fish. Rev. 40* (8), 24–28.

BOEREMA, L K. Provisional note on shrimp assessment and
1974   management. Paper presented at the 'Government consultation on shrimp resources in the CICAR area' Caracas, Venezuela, Sept. 1974. *FAO FIR:SR/WP12*, 13 p.

BYRNE, J L. The South Australian prawn fishery (A case study
1980   in limited entry regulation). Presented to 'Australian National Seminar on economic aspects of limited entry and associated fisheries management measures. University of Melbourne. (Mimeo). 21 p.

CADDY, J F. Some considerations underlying definitions of
1979   catchability and fishing effort in shellfish fisheries, and their relevance for stock assessment purposes. *ICES C.M. 1977/K:18*, 22 p (Mimeo).

CUSHING, D H. Dependence of recruitment on parent stock. *J.*
1973   *Fish. Res. Bd. Can. 30* (12) Pt 2, 1965–1976.

DALL, W. Northern prawn fishermen–pray for rain. *Aust.*
1980   *Fisheries, 39* (12), 3–4.

EDWARDS, R R C. The fishery and fisheries biology of penaeid
1978   shrimp on the Pacific Coast of Mexico. *Oceanogr. Mar. Biol. Ann. Rev.*, 145–180.

ELLIS, R W. An analysis of the state of the shrimp stocks in the
1975   Gulf between Iran and the Arabian Peninsular. (Provisional Report). *FAO IOFC/75/Inf. 10*, 17 p.

FORD, T B and ST AMANT, L S. Management guidelines for
1971   predicting brown shrimp, *Penaeus aztecus*, production in Louisiana. *Proc. Gulf Caribb. Fish. Inst. 23rd Annu. Sess. Nov. 1970*, p 149–161.

FUSS, C M. Shrimp behavior as related to gear research and
1964   development. 1. Burrowing behavior and responses to mechanical stimulus. In 'Modern fishing gear of the world'. Fishing News (Books) Ltd, London. *Vol. 2*, 563–66.

FUSS, C M and OGREN, L H. Factors affecting activity and
1966   burrowing habits of the Pink Shrimp *Penaeus duorarum* burkenroad. *Biol. bull. 130* (2) 170–191.

GLAISTER, J P. The impact of river discharge on distribution and
1978   production of the school prawn *Metapenaeus macleayi* (Haswell). (Crustacea: Penaeidae) in the Clarence River. *Aust. J. Mar. Freshwater Res. 29*, 311–23.

GULLAND, J A. Fishery Management: New strategies for new
1978   conditions. *Trans. Ann. Fish. Soc. 107*, 1–11.

GUNTER, G and EDWARDS, J C. The relation of rainfall and
1969   freshwater drainage to the production of penaeid shrimp (*Penaeus fluviatilis* Say and *Penaeus aztecus* Ives) in Texas and Louisiana waters. *FAO Fish. Rep. 57* (3) 875–92.

HALL, N G and PENN, J W. Preliminary assessment of effective
1979   effort in a two species trawl fishery for Penaeid prawns

in Shark Bay, Western Australia. *Rapp. P.-V. Reun. Cons. Int. Explor. Mer, 175*, 147–154.

HANCOCK, D A. The relationship between stock and recruit-
1973 ment in exploited invertebrates. *Rapp. P.-V. Reun. Cons. Int. Explor. Mer. 164*, 112–131.

HAYSOM, N M. The Moreton Bay permit system. An exercise in
1975 licence limitation. In 'First Australian National prawn seminar', Maroochydore, Queensland, November 1973. *A.G.P.S.* Canberra, 240–245.

HINDLEY, J P R. Effects of endogenous and some exogenous
1975 factors on activity of these juvenile banana prawn, *Penaeus merguiensis. Mar. Biol. 29*, 1–8.

HINDLEY, J P R and PENN, J W. Activity measurement of
1975 decapod crustaceans: A comparison of white and infrared photographic illumination. *Aust. J. Mar. Freshwater Res., 26*, 281–5.

HUDINAGA, M. Reproduction, development and rearing of
1942 *Penaeus japonicus* Bate, *Jpn. J. Zool. 10*, 305–93.

HUGHES, D A. Factors controlling the time of emergence of
1969 pink shrimp *Penaeus duorarum. FAO Fish. Rep. No. 57, 3*, 971–81.

JONES, A C and DRAGOVICH, A. The United States shrimp
1977 fishery of northeastern South America (1972–74). *Fish. Bull. 75*, 703–716.

JOYCE, E A and ELDRED, B. The Florida shrimping industry.
1976 Florida Board of Conservation, Educational Series, No. 15, 47 p.

KRISTJONSSON, H. Techniques of finding and catching shrimp in
1969 commercial fishing. *FAO Fish. Rep. 57* (2), 125–192.

KUTTY, M N and MURUGOPOOPATHY, G. Diurnal activity of the
1968 prawn *Penaeus semisulcatus* de Haan. *J. Mar. biol. Ass. India, 10*, 95–98.

LE RESTE, L and MARCILLE, J. Reflexions sur les possibilités
1973 d'amenagement de la peche crevettiere a Madagascar. *Bull.* Madagascar, (320), 15 p.

LUCAS, C. Preliminary estimates of stocks of king prawn,
1974 *Penaeus plebejus*, in South Eastern Queensland. *Aust. J. Mar. Freshwater Res. 25*, 35–47.

LUCAS, C, KIRKWOOD, G and SOMERS, I. Assessment of the
1979 stocks of the banana prawn *Penaeus merguiensis* in the Gulf of Carpentaria. *Aust. J. Mar. Freshwater Res. 30* (5), 639–652.

MARCELLE, J. Dynamique des populations de crevettes
1978 penaeides exploitees a Madagascar. *Trav. doc. ORSTROM*, (92), 127 p.

MOFFET, A W. The shrimp fishery in Texas. Texas Parks and
1967 Wildlife Dept. *Bull. No. 50*, 28 p.

MOLLER, T H and JONES, D A. Locomotory rhythms and
1975 burrowing habits of *Penaeus semisulcatus* (de Haan) and *P. monodon* (Fabricius) (Crustacea: Penaeidae). *J. Exp. Mar. Biol. Ecol., 18*, 61–77.

MOTOH, H. Studies on the fisheries biology of the giant tiger
1981 prawn, *Penaeus monodon* in the Philippines. *SEAF-DEC, Technical report No. 7*, 128 p.

MUNRO, I S R. Biology of the banana prawn (*Penaeus merguien-*
1975 *sis*) in the south-east corner of the Gulf of Carpentaria. In 'First Australian National prawn seminar',
Maroochydore, Queensland, November 1973. *A.G.P.S.*, Canberra: 60–78.

NEAL, R A. The Gulf of Mexico research and fishery on Penaeid
1975 prawns. In 'First Australian National prawn seminar', Maroochydore, Queensland, November 1973. *A.G.P.S.*, Canberra: 60–78.

NICHOLS, J P and GRIFFIN, W L. Trends in catch-effort relation-
1975 ships with economic implications: Gulf of Mexico shrimp fishery. *Mar. Fish. Rev. 37* (2), 1–4.

PENN, J W. Tagging experiments with western king prawn,
1976 *Penaeus latisulcatus* Kishinouye II. Estimation of population parameters. *Aust. J. Mar. Freshwater Res., 27*, 239–50.

PENN, J W. Spawning and fecundity of the western king prawn,
1980 *Penaeus latisulcatus* Kishinouye, in Western Australian waters. *Aust. J. Mar. Freshwater Res., 31*, 21–35.

PENN, J W and HALL, N G (MS). Stock assessment of the Western Australian limited entry prawn fisheries with special reference to the 1975–77 triennium. *Fish. Res. Bull. West. Aust.* (in prep.).

PENN, J W and STALKER, R W. The Shark Bay prawn fishery
1979 (1970–76) Dept. Fish. Wild. West. Aust. *Rept., 38*, 38 p.

RACEK, A A. Prawn investigations in eastern Australia. *Res.*
1959 *Bull. St. Fish. N.S.W., 6*, pp 57.

RICKER, W E. Computation and interpretation of biological
1975 statistics of fish populations. *Bull. Fish. Res. Bd. Can. 191*, 328 p.

RUELLO, N V. The influence of rainfall on the distribution and
1973 abundance of the school prawn *Metapenaeus macleayi* in the Hunter River Region (Australia). *Mar. Biol. 23*, 221–228.

RUELLO, N V. Geographical distribution, growth and breeding
1975 migration of the eastern king prawn *Penaeus plebejus* Haas. *Aust. J. Mar. Freshwater Res. 26*, 343–54.

SOMERS, I F and TAYLOR, B R. Fishery statistics relating to the
1981 declared management zone of the Australian northern prawn fishery, 1968–1979. *Rep. CSIRO Mar. Lab.*

WHITE, T F C. Population dynamics of the tiger prawn *Penaeus*
1975 *esculentus* in the Exmouth Gulf prawn fishery, and implications for the management of the fishery. Ph.D. Thesis, University of Western Australia.

WICKHAM, D A and MINKLER, F C. Laboratory observations on
1975 daily patterns of burrowing and locomotor activity of pink shrimp, *Penaeus duorarum*, brown shrimp *Penaeus aztecus*, and white shrimp, *Penaeus setiferus*. *Contr. mar. Sci, 19*, 21–35.

WILLIAMS, A B. Substrates as a factor in shrimp distribution.
1958 *Limnol. Oceanog., 3*, 283–90.

VAN ZALINGE, N P. Report on the shrimp resources of the west
1980 coast of the Gulf (Bahrain, Kuwait, Saudi Arabia.) *FAO, IOFC: DMG/80/W.P. 1*, 22 p.

YOUNG, P C and CARPENTER, S. The recruitment of postlarval
1977 penaeid prawns to nursery areas in Moreton Bay, Queensland, Australia. *Aust. J. Mar. Freshwat. Res. 28*, 755–83.

186

Methods of analysis

Standardization of fishing effort and production models for brown, white and pink shrimp stocks fished in U S waters of the Gulf of Mexico

Susan L Brunenmeister

Abstract

Standardized fishing effort estimates pertaining to the offshore fishery on brown shrimp (*Penaeus aztecus* Ives), white shrimp [*Penaeus setiferus* (L.)], and pink shrimp (*Penaeus duorarum* Burkenroad) in U S waters of the Gulf of Mexico from 1965 to 1980 were developed from multiple regression models. These models predicted catch-per-unit effort (CPUE) of individual vessels fishing each shrimp species within years based on month, area, depth, bycatch of other shrimp species, and composite vessel characteristics variables (estimated from principal components analysis). Month effects were most prominent reflecting the seasonal nature of the fisheries; in general, vessel characteristics were less important than seasonal, areal and bycatch variables concerning within year variation in CPUE. Seasonal variation in effort and CPUE pertaining to each shrimp species were staggered and illustrated the ecological distinction of the species and the annually protracted nature of the shrimp fishery. Average vessel size, net number, and horsepower of vessels fishing each shrimp species increased such that the average fishing power of vessels increased by about 20% from 1965 to 1977 (vessel data were not available after 1977).

Increases in annual standardized fishing effort pertaining to each species were substantial between 1965 and 1980 and were greatest for brown shrimp and lowest for pink shrimp. During this period, the Gulf-wide CPUE of brown shrimp declined by 55% while that of white shrimp declined by 68%. Rates of increase in effort and changes in CPUE varied among catch regions of brown shrimp and pink shrimp that were defined on the basis of cluster analysis. For each shrimp species, estimates of 'm' in the generalized stock production model were close to zero in most cases.

However, variability about the Gompertz model ($m = 1 \cdot 001$) and about the logistic model ($m = 2 \cdot 000$) were about the same as that for the $m \cong 0$ case and the range of data made it difficult to choose the model that best represented each fishery. These results were based on biological years, defined on the basis of species recruitment patterns, rather than calendar years which have been used in previous analyses.

Introduction

The shrimp fisheries of the Gulf of Mexico are among the most economically important United States fisheries. Despite their importance, relatively little is known about shrimp population dynamics and the extent to which fishing affects the stocks. The bulk of the catch is comprised of pink shrimp, *Penaeus duorarum* Burkenroad; white shrimp, *Penaeus setiferus* (L.); and brown shrimp, *Penaeus aztecus* Ives. Each species is differentially available to the fishery as a whole with respect to area, depth and season. They are fished by inshore fisheries as well as by a variety of trawlers offshore whose differences in length, horsepower, range, net size and net number also reflect innovations in fishing technology within the last two decades.

Effective management of a fishery involves the reasonably accurate assessment of the impact of fishing mortality. Since the magnitude of fishing mortality is related to the magnitude of fishing effort, conservation of a stock at an optimal level necessitates estimating effective fishing effort and its relation to catch. In dealing with fisheries comprised of diverse vessel types that also vary temporally and spatially in their capability to catch a given fraction of the population, it is necessary to standardize fishing effort in order to assess its magnitude and to determine catch per unit of effort with a reasonable degree of accuracy. Standardization is also necessary in order to analyze changes in fishing technology which may affect estimation of fishing mortality during the period of interest.

This paper develops estimates of fishing power for vessels fishing shrimp in the Gulf of Mexico and employs these estimates to calculate 'surplus production' for the fisheries. Multiple regression techniques as outlined by Robson (1966) were used to estimate fishing power. The estimates of fishing power were used to analyze various aspects of the spatiotemporal distribution of fishing effort of the offshore component of the Gulf of Mexico shrimp fishery and to develop production models referring to the offshore fishery for each species of shrimp.

Materials and methods

Data sources

Landings, nominal fishing effort (hours fished), and vessel characteristics utilized in this study were drawn from data files maintained by the Southeast Fisheries Center, Technical Information Management System (SEFC, TIMS) of the National Marine Fisheries Service. These were (a) *Interview Records,* (b) *Dealer Records*, and (c) *Vessel Records*.

(a) *Interview Records* contain individual fishing trip information obtained by TIMS port agents from vessel-captain interviews. Data include pounds of shrimp (heads-off) landed by species, date landed, depth fished, number of 24-hour days fished to tenths of a day, area of catch, and vessel identification number. The species recorded are brown shrimp (*Penaeus aztecus*), white shrimp (*Penaeus setiferus*), pink shrimp (*Penaeus duorarum*), seabob (*Xiphopenaeus kroyeri* Heller), rock shrimp (*Sicyonia brevirostris* Stimpton) and royal red shrimp (*Hymenopenaeus robustus* Smith). Depth fished is coded in five fathom interverals from zero to 250 fathoms. Area of catch is categorized into one of 21 zones, numbered sequentially from the west coast of Florida to the south coast of Texas (*v* Figure 2 in Rothschild and Brunenmeister, 1984).

(b) *Dealer records* represent the entire landings of shrimp sold by commercial fishermen to dealers at U S ports and usually include much of the same information as interview records except for the magnitude of fishing effort in hours fished. The data comprising these records is recorded by the fish dealer to whom the fishermen sells his catch. Two types of Dealer Records exist. Those available for the years 1965–1975 and 1981, contain actual statistical area of catch. Those available for the years 1976–1980 contain shrimp catches as estimated by TIMS using interview records and information from port agents, since the port sampling scheme in effect from 1976 to 1980 did not generate specific catch area information.

(c) *Vessel Records* refer to fishing craft of at least five gross tons that were registered for operation within a calendar year. These records were only available for 1965 through 1977. Vessel characteristics recorded include gross tonnage, vessel length, engine horsepower, number of nets, total headrope length of nets in yards, number of full-time crew, year of vessel construction, type of hull material and type of gear. Type of gear for all vessels considered here was the otter trawl. Vessel Records for the appropriate year were matched to Interview Records according to the vessel identification number in order to generate data files containing catch, effort and vessel characteristics data by individual vessel for the years 1965 to 1977.

Data treatment

The number of interviews conducted per day in a statistical area was strongly periodical within months and the frequency in which individual vessels were interviewed during a month varied. The periodicity in number of interviews can be ascribed to the daily routine of the port agent, and to movements of the local fleet. The varying frequency in which individual vessels were interviewed reflects the cooperation of individual captains and the range in size of the vessels landing at the port since the length of time a vessel is out of port varies with vessel size. Thus, to facilitate analysis with this complex data set, landings and effort data for a single vessel recorded from the same statistical area and depth zone were summed over trips within a month. In this scheme, then, a single vessel was multiply represented in a month and statistical area sample-block of the data set only if it landed catches from more than one depth zone.

Five vessel characteristics were chosen for fishing-power analysis: (1) gross tonnage, (2) vessel length in feet, (3) horsepower, (4) net number and (5) average net size in feet (headrope in feet/net number). Type of hull material (wood or steel) was excluded since relationships between the selected variables did not change according to hull type. It can be noted, however, that most small vessels were wooden and that most large vessels were steel. Number of full-time crew was excluded as a variable since it tends to be related to vessel size and actual crew number per trip was not available. Year of vessel construction was excluded since change in configuration of new vessels due to improvements in vessel design would be reflected by changes in relationships of the selected variables as would modifications or re-outfitting accomplished by owners of older vessels.

Because there were correlations among vessel characteristics, the number of vessel variables was reduced using principal components analysis (*v eg*, Green and Carroll, 1976). The correlations among vessel characteristics in 1977 (*Table 1*) was deemed representative of other years data; thus, composite variables for vessels operating in all years were calculated using vessel characteristic loadings referring to the 1977 data set (*Table 2*). The utility of expressing vessel characteristics by these composite variables lay in the reduction, from five to three, of the number of variables employed to describe a vessel without signficant loss in information (less than 4.5%) and in the creation of a set of relatively independent variables which pertained to their utilization in the regression models developed below.

Table 1

CORRELATIONS BETWEEN CHARACTERISTICS OF VESSELS OPERATING IN THE OFFSHORE SHRIMP FISHERY IN 1977

	Gross tonnage	Net number	Net size	Vessel length
Gross tonnage	1·000	—	—	—
Net number	0·457	1·000	—	—
Net size	0·290	−0·519	1·000	—
Vessel length	0·938	0·437	0·312	1·000
Horsepower	0·803	0·368	0·318	0·758
Sample size	2 209			

Table 2

VESSEL CHARACTERISTICS LOADINGS ON PRINCIPAL COMPONENT AXES

	1	2	3
Gross tonnage	0·729	−0·271	−0·236
Net number	−0·404	0·626	−0·177
Net size	−0·334	1·441	−0·212
Vessel length	0·866	−0·168	−0·554
Horsepower	−0·468	−0·190	1·567
Cumulative % of total variance	60·0	90·2	95·6

Effort standardization
The models Data were fitted to models of the form,

$$Y_{ijkv} = B_0 + B_1 d_{ijkv} + B_2 x_v + B_3 w_v + B_4 z_v$$
$$+ B_5 p_{ijkv} + B_6 q_{ijkv} + B_7 r_{ijkv}$$
$$+ \sum_{i=1}^{12} B_{i+7} m_i + \sum_{j=1}^{21} B_{j+19} s_j + e_{ijvk}$$

where

Y_{ijkv} is the log (all logarithms are natural logarithms) of the catch-per-unit-effort (pounds/hour) of brown, white or pink shrimp of vessel (v) in month (i), statistical area (j) and depth (k)

d_{ijkv} is the log (midpoint of depth class in fathoms)

x is the log (vessel score on first principal component axis, +10·0)

w is the log (vessel score on second principal component axis, +10·0)

z is the log (vessel score on third principal component axis, +10·0)

m_i is 1·0, if the month of the observation = i, 0, otherwise; i = 1 to 12

s_j is 1·0, if the statistical area of the observation = j, 0, otherwise; j = 1 to 21

p_{ijkv} is the log (catch of brown shrimp when Y_{ijkv} refers to white shrimp, +1·0)
is the log (catch of white shrimp when Y_{ijkv} refers to brown shrimp, +1·0)
is the log (catch of brown shrimp when Y_{ijkv} refers to pink shrimp, +1·0)

q_{ijkv} is the log (catch of pink shrimp when Y_{ijkv} refers to white shrimp, +1·0)
is the log (catch of pink shrimp when Y_{ijkv} refers to brown shrimp, +1·0)
is the log (catch of white shrimp when Y_{ijkv} refers to pink shrimp, +1·0)

r_{ijkv} is the log (catch of seabob + rockshrimp, +1·0)

and the B_i are the estimated parameters and e is the error term.

Although sample sizes were adequate to estimate coefficients for each month in each species case, sample sizes were not adequate to estimate coefficients for every statistical area in which catches of brown, white or pink shrimp were recorded. However, these were cases in which relatively small catches of the species occurred and data pertaining to these cases were omitted from the regression analyses.

A complete data set was not available for the entire range of years, 1965 to 1981 studied here. Catch data were available through August 1981. Vessel characteristics data were not available for the years 1978 through 1981. Thus, the vessel characteristics terms in the model were omitted from the regression analyses pertaining to 1978 through 1981. In what follows, *complete model* refers to that containing all the terms indicated above and *reduced model* refers to that in which the vessel characteristics terms were omitted.

In each model, covariates and factors were estimated simultaneously. In order to obtain a solution, one month and one statistical area were chosen as references. The reference statistical area and reference month chosen for each species model were those containing the highest sample sizes in the entire data set of each species and were the same as those chosen as standards for standardized regional annual effort standardization (v Table 5). No interactions between variables were assumed. Incorporation of interaction terms in this model, given the unbalanced nature of the data would confound estimation of statistical area and month effects (v Searle, Speed and Henderson, 1981).

Adequacy of models and estimated coefficients
The percentage of total variability accounted for by the regression models differed among the species (*Table 3*). In both the complete and reduced models, average R^2 was highest for white shrimp and lowest for pink shrimp. No trends in values of R^2 from 1965 to 1977 in the complete models, or from 1978 to 1980 in the reduced models, were evident for any of the species cases suggesting that the models were equally adequate over the years in each case.

The relative contribution of the different variables to R^2 is given in *Table 3*. For brown shrimp, month effects on the average were most prominent followed by catch of other species, statistical area, vessel characteristics (in the complete models) and depth variables. Over-all trends among years in

the importance of variables to the complete models were consonant with those in the reduced models.

For white shrimp, month was also the most important variable in the complete and reduced models. However, depth of catch was secondary in importance followed by catch of other species, statistical area and, in the complete models, vessel characteristics. Trends over years in the relative importance of the variables to the complete and reduced models appeared similar.

For pink shrimp, month in the complete models was also the most important variable with statistical area, vessel characteristics, catch of other species and depth variables following in importance. In the reduced models, month was less important on the average, with statistical area and catch of other species variables being of more importance than in the complete models.

The most apparent effects that the lack of vessel characteristics data had on model predictability were observed in the pink shrimp model. The difference in explained variability between the complete and reduced models fitted for comparison to 1977 data showed that omitting vessel variables had little effect on the brown shrimp fit (−0·67%), more effect on the white shrimp fit (−4·48%), and most effect on the pink shrimp fit (−13·75%). That no large interactions with vessel variables existed and that the variability in catch per unit effort (CPUE) contributed by vessel type could not be partially accounted for by a correlated variable were indicated from an examination of

Table 3

AVERAGED MODEL FITS OBTAINED FOR EACH SHRIMP SPECIES AND AVERAGED RELATIVE IMPORTANCE OF VARIABLE TYPES TO REDUCTION OF UNEXPLAINED VARIABILITY IN CPUE OF VESSELS EXPRESSED AS PERCENTAGE OF R^2, THE PROPORTION OF VARIABILITY EXPLAINED BY THE REGRESSION MODEL

| Species/Years | Average R^2 | Average % of R^2 | | | | |
		Vessels	Months	Statistical Areas	Depth	Catch of other species
Brown shrimp						
complete model (1965–1977)	0·43	10·4	54·7	11·5	5·3	17·6
reduced model (1978–1981)	0·37	—	63·5	6·5	2·8	27·6
White shrimp						
complete model (1965–1977)	0·62	5·4	35·3	10·5	27·2	21·9
reduced model (1978–1981)	0·53	—	35·6	16·2	21·2	27·0
Pink shrimp						
complete model (1965–1977)	0·36	13.1	45·8	24·2	5·9	10·7
reduced model (1978–1981)	0·26	—	27·8	33·8	8·5	31·6

pink shrimp models. This examination showed that recomputed variable contributions were similar between the 1977 complete model minus the vessel contributions and the 1977 reduced model.

Estimates of the month parameters obtained for each year by the models for brown shrimp were more concordant over the years than those obtained for pink or white shrimp and suggested clear seasonal changes in shrimp abundance (*Figure 1a*). Periodicity was also discernible in the white and pink shrimp month coefficients; however, these were less consistently patterned (*Figures 1b* and *1c*, respectively). Estimates of the month parameters obtained for each species in 1977 from the reduced model form did not differ greatly from those obtained from the complete model. This suggested that the month coefficients estimated for each species for the years 1978 to 1981 in the reduced model were not greatly biased. Moreover, patterns in the coefficients obtained for each species from 1978 to 1981 were similar to those observed from 1965 to 1977.

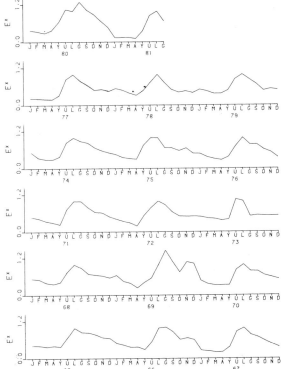

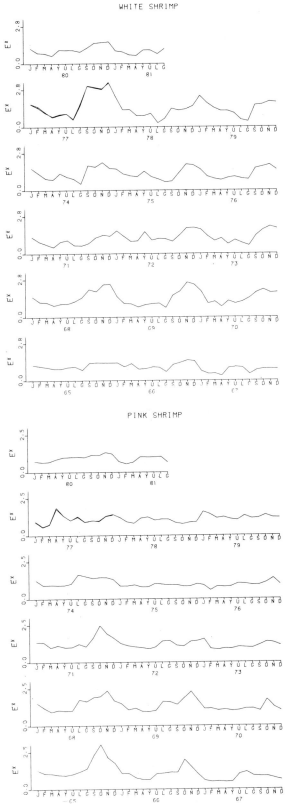

Fig 1 Month coefficients estimated by complete models (1975 to 1977) and reduced models (1977 for comparison; 1978– August 1981) for (a) brown shrimp, (b) white shrimp, and (c) pink shrimp. Estimates obtained for 1977 by the reduced models in each case are plotted over the estimates obtained from the complete models. Abscissa labels Y, U, L and G refer to months May, June, July and August, respectively

191

The influence of each statistical area on predicted CPUE of brown, white, or pink shrimp, respectively, relative to the standard statistical area varied from year to year in both the complete and reduced models. Although the effect of statistical area was of less than secondary importance in predicting CPUE's of brown or white shrimp in both model types, concordant trends among coefficients of particular statistical areas were observed over years in all species cases (*Figures 2a–2c*). As in the case of month effects, statistical area effects estimated for each species for 1977 by the reduced model did not differ dramatically from those estimated by the original models.

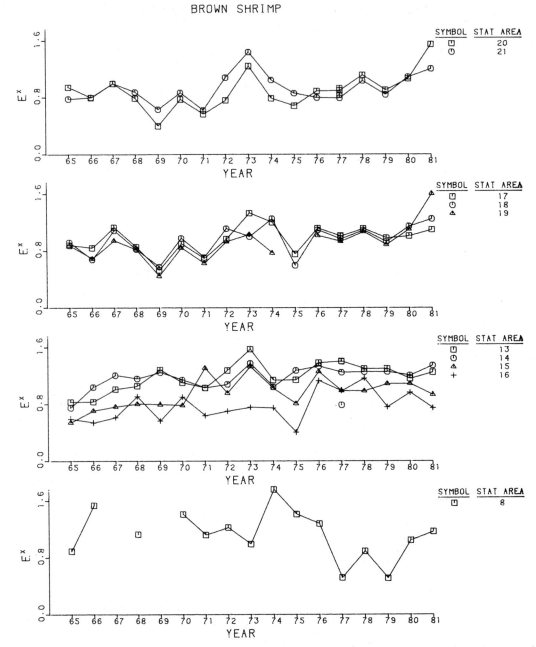

Fig 2 Statistical area coefficients estimated by complete models (1965 to 1977) and reduced models (1977 for comparison; 1978 – August 1981) for (a) brown shrimp, (b) white shrimp and (c) pink shrimp. Estimates obtained for 1977 by the reduced models in each case are plotted free-standing for comparison

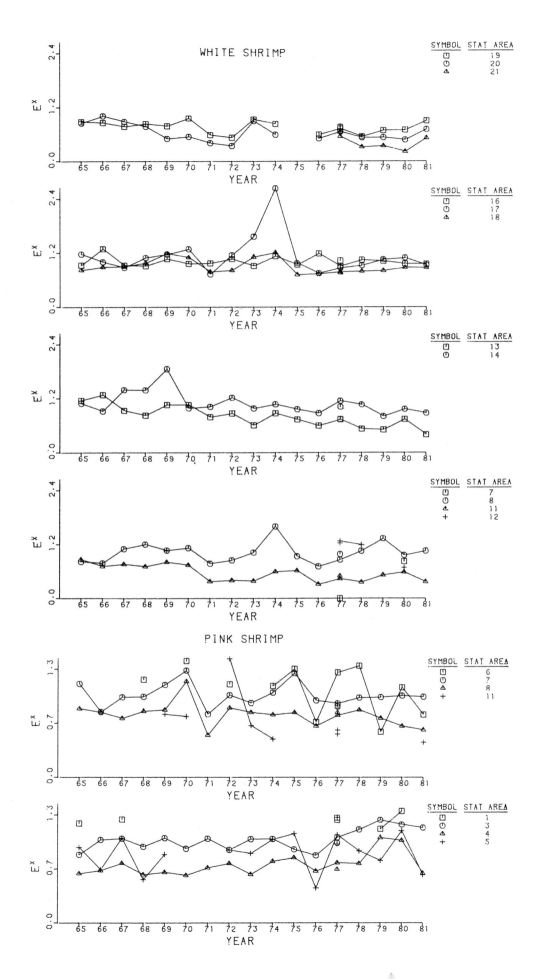

The depth coefficients estimated by both complete and reduced models indicated that CPUE of white or pink shrimp decreased wth depth. However, depth coefficients estimated for brown shrimp by the complete model indicated CPUE increased slightly with depth whereas those estimated by the reduced models for 1978 and 1979 estimated declines in CPUE with depth.

Significant coefficients of bycatch variables estimated by both complete and reduced models were negative for each species in each year as would be expected. Hence, expected CPUE for each species was adjusted downward according to the amount of catch of other shrimp species.

Estimation of standardized effort Three types of standardized effort were computed. These were (*1*) annual standardized regional (or subregional) effort in which vessel effects (years 1965–1977 only), month, statistical area and depth effects were standardized; (*2*) monthly standardized regional effort, in which the same effects except for month effects were standardized; and (*3*) annual standardized statistical area effort, in which all effects save statistical area effects were standardized.

The procedure for estimating standardized effort was as follows:
(*1*) Compute a measure of fishing power, $\hat{P}_{ijkv}$, for each individual interviewed vessel observation (v) at depth (k), statistical area (j) and month (i) (see *Table 4* for definitions of $\hat{Y}_{ijkv}$ and $\hat{Y}_{ijks}$ and *Table 5* for standards)

$$\hat{P}_{ijkv} = \frac{e^{\hat{Y}_{ijkv}}}{e^{\hat{Y}_{ijks}}}$$

Table 4
ESTIMATION EQUATIONS USED IN FISHING POWER CALCULATIONS

Interviewed vessel	$\hat{Y}_{ijkv} = B_0 + B_i d_k + B_2 v_v + B_3 w_v + B_4 w_v$ $+ B_5 P_{ijkv} + B_6 q_{ijkv} + B_7 r_{ijkv}$ $+ B_i m_i + B_j s_j$ *
Standard vessel Standardized annual effort for the entire Gulf of Mexico	$\hat{Y}_{ijks} = B_0 + B_1 d_s + B_2 v_s + B_3 w_s + B_4 z_s$ $+ B_5 p_{ijkv} + B_6 q_{ijkv} + B_7 r_{ijkv}$ $+ B_s m_s + B_s s_s$
standardized monthly effort	$\hat{Y}_{ijks} = B_0 + B_1 d_s + B_2 v_s + B_3 w_s + B_4 z_s$ $+ B_5 P_{ijkv} + B_6 q_{ijkv} + B_7 r_{ijkv}$ $+ B_i m_i + B_s s_s$
standardized statistical area effort	$\hat{Y}_{ijks} = B_0 + B_1 d_s + B_2 v_s + B_3 w_s + B_4 z_s$ $+ B_5 p_{ijkv} + B_6 q_{ijkv} + B_7 r_{ijkv}$ $+ B_s m_s + B_j s_j^*$

* In those cases where no coefficient for the pertinent statistical area had been estimated due to small sample size, the coefficient of another statistical area was substituted in the estimation equation. The substituted statistical area either showed the highest correlation in estimated coefficients with the statistical area in question in other years, or, if no ancillary information was available, was simply the closest area geographically for which a coefficient had been estimated.

(*2*) Standardize the reported effort, f_{ijkv}, of each interviewed vessel to f^*_{ijkv}, as

$$f^*_{ijkv} = \hat{P}_{ijkv} \cdot f_{ijkv}$$

(*3*) Estimate standardized CPUE statistics for the interviewed fleet, as
(*a*) annual standardized regional (or subregional) CPUE a_t

$$a_t = \frac{\sum^i \sum^j \sum^k \sum^v C_{ijkv}}{\sum^i \sum^j \sum^k \sum^v f^*_{ijkv}}$$

Table 5
STANDARD VESSEL CHARACTERISTICS, MONTHS, STATISTICAL AREAS AND DEPTHS USED IN CALCULATING STANDARDIZED EFFORT FOR BROWN, WHITE AND PINK SHRIMP.

Standard vessel characteristics and standard depth were the grand means of vessels and depth variables in the data sets pertaining to each species for 1965 to 1977. Standard month and standard statistical areas were those containing the highest sample size in the data sets pertaining to each species.

	Brown shrimp	White shrimp	Pink shrimp
Vessel characteristics			
Vessel gross tonnage	79·55	64·06	64·06
Horsepower	272·07	235·47	231·90
Vessel length in feet	62·37	57·73	58·49
Net number	2·04	1·96	1·99
Average net size in feet	54·58	52·95	53·71
Month	July	June	June
Statistical Areas			
For regions	Nos. 1–21: No. 11	Nos. 1–21: No. 15	Nos. 1–21: No. 2
For subregions	Nos. 11–16: No. 11	—	Nos. 1–4: No. 2
For subregion	Nos. 17–21: No. 20	—	Nos. 5–9: No. 8
Depth in fathoms	17	16	6

194

where j is summed over the statistical areas of interest

(b) monthly standardized regional CPUE, a_i

$$a_i = \frac{\sum^j \sum^k \sum^v C_{ijkv}}{\sum^j \sum^k \sum^v f^*_{ijkv}}$$

(c) annual standardized statistical area CPUE, a_j

$$a_j = \frac{\sum^i \sum^k \sum^v C_{ijkv}}{\sum^i \sum^k \sum^v f^*_{ijkv}}$$

and C_{ijkv} in each case is the catch of the species referred to by $\hat{Y}_{ijkv}$

(4) Estimate the standardized CPUE for the catch, C^*_{ijk} of the entire fleet, as

(a) annual standardized regional (or sub-regional) effort, f^*_t

$$f^*_t = \frac{\sum^i \sum^j \sum^k C^*_{ijk}}{a_t}$$

where j is summed over the statistical area of interest

(b) monthly standardized regional CPUE, f^*_i

$$f^*_i = \frac{\sum^j \sum^k C^*_{ijk}}{a_i}$$

(c) annual standardized statistical area CPUE, f^*_j

$$f^*_j = \frac{\sum^i \sum^k C^*_{ijk}}{a_j}$$

Results

Changes in fishing power

For the years 1965 to 1977, trends in the vessel characteristics of the shrimp fleet, as indicated by average vessel scores of interviewed vessels, were very evident and were relatively concordant among vessels catching brown, white, or pink shrimp. The average gross tonnage and length of vessels that caught white or brown shrimp increased from 1965 and 1975 and declined thereafter (*Figure 3a*). This decline was more marked for those vessels that caught brown shrimp, which were larger vessels on the average than those which caught white shrimp. Gross tonnage and vessel length also increased over the years among vessels that caught pink shrimp (*Figure 3a*). Average net number and net size of vessels that caught each species increased through 1974 or 1975, with declines thereafter concomitant with declines in vessel size (*Figure 3b*). Lastly, steady increases in engine horsepower were evident in vessels that caught each species suggesting that not only were these increases due to increased average vessel size over the years, but also to increased powering of vessels (*Figure 3c*).

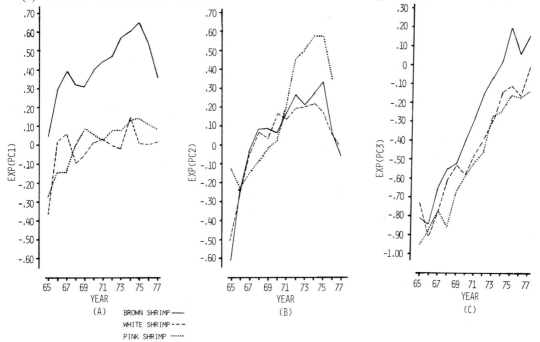

Fig 3 Annual variation in characteristics of the average type of vessel fishing for brown, white or pink shrimp as evidenced by vessel composite variables reflecting primarily (a) gross tonnage and length, (b) net number and net size, and (c) engine horsepower

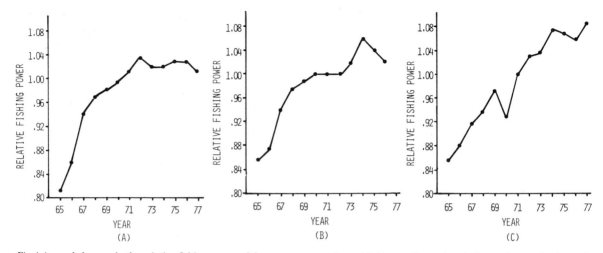

Fig 4 Annual changes in the relative fishing power of the average type of vessel fishing each year in relation to the standard vessel fishing (a) brown shrimp, (b) white shrimp, and (c) pink shrimp

Changes in the fishing power of the average vessel type operating in a year relative to the standard vessel type (calculated for the average vessel and the standard vessel at the species' standard depth, month and statistical area) showed an increase of approximately 20% occurred in the fishing power of vessels that caught each shrimp species between 1965 and 1977 (*Figure 4a–4c*).

How changes in the complexion of the fleets fishing each species were related to changes in the distribution of fishing mortality exerted by vessels in the fleets is illustrated in *Figures 5a, 5b,* and *5c.* In these plots, annual landings of each interviewed vessel in 1965 and 1977, respectively, expressed as a percentage of the total landings of vessels of the interviewed fleet in each year is used as a reflection of vessels in the entire fleet. In the brown shrimp fishery, vessel size varied in proportion to mag-

nitude of landings in both example years. This was true of vessels fishing white shrimp in 1977 but was not a pattern that was evident in 1965. In the pink shrimp fishery in both 1965 and 1977, average vessel size varied little with catch magnitude.

In terms of the distribution of the annual catch among individual vessels, the distributions pertaining to vessels landing brown or white shrimp were more skewed in 1977 than in 1965 and in both years were more skewed than those pertaining to pink shrimp in 1965 or 1977, which were similar (*Figures*

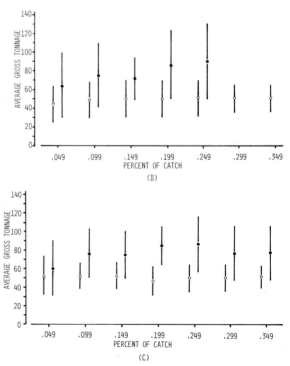

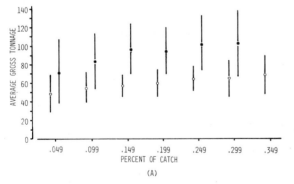

Fig 5 Average gross tonnage of vessels landing various proportions of the catch of (a) brown shrimp, (b) white shrimp and (c) pink shrimp as reflected by interviewed vessels in two sample years, 1965 (open circles) and 1977 (solid circles). Vertical lines represent standard deviations about means

196

6a, 6b, and 6c). Thus, in 1977, 74% and 71% of the vessels landing brown or white shrimp, respectively, landed less than 0·10% of the total catch, whereas 51% of the vessels landing pink shrimp fell into this category.

These patterns suggest, then, that fishery management options concerning catch limitations, nominal effort limitation, or fishing power constraints designed to affect fishing mortality levels would exert different effects in the fisheries for brown, white and pink shrimp, respectively.

Temporal patterns

Fluctuations in monthly standardized regional effort and in catch per unit of monthly standardized region effort for brown, white and pink shrimp are shown in *Figures 7, 8,* and *9,* respectively. Monthly

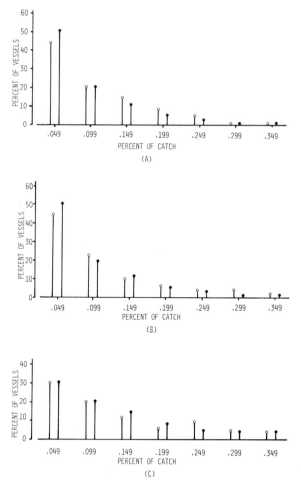

Fig 6 Number of vessels landing various proportions of the catch of (a) brown shrimp, (b) white shrimp and (c) pink shrimp as reflected by interviewed vessels in two sample years, 1965 (open circles) and 1977 (solid circles). Vertical lines represent standard deviations about means

effort for brown shrimp was cyclic. Effort was lowest in January, February and March and highest in July and August. The highest magnitude of effort observed during the study period occurred in August 1981. CPUE was highest in July or August. That the July 1981 CPUE level was as high as those occurring in earlier years when effort levels were much lower may be related to novel regulations imposed in 1981. These regulations effected a closure on brown shrimp fishing off the Texas coast from 4 fathoms to 100 fathoms from May 22 to July 15.

Effort for white shrimp tended to be bimodal in the early years of the study period, with the first peak occurring during May or June and the second peak occurring in the fall, during October or November. CPUE tended to be high in the fall, and low in late summer.

Monthly effort levels estimated for pink shrimp also varied seasonally. Lowest effort was expended from July to September; effort was highest during the winter and early spring months, with peaks occurring between January and April. CPUE also fluctuated broadly throughout each year. Highest CPUE occurred during the summer and early fall; lowest returns were observed from January or February through April or May. These cyclic patterns were most pronounced and consistent prior to 1975. These patterns suggest, then, that annual cycles in effort and CPUE occur in each species fishery and that, as a whole, the fishery is temporally partitioned among the three species.

Spatial patterns

To examine spatial heterogeneity, trends in catch per unit of annual standardized statistical area effort were compared among statistical areas from which the greatest portions of the catches of each species were landed which were those areas for which statistical area coefficients had been estimated by the models. In each species case, correlations in catch per unit of annual standardized statistical area effort between vessels operating in different statistical areas were highest between those operating in adjacent areas. These patterns of correlations were examined using cluster analysis. The average linkage method, which formed clusters on the basis of the average similarity between all variables in each cluster was employed.

The cluster analysis for brown shrimp indicated that Area 8 differed greatly from the Areas off the Louisiana and Texas coasts (*Figure 10a*). The method grouped Areas 18–21 (off the lower Texas coast) together and grouped the adjacent Area 17

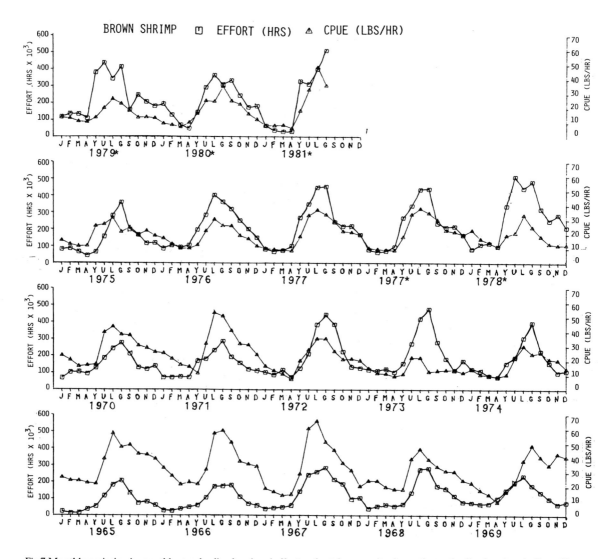

Fig 7 Monthly variation in monthly standardized regional effort and catch-per-unit-of-month standardized regional effort of brown shrimp. Starred years represent estimates obtained using reduced models; other years refer to estimates obtained from complete models

198

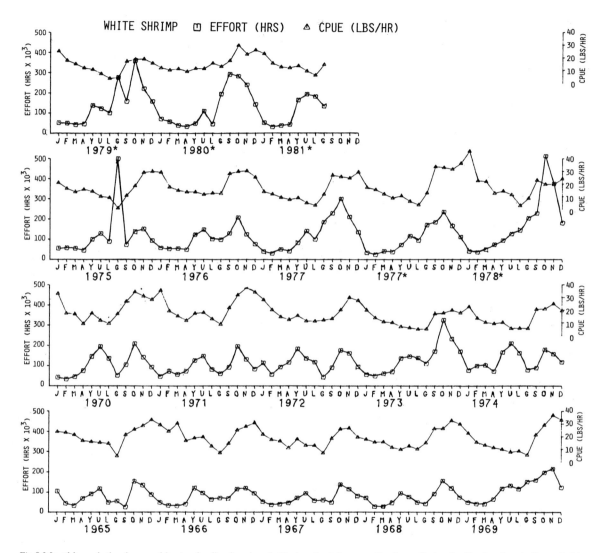

Fig 8 Monthly variation in monthly standardized regional effort and catch-per-unit-of-month standardized regional effort of white shrimp. Starred years represent estimates obtained using reduced models; other years refer to estimates obtained from complete models

199

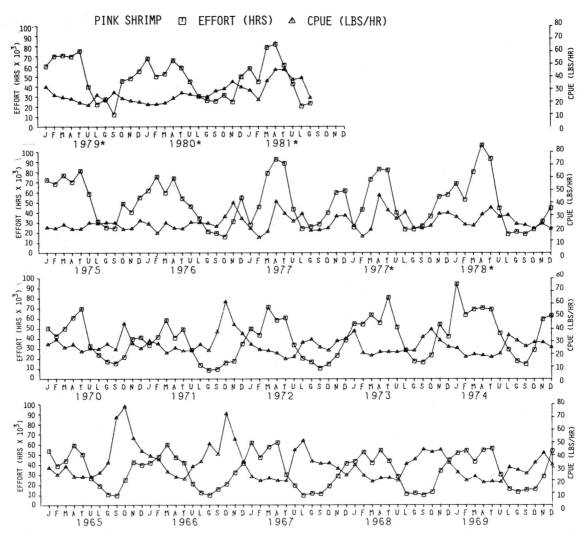

Fig 9 Monthly variation in monthly standardized regional effort and catch-per-unit-of-month standardized regional effort of pink shrimp. Starred years represent estimates obtained using reduced models; other years refer to estimates obtained from complete models

closely to Areas 18–21. The remaining areas, Nos. 11–16, all of which lie off the central and eastern regions of Louisiana, were grouped as a distinct unit.

Analysis of the pattern of correlations in CPUE among catch areas of white shrimp indicated three groups (*Figure 10b*). Since these groups were not adjacent, they did not appear to reflect a substock structure as was observed for brown shrimp. However, the patterns do suggest that CPUE trends were similar in the marginal areas of fishing for this species for which interview samples were available, *ie*, No. 8 and No. 20. The next cluster grouped the adjacent marginal areas for which samples were available, No. 11 and No.'s 18 and 19. Hence, the

cluster analysis of white shrimp was not suggestive of substock structures.

Cluster analysis of pink shrimp CPUE correlations among statistical areas revealed two distinct groupings; these were Area 11 and Areas 2 through 8 (*Figure 10c*). The latter group was separated into Areas 2 and 7, Areas 3 and 4, and Areas 5, 6, and 8 respectively. Areas 2 and 7 were most distinct of the areas off Florida. However, it should be noted that these groups were separated at relatively high correlations and thus CPUE trends among the sectors in Areas 2–8 were highly concordant. These patterns suggested, then, that significant areal patterns in CPUE of brown and pink shrimp within the Gulf of Mexico existed.

200

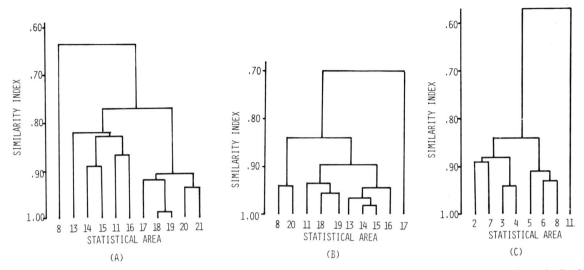

Fig 10 Correlograms obtained using the average linkage method of cluster analysis on catch-per-unit-of-annual standardized statistical area effort among statistical areas for (a) brown shrimp, (b) white shrimp and (c) pink shrimp

Annual regional catch and effort trends

The temporal and spatial patterns in CPUE of each shrimp species suggested two approaches to characterizing annual changes in the fisheries. The first was to analyze catch and effort trends in subregions suggested by the cluster analyses results. The second was to partition the data by biological years as well as by calendar years.

Regional groupings chosen for brown shrimp were the subregions Areas 11–16 (standardized to Area 11), and Areas 17–21 (standardized to Area 20), and the Gulf of Mexico as a whole (Areas 1–21, standardized to Area 11). Because no interpretable areal patterns resulted from the cluster analysis pertaining to white shrimp, only the Gulf of Mexico as whole (Areas 1–21, standardized to Area 15) was chosen for study. Regional groupings for pink shrimp catches in the eastern Gulf of Mexico were made considering the patterns formed by the cluster analyses, the contiguity of statistical areas, and patterns in the average size of shrimp landed per month among areas (Brunenmeister, unpublished data). Thus, Areas 1–9 (standardized to Area 2) were grouped together and Areas 1–4 (standardized to Area 2) and Areas 5–9 (standardized to Area 8) were treated as subregions. No other statistical areas were included in the analysis of pink shrimp since recorded catch levels of them in other areas are comparatively low.

Biological years for each species were defined on the basis of recruitment patterns. For brown shrimp, a biological year was defined from May through April since these months demarcated best the annual cycle in average size of shrimp landed

and in CPUE expressed in number of shrimp (*v* Rothschild and Brunenmeister, 1983, Figures 3 and 7). Similarly, for white shrimp patterns in these descriptors (*v* Rothschild and Brunenmeister, 1983, Figures 4 and 8) suggested that a biological year began in September and terminated in August of the following year. The period September–August was selected as the biological year for pink shrimp assessment since Kutkuhn (1966) noted that the period from September–March marked the greatest migrations of maturing shrimp to the Tortugas trawling grounds and the period of heaviest fishing. Also, recruitment patterns to the offshore fishery were defined by these bounds (*v* Rothschild and Brunenmeister, 1983, Figures 5 and 9). Also, monthly variation in effort levels depicted in *Figure 9* shows that September marks the onset of a fishing cycle.

Because of the largely annual cycle of these shrimp species, fluctuations in the shrimp fisheries would appear to be better described by biological years rather than by arbitrarily delimited calendar years. Hence, trends on the basis of biological year are described here. Previous work on these shrimp fisheries have utilized the calendar year (*eg*, Lopik *et al.*).

Effort levels for brown shrimp Gulf-wide showed a steady increase from 1965 to 1980 of approximately 8·4% per year (*Figure 11a*). Effort trends in the subregions, were disparate with effort in Areas 11–16 increasing more rapidly, at an average rate of ca. 15% per year, than in Areas 17–21 in which it increased at an average rate of ca. 2·4% per year (*Figures 11b* and *11c*). The continu-

201

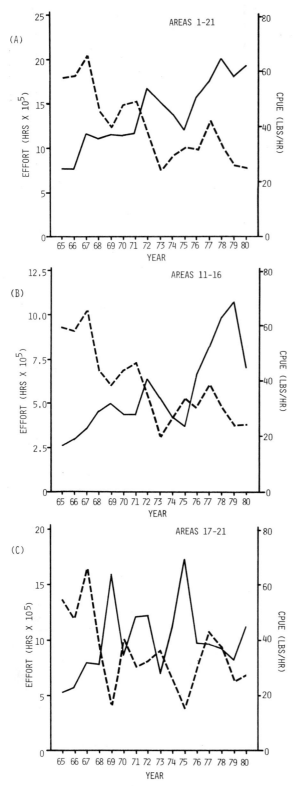

Fig 11 Trends in annual (biological year) regional/subregional standardized effort (solid line) and catch-per-unit-of-annual standardized regional/subregional effort (dashed line) for brown shrimp in (a) Areas 1–21, (b) Areas 11–16 and (c) Areas 17–21

ous decline in effort observed in Areas 11–16 from 1973 to 1975 was not evident in Areas 17–21, although the effort level in the latter area in 1973 was substantially lower than in 1972. These effort patterns calculated on the basis of a 'biological' year are apparently more reflective of prevailing conditions during the fishing season than those derived from calendar year calculations, especially in Areas 11–16. Mitchell (1976) states that most shrimp vessels were operated at a loss from mid-1973 to mid-1975 and that a number of vessels were sold for conversion to other U S fisheries or for entry into foreign shrimp fisheries as a result of dramatically increased fuel and equipment costs during that period. That effort levels in Areas 17–21 did not remain depressed through 1974 and 1975 may be due to differential characteristics of the fleets operating in the subregions areas, as Mitchell also noted that despite losses, many producers continued to operate their boats in order to keep their crews working. The effort level in Areas 11–16 increased from 1975 to 1979 and declined sharply in 1980. The effort level in Areas 17–21 showed no remarkable increase from 1976 to 1980, sustaining an average level observed since 1970.

Annual effort levels for white shrimp Gulf-wide increased overall from 1965 to 1980 (*Figure 12*). Lowest effort levels occurred in 1967, 1972, 1975 and 1979; highest levels occurred at approximately 4–5 year intervals, during 1969, 1973 and 1977. Average increases in effort were approximately 7·6% per annum.

Effort levels for pink shrimp in Areas 1–9 increased during the period from 1965 to 1980 at an

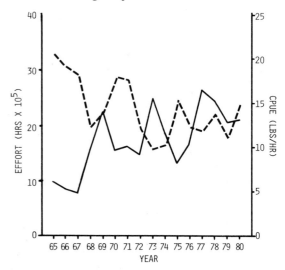

Fig 12 Trends in annual (biological year) regional standardized effort (solid line) and catch-per-unit-of-annual standardized regional effort for white shrimp

202

average annual rate of approximately 1·2% (*Figure 13a*). However, trends in increases in effort levels differed markedly between subregions (*Figures 13b* and *13c*). While effort levels in areas 1–4 increased an average of 1·3% per year, effort levels in Areas 5–9 increased sharply in an exponential fashion from 1965 through 1974/1976 and thereafter declined to levels approximately those of the early 1970's.

The CPUE of brown shrimp in the entire Gulf and in each subregion showed declines to 1980 from the peak years of 1965–1967 (*Figures 11a–11c*). While CPUE in both subregions declined, lowest CPUE levels in Areas 11–16 occurred during 1969, 1973 and 1979 whereas those in areas 17–21 occurred during 1969, 1975 and 1979. These trends differed somewhat from the Gulf-wide estimates which indicated lowest CPUE's occurred in 1969, 1973 and 1980.

The CPUE of white shrimp during this period showed an overall decline despite peaks in 1965, 1970 and 1975 (*v Figure 12*). After 1975, CPUE appears to have fluctuated around a low average level showing a slowing of the decline observed from 1965 to 1976. The average annual decline in CPUE was approximately 3·8% between 1965 and 1980.

The CPUE of pink shrimp in Areas 1–9 declined from 1965/1966 to 1967/1968 and thereafter fluctuated around an average low level up to 1979 (*v Figure 13a*). CPUE increased dramatically in the 1980 season to levels previously observed in 1965/1966. With subregions, CPUE in Areas 1–4 mirrored those seen overall in Areas 1–9, and not surprisingly, since the bulk of the catch is derived from this subregion (*v Figure 13b*). CPUE in Areas 5–9 evidenced a slow exponential decline, despite increasing catch levels (*v Figure 13c*).

Production models Production models pertaining to each shrimp species were estimated by fitting the generalized stock production model to annual regional (or subregional) catch and standardized effort data using PRODFIT (Fox, 1975). For brown shrimp on a Gulf-wide basis, the best fit was obtained with $m = 0·31$ ($R^2 = 0·572$), but similarly good fits were obtained with the Gompertz form ($R^2 = 0.568$) and the logistic form ($R^2 = 0·551$) (*Table 6*; *Figure 14a*). Estimated maximum sustainable yields ranged from 55·8 to 59·0 million pounds with 2·5 and 1·6 million hours of effort, respectively. The optimum effort level estimated by the best fit model exceeded that of the highest observed level (which occurred in 1978) by 25·5%, whereas that estimated by the logistic model

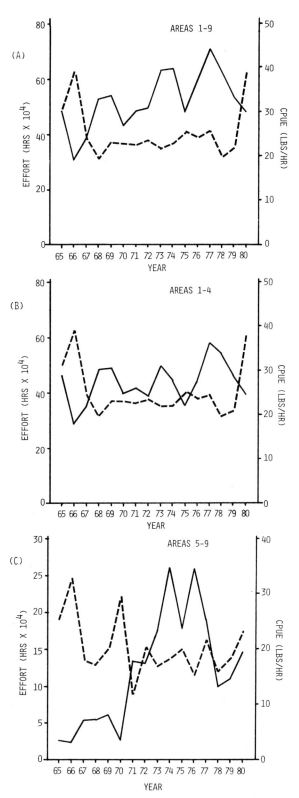

Fig 13 Trends in annual (biological year) regional/subregional effort (solid line) and catch-per-unit-of-annual standardized regional/subregional effort (dashed line) for pink shrimp in (a) Areas 1–9, (b) Areas 1–4 and (c) Areas 5–9

203

approximated the 1976 observed effort level. MSY's estimated by each model were slightly above the average offshore catch level of 52·0 million pounds observed during the study period.

With respect to the subregions, the best fit to catch and effort levels of brown shrimp in Areas 11–16 was obtained with $m = 0$ (*Table 6*; *Figure 14b*; $R^2 = 0·551$). This model implies over-fishing cannot occur, *ie*, regardless of increasing effort levels, a maximum yield of 36·4 million pounds

Table 6

BROWN SHRIMP PRODUCTION MODEL ANALYSIS FOR BIOLOGICAL YEARS 1965–1980

R^2 = proportion of variance explained
CPUE = catch per unit effort in pounds (heads-off)
f = standardized regional annual fishing effort in hours fished
MSY = maximum sustainable yield in pounds (heads-off)

	m not fixed		*m fixed*
All Statistical Areas	$m = 0·311$	$m = 1·001$	$m = 2·000$
R^2	0·572	0·568	0·551
Optimum CPUE	23·233	33·226	36·252
Optimum f	2 530 830	1 735 100	1 628 010
MSY	58 800 000	57 651 400	59 018 000
Statistical Areas 11–16	$m = 0·000$	$m = 1·001$	$m = 2·000$
R^2	0·551	0·517	0·485
Optimum CPUE		25·033	29·723
Optimum f		1 054 640	905 433
MSY	36 484 400	26 400 000	26 912 500
Statistical Areas 17–21	$m = 1·521$	$m = 1·001$	$m = 2·000$
R^2	0·693	0·682	0·684
Optimum CPUE	35·483	37·778	33·863
Optimum f	1 014 760	929 831	1 094 260
MSY	36 006 700	35 126 700	37 055 200

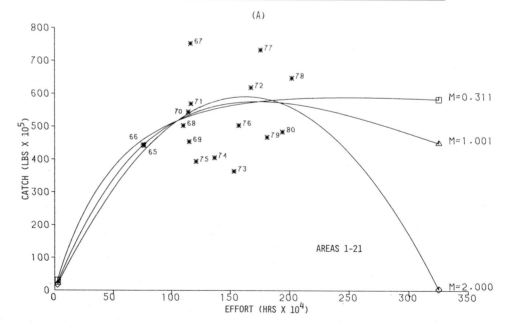

Fig 14 The best model, the Gompertz model ($m = 1·001$) and the logistic model ($m = 2·0$) fitted by the generalized stock production model to annual (biological year) catches and annual (biological year) standardized regional/subregional effort levels of brown shrimp in (a) Areas 1–21, (b) Areas 11–16 and (c) Areas 17–21

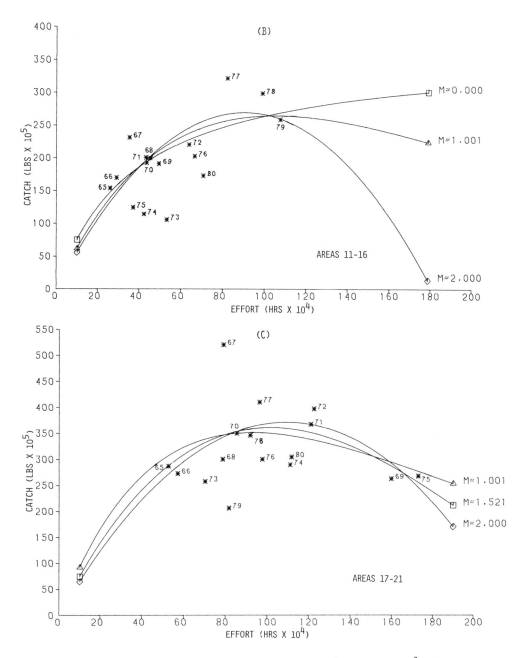

would be realized. However, a more appropriate conclusion to draw from this data is that effort has not substantially exceeded that which produces maximum sustainable yield. Very similar MSY estimates of 26·4 and 26·9 million pounds with 1·1 and 0·9 million effort hours, respectively was obtained with the Gompertz and the logistic models.

In contrast to Areas 11–16, the best model fit for brown shrimp catch and effort data pertaining to Areas 17–21 indicated a more well defined maximum ($m = 1·521$; $R^2 = 0·693$; *Figure 16c*). However, the values of R^2 for the Gompertz and the logistic models were quite similar. The parabolic form estimated MSY at 36·0 million pounds at 1·0 million effort hours. These values approximated average observed catch and effort levels of 32·1 million pounds and 0·99 million hours, respectively. The fits of these production models to the data suggest that 1969 and 1975 were anomalous years; indeed, recruitment levels to the fishery in this region from 1965 to 1981 as reflected by the CPUE in numbers of shrimp during July appeared to be lowest during 1969 and 1975 (Brunenmeister, unpublished data).

The Generalized Stock Production Model applied to the biological year catch and standardized effort data of white shrimp estimated the asymptotic yield form as the best fit ($R^2 = 0.508$; Table 7, Figure 15). The Gompertz form ($R^2 = 0.499$) and the logistic from ($R^2 = 0.468$) gave successively poorer fits. As can be seen in the figure, the present range of data is not sufficient to determine which model most appropriately describes the fishery's dynamics. However, effort levels have not exceeded that which produces a maximum sustainable yield. Maximum sustainable yield estimated by the Gompertz model and the logistic model approximate the range in yields observed at the highest effort levels during the study period.

Best fits of the Generalized Stock Production Model to pink shrimp biological year catch and standardized effort data for Statistical Areas 1–9 and Areas 1–4 were to the asymptotic yield model ($R^2 = 0.380$; $R^2 = 0.408$, respectively; Table 8; Figures 16a, 16b). In each case, the Gompertz and logistic model forms produced successively poorer fits. Reference to the figures for each region, however, indicates that the most appropriate model for each area cannot be discerned with the present ranges of data. Model fits to Areas 5–9 were equally poor and did not differ appreciably among one another (Table 8; Figure 16c). Here too, the range of data was not sufficient to distingish which among the models provided the best description of

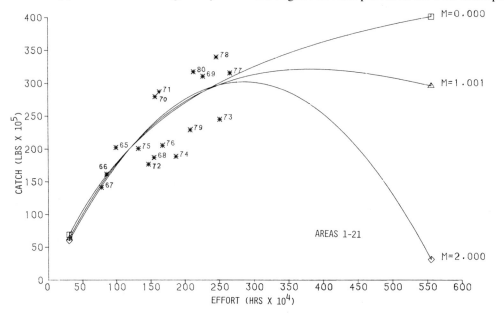

Fig 15 The best model, the Gompertz model ($m = 1.001$) and the logistic model ($m = 2.0$) fitted by the generalized stock production model to annual (biological year) catches and annual (biological year) standardized regional effort levels of white shrimp in Areas 1–21

Table 7

WHITE SHRIMP PRODUCTION MODEL ANALYSIS FOR BIOLOGICAL YEARS 1965–1980

R^2 = proportion of variance explained
CPUE = catch per unit effort in pounds (heads-off)
f = standardized regional annual fishing effort in hours fished
MSY = maximum sustainable yield in pounds (heads-off)

	m not fixed	m fixed	
All Statistical Areas	$m = 0.000$	$m = 1.001$	$m = 2.000$
R^2	0.508	0.499	0.486
Optimum CPUE		8.470	10.601
Optimum f		3 798 630	2 850 570
MSY	55 362 600	32 174 800	30 219 100

the underlying dynamics of the fishery. Finally, while production model fits of pink shrimp were poorer than those obtained for white or brown shrimp, it can be noted that treating Areas 1–4 and 5–9 separately resulted in improved fits of the models relative to those dealing with both subregions combined.

Discussion

Two aspects of uncertainty in fisheries have been distinguished: the behavior of fishermen and the behavior of Nature (Rothschild, 1972). There are several reasons why these factors could vary among the shrimp fisheries, and which may account for the average differences in the models to predict the

Table 8

PINK SHRIMP PRODUCTION MODEL ANALYSIS FOR BIOLOGICAL YEARS 1965–1980

R^2 = proportion of variance explained
CPUE = catch per unit effort in pounds (heads-off)
f = standardized regional annual fishing effort in hours fished
MSY = maximum sustainable yield in pounds (heads-off)

	m not fixed		m fixed
Statistical Areas 1–9	$m = 0.000$	$m = 1.001$	$m = 2.000$
R^2	0·380	0·358	0·335
Optimum CPUE		17·244	20·282
Optimum f		894 058	741 759
MSY	22 002 300	15 417 600	15 044 700
Statistical Areas 1–4	$m = 0.000$	$m = 1.001$	$m = 2.000$
R^2	0·408	0·387	0·364
Optimum CPUE		19·289	21·678
Optimum f		614 168	544 334
MSY	14 612 600	11 846 800	11 800 100
Statistical Areas 5–9	$m = 1.101$	$m = 1.001$	$m = 2.000$
R^2	0·391	0·372	0·354
Optimum CPUE	2·097	9·568	12·701
Optimum f	4 103 360	559 628	362 662
MSY	8 603 640	5 354 560	4 606 280

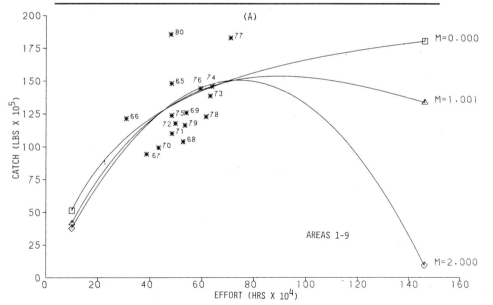

Fig 16 The best model, the Gompertz model ($m = 1.001$) and the logistic model ($m = 2.0$) fitted by the generalized stock production model to annual (biological year) catches and annual (biological year) standardized regional/subregional effort levels of pink shrimp in (a) Areas 1–9, (b) Areas 1–4 and (c) Areas 5–9. (*Continued overleaf*)

207

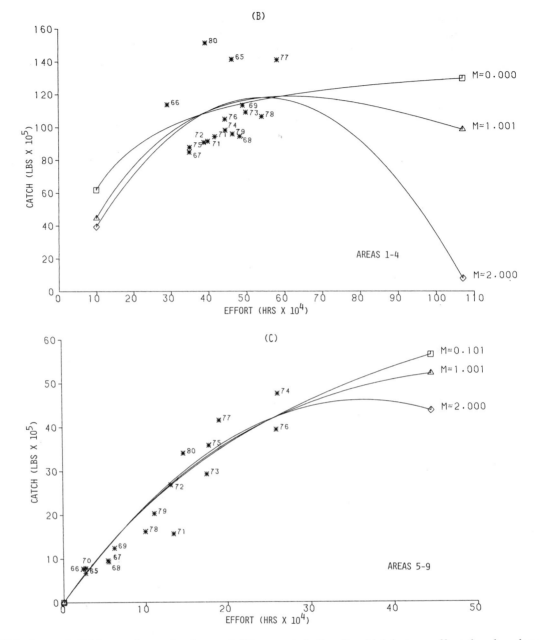

CPUE of vessels fishing each species, beyond differences in the suitability of the model construct (*eg* lack of interaction terms). With respect to the behavior of Nature, life history patterns of the species differ as do their environmental characteristics and tolerances. For example, brown shrimp prefer higher salinities than white shrimp and appear to tolerate low temperatures better than white shrimp (Farfante, 1969). Such factors affect the vulnerability of the populations to environmental perturbations and the variability and amplitude of population responses to environmental stimuli, which in turn affect the abundance and distribution of populations and the conformity (and predictability) of distribution patterns in space or time.

With respect to the behavior of the fishermen, important factors are fishing goals and constraints. For example, since economic costs incumbent in vessel operation vary according to vessel size, different vessel sizes may be expected to impart different kinds of fishing constraints. The reduction in effort and concomitant decline in average vessel size observed in the brown shrimp fishery in response to the mid-1973 fuel price escalation which was not observed in the pink or white shrimp fisheries is suggestive of such disparity. Variation in skill among captains also could have imparted a degree of unpredictability to the models since this factor was unaccounted for. The importance of the fisherman's skill over that of vessel characteristics *per se* to fishing power has been pointed out (*eg*, Rothschild, 1972). The degree to which the distributions of fishing skill varied among captains of vessels fishing brown, white or pink shrimp was not estimable; however, it is worth noting that these distributions could have changed over the years studied here according to changes in fishermen's attitudes regarding fishing success, their incentives to fish as a vocation, and economics of fishing. Smith (1974), for example, lists a variety of needs cited by salmon fishermen that constituted fishing satisfaction and which compensated for low fishing success in economic terms. That changes in the percentage of the catches of each species landed by individual vessels occurred between 1965 and 1977 suggests the fisheries have been dynamic in these respects.

The relative importance of the various variables to the variability explained by the models pertaining to vessels fishing each species showed that factors affecting the CPUE of vessels varied somewhat among the fisheries. That month effects were generally most important for each species emphasizes the common feature of temporal variation in shrimp abundance and size. However, the relatively greater importance of the catch area to predicted CPUE of pink shrimp than to predicted CPUE of brown or white shrimp is related to the heterogeneity in size of shrimp (and hence abundance) among statistical areas due to juvenile migrations offshore which are not evidenced to the same degree by brown or white shrimp. The influence of the catch of other species on CPUE was highest for vessels fishing brown and white shrimp, and both these valuable species frequently comprised catches of individual vessels. In contrast, lesser-valued rock shrimp usually comprised the bycatch of pink shrimp and therefore catch of other species emerged as a less important factor in predicting CPUE of this target species. Lastly, the greater importance of depth on CPUE of white shrimp can be related to the more restricted depth distribution of this species compared to brown or pink shrimp, as catches of white shrimp fall off dramatically beyond 15 fathoms.

The estimation of the fishing power of the average type of vessel operating offshore during each year of the study compared to the standard vessel (which was the average type of vessel operating over the entire span of years) for which vessel data were available suggested that the fishing power of vessels fishing each shrimp species increased on the average about 20% between 1965 and 1977. Changes in fishing power in the brown and white shrimp fisheries were most dramatic between 1965 and 1969 or 1970, and the rate of change in increases in fishing power declined thereafter. This was not observed in the pink shrimp fishery, however, in which increases were fairly sustained throughout the period. Lack of adjustment in effort levels for the increases in vessel size and powering and changes in gear that occurred in this 13 year period would thus bias estimates of shrimp population trends, as reflected by CPUE, upward, although the degree of bias in the later years would be less for brown and white shrimp than for pink shrimp. In this regard, it should be noted, that the average characteristics of vessel operating from 1965 to 1977 as evidenced by the interviewed portion of the fleet fishing each species, changed by more than 20% between 1965 and 1977. Thus, the rate of increase in fishing power lagged increases in vessel size, powering and gear, and in the case of brown and white shrimp, appeared in more recent years to increase in the rate of decline. These aspects emphasize the importance of quantifying actual measures of fishing power, since in the case of brown and white shrimp seen here, increases in the apparent characteristics of vessels did not correspond to increases in fishing power in a consistent fashion.

Effort levels estimated for brown and white shrimp Gulf-wide and in the eastern Gulf of Mexico for pink shrimp on the basis of biological years all increased substantially between 1965 and 1980. Average annual increases were greatest for brown shrimp with those for white and pink shrimp following in magnitude.

CPUE levels of brown shrimp and white shrimp, respectively, calculated on the basis of biological years decreased from 1965 to 1980 while those of pink shrimp showed a recovery in 1980. The general decline in CPUE from the years 1965–1967 to 1978–1980 observed for brown shrimp (55%) was slightly lower than that observed for white shrimp (68%). CPUE of pink shrimp in the major production region (Areas 1–9) was relatively stable from 1967 to 1976 after falling off from high levels in 1965 and 1966; however, CPUE increased in 1980 up to the high level observed in 1965/1966. Gulf-wide catch trends of brown shrimp showed overall little increase in production occurred between 1965 and 1980. Catches of white shrimp and pink shrimp, however, did increase during these years. The average annual catch of pink shrimp in the eastern Gulf of Mexico (Areas 1–9) increased approximately by 38% while that of white shrimp increased by 49% (v Rothschild and Brunenmeister, 1983, Figure 1).

The best production model fit to Gulf-wide biological year catch and effort data of brown shrimp was by an exponential form ($m = 0.311$) whereas that to white shrimp was by the asymptotic yield form. The asymptotic form also best fit data for pink shrimp in the eastern Gulf (Areas 1–9). These patterns indicated that effort levels during fishing seasons over the years 1965 to 1980 did not substantially exceed those estimated to produce maximum sustainable yields.

These general patterns were not completely reflective of trends in effort and CPUE of pink or brown shrimp in subregions These differences were most striking for pink shrimp in which effort levels in the Appalachicola grounds (Areas 5–9) increased exponentially from 1965 to 1977 by approximately 1432% whereas effort levels in the Sanibel-Tortugas grounds (Areas 1–4) increased on the average only 1.3% in a linear fashion during the study period. The utility of treating these areas separately in stock production analysis was indicated by the increased fit of the models for both the Sanibel-Tortugas and Appalachicola grounds, relative to those fitted to the subregions combined.

With respect to brown shrimp, trends were also different in the subregions. Effort levels off central and eastern Louisiana (Areas 11–16) increased at a high rate resulting in an approximate increase in effort of 245% between 1965 and 1980. Effort levels off the Texas coast and western Louisiana (Areas 17–21) increased from 1965 to 1969 and thereafter fluctuated at a relatively stable level up to 1976. Production models estimated for Areas 17–21 were closer fits than those to the Gulf region as a whole. The best fit form of the models differed greatly between the subregions. A more parabolic form best described the catch and effort trends observed for Areas 17–21 and suggested that the optimum effort level was close to that observed in the 1976 season. Effort levels in Areas 11–16 apparently had not reached or exceeded optimum levels as catch levels varied positively with effort throughout the time series.

The significance of such geographical partitions to management of the stocks should be carefully evaluated. Mark-recapture studies on brown shrimp have shown east to west migration from central Louisiana towards Texas and north–south migration along the Texas coast, although no shrimp tagged in central Louisiana waters has been recovered off south Texas. The most westward movement observed has been of a single white shrimp which was released in Caillou Lake, Louisiana and recaptured in Galveston Bay, Texas (J Lyon, pers. comm.). The relationship between shrimp east and west of the Mississippi River is poorly known. One shrimp tagged in September 1979 east of the River was recaptured west of the River (J Lyon, pers. comm.). How reflective (apparent) movements of tagged shrimp are of shrimp in general is difficult to assess. Certainly the high mortality incurred by some release groups has precluded observation of long-term or long-distance movements. However, should such mortality levels reflect levels on the stocks as a whole, then local high fishing effort levels could reduce migration rates between populations to the extent that catch and effort trends in coastal areas are heterogeneous and thereby exert differential effects on the persistence of the stock as a whole.

These analyses have dealt only with the offshore component of each fishery. This owed primarily to a relative lack of data on the inshore fisheries. However, a high degree of correlation appears to exist between inshore and offshore catches of these species and the proportion of the total catch that is landed by the offshore fishery on each species is high.

Acknowledgements

William Fox, Jr. and Joseph Powers of the Southwest Fisheries Center, National Marine Fisheries Service contributed several helpful comments and suggestions. Brian Rothschild of the Chesapeake Biological Laboratory, University of Maryland provided continued encouragement and guidance during the study. Special thanks are also extended to Carl Kinerd of the Southwest Fisheries Center for programming assistance and support and to Philip Jones, Janice Silverstein and Harry Hornick of the Chesapeake Biological Laboratory for their help in the production of figures.

References

FARFANTE, I P. Western Atlantic shrimps of the genus *Penaeus*.
1969   *Fish. Bull.* 67(3): 1–591.
FOX, W W, Jr. Fitting the generalized stock production model
1975   by least squares and equilibrium approximation. *Fish. Bull.* 73(1): 23–37.
GREEN, P E and CARROLL, J D. Mathematical tools for applied
1976   multivariate analysis. Academic Press, N.Y., 376 pp.
KUTKUHN, J H. Dynamics of a penaeid shrimp population and
1966   mangement implications. *Fish. Bull.* 65(2): 313–338.

LOPIK, J R, DRUMMOND, K H and CONDREY, R E. Management
1980   plan and final environmental impact statement for the
        shrimp fishery of the Gulf of Mexico, United States
        waters. Center for Wetland Resources, Louisiana State
        University, Baton Rouge, La.
LYON, J. Personal communication. Galveston Laboratory,
1981   Southeast Fisheries Center, National Marine Fisheries
        Service, Galveston, Texas.
MITCHELL, J P. The present status of the shrimping industry.
1976   *Proc. 1st Ann. Tropical and Subtropical Fisheries Tech.*
        Conf. Texas A&M Univ. 1: 412–416.
ROBSON, D S. Estimation of the relative fishing power of

1966   individual ships. *Res. Bull. ICNAF*, 3: 5–14.
ROTHSCHILD, B J. An exposition on the definition of fishing
1972   effort. *Fish. Bull.* 70(3): 671–679.
ROTHSCHILD, B J and BRUNENMEISTER, S L. The dynamics and
1984   management of shrimp in the northern Gulf of Mexico –
        U.S. Overview Paper. In this volume.
SEARLE, S R, SPEED, F M and HENDERSON, H V. Some computa-
1981   tional and model equivalences in analyses of variance of
        unequal-subclass-numbers data. *Amer. Stat.* 35(1): 16–
        33.
SMITH, C R. Fishing success in a regulated commons. *Ocean.*
1974   *Devel. and Internat. Laws J.* 1(4): 369–381.

Modelling of the Gulf of Carpentaria prawn fisheries

G P Kirkwood

Abstract

The Gulf of Carpentaria and nearby northern Australian waters is the site of a major fishery for penaeid prawns. Prawn catches are taken in essentially two distinct fisheries; a single species fishery for banana prawns (*Penaeus merguiensis*), and a mixed species fishery for tiger (*P. esculentus* and *P. semisulcatus*), endeavour (*Metapenaeus endeavouri*, *M. ensis*) and king (*P. latisulcatus*) prawns, with incidental catches of banana prawns. Commercial fishing for banana prawns commenced in 1966, and during the 1970's there have been marked fluctuations in annual catch. The mixed species fishery, which generally commences each year when banana prawn catch rates have fallen to unprofitable levels, has seen a quadrupling in fishing effort over the last 5 years. A limited entry management scheme has been in operation for fisheries in the Gulf of Carpentaria and adjacent areas since 1977. Significant changes in fleet composition have led to large increases in the fishing power of the fleet.

In this paper, the history of the fishery and the management problems that have arisen throughout its development are outlined. The perceived management problems have motivated several modelling studies, and description of them and the management implications of their results form the backbone of the paper. For the banana prawn fishery, the within season dynamics are modelled both by a simple difference equation and a detailed simulation model, with the aim of estimating exploitation rates and optimal yields per recruit. Research is only beginning on the mixed species fishery, and indications of the problems and intended lines of research are given. Optimum fleet sizes and compositions are examined by means of simulation and detailed bioeconomic models.

1 Introduction and history of the fisheries

Following the discovery of commercially exploitable stocks of prawns in the south eastern section of the Gulf of Carpentaria during a government financed survey in 1963–65, the prawn fisheries in northern Australian waters collectively have become one of Australia's largest fisheries in terms of both gross production and value. The commercial catch is composed of three major species groups: banana prawns (*Penaeus merguiensis*), tiger prawns (*P. esculentus* and *P. semisulcatus*) and endeavour prawns (*Metapenaeus endeavouri* and *M. ensis*). More recently, increased catches of western king prawns (*P. latisulcatus*) have also been taken.

These species are caught in two quite distinct fisheries; a daytime single species fishery for banana prawns, and a predominantly night time mixed species fishery for tiger, endeavour and king prawns. Incidental catches of banana prawns are taken in the mixed species fishery. In late summer/early autumn of each year, generally in mid-March, fishing commences on banana prawns, which characteristically yield large but variable catches. When catch rates of banana prawns have fallen to unprofitable levels (Clark and Kirkwood, 1979), attention is switched to the mixed species fishery. In 1977, a limited entry management scheme was introduced for these fisheries on an interim basis, and it was subsequently extended indefinitely in slightly modified form in 1979. The scheme restricted entry to prawn fisheries in a Declared Management Zone (DMZ), encompassing the Gulf of Carpentaria and waters to the west of the Gulf of Northern Territory (see Figure 1 of Walker, this volume). Detailed statistics for the prawn fisheries in the DMZ have been collated by Somers and Taylor (1981), and the catch history for each species group is shown in *Table 1*.

Research into the commercial fishery and the species taken by it commenced in late 1969, and is still continuing. Throughout the 1970's the primary concentration was on banana prawns. This research has covered the full range from basic biological research into the life history of banana prawns, through to applied management and assessment modelling and catch predictions. In this paper, attention is focussed on the modelling studies that have been undertaken. Bearing in mind the aims of the Workshop, specific note is made of the contexts in which the needs for such modelling arose. The management implications of the results of these studies are also discussed. In taking this approach, otherwise important aspects of the basic biology, and history and management of the fishery have been glossed over. Further details on these topics may be found in Hynd (1974), Munro (1975), Ruello (1975), Somers (1977), Staples *et al* (1981), Walker (1975, 1981) and references therein. In the remainder of this section, a brief account is given of the evolution of the fishery and fleet composition, highlighting the management concerns that have motivated the modelling studies.

The commercial fishery for banana prawns commenced in the Gulf of Carpentaria in 1966, and annual catches rose steadily until 1968. However in 1969, coincident with a substantial increase in vessel numbers, the catch fell markedly. This drop in catch gave rise to fears of possible recruitment overfishing (Hynd, 1974), however these initial fears were allayed by the subsequent catch history. Catch and effort data for the banana prawn fishery from 1971 are shown in *Table 1*. These data indicate that, while catches have increased substantially from the 1968 level, there have been very large year to year fluctuations. In 1974, a record banana prawn catch of 12 711 tonnes was taken, with this year being characterized by extensive flooding of the coastal river system that forms the nursery area for post larval and juvenile banana prawns. However the 1975 catch was less than one third of that of the preceding year, and while catches since then have continued to fluctuate, viewed simply as a time series there is some indication of a decline during the late 1970's. Understandably, the marked fluctuations in banana prawn catches have been a subject of continuing concern to management authorities.

Prior to 1977, the mixed species tiger, endeavour and king prawn fishery was very much subordinate to banana the prawn fishery, and both annual catch and effort remained relatively stable, as indicated by the catch and effort data for this fishery shown in *Table 2*. However, from 1977 there has been a rapid escalation in effort expended in this fishery, and in 1980 the annual number of boat days was some four times the average expended up to 1976. Despite the limited entry management regime, there exists within the licensed fleet the capacity for further large increases in effort. Consequently research aimed at assessing the current state of these stocks of prawns has assumed a high priority.

Table 1

ANNUAL BANANA PRAWN CATCH, INCLUDING (TOTAL) AND EXCLUDING (HOMOGENEOUS) INCIDENTAL CATCHES DURING MIXED SPECIES FISHERY, AND EFFORT DATA FOR THE DMZ, 1968–80 FROM SOMERS AND TAYLOR (1981)

Year	Total catch (tonnes)	Homogeneous catch (tonnes)	Homogeneous effort (boat days)	CPUE (tonnes/boat day)	No. vessels
1968	1 978	—	—	—	19
1969	1 078	—	—	—	115
1970	1 702	1 527	—	—	142
1971	7 365	6 597	5 024	1·313	165
1972	4 804	4 131	3 861	1·070	154
1973	4 226	3 524	4 956	0·711	166
1974	12 711	12 291	6 916	1·777	165
1975	2 980	2 855	4 690	0·609	105
1976	4 436	4 164	6 427	0·647	145
1977	6 216	5 956	6 535	0·911	175
1978	2 535	2 263	4 977	0·455	193
1979	4 775	4 335	6 549	0·662	199
1980*	2 681	2 092	6 826	0·305	

* Preliminary data

Table 2
ANNUAL CATCH (ALL SPECIES COMBINED) AND EFFORT DATA FOR
THE MIXED SPECIES FISHERY IN THE DMZ, 1968–80. FROM SOMERS
AND TAYLOR (1981)

Year	Catch (tonnes)	Effort (boat/days)	CPUE	No. vessels
1968	127	—	—	65
1969	845	—	—	144
1970	1 555	5 939	0·262	191
1971	1 583	7 725	0·205	169
1972	1 854	8 060	0·230	180
1973	2 266	7 361	0·294	217
1974	1 153	3 016	0·382	196
1975	1 425	5 817	0·245	107
1976	1 805	6 915	0·261	145
1977	4 071	11 637	0·350	193
1978	4 937	18 746	0·263	237
1979	5 576	18 618	0·299	240
1980*	6 543	28 858	0·227	

* Preliminary data

For both the banana prawn and mixed species fisheries, the fleet vessel numbers throughout the 1970's have shown at most a modest increase, and with the exception of the last few years in the mixed species fishery, the annual total effort expended has changed similarly. However, these data mask the substantial changes that have occurred in the composition of the fleet.

Although the known fishing grounds cover almost all coastal waters within the DMZ, there are few ports with adequate berthing and unloading facilities in this remote area of Australia. In the early years of the banana prawn fishery, the fleet consisted mainly of vessels that operated locally out of ports on the east coast of Australia. In that region, prawn fishing grounds are quite close to the fishing ports, and trawlers normally spend at most a few nights at sea on the grounds before returning to port for unloading. Consequently the vessels tended to be rather small, with limited catching and carrying capacities. Catches on these vessels were stored in chilled brine, and this factor alone restricted the time the vessels could spend at sea before returning to port and unloading their catch.

As the banana prawn fishery in the Gulf expanded to encompass areas at some distance from the ports, it soon became apparent that vessels of this design were not ideal. Consequently, there has been continuous and increasing trend towards replacement of these vessels with larger vessels more suited to the conditions in the DMZ. The replacement vessels generally have been substantially larger, with high catching and carrying capacities. These vessels were equipped with efficient snap freezing facilities, and at present almost all the catch taken by these vessels is processed on board, allowing them to stay at sea for months at a time. An indication of the change in fleet composition is given by the fact that in 1972 more than 70% of the vessels in the fleet were less than 16·5 m in length, with 39% less than 13·7 m, and only 23% were longer than 18 m, with 12% longer than 21 m. In 1980, only 15% were less than 16 m in length, with 8% less than 14 m, while 71% were longer than 18 m and 44% longer than 22 m. Bearing in mind the greater sophistication in design, construction and equipment of the new larger trawlers, these data indicate a very large increase in capitalization in the fleet.

At present, the capacity of the fleet far exceeds that necessary to take the largest recorded catches taken since commencement of the fishery. Furthermore the inevitable penalty paid for introduction of these newer larger vessels is a substantial increase in fuel and general running costs, and thus in the break-even catch rates required by such vessels. During poor banana prawn seasons, such as in 1975, there is evidence that a number of the vessels in the fleet failed even to cover their fuel costs. These symptoms indicate possible severe overcapitalization in the fleet, and in part recognition of this led to the imposition of the limited entry management regime.

The modelling studies described in this paper were motivated by and designed to provide advice on the management concerns outlined above. These studies generally have been restricted to those fisheries in the Gulf of Carpentaria, rather than the entire DMZ, however the major individual fisheries of the DMZ are contained in the Gulf of Carpentaria. In Section 2 an assessment of banana prawn stocks in the Gulf of Carpentaria is described. In this assessment, estimates of annual rates of exploitation were obtained, and, via a yield per recruit analysis, the optimal rates of exploitation and lengths at first capture were estimated. These allow discussion as to whether current levels of effort are excessive, and on optimal timing of seasonal closures. In Section 3, preliminary results of analyses of data from the mixed species fishery are outlined, while in Section 4 several modelling studies aimed at determining the optimal fleet composition are described. In the final Section, the implications of the investigation into banana prawn catch prediction by Staples *et al* (1981) with respect to possible recruitment overfishing and the presence or absence of a discernible stock recruitment relation are briefly examined, and directions of future research are discussed.

2 The banana prawn fishery in the Gulf of Carpentaria

2.1 *Fishery characteristics*

Adult banana prawns form isolated, densely packed schools during daylight hours, and the banana prawn fishery is based on catching from these schools. Once located, a single school may yield very large catches. For example, a survey of trawler skippers indicated school sizes in the south eastern Gulf of Carpentaria of up to 180 tonnes (Somers, 1977), and much larger schools occasionally have been reported in other areas. However, the distribution of schools over the fishing grounds is sufficiently sparse that most of the time on the grounds (of the order of 70–95%) is spent searching for these schools (Lucas *et al*, 1979). Some banana prawn schools are associated with patches of discoloured water (hence the term 'boils' to describe a banana prawn school) and thus may be located by aerial spotting, but the majority are located in clear water by echosounder.

The fishery for banana prawns normally commences each year in mid-March (early autumn). Only once a school has been found does trawling commence, and the school normally is fished by making a series of short trawls through it. The remainder of the time is spent searching for schools. Banana prawn fishing generally ceases when catch rates fall to unprofitable levels, and attention is then switched to the mixed species tiger and endeavour prawn fishery (Clark and Kirkwood, 1979).

2.2 *Estimation of exploitation rates*

Although the early effort data collected were in terms of hours trawled (Hynd, 1974), the above described characteristics of the fishery clearly indicate that for catches per unit effort to provide a reliable index of abundance, account must be taken of the time spent searching. Subject to weather conditions, breakdowns and unloading, trawlers customarily spend every available daylight hour on the fishing grounds searching and fishing for banana prawns. Since 1974, data on catches by weight for each day spent on the fishing grounds have been collected. These data formed the basis for the stock assessment carried out by Lucas *et al* (1979).

For any one fishing ground, the number of banana prawns available at the start of each time period was assumed to follow the simple model

$$N_i = (N_{i-1} + R_{i-1})e^{-Z}, \qquad \text{for } i = 1, 2, \ldots (1)$$

In this expression,

N_i = number of prawns available at the start of period i,

R_i = recruitment during period i, assumed for simplicity to occur at the beginning of period i,

$N_o = R_o$

and $Z = F + M + X$,

where Z = average instantaneous total mortality rate, with components M, the natural mortality rate, F the average fishing mortality rate and the average emigration rate from the fishing ground.

Using length frequency samples from the catches for each ground and each time period, catches by weight were converted to catches by number, allowing calculation of time series of catches in number per day on the ground. These were taken to be proportional to the abundance on the ground (Lucas *et al*, 1979).

Along with the collection of the above catch per unit effort and length frequency data, tag recapture experiments were undertaken to study migration, growth and mortality of banana prawns. From the results of these experiments Lucas *et al* (1979) concluded that there was at most negligible emigration from the fishing grounds. Therefore X may be taken as zero, and equation (1) then resembles a model for the familiar catch curves of Chapman and Robson (1960).

From the catch length frequency samples collected, Lucas *et al* (1979) were able to ascertain the time period towards the end of each time series during which recruitment of small prawns to the fishing grounds had effectively ceased ($R_i = o$). The data following that time period then allowed estimation of the average total mortality rate Z. Although no reliable independent estimate of the natural mortality rate M could be obtained, catch and effort data were available in appropriate form for the 1968 banana prawn fishing season off Weipa. In that year, effective effort and therefore the fishing mortality rate was very low (see *Table 2*). Application to these data of the above techniques led to an estimate of 0·05 week^{-1} for Z in that year. Taking this value as an (over-) estimate of M allowed estimation of F, and thus of the average exploitation rate for each ground for the 1974–76 data. From these estimates, Lucas *et al* (1979) concluded that the average exploitation rates for the entire eastern Gulf of Carpentaria were 0·78, 0·86 and 0·85 on the years 1974–76.

Lucas *et al* (1979) further noted that these estimates were consistent with those obtained from tag recapture experiments, once allowance had been made for initial and long term tagging mortality, and with the observed shortening in the length of the season from around 35 weeks at Weipa in 1968, to around 11 weeks on average in any one area during 1974.

2.3 Simulation modelling

Although the unit of effort adopted by Lucas *et al* (1979) is clearly preferable to one based solely on hours trawled, the wide range of fishing powers of the vessels in the fleet and the complex nature of the catching process indicated that the assumption that this unit of effort will lead to an adequate index of abundance should be further examined. Partly to this end, Somers (1977) developed a detailed simulation model of the banana prawn fishery which specifically accounted for those factors. In this model, the searching, catching and unloading operations of vessels working out of Karumba in the south eastern Gulf were simulated. Starting with a given number of available banana prawns, the prawns were assumed to be distributed randomly over the fishing ground in schools. The distribution of school sizes and the areas occupied by each school were estimated from experimental data collected via a survey of vessel skippers (Somers, 1977). The probability of an encounter of a particular school of banana prawns by a searching vessel was taken to be the ratio of the effective area searched to the total area of the ground (cf Berry, 1970), where the effective area searched is a function of the vessel speed and the dimensions of the school.

Once a school had been located, the successful vessel immediately commenced fishing. However, mimicking the actual situation in which all vessels in the vicinity soon hear of the location of a school, vessels in the vicinity were assumed to steam towards the located school and commence fishing on arrival. The catch taken from a school by each vessel was assumed to be related to the size of the school, the number of vessels fishing the school and individual vessel characteristics, such as maximum handling/freezing rates, hold capacity, endurance and fuel consumption. Once a vessel had reached the limit of its hold capacity or endurance, it returned to port. When running the model, data on actual fleet characteristics were used.

As described by Somers (1977), when the model was applied to 1977 data, time series of predicted catches per unit effort were obtained that closely matched the observed data, and the total mortality rate Z was estimated to be 0.27 week^{-1}. Application of the Lucas *et al* (1979) methods to the observed 1977 catch and effort data led to an estimated Z of 0.32 ± 0.20 week^{-1}. It thus appeared that, at least for fleet compositions similar to those at present, the methods of Lucas *et al* (1979) will not lead to serious bias.

2.4 Yield per recruit analyses

Having obtained reliable estimates of the exploitation rates, the relationship between the current levels of exploitation or lengths at first capture and those that are optimal on a per recruit basis may be determined by utilizing the familiar Beverton and Holt (1965) yield per recruit equation. This equation relates the yield per recruit to parameters E, the exploitation rate; M/K, the ratio of the natural mortality rate to the von Bertalanffy growth rate parameter K; and L_c/L_∞, where L_c is the length at first capture and L_∞ the average maximum length.

Using estimates of L_c, L_∞ and K obtained from length frequency catch samples, Lucas *et al* (1979) showed that for a wide range of values of M/K and L_c/L_∞ spanning the estimated values of these parameters, the yields in weight per recruit for 1974–76 estimates of E lay within 10% of the maximum, with the maximum probably occurring within the range 0.75–0.85. It was concluded that an increase in effective effort from 1975–76 levels would not lead to an appreciable increase in total catch, and that even a one-third to one-half reduction in effective effort from those levels would be unlikely to lead to substantial losses in yield.

As most of the catch is exported, with higher export prices (and lower processing costs) associated with larger prawns, it is more appropriate to determine optimum levels of exploitation in terms of value rather than weight. Calculations along these lines served to emphasize the conclusion that exploitation rates similar to those in 1975–76 may be excessive.

The yield (or value) per recruit relationship may also be used to determine optimum lengths at first capture for a given level of exploitation. Corresponding to the 1975–76 estimates of exploitation rates, Lucas *et al* (1979) found that the optimum value per recruit was obtained for L_c within the range 30.6–32.6 mm carapace length. Both from the estimated von Bertalanffy growth curve and from catch length frequency samples taken in various areas during March over a number of years, it was found that the average carapace length of prawns in the catch in mid to late March lay within this optimum range. Prawns taken in early March fell just below this range. At least for banana prawns, seasonal closures appear to be the only management tool available for ensuring an optimal length at first capture, as the fishing gear is quite unselective. However observed annual and between ground variations in length compositions precluded accurate selection of a fixed seasonal opening date solely on the basis of yield per recruit calculations.

Since 1971, certain areas of the eastern Gulf of Carpentaria have been subjected to a seasonal closure to the taking of banana prawns, with the opening date for the season generally being March 15. Historically, the principal basis for such closures appears to have been socioeconomic in nature. The results described above, while not suggesting a change in the customary opening date, suggest that the value per recruit is fairly constant for average lengths at first capture during March. However, such analyses assume a single pulse of knife edge recruitment, while in fact banana prawns recruitment occurs over several months, with peaks occurring at different times in different areas. The biasing effects of relying upon this simplifying assumption were graphically seen when, in 1976, the opening date of the season was shifted to March 1. In that year, due to the restricted nature of the areas closed to the taking of banana prawns, some fishing commenced in late February. With the high exploitation rate during 1976, over 40% of the total catch was taken by March 15. The overall size composition of the 1976 catch was markedly different from the average of the previous three years, when fishing commenced on March 15, as seen in *Table 3*.

Table 3
BANANA PRAWN CATCH SIZE COMPOSITIONS 1973–76

Count per lb	Percentages			Average 1973–75	1976
	1973	1974	1975		
11–15	3	1	1	2	2
16–20	17	16	11	13	8
21–25	37	45	42	40	24
26–30	33	33	35	35	32
31–35	10*	5*	11*	10*	21
36–40					9
40+					4

* 31+

If it is attempted to account for the change in opening date by calculating the 1976 L_c using the growth curve estimated by Lucas *et al* (1979), a value per recruit analysis suggests only a negligible loss in value. However, on the not unreasonable assumption that the 1976 catch size composition would have been similar to the 1973–75 average had fishing commenced on March 15, one is led to the conclusion that the loss in value of catch would have been around 7% from (1977). As this repre-

sents several million dollars in lost revenue, it is not surprising that in subsequent years, the opening date reverted to March 15.

3 The mixed species fishery

The mixed species prawn fishery commences when catch rates for banana prawns have fallen to unprofitable levels. The primary targets of this fishery are tiger prawns (*P. esculentus*, *P. semisulcatus*) and endeavour prawns (*M. endeavouri*, *M. ensis*). In recent years, increased catches of king prawns (*P. latisulcatus*) have been taken, and banana prawns form an incidental component of the catch. Unlike banana prawns, tiger, endeavour and king prawns do not form dense aggregations, and as a consequence little or no searching is undertaken. They are fished generally at night. The catch from each two to three hour trawl usually contains significant quantities of each species group.

Not unexpectedly, the mixed species nature of this fishery presents problems for analysis of a much greater magnitude than those present for the banana prawn fishery, at least in terms of estimation of individual population parameters and yield per recruit analyses. The catch and effort data collected for this fishery indicates that the relative proportions of tiger, endeavour and king prawns vary not only from fishing ground to fishing ground, but also within each ground. Unfortunately the individual tiger and endeavour prawn species are not distinguished by fishermen, and to date catch sampling has not been carried out on a sufficiently large scale for the catch to be determined reliably on an individual species basis. However the samples that have been taken suggest that within each species group, the relative proportions of the component species also vary from area to area (I Somers, N Carrol, pers. comm). Detailed research into the life histories of initially the two tiger prawn species is planned to commence within the near future. However until at least detailed data on the geographical distribution, recruitment patterns and catch of each species taken in the fishery are available, within season modelling of the fishery appears not to be feasible.

The most profitable approach to take at present appears to be that of production modelling along the lines described by Pella and Tomlinson (1969) and Fox (1975), with all species combined. Such an approach may provide, at least in the short term, an indication of the effects of the recently increasing effort. *Table 2* shows the relationship between

216

the total mixed species catch and nominal fishing effort in boat days from 1970. However these preliminary data cannot be interpreted reliably as they stand, as the effort data have not been standardized as yet to allow for the large increases in fishing power of the vessels that were documented earlier. Furthermore, there have been a number of important, previously unfished grounds discovered in recent years, notably in 1977 and 1979. Work is continuing to take account of these problems.

4 Optimum fleet size and composition

The dramatic changes in fleet composition in the Gulf of Carpentaria prawn fisheries described earlier heralded on the one hand a large increase in catching and handling capacity and therefore fishing power, and on the other hand substantial increases in capital and running costs. Fears of possible overcapitalization of the fleet were a major underlying factor for the introduction of the limited entry management regime introduced in 1977. Several modelling studies have been directed at throwing light on the effect of increases in the fleet fishing power, and changes in the fleet composition and its optimization.

Looking first at the banana prawn fishing fleet, the yield per recruit analysis described earlier suggested that any increase from the fishing mortality exerted in 1975–76 would be unlikely to lead to appreciable increases in the weight or value of the total catch. Further, it was estimated that a decrease up to 50% in fishing mortality would not lead to significant catch losses. Bearing in mind that this implies that rather higher per vessel catch rates could be obtained, this suggests that the current fleet is overcapitalized.

As the number of vessels operating in this fishery has at most increased modestly, increases in the fleet fishing power were due almost solely to its changing composition. Crudely, this change may be characterized as being from small 'brine' vessels to larger, sophisticated 'freezer' vessels. Using his simulation model, Somers (1977) investigated the interactions and energy efficiencies of these two types of vessel in the banana prawn fishery. He found that while on average brine vessels located as many banana prawns as did freezer vessels, they actually caught substantially less. However when the fuel consumption of each vessel type was taken into account, brine vessels were slightly superior even in catch per litre of fuel, and were much superior in amount located per litre of fuel. In this sense, Somers (1977) found the brine vessels to be more efficient than the freezer vessels. However,

when processing of the catch was taken into account, the overall energy efficiency of the two vessel types was found to be similar, due to the on board processing of catch on freezer vessels as opposed to the brine vessels which must steam to port for onshore processing of their catch.

At least until the late 1970's, the operating strategies of brine and freezer vessels differed. Brine vessels tended to operate in the Gulf of Carpentaria mainly during the banana prawn season only with most returning to the east coast of Australia and its local prawn fisheries once the banana prawn season had concluded. Freezer vessels however, were specifically designed to operate within the DMZ throughout the year, switching to the mixed species fishery when banana prawn fishing became unprofitable. Using a model of moderate complexity and realism, Clark and Kirkwood (1979) examined the performance of each vessel type exploiting the several prawn stocks.

Looking first at the dynamics of each of the prawn stocks the dynamics of the banana prawn stocks were assumed to be described by

$$dx(t)/dt = -(M + F(t))x(t), \qquad 0 \le t \le T$$

where $x(t)$ = number of banana prawns t weeks
from the start of the season,
M = natural mortality rate
and $F(t)$ = fishing mortality rate during week t.

Recruitment to the stock was assumed to occur at the start of the season, and was denoted by $R = x(o)$; T was taken as 52 weeks. Growth of individuals was assumed to follow the von Bertalanffy growth curve estimated by Lucas et al (1979).

The biological dynamics of the alternative stocks for each vessel type were suppressed, and they were treated as a fixed pool that can be exploited with constant catch per unit effort. As Clark and Kirkwood (1979) noted, this assumption was not too unreasonable for the mixed species fishery at the time. However, it may have been unrealistic for the east Australian coast fisheries, but data were lacking to do otherwise.

The fishing mortality $F(t)$ in the banana prawn fishery was assumed to be given by

$$F(t) = q_f E_1^f(t) + q_b E_1^b(t)$$

here q_f, q_b are the catchability coefficients for freezer and brine vessels, and E_1^f, E_1^b denote the effort levels for each vessel type. Then, if E_2^f, E_2^b denote the effort levels for the two vessel types in their alternative fisheries, there are constraints

$$E_i^f(t) \geq o, \quad E_i^b(t) \geq o, \quad i = 1, 2$$
$$E_1^f(t) + E_2^f(t) \leq E_{\max}^f = S_f Z^f$$

and

$$E_1^b(t) + E_2^b(t) \leq E_{\max}^b = S_b Z^b$$

In these constraints, $E_{\max}^f$ and $E_{\max}^b$ represent the total effort capacity for the two vessel types, Z^f and Z^b denote total number of vessels of each type, and S_f and S_b are serviceability factors, representing the average fraction of time that vessels are actually available for fishing. By their design, the serviceability factor for freezer vessels is rather greater than that for brine vessels.

With these definitions, the net annual economic revenue accruing from the combined prawn fisheries, neglecting fixed costs, is then

$$V(E_1^f, E_2^f; E_1^b, E_2^b)$$
$$= \int_o^T \{[p_1^f q_f E_1^f(t) + p_1^b q_b E_1^b(t)]x(t)w(t)$$
$$- c_1^f E_1^f(t) - c_1^b E_1^b(t)$$
$$+ (p_2^f y_2^f - c_2^f)E_2^{f(t)} + (p_2^b y_2^b - c_2^b)E_2^b(t)\}dt.$$

where p_1^f, p_1^b = landed prices of banana prawns for each vessel type,

p_2^b, b_2^b = landed prices of alternative prawns,

y_2^f, y_2^b = average catch per unit effort for alternative stocks

and c_i^f, c_i^b = operating costs of effort for each vessel type and fishery, $i = 1, 2$.

For given fleet sizes Z^f and Z^b, Clark and Kirkwood (1979) examined the economically optimum allocation over time of effort between the various stocks. In particular, they estimated the optimal opening date for the banana prawn season as a function of model parameters and fleet sizes; optimal timing of the switch from banana prawns to the alternative stocks; and taking into account fixed costs, determined the long run fleet sizes corresponding to open entry and the economically optimum fleet sizes.

On the basis of the above model, it was found that at any time t, each fleet should be completely allocated to the banana prawn stock or its particular alternative, and each fleet should (and will, without regulation) switch from banana prawns to its alternative fishery at the moment when the revenue flow declines to the opportunity cost of harvesting (ie, as soon as banana prawn returns fall below expected returns from the alternative fisheries). The optimal opening date for the banana prawn fishery was found to vary both as a function of fleet size and of vessel type, although taking

account of the latter was felt to be rather impracticable. Optimal opening dates varied from mid to late March as fleet sizes increased.

Taking into account fixed costs C_F^f, C_F^b, the net annual revenue can be calculated as

$$V_{net}(Z^f, Z^b) = V(Z^f, Z^b) - C_F^f Z^f - C_F^b Z^b$$

This equation was used by Clark and Kirkwood (1979) to estimate both the long run fleet sizes Z^f, Z^b resulting from unrestricted entry, and those fleet sizes that are economically optimum in that they maximise net annual revenue. Unfortunately, these fleet sizes are sensitive to the values of the fixed cost parameters C_F^f, C_F^b, which are difficult to specify with precision. However, Clark and Kirkwood (1979) were able to estimate them by fine tuning observed and expected vessel numbers under unrestricted entry prior to 1977. This enabled estimation of economically optimal fleet sizes for each vessel type.

Clark and Kirkwood (1979) noted that owing to the (necessary) unrealism in the model and its assumptions, the actual estimates obtained for economically optimal fleet compositions and sizes are subject to some uncertainty. In particular, the computed optimum fleet consists exclusively of one or other vessel type, with the type to be preferred depending critically on the value of C_F^b. More recently constructed vessels indicate that the currently preferred vessel type is a less expensive, smaller freezer vessel than that modelled by Clark and Kirkwood (1979), and there has been also a tendency for brine trawlers to be fitted with freezing facilities. Clark and Kirkwood (1979) note, however, echoing Somers (1977), that it may be advisable to retain some vessels of each extreme type in the fishery.

Even allowing for the caveats expressed by Clark and Kirkwood (1979) the results of the studies described above clearly indicate that the current fleet is larger than that which would be economically optimal in terms of net revenue earned from catching.

5 Discussion

The theme throughout this paper has been the development and application of models in reaction to problems perceived in the management of the Gulf of Carpentaria prawn fisheries. No attempt has been made to provide a compendium of prawn modelling techniques. However, it is apparent that there are several notable omissions in the preceding account; the most important of which is discussion of the relationship between stock and recruitment.

Annual banana prawn catches, and by inference stock abundances due to the high exploitation rate, have fluctuated markedly over the last decade, and there is even some suggestion of a downward trend over the later years. It is clearly of primary importance to investigate the causes of these fluctuations and to determine whether the stock has been subjected to recruitment over-fishing. Once it had been found that the extensive estuarine river systems around the Gulf of Carpentaria provided nursery grounds for post larval banana prawns, one of the aims of the studies of Hynd (1974) and Staples (1979), Staples *et al* (1981) was prediction of banana prawn catches. Staples *et al* (1981) have found that most of the variation in annual catch could be explained by the spring and summer rainfall, and their work provided an excellent basis for catch prediction. In addition, the residuals in annual catch once the rainfall effect has been removed should provide useful information on possible stock recruitment relations. For example, while there may appear to be a downward trend in catches since 1974, there has been a similar trend in rainfall (Staples *et al* 1981, Fig. 7). These residuals have been examined by Vance *et al* (in prep.), who found for each major banana prawn fishing ground no time trend in residual catches. There would appear to be no evidence of discernible stock related trends in recruitment, and thus of recruitment overfishing, despite the very high exploitation rates.

Another problem area common to many of the world's penaeid prawn fisheries is that of inter-actions between prawn species and interactions with other co-occurring fish species. Such problems clearly arise in the mixed species prawn fishery in the Gulf of Carpentaria, and as described earlier research is planned to tackle these problems. However, in respect to the fishery in the south-eastern Gulf of Carpentaria, a possible rare opportunity exists to examine the effects of an intense trawl fishery for prawns on the fish and benthic fauna of the area, due to the detailed surveys carried out prior to the commencement of the commercial fishery. These surveys were described in part by Munro (1975). A similar survey carried out in the near future would allow examination of the effects of almost ten years of very intense fishing pressure. Until fairly recently, the prawn fishery was virtually the sole fishery in the Gulf of Carpentaria of

any note. However, there is now increasing interest in harvesting other species, and in those circumstances a study such as suggested above would assume rather greater importance.

References

ANON. Gulf of Carpentaria prawn fishery complex and intrigu-
1977 ing. *Aust. Fish.* 36(6): 4–11.
BERRY, R J. Shrimp mortality rates derived from fishery statis-
1970 tics. *Gulf Caribb. Fish. Inst.,* Proc. 22nd Annual Ses-
sion, 66–78.
BEVERTON, R J H and HOLT, S J. Tables of yield functions for
1965 fishery assessment. FAO Fish. Tech. Pap. No. 38.
CLARK, C W and KIRKWOOD, G P. Bioeconomic model of the
1979 Gulf od Carpentaria prawn fishery. *J. Fish. Res. Bd.
Can.* 36(11): 1304–12.
FOX, W W Jr. Fitting the generalized stock production model by
1975 least squares and equilibrium approximation. *Fish.
Bull. US*, 73(1): 23–37.
HYND, J S. Year-round prawn fishery possible in the Gulf. *Aust.
1974 Fish.* 33(5): 2–5.
LUCAS, C, KIRKWOOD, G and SOMERS, I. An assessment of the
1979 stocks of the banana prawn *Penaeus merguiensis* in the
Gulf of Carpentaria. *Aust. J. Mar. Freshwater Res.* 30:
639–52.
MUNRO, I S R. Biology of the banana prawn (*Penaeus merguien-
1975 sis*) in the south-east corner of the Gulf of Carpentaria.
In 'Australian National Prawn Seminar (1st),
Maroochydore, 1973'. (Ed. Young, P C), pp. 60–78.
Aust. Govt. Printing Service, Canberra.
PELLA, J J and TOMLINSON, P K. A generalized stock production
1969 model. *Inter. Amer. Trop. Tuna Comm. Bull.*, 13:
421–96.
RUELLO, N V. An historical review and annotated bibliography
1975 of prawns and the prawning industry in Australia. In
'Australian National Prawn Seminar (1st), Maroochy-
dore, 1973'. (Ed. Young, P C), pp. 305–41. Aust. Govt.
Printing Service, Canberra.
SOMERS, I. Management of the Australian Northern Prawn
1977 Fishery. M.Sc. thesis, Griffith University, Qld. 68 pp.
SOMERS, I F and TAYLOR, B R. Fishery statistics relating to the
1981 Declared Management Zone of the Australian Northern Prawn Fishery, 1968–79. C.S.I.R.O. Division of
Fisheries Research, Report No. 123. In press.
STAPLES, D J. Seasonal migration patterns of postlarval and
1979 juvenile banana prawns, *Penaeus merguiensis* de Man,
in the major rivers of the Gulf of Carpentaria,
Australia. *Aust. J. Mar. Freshwater Res.* 30: 143–7.
STAPLES, D J, DALL, W and VANCE, D J. Catch predictions of
1981 the banana prawn, *Penaeus merguiensis* in the south-
eastern Gulf of Carpentaria. Paper prepared for Workshop on the Scientific Basis for the Management of
Penaeid Shrimp.
VANCE, D J, STAPLES, D J and KERR, J D. Factors affecting
(in year-to-year variations in the catch of banana prawn,
prep.) *Penaeus merguiensis*, in the Gulf of Carpantaria.
WALKER, R H. Australian prawn fisheries. In 'Australian
1975 National Prawn Seminar (1st), Maroochydore, 1973'.
(Ed. Young, P C), pp. 284–304. Aust. Govt. Printing
Service, Canberra.
WALKER, R H. Australian prawn fisheries. Paper prepared for
1981 Workshop on the Scientific Basis for the Management
of Penaeid Shrimp.

Application to shrimp stocks of objective methods for the estimation of growth, mortality and recruitment-related parameters from length-frequency data (ELEFAN I and II)*

D Pauly, J Ingles and R Neal

Abstract

This paper presents applications to several shrimp stocks of a new, computer-based set of methods for the detailed analysis of length-frequency data. These applications include the objective estimation of growth parameters, including seasonally oscillating growth, and of total mortality as estimated from length-converted catch curves and from mean lengths. A method for obtaining estimates of natural mortality in shrimps is discussed. Methods are also presented by which inferences can be drawn on the pattern of mesh selection prevailing in the fishery from which the length-frequency data were obtained. Finally, a method is presented which helps assessment of the seasonality of the recruitment of young shrimps into a fishery.

Introduction

Reasonably reliable estimates of the growth and mortality parameters of exploited shrimp populations are essential for their proper management. However, as opposed to the situation prevailing in fish where periodic markings can be used for precise and accurate aging (Brothers, 1980), the growth of shrimps is very difficult to estimate reliably, because no calcareous structure survives the periodic shedding of the exoskleleton.

Methods that have been used involved the establishment of calibrating growth curves from shrimp grown in captivity (Zein-Eldin and Griffith 1969, Forster 1970), the study of the growth increments of tagged and recaptured shrimps (Lindner and Anderson 1956, Berry 1967), the detailed study of the growth increments associated with moultings (Forster, 1970) as well as the study of size-frequency data (Kutkuhn 1962, Boschi 1969). Unfortunately, none of these methods approaches the degree of reliability generally obtained in fish aging by means of their otoliths, scales or other bones. This applies to techniques for the analysis of length-frequency data, which have remained essentially unmodified since they were proposed

by Petersen (1872), as well as to the analysis of mark-recapture data, which have been the principal method for estimation of growth parameters to date. The latter method is limited by the fact that many marks interfere with successful completion of moulting. Thus growth estimates, at best, are for short period – requiring considerable extrapolation – and at worst, do not represent growth because growth of shrimps occurs only at the time of moulting. Some marks have been developed that are retained through moulting (Neal, 1969) although they probably inhibit growth to some extent. Another problem is the effect of seasonal temperature oscillations, which until recently were difficult to account for (see Pauly and Ingles, 1981, and Nichols, this volume).

The methods introduced by Petersen (1892) for the analysis of length-frequency data consists of two approaches:

– the Petersen method (*sensu stricto*), and
– the *modal* class progression analysis.

The first of these approaches involves the attribution of assumed (relative) ages to the distinct peaks of a single, multi-peaked length-frequency sample. The problems here are of identifying the 'real' peaks, representing distinct broods, and attributing the proper relative age to the peaks representing broods (*Fig 1*).

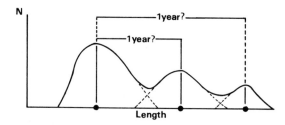

Fig 1 Application of the 'Petersen method' (*sensu stricto*) to a hypothetical length-frequency sample. Note that the *time* separating various peaks must be assumed, a difficult task in animals which may, or may not, spawn several times a year.

Much thought has been devoted to the problem of identifying 'real' peaks and splitting up single, multi-peaked samples into their constituent broods; milestones here include papers by Harding (1949), Cassie (1954), and Tanaka (1956), as well

*ICLARM Contribution No 122.

as the computer programs NORMSEP (Tomlinson, 1971), and ENORMSEP (Young and Skillman, 1975). All of these methods assume the constituent broods of a multi-peaked sample to be normally distributed.

On the other hand, relatively little attention has generally been devoted to improving what George and Banerji (1964) called 'modal class progression analysis' in which individual peaks are followed through a time series of length-frequency samples.

The problem with this method (in addition to the separation of broods) is the identification of those peaks (=broods) that are to be connected with each other – by no means a trivial problem (see *Fig 2*). Thus, as in the case of *Fig 1*, it appears that the use of complex methods (*eg* ENORMSEP) for the separation of multipeaked samples into normally distributed subsets ('broods') and the subsequent computation of their means and standard deviations helps little in identifying the peaks that should be interconnected or, as in *Fig 1*, in defining the time separating peaks.

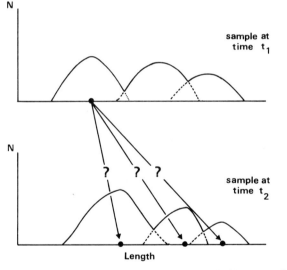

Fig 2 Application of the 'modal progression analysis' to a set of two samples obtained at known times (t_1, t_2). Note that the problem is the proper identification of peaks to be interconnected, and not the problem of time (as was the case in Fig. 1).

Pauly (1978, 1983) suggested a method involving an explicit criterion by which the 'Petersen method' and the 'modal class progression analysis' can be used to validate each other (*Fig 3*). This method, called 'integrated method', although it represents an improvement over the separate use of the two earlier approaches, is still subjective in that different workers can obtain different results from the same set of data.

In this paper, the applicability to shrimp stocks

of a set of new methods for the detailed analysis of length-frequency data is demonstrated. Also, it is shown that the new methods allow for the extraction from the length-frequency data available of a large amount of information on the biology of shrimps, information which normally is embedded in any set of length-frequency data but cannot be extracted by the methods used to date. All rates (growth, mortality) discussed in this paper are put on an annual basis, unless mentioned otherwise.

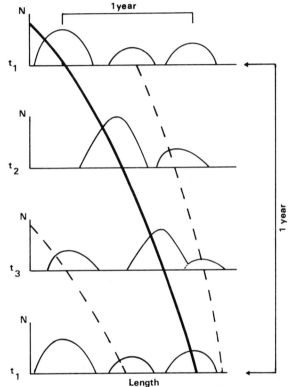

Fig 3 An application of the 'integrated method' to a hypothetical set of length-frequency samples. Note that the attribution of a relative age to the third peak of sample t_1 is confirmed by the modal class progression, which suggests a growth curve passing through the major peaks of samples t_1, t_2, t_3 and through the third peak of sample t_1 *repeated* after one year (*ie*, placed at the appropriate place on the time scale, after sample t_3). Thus a smooth growth curve can be traced which explains most of the peaks of a set of length-frequency samples, including those of earlier samples repeated once, twice or more along the time axis. A certain degree of reliability is achieved which could not be achieved by applying either of the two earlier methods.

Methods

Estimation of growth parameters

The method presented here for the extraction of growth parameters from length-frequency data may be considered an extension of the integrated method, yet is wholly objective, *ie*, allows for fully reproducible results (Pauly and David, 1980, 1981). The method, which is called ELEFAN I (*Electronic LEngth Frequency ANalysis*) involves the following steps:

– identify peaks, and the troughs separating peaks in terms of the deviation of each length class frequency from the corresponding running average frequency (peaks are positive, troughs negative deviations) – note that this definition involves no assumption of normality for the distribution of the broods in each sample,
– attribute to each positive deviation a number of positive points, proportional to its deviation from the running average and attribute, similarly, a certain number of negative points to each trough,
– identify the set of growth parameters, which, by generating a growth curve which passes through a maximum number of peaks and avoiding troughs as much as possible, accumulates the largest number of points, termed 'Explained Sum of Peaks' or ESP, and
– divide the ESP by the sum of points 'available' in a set of length-frequency samples, *ie*, by the 'Available Sum of Peaks' (ASP) to obtain an estimator of the goodness of fit, the ESP/ASP ratio, which generally ranges between 0 and 1 may be considered analogous to a coefficient of determination (r^2).

Figure 4 depicts a set of restructured length-frequency samples of penaeid shrimps, showing the peaks and troughs as defined above, together with the best growth curve estimated from the data

set. This procedure assumes that:

– the length-frequency data are representative of the population,
– the growth patterns are repeated from year to year,
– the von Bertalanffy Growth Formula (VBGF) describes the mean growth in the population, and
– all shrimps in the samples have the same growth parameters, *ie*, all differences in size reflect difference in age.

Of these four assumptions, the fourth is the least realistic, since it is known that shrimps of different lengths can have the same age. However, the error involved by making this (essential) assumption is probably small. The procedure for ELEFAN I is fully described in Pauly *et al* (1980).

An aspect of ELEFAN I which adds considerably to the versatility of the program is that the growth equation used for generating the growth curves is a seasonally oscillating version of the VBGF which has the form

$$L_t = L_\infty(1 - e^{-[K(t-t_0)+C(K/2\pi)\sin 2\pi(t-t_s)]}) \qquad (1)$$

where L_t is the length at age t, L_∞ is the length the fish would reach if they were to grow indefinitely, and t_0 is the computed length at age zero while t_s and C are parameters which control seasonal growth oscillations of period one year; t_s is the start of a sinusoid growth oscillation with respect to $t = 0$ and C is a parameter exposing the intensity of the seasonal growth oscillation. C has values generally ranging from zero (when there are no oscillations) to a value of unity (in which case dL/dt has one

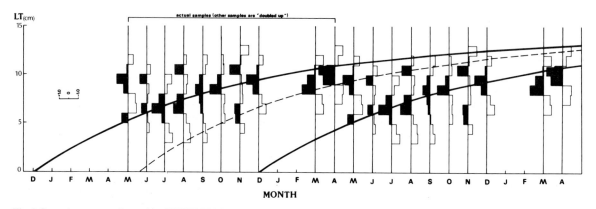

Fig 4 Growth curves estimated by ELEFAN I from a 'restructured' set of shrimp length-frequency data (stock #6, see *Table 1*). Note that 'peaks' are here expressed as positive, and troughs as negative points. The growth curve (a von Bertalanffy curve) as fitted by ELEFAN I goes through as many peaks as possible, while avoiding as many troughs as possible resulting in ESP/ASP = 0·418 for the main curve (solid line) and ESP/ASP = 0·209 for the secondary curve (dotted line), *ie* a sum of 62·7% of the available peaks are expressed by the curves. It will be noted that, as in *Fig 3*, there is a multiple use of some samples, to allow for a better visualization of the shape of the growth curves, and of the peaks they explain.

zero value per year). Equation (1) can be reduced to the form of the VBGF commonly used in fishery biology by setting $C = 0$, ie,

$$L_t = L_\infty(1 - e^{-K(t-t_0)}) \qquad (2)$$

The seasonally oscillating growth model built into ELEFAN I is a feature which is essential to the use of this program in conjunction with sets of data pertaining to temperate or subtropical shrimp stock, as will be shown below.

Finally, a parameter, the Winter Point (WP) is defined as

$$WP = t_s + 0.5 \qquad (3)$$

or as the time of the year (expressed as a decimal fraction) where growth is slowest ie, in 'winter' (Pauly and Gaschütz, 1979).

Estimation of total mortality from length-frequency samples

Although the estimation of total mortality based on a 'catch curve' was used by Edser (1908) in conjunction with a length-frequency sample, the catch-curve method of estimating total mortality is generally applied to animals that have been aged, using the relationship

$$\log_e N = a + bt \qquad (4)$$

where N is the number of fully recruited and vulnerable animals of a given age t, and $-b = Z$, the exponential rate of total mortality, as commonly used in fishery biology (see Robson and Chapman, 1961 or Ricker, 1975 for reviews).

When animals cannot be aged individually, as in shrimps, one could conceive of replacing N in equation (4) by the number of animals in a given length class, and replacing t by the relative age (t') of the animals at the mid-length of that class ($L_{t'}$), where t' is obtained by solving the VBGF

$$t' = \frac{\log_e \{1 - (L_{t'}/L_\infty)\}}{-K} \qquad (5)$$

This method was applied to shrimp, *eg*, by Berry (1970). However, the method is subject to a large bias, because larger shrimps need a longer time to grow through a length class than small shrimps, this being due to the fact that shrimp growth in length is not linear. Thus, large, old shrimps 'pile up' in the larger size groups and Z is underestimated. A correction for this effect can be achieved by dividing each value of N by the time needed to grow through a length class (Δt) computed as

$$\Delta t = \frac{\log_e \{(L_\infty - L_1)/(L_\infty - L_2)\}}{K} \qquad (6)$$

where L_1 and L_2 are the lower and upper limits of the length class, respectively. Thus, a length-converted catch curve should (as suggested by J A Gulland, pers. comm.) have the form

$$\log_e (N/\Delta t) = a + bt' \qquad (7)$$

The differences between Z values obtained from equation (7) and those obtained when not accounting for the non-linearity of shrimp growth can be considerable (see *Fig 5*).

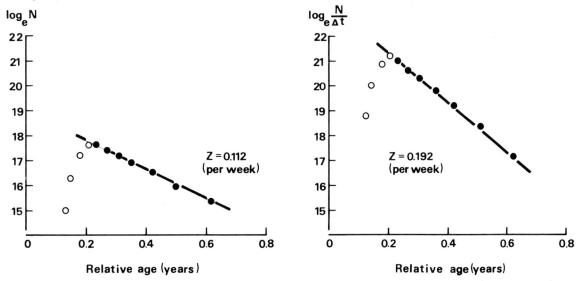

Fig 5 Two methods for the estimation of total mortality (Z) from length-converted catch curves. Left: straightforward conversion of lengths to relative ages. Right: conversion of lengths to ages with compensation for non-linearity of growth. Note large deviation (58%) between the estimates of Z. The catch data (*P. duorarum* ♀) are from Berry (1970, Fig. 1) whose growth parameters ($L_\infty = 20.5$, $k = 0.08$ per week) were also used.

The computer program ELEFAN II (a sequel to ELEFAN I) can be used to estimate values of Z using the method presented here, based on any combination of length-frequency samples believed to be representative of the population, as well as any set of growth parameters (Pauly *et al*, 1981).

A feature of this program is that it allows for the selection of data points to use in the analysis. Thus, it is possible to disregard both the points pertaining to the 'ascending' part of the curve and those points (generally one only) derived from lengths so close to L_∞ that unrealistically high 'age' is generated by equation (5) (see *Figs 7–9* for examples).

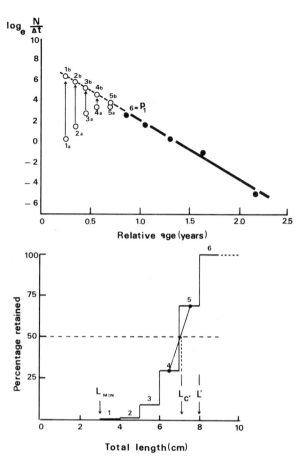

Fig 6 Construction of 'selection patterns' from catch curves. Above: backward projection of straight portion of catch curve, and computation of 'expected' numbers (1b to 5b). Below: percentage retained, as obtained by performing 1a/1b, 2a/2b *etc*, and plotting the resulting fractions (expressed in %) against the corresponding lengths. L_{min}, L'_c and L' are defined. Catch curve and selection patterns shown here pertain to stock #8 in *Tables 1* to *4*.

Identification of selection patterns

As will be noted in the results section, the descending limb of length-converted catch curve is generally remarkably straight. This suggests that Z would also be constant for at least part of the ascending part of the curve, were it not for the effects of incomplete selection and/or recruitment. Thus, by dividing, for each length class in the ascending part of the curve, the numbers actually sampled by the expected numbers (as obtained by projecting backward the straight portion of the catch curve) as series of ratios are obtained which resemble a selection curve, but is in fact akin to a 'resultant curve' (Gulland, 1969), *ie* the result of the interactions of a recruitment with a selection curve (*Fig 6*).

The key assumption of this method, which generates what will be called "selection patterns", is that the straight portion of the catch curve could legitimately be extrapolated backward, *ie* that Z would be constant over a range of values larger than the range from which it is estimated, if it were not for the effect of mesh selection. (It will be realized that this assumption is fully met only when $Z = M$.)

Estimation of natural and fishing mortality

When the growth of animals can be described by the VBGF, and if it can be assumed that the oldest animals of a given stock reach approximately 95% of their L_∞ value, then the VBGF can be solved such that

$$t_{max} \approx 3/K \qquad (8)$$

where K is the growth coefficient of the VBGF and t_{max} the longevity. That natural mortality should be related to longevity, hence to K is thus obvious, and has been demonstrated on the basis of a vast amount of data by Beverton and Holt (1959). Similarly, animals with small L_∞ (*ie* animals that stay small) should have more predators than

animals having a large size. Hence, small animals should generally have, for constant K, a higher natural mortality than large animals. Mean environmental temperature – at least in fishes – can also be demonstrated to have a direct relationship with natural mortality (Pauly, 1980a). On the basis of 175 independent data sets, it was thus possible to establish for fishes the relationship

$$\log M = -0.0066 - 0.279 \log_{10} L_\infty + 0.6543 \log_{10} K$$
$$+ 0.4634 \log_{10} T$$
$$(9)$$

where L_∞ is expressed in cm (total length), where K is put on an annual basis and T is expressed in °C. This relationship, which can be used to predict reasonable values of M in any species of fish can be expected to generate equally reasonable estimates

224

of M in shrimps for the reasons that shrimps and fish generally share the same habitats, resources and predators, and that therefore, they are not likely to differ widely in their vital parameters (see Discussion).

Equation (9) thus allows a rough estimation of M in shrimp stocks where L_∞, K and T are known. Subtraction of these estimates of M from the estimates of Z (obtained by using one of the methods presented above) gives values of fishing mortality via the definition

$$Z = M + F \qquad (10)$$

Also, the exploitation rate $E = F/Z$ can be computed for a preliminary assessment of whether a stock is lightly ($E < 0\cdot5$) or strongly exploited ($E > 0\cdot5$), based on the assumption that a stock is optimally exploited when $F = M$ or $E = 0\cdot5$ (Gulland, 1971).

Derivation of recruitment patterns
It has been a common practice, when using the traditional methods for the analysis of length-frequency data to make inferences, once patterns of growth have emerged, as to

- the length of the spawning season (more precisely the length of the period during which animals are recruited into the set of length-frequency samples under investigation),
- the number of times 'spawning' occurs per year (recruitment again is actually considered),
- the relative magnitude of the various 'spawning' (recruitment) events.

However, these inferences, although helpful, are as subjective as the growth estimates obtained by these methods. ELEFAN II incorporates a routine which, by projecting the length-frequency data available backward onto the time axis (by means of a set of growth parameters) generates 'recruitment patterns' which can be used to obtain objective information pertaining to the recruitment processes (*Fig 7*).

Material

Table 1 indicates the source of the data used here to demonstrate the applicability of ELEFAN I and II to shrimps.

Except for data collected in the Philippines (Ingles, 1980) all length-frequency data used in the present analysis stem from the literature. There are two reasons for this:

- we do not have enough original data to illus-

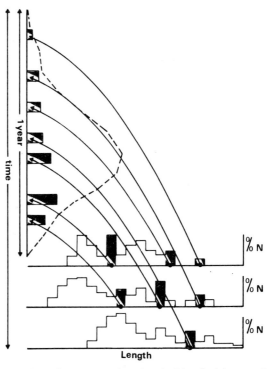

Fig 7 Schematic representation of method for obtaining recruitment patterns. The steps involved are: (1) projection onto the time axis of the frequencies of a set of length-frequency data (this process includes a correction for non-linearity of growth, not shown here), (2) summation for each month (and irrespective of year) of the frequencies projected onto each month, (3) subtraction, from each monthly sum, of the lowest monthly sum to obtain a zero value where apparent recruitment is lowest, (4) output of monthly recruitment, in percent of annual recruitment.

Concerning points 3 and 4, it must be noted that the monthly values pertain to real times of the year only when t_0 is available. Otherwise, the times of maximum and minimum recruitment are not known – hence the use of '1 year' as a time scale in the various figures presented below.

trate sufficiently all possible applications of the ELEFAN programs, and
- one of our aims is to demonstrate that length-frequency data in the literature have generally been underutilized, *ie*, the data allow for a much deeper analysis.

Results

Growth parameter estimates
The growth parameters estimated by ELEFAN I from the data in *Table 1* are summarized in *Table 2*. Some of the growth curves involved are illustrated in *Figures 8* and *9*.

An 'auximetric' grid (Pauly, 1970, 1980c) is presented which allows comparison of the growth patterns of the shrimp stocks investigated here (*Fig. 10*). It will be noted that for given K-values

225

Table 1

SUMMARY OF DATA ON THE SHRIMP STOCKS USED FOR THE ANALYSIS

Fig. No.	Stock No.	Species, sex	Locality and date	Water temperature (monthly means) Min	Max	Mean	ΔT	Source of length-frequency data
8	1	*Hymenopenaeus robustus* ♀	Northeast Florida, 100 fathoms, 1957	—	—	15[a]	—	Anderson and Lindner (1971, App. Table 6)
	2	*Metapenaeopsis durus* ♀ and ♂	Visayan Sea, Philippines, 1976/77	25·1	28·3	26·8[b]	3·2	Ingles (1980, Appendix IIIc)
9	3	*Metapenaeus brevicornis* ♀ and ♂	Off Kutch, India, 1960/61	17·7	29·3	25·0[c]	11·6	Ramamurthy (1965, Fig 6)
	4	*Metapenaeus brevicornis* ♀ and ♂	Off Kutch, India, 1961/62	16·5	29·1	24·0[c]	12·6	Ramamurthy (1965, Fig 6)
	5	*Metapenaeus affinis* ♀ and ♂	Off Versoba, India, 1958/59	24·6	28·1	26·7[c]	3·5	Mohamed (1967, Fig 3)
	6a	*Metapenaeus kuchensis* ♀ and ♂	Off Kandla, India, 1959/60	21·3	29·0	25·0[d]	7·7	Ramamurthy (1965, Fig 3)
	6b	*Metapenaeus kuchensis* ♀ and ♂	Off Kandla, India, 1959/60	21·3	29·0	25·0[d]	7·7	Ramamurthy (1965, Fig 3)
	7	*Metapenaeus kuchensis* ♀ and ♂	Off Kandla, India, 1960/61	17·7	29·3	25·0[d]	11·6	Ramamurthy (1965, Fig 3)
	8	*Metapenaeus kuchensis* ♀ and ♂	Off Kandla, India, 1961/62	16·5	29·1	24·0[d]	12·6	Ramamurthy (1965, Fig 3)
	9a	*Parapenaeus longipes* ♀ and ♂	Visayan Sea, Philippines, 1976/77	25·1	28·3	26·8[b]	3·2	Ingles (1980, Appendix IIIb)
	9b	*Parapenaeus longipes* ♀ and ♂	Visayan Sea, Philippines, 1976/77	25·1	28·3	26·8[b]	3·2	Ingles (1980, Appendix IIIb)
	10a	*Penaeus duorarum* ♂	Tortugas, Florida, 1957/58	22·8	28·6	25·6[e]	5·8	Iversen *et al* (1960, Appendix Table 2)
	10b	*Penaeus duorarum* ♂	Tortugas, Florida, 1957/58	22·8	28·6	25·6[e]	5·8	Iversen *et al* (1960, Appendix Table 2)
	11	*Penaeus kerathurus* ♀	Gulf of Cadiz, Spain, 1971/72	14·2	20·3	18·0[f]	6·1	Rodriguez (1977, Fig. 12)
	12	*Penaeus kerathurus* ♂	Gulf of Cadiz, Spain, 1971/72	14·2	20·3	18·0[f]	6·1	Rodriguez (1977, Fig. 12)
	13	*Penaeus setiferus* ♀	Off Texas, USA	16·9	28·2	22·7[e]	11·3	Anderson and Lindner (1958, Table 1)
	14	*Penaeus setiferus* ♂	Off Texas, USA	16·9	28·2	22·7[e]	11·3	Anderson and Lindner (1958, Table 1)
	15	*Trachypenaeus fulvus* ♀	Visayan Sea, Philippines, 1976/77	25·1	28·3	26·8[b]	3·2	Ingles (1980, Appendix IIIa)
	16	*Trachypenaeus fulvus* ♂	Visayan Sea, Philippines, 1976/77	25·1	28·3	26·8[b]	3·2	Ingles (1980, Appendix IIIa)

[a] Bottom temperature not available. Value of 15°C is an estimate.
[b] Actual bottom temperature, as measured by Ingles (pers. obs.).
[c] Anon. (1944).
[d] Ramamurthy (1967, Table VII).
[e] Rivas (1968).
[f] Rodriguez (1977, Fig 20).

the *Penaeus spp.* have higher values of W_∞ than the other Penaeidae *ie*, they grow faster. Pauly (1979, 1980c) notes other inferences and comparisons, *eg*, with fish, which can be made using an auximetric grid.

As will be noted from *Table 2*, seasonally oscillating growth curves have not been fitted to tropical stocks, but only to temperate and subtropical stocks, *ie*, to 7 of the 19 cases considered here (see *eg, Fig 9*). Thus, for these stocks, the value of *C* in

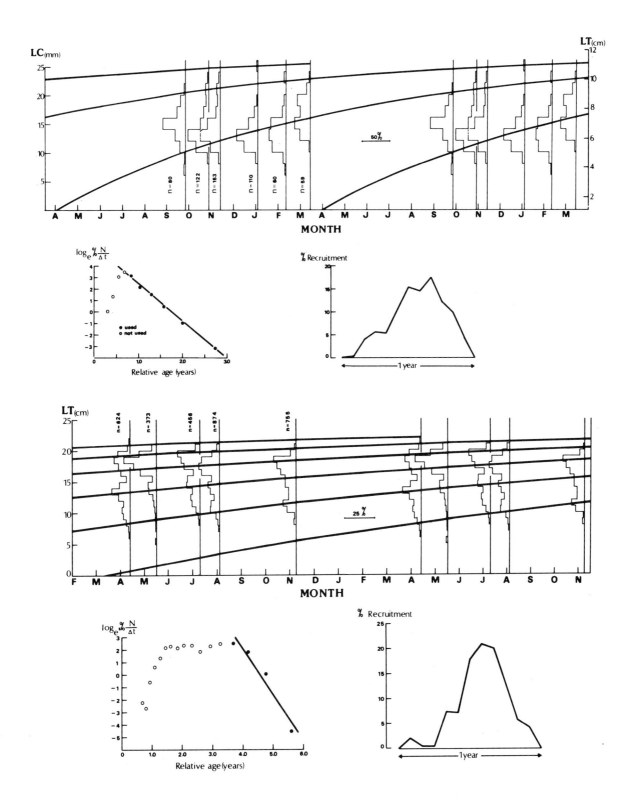

Fig 8 Above: growth, catch curve and recruitment pattern of *Hymenopenaeus robustus* ♀ off Florida. Below: growth, catch curve and recruitment pattern for *Metapenaeopsis durus* ♀ and ♂ from the Visayan Sea, Philippines. See *Tables 1* to *4* for details on these stocks.

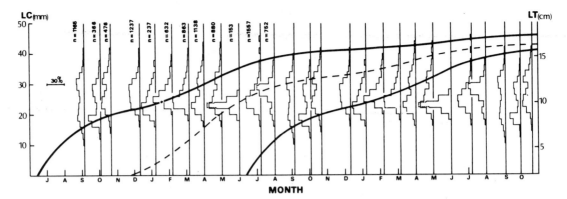

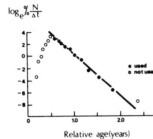

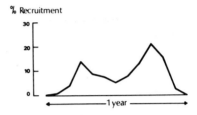

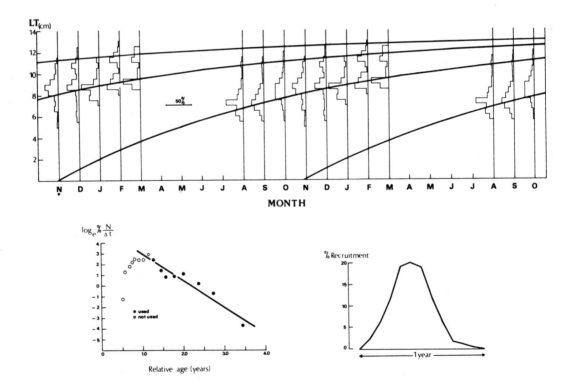

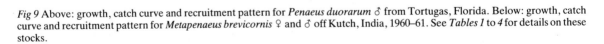

Fig 9 Above: growth, catch curve and recruitment pattern for *Penaeus duorarum* ♂ from Tortugas, Florida. Below: growth, catch curve and recruitment pattern for *Metapenaeus brevicornis* ♀ and ♂ off Kutch, India, 1960–61. See *Tables 1* to *4* for details on these stocks.

228

Table 2
SUMMARY OF RESULTS PERTAINING TO THE GROWTH OF SHRIMP

Fig. No.	Stock No.	Species, sex	Growth parameters					
			L_∞	'W_∞'[a]	K	C	Winter Point	ESP/ASP[b]
8	1	*Hymenopenaeus robustus* ♀	24·25	114	0·39	0·10	0·4 (May)	0·531
	2	*Metapenaeopsis durus* ♀ and ♂	11·6	12·5	0·95	—	—	0·446
9	3	*Metapenaeus brevicornis* ♀ and ♂	13·3	18·8	0·93	—	—	0·465
	4	*Metapenaeus brevicornis* ♀ and ♂	14·25	23·1	0·90	—	—	0·490
	5	*Metapenaeus affinis* ♀ and ♂	17·5	42·9	1·20	—	—	0·392
	6a	*Metapenaeus kutchensis* ♀ and ♂	14·0	22·0	1·15	—	—	0·418
	6b	*Metapenaeus kutchensis* ♀ and ♂	14·0	22·0	1·2	—	—	0·209
	7	*Metapenaeus kutchensis* ♀ and ♂	13·5	19·7	1·05	—	—	0·456
	8	*Metapenaeus kutchensis* ♀ and ♂	13·75	20·8	1·10	—	—	0·328
	9a	*Parapenaeus longipes* ♂	10·0	8·00	1·4	—	—	0·379
	9b	*Parapenaeus longipes* ♀	10·25	8·62	1·15	—	—	0·251
	10a	*Penaeus duorarum* ♂	17·6	43·6	1·45	0·60	0·93 (Dec)	0·347
	10b	*Penaeus duorarum* ♂	17·6	43·6	1·2	0·54	0·87 (Nov)	0·264
	11	*Penaeus kerathurus* ♀	21·0	74·1	0·8	0·9	0·8 (Oct)	0·457
	12	*Penaeus kerathurus* ♂	18·0	46·7	0·9	0·85	0·75 (Sept)	0·566
	13	*Penaeus setiferus* ♀	22·5	91·1	1·25	0·61	0·11 (Feb)	0·433
	14	*Penaeus setiferus* ♂	19·25	61·6	1·55	0·675	0·15 (Feb)	0·476
	15	*Trachypenaeus fulvus* ♀	13·0	17·6	1·4	—	—	0·481
	16	*Trachypenaeus fulvus* ♂	11·4	11·9	1·6	—	—	0·451

[a] Computed throughout by setting $W = 0.008 \, L^3$, where L is the total length in cm, and W the weight in grams.
[b] Explained sum of peaks/available sum of peaks; see text for definitions.

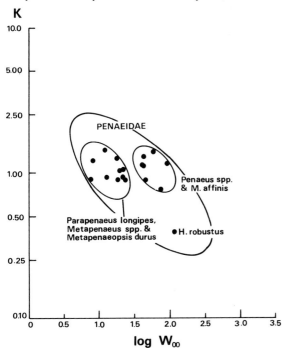

Fig 10 Comparison of growth performance of various penaeid species. The stocks with a highest value of K for a given value of W_∞ (or conversely) grow fastest (*ie* reach faster any given size). Note that various groups fall into ellipsoid clusters, suggesting ranges of $W_\infty K$, combinations possible in Penaeidae. Based on data in *Table 2* and 'auximetric grid' method of Pauly (1979).

equation (1) has simply been set at zero. For all other cases, values of C and WP were estimated from the data. As might be seen, the various 'Winter Points' indeed fall within the periods where the temperature is reduced. This may also apply to *Hymenopenaeus robustus* (stock #1 in *Tables 1* to *4*) whose 'Winter Point' falls in May, and which occurs in deep water, where the annual temperature minimum probably occurs later than in shallow waters. The values of C correspond, for a given difference between highest mean monthly summer and winter temperature (ΔT), very well with values obtained from seasonally oscillating growth curves of fish (*Fig 11*). Thus, as in the case of fish, it appears in shrimp that the intensity of seasonal growth oscillations correlates well with the intensity of the annual temperature fluctuations.

Total mortality estimates
Table 3 gives the estimates of Z obtained from the length converted catch curves. Overall, they are similar to those obtained using Beverton and Holt's (1956) equation for the estimation of Z from mean lengths.

Selection patterns
In *Fig 12*, four selection patterns are presented. The limited number of examples were selected:

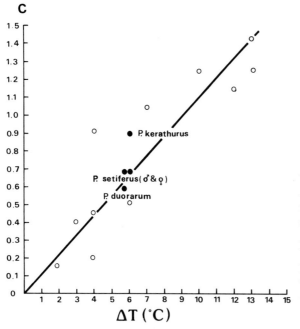

C

ΔT (°C)

Fig 11 Relationship between intensity of seasonal growth oscillation, as expressed by the parameter *C* of equation (1) and the difference between highest and lowest mean monthly water temperature in the course of one year (ΔT). Open dots pertain to fish stocks, documented in Pauly (1982) and Pauly and Ingles (1981), black dots pertain to shrimp stocks, documented in *Tables 1* and *2*.

– to include the females and males of a given species, in order to show that similar results are indeed obtained where similar results may be expected, and

– to include a stock with a 'knife-edge' selection pattern as well as a stock with selection occurring over a wider length range, to show that the method does not, due to some inherent bias, always generate similar recruitment patterns.

Selection patterns as defined here thus can be used to extract useful information, *eg* length at 50% retention, or $L_{c'}$,* from a given set of length-frequency data. They may serve as complement to selection experiments, the results of which they generally should confirm. Also, they might be used to replace selection experiments where such experiments cannot be conducted, *eg*, because of the costs involved in such experiments, or when analyzing old data.

Estimates of natural and fishing mortalities
Table 3, which summarizes the estimates of mortalities, shows the estimates of *M* obtained by

Footnote: The 50% retention length estimated from selection patterns may be coded $L_{c'}$ to distinguish it from a 'real' estimate of L_c, as obtained from a selection experiment.

Table 3
SUMMARY OF RESULTS PERTAINING TO THE MORTALITY OF SHRIMP

Fig.	Stock	Species, sex	Catch curve						Remarks
			r^2	n	Z_1	$\hat{M}$	F	E	
8	1	*Hymenopenaeus robustus* ♀	0·994	4	3·77	0·77	3·00	0·80	active fishery
	2	*Metapenaeopsis durus* ♀ and ♂	0·995	6	3·28	2·21	1·07	0·33	incidental catch in demersal fisheries
9	3	*Metapenaeus brevicornis* ♀ and ♂	0·910	8	2·46	2·03	0·43	0·18	active fishery
	4	*Metapenaeus brevicornis* ♀ and ♂	0·960	11	2·55	1·91	0·64	0·25	
10	5	*Metapenaeus affinis* ♀ and ♂	0·983	5	5·29	2·29	3·00	0·57	'no overfishing in the areas' (George 1970)
	6a	*Metapenaeus kutchensis* ♀ and ♂	0·970	4	6·49	2·33	4·16	0·64	
	6b	*Metapenaeus kutchensis* ♀ and ♂							active fishery
11	7	*Metapenaeus kutchensis* ♀ and ♂	0·999	6	3·55	2·19	1·36	0·38	
	8	*Metapenaeus kutchensis* ♀ and ♂	0·988	5	5·83	2·20	3·63	0·62	
12	9a	*Parapenaeus longipes* ♀	0·997	6	3·83	2·79	1·04	0·27	not accessible to most commercial trawlers (too deep)
	9b	*Parapenaeus longipes* ♀							
	10a	*Penaeus duorarum* ♂	0·991	10	7·07	2·40	4·67	0·66	known to be heavily fished
	10b	*Penaeus duorarum* ♂							
13	11	*Penaeus kerathurus* ♀	0·967	6	1·96	1·39	0·57	0·29	'the catch and yields decreased in an alarming manner in the last years' (Rodriguez 1977)
	12	*Penaeus kerathurus* ♂	0·988	4	2·76	1·57	1·19	0·43	
14	13	*Penaeus setiferus* ♀	0·881	6	6·71	2·03	4·68	0·70	known to be heavily fished
	14	*Penaeus setiferus* ♂	0·863	4	5·43	2·55	2·88	0·53	
15	15	*Trachypenaeus fulvus* ♀	0·981	7	4·59	2·75	1·84	0·40	active fishery
	16	*Trachypenaeus fulvus* ♂	0·834	8	5·9	3·12	2·78	0·47	

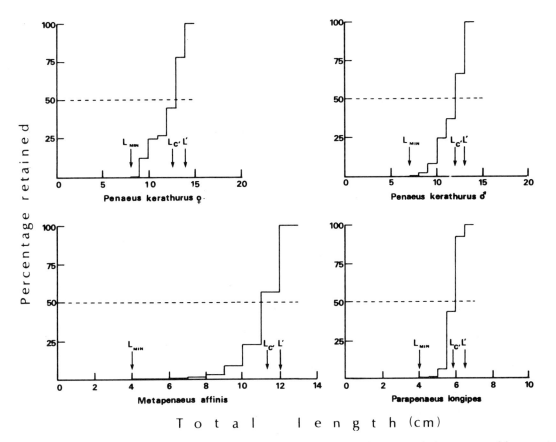

Fig 12 Four examples of 'selection patterns'. Above: showing the similarity of selection patterns in the two sexes of the same stock. Below left: a selection pattern covering a wide range of lengths. Below right: an almost 'knife-edge' selection pattern.

applying equation (9) to the growth parameter estimates of *Table 2* and the mean temperatures of *Table 1*. These values of M range from 0·77 in *Hymenopenaeus robustus* to 3·12 in male *Trachypenaeus fulvus*, a small Philippine shrimp.

When subtracted from the value of Z in *Table 3*, estimates of fishing mortality are obtained which range from 4·67 in male *Penaeus duorarum* to 0·43 in *Metapenaeus brevicornis*. Values of $E = F/Z$, are listed in *Table 3* which are complemented by brief comments gleaned from the literature as to the status of the various fisheries. There is broad agreement between the comments on the status of the fishery and the 'message' represented by a value of E ('underfishing' when $E < 0.5$; 'overfishing' when $E > 0.5$).

Recruitment patterns
The recruitment patterns obtained here fall into 3 categories:

– recruitment patterns suggestive of one recruitment event per year (or of two events, with one completely dominating the other);

– recruitment pattern suggestive of two recruitment events per year;

– recruitment patterns from which no secure inference can be drawn.

Table 4 summarizes and discusses the information extracted from the recruitment patterns here (see also *Figs 8* and *9* for examples).

Discussion

The set of methods presented here to extract growth parameters from length-frequency data (ELEFAN) is not fail safe. Given a set of highly unrepresentative length-frequency data (*eg*, as might be obtained from fishing only in the extremes of the depth range of *Penaeus* spp.) ELEFAN I will output ridiculous growth parameter values. However, when this happens, the goodness of fit values will generally be very low, suggesting themselves that the results are unreliable.

In fact, the goodness of fit values of the examples provided here (*Table 2*) are low compared with values for fish (Pauly and David 1981, Pauly and

231

Stock No.	Species, sex	Structure of recruitment pattern	Biological interpretation and remarks
1	*Hymenopenaeus robustus* ♀	one normally distributed group, or two, with one much larger than the other	Probably one recruitment event per year, with smaller 'mode' reflecting only sampling variability (see *Fig 8*)
2	*Metapenaeopsis durus* ♀ and ♂	two overlapping, normally distributed groups	Two recruitment events per year, one larger than the other (see *Fig 8*)
3	*Metapenaeus brevicornis* ♀ and ♂	one normally distributed group, or two, with one much larger than the other	Pattern suggestive of one recruitment event per year. Comparison with stock No. 4 suggests some broods may fail to recruit (see *Fig 9*)
4	*Metapenaeus brevicornis* ♀ and ♂	two overlapping, normally distributed groups	Two recruitment events per year confirming George (1970) who reports 'breeding' twice a year
5	*Metapenaeus affinis* ♀ and ♂	one normally distributed group, or two, widely overlapping groups	No clear inference can be drawn; Mohamed (1967) suggests two spawning seasons, while George (1970), reviewing several papers, suggests one spawning season per year
6	*Metapenaeus kutchensis* ♀ and ♂	two readily identifiable groups	Two recruitment events per year
7	*Metapenaeus kutchensis* ♀ and ♂	two sidely overlapping groups	Two recruitment events per year
8	*Metapenaeus kutchensis* ♀ and ♂	four modes in pattern, but two groups apparent	Probably two recruitment events per year
9	*Parapenaeus longipes* ♀	three modes in pattern, but two groups apparent	Probably two recruitment events per year
10	*Penaeus duorarum* ♂	Two readily identifiable groups	Two recruitment events per year (see *Fig 9*) confirming observations reported by Castello and Allen (1970)
11 12	*Penaeus kerathurus* ♀ *Penaeus kerathurus* ♂	one normally distributed group, or two, with one much larger than the other	Patterns equivocal, but not contradicting solid evidence Presented by Rodriguez (1977) for one spawning/recruitment season per year
13 14	*Penaeus setiferus* ♀ *Penaeus setiferus* ♂	one single, normally distributed group	One recruitment event per year, as reported by Lindner and Cook (1970)
15 16	*Trachypenaeus fulvus* ♀ *Trachypenaeus fulvus* ♂	two readily identifiable groups, one stronger than the other	Two recruitment events per year

Ingles 1981). However, this is probably due to the feature that these previous demonstrations of the use of ELEFAN I were selected for their 'good fit' (the textbook syndrome), whereas availability of data was the main criterion for including the shrimp examples provided here.

To date, use of length-frequency analysis for determination of growth parameters has been of limited use with penaeids, due to extended and fluctuating recruitment and mixing of broods. Thus, it has most often been impossible to 'follow' a selected brood. The method described here is a viable alternative if a broadly representative set of samples of the population can be obtained. When this is the case, the method will be superior to mark-recapture experiments.

Deriving catch curves from length-frequency data has been done only rarely in the past, and when done, a serious bias was involved which generated erroneous results (see *Fig 5*). Correction of this bias, however, is simple, and length-converted catch curves can thus be used to obtain reasonable estimates of Z.

However, representativeness of the samples used in the analysis remains a problem, because the distribution of shrimps is correlated with size. A related problem is that samples taken in an area of very active fishing (also an area of active recruitment) may lead to an over-estimation of total mortality for the population as a whole. Thus, it is imperative that the samples used for this part of the analysis adequately represent the population for a period of one (or more) year(s).

Given representative samples, length-converted catch curves are, most often, remarkably straight, and allow the identification of the point on the catch curve from which recruitment and selection are completed (P_1).

Only a very short comment seems appropriate concerning the concept of selection patterns as introduced here. The concept is apparently new, and there were no data available for a comparison of a selection pattern with a real selection curve, as obtained from a selection experiment, or better with a 'resultant curve', as obtained by combining a selection with a recruitment curve.

The backward projection of the catch curve as used here to generate selection patterns evidently involves a bias ($L'_c > L_c$) since fishing mortality – hence also Z – gradually decreases as one progresses from right to left on the ascending part of a catch curve. However, two factors still contribute toward making L'_c as defined here a reasonable estimator of L_c:

- among smaller shrimps, natural mortality could be higher, thus compensating (at least in part) for the reduction in fishing mortality
- both L'_c and L_c are constrained by the same values of L_{min} (smallest size caught) and L' (length at full retention, corresponding to P_1, see *Figs 6, 12* and text).

The method presented here to estimate the natural mortality of shrimps is risky in that it is based on data obtained from fish stocks. However, even if shrimp differed from fish in having – for a given set of growth parameters and environmental temperature – a slightly higher or lower natural mortality, the method would still provide estimates of M that are more reasonable than many of those to be found in the literature, in which values of M, *eg*, for *P. duorarum*, ranging from 28·6 (Kutkuhn, 1966) to 1·04 (Berry, 1970) may be found. Also, the Z and M data in *Table 3* provide indirect evidence that M in shrimp generally cannot be much lower than in fish. If they were, some of the Z values associated with low fishing intensities would have turned out to be lower than the M value estimated from equation (12).

Finally, it should be pointed out that the values of E for the various shrimp stocks, as obtained from the estimates of Z, M and F, as a whole, do not suggest anything different from what is known from the various fisheries referred to in *Table 3*. Thus, for example, in the case of heavily exploited stocks such as *Penaeus duorarum* off Tortugas, high values of E were obtained, while a relatively

low value of E was obtained for *Parapenaeus longipes*, which occurs at depths between 40 and 200 fathoms and is therefore virtually inaccessible to the trawlers operating in the Visayan Sea area of the Philippines. Thus, the values of M obtained from equation (12) seem to provide reasonable results for shrimps as well as fish. These values allow, in conjunction with values of Z estimated from length-converted catch curve, the estimation of values of E and therefore a direct assessment of the status of a fishery.

The information obtained from recruitment patterns (*Table 4*) is in general agreement with studies on the reproduction of the shrimp stocks in question. It may be concluded that this method of examining recruitment patterns is a useful tool.

In summary the methodology proposed here (the application of ELEFAN I and II) for the study of shrimp stocks has the following features:

- given suitably representative length-frequency data, information is extracted which is important both for practical stock assessment purposes and for understanding the basic biology of shrimp,
- all results obtained by these methods are reproducible, *ie*, different authors using the same set of data will get the same results,
- the methods can be applied to both tropical and temperate shrimp stocks and implemented on cheap microcomputer, affordable even by the fishery research laboratories of developing countries, and
- the methods can be applied both to new or existing (published or unpublished) data. This permits comparative studies and the reconstruction of time series (*eg*, of Z, growth parameters) which should prove invaluable in understanding the dynamics of shrimp stocks.

References

ANDERSON, W W and LINDNER, M J. Contribution to the
1971   biology of the royal red shrimp, *Hymenopenaeus robustus* Smith. *U.S. Fish. Bull.* 69: 313–336.

ANON. World Atlas of sea surface temperatures. *2nd* Ed.
1944   Hydrogr. Office. U.S. Navy H.O. No. 225.

BERRY, R J. Dynamics of the Tortugas (Florida) pink shrimp
1967   populations. Ph.D. Thesis. University of Rhode Island. University Microfilms, Inc. Ann Arbor, Michigan. 160 p.

BERRY, R J. Shrimp mortality rates derived from fishery statis-
1970   tics. *Proc. Gulf Caribb. Fish. Inst.* 22: 66–78.

BEVERTON, R J H and HOLT, S J. A review of methods for
1956   estimating mortality rates in fish populations, with special references to sources of bias in catch sampling. *Rapp. P.-V. Réun. Cons. Perm. Int. Explor. Mer.* 140: 67–83.

BEVERTON, R J H and HOLT, S J. A review of the life-spans and
1959   mortality rates of fish in nature, and their relation to growth and other physiological characteristics. *Ciba Found. Colloq. Ageing.* 5: 142–180.

BOSCHI, E E. Crecimiento, migracion y ecologia del camaron
1969 commercial *Artemesia longinaris* Bate, p. 833–846. *In* Mistakidis, M N (ed.) Procedings of the world scientific conference on the biology and culture of shrimps and prawns. FAO Fish. Rep. 57 Vol. 3.

BROTHERS, E B. Age and growth studies on tropical fishes, p.
1980 119–136. *In* Saila, S and Roedel, P (eds.) Stock assessment for tropical small-scale fisheries. Proceedings of an international workshop held September 19–21, 1979, at the University of Rhode Island. Int. Cent. Mar. Res. Dev., Univ. of Rhode Island. 198 p.

CASSIE, R M. Some uses of probability paper in the analysis of
1954 size frequency distribution. *Aust. J. Mar. Freshwater Res.* 5: 513–522.

DIXON, W J and MOOD, A M. The statistical sign test. *J. Amer.*
1946 *Stat. Assoc.* 41: 557–566.

EDSER, T. Note on the number of plaice at each length, in
1980 certain samples from the southern part of the North Sea, 1906. *J. R. Stat. Soc.* 71: 686–690.

FORSTER, J R M. Further studies on the culture of the prawn,
1970 *Pelaemon serratus* Pennant, with emphasis on the post-larval stages. *Fish. Invest. Ser.* 11, 26(6). 40 p.

GEORGE, M J. Synopsis of biological data on the penaeid prawn
1970 *Metapenaeus affinis* (Miers, 1978), p. 1359–1375. *In* Mistakidis, M N (ed.) Procedings of the world scientific conference on the biology and culture of shrimps and prawns. FAO Fish. Rep. 57 Vol. 4.

GEORGE, K and BANERJI, S K. Age and growth studies on the
1964 Indian mackerel *Rastrelliger kanagurta* (Cuvier) with special reference to length-frequency data collected at Cochin. *Indian. J. Fish.* 11(2): 621–638.

GULLAND, J A. Manual of methods for fish stock assessment.
1969 Part I. Fish population analysis. FAO Man. Fish. Sci. 4, 154 p. FAO, Rome.

GULLAND, J A, editor. The fish resources of the ocean. Fishing
1971 News (Books) Ltd., Surrey, England. 255 p.

HARDING, J P. The use of probability paper for graphical
1949 analysis of polymodal frequency distributions. *J. Mar. Biol. Assoc. U.K.* 28: 141–153.

INGLES, J. Distribution and relative abundance of penaeid
1980 shrimps (subfamily Penaeinae) in the Visayan Sea. M.Sc. Thesis. University of the Philippines, Manila, Philippines. 78 p.

IVERSEN, E S, JONES, A E and IDYLL, C P. Size distribution of
1960 pink shrimp, *Penaeus duorarum* and fleet concentrations on the Tortugas fishing grounds. *U.S. Dept. Int. Fish. Wildl. Serv. Spec. Sci. Rep. Fish.* 356. 62 p.

KUTKUHN, J H. Gulf of Mexico commercial shrimp popula-
1962 tions – trends and characteristics, 1956–59. *U.S. Fish. Bull.* 62: 343–402.

KUTKUHN, J H. Dynamics of a penaeid shrimp population and
1966 management implications. *U.S. Fish. Bull.* 65: 313–338.

LINDNER, M J and ANDERSON, W W. Growth, migration,
1956 spawning and size distribution of shrimp *Penaeus setiferus* (Linnaeus) 1767, p. 1439–1469. *In* Mistakidis, M N (ed.) Procedings of the world scientific conference on the biology and culture of shrimps and prawns. FAO Fish. Rep. 57 Vol. 4.

MOHAMED, K H. Penaeid prawns in the commercial shrimp
1967 fisheries of Bombay with notes on species and size fluctuations, p. 1408–1418. *In* Proceedings of the symposium on Crustacea held at Ernaculam, January 1965. Mar. Biol. Assoc. India. Mandapam Camp., India. Part IV.

NEAL, R A. Methods of marking shrimp, p. 1149–1165. *In*
1969 Mistakidis, M N (ed.) Proceedings of the world scientific conference on the biology and culture of shrimps and prawns. FAO Fish. Rep. 59 Vol. 4.

PAULY, D. A preliminary compilation of fish length growth
1978 parameters. *Ber. Inst. Meereskunde (Kiel) No. 55.* 200 p.

PAULY, D. Gill size and temperature as governing factors in fish
1979 growth: a generalization of von Bertalanffy's growth formula. *Ber. Inst. Meereskunde (Kiel) No. 63.* 156 p.

PAULY, D. On the interrelationships between natural mortality,
1980a growth parameters and mean environmental temperature in 175 fish stocks. *J. du Conseil* 39(3); 175–192.

PAULY, D. A new methodology for rapidly acquiring basic
1980b information on tropical fish stocks: growth mortality and stock-recruitment relationships, p. 154–172. *In* Saila, S and Roedel, P (eds.) Stock assessment for tropical small-scale fisheries. Proceedings of an international workshop held September 19–21, 1979, at the University of Rhode Island. Intern. Cent. Mar. Res. Dev. Univ. of Rhode Island. 198 p.

PAULY, D. Studying single-species stocks in a multispecies
1982 context. p. 33–70. *In* Pauly, D and Murphy, G. I. (eds.) Proceedings of the ICLARM/CSIRO Workshop on the Theory and Management of Tropical Fisheries. ICLARM Conference Proceedings 9. 360 p.

PAULY, D. Some simple methods for the assessment of tropical
1983 fish stocks. FAO Fish Tech. Pap. No 234, 52 p.

PAULY, D and DAVID, N. An objective method for determining
1980 growth from length-frequency data. ICLARM Newsletter 3(3): 13–15.

PAULY, D and DAVID, N. ELEFAN I, a BASIC program for the
1981 objective extraction of growth parameters from length-frequency data. *Meeresforsch.* 28(4): 205–211.

PAULY, D, DAVID, N and INGLES, J. ELEFAN I. Users' instruc-
1980 tions and program listings. (Mimeo, pag. var.)

PAULY, D, DAVID, N and INGLES, J. ELEFAN II. Users'
1981 instructions and program listings. (Mimeo, pag. var.)

PAULY, D and GASCHÜTZ, G. A simple method for fitting
1979 oscillating length growth data, with a program for pocket calculations. Int. Counc. Explor. Sea. CM 1979/ G: 24, 26 p. (Mimeo.)

PAULY, D and INGLES, J. Aspects of the growth and natural
1981 mortality of exploited coral reef fishes, p. 89–98. *In* Proceedings of the 4th International Coral Reef Symposium, 17–22 May 1981. Vol. 1. Manila.

PETERSEN, C G J. Fiskensbiologiske forhold i Holboek Fjord,
1892 1890–91. *Beret. Danske Biol. Sta.* 1890 (91). 1: 121–183.

RAMAMURTHY, S. Studies on the prawn fisheries of Kutch, p.
1967 1424–1436. *In* Proceedings of the symposium on Crustacea held at Ernakulam, January, 1965. Pat IV. Mar. Biol. Assoc. India. Mandapam Camp., India.

RICKER, W E. Computation and interpretation of biological
1975 statistics of fish populations. *Bull. Fish. Res. Board. Can.* 191. 382 p.

RIVAS, L R. Fishermen's atlas of monthly sea surface tempera-
1968 tures for the Gulf of Mexico. *U.S. Dept. Int. Fish. Wildl. Surv. Bur. Comm. Fish. Circ.* 300. 33 p.

ROBSON, D S and CHAPMAN, D G. Catch curves and mortality
1961 rates. *Trans. Amer. Fish. Soc.* 90: 181–189.

RODRIGUEZ, A. Contribución al conocimiento de la biologia y
1977 pesca del langostino, *Penaeus kerathurus* (Forskål 1776) del Golfo de Cádiz (Region Sudatlántica española) Invest. Pesq. 41(3): 603–635.

TANAKA, S. A method of analyzing the polymodal frequency
1956 distribution and its application to the length distribution of porgy, *Taius tumifrons* (T. and S.). *Bull. Tokai Reg. Fish. Res. Lab.* 14: 1–2 (in Japanese, English summary).

TOMLINSON, P K. Program Name: NORMSEP. Programmed
1971 by Victor Hasselblad. *In* Computer Programs for Fish Stock Assessment, N J Abramson, comp. FAO Fish. Tech. Pap. 10 p.

YONG, M Y and SKILLMAN, R A. A computer program for
1975 analysis of polymodal frequency distributions (ENORMSEP), FORTRAN IV. *U.S. Fish. Bull.* 73(3): 681.

ZEIN-ELDIN, Z P and GRIFFITH, G W. An appraisal of the
1969 effects of salinity and temperature on growth and survival of postlarval penaeids, p. 1016–1026. *In* Mistakidis, M N (ed.) Proceedings of the world scientific conference on the biology and culture of shrimps and prawns. FAO Fish. Rep. 57 Vol. 3.

Interaction with other species

Ecological interactions between penaeid shrimp and bottomfish assemblages

Peter F Sheridan
Joan A Browder
and Joseph E Powers

Abstract

A biological review identifies potential ecological interactions between commercial penaeid shrimp and bottomfish that share the same habitat and are caught with shrimp in trawls, then returned to the sea as carrion. Two models are used to evaluate the possible impact on shrimp stocks of reducing the quantity of fish discarded, assuming certain ecological interactions exist. The first model, using a classical population dynamics approach, treats the problem as one of two stocks exploited by a common fishery, each partially supported by the discards of the other. Conclusions from this model are that the elimination of bottomfish discards could reduce shrimp production by only a small amount. The energy-flow ecosystem model treats the problem as one of 11 trophic compartments linked by the flow of energy and the cycling of nitrogen. In this model system, a reduction in discards through utilization of one-half the by-catch reduces shrimp stocks by 25 percent, but a reduction in discards through the use of special trawls with one-half the catch-efficiency for fish reduces shrimp stocks by only 8 percent, even in the 'worst case' of bottomfish predation on shrimp. In the model system, there is no long-term impact on shrimp of reducing discards by means of special trawls, if bottomfish are moderately selective against shrimp relative to their biomass in the environment.

Introduction

Penaeid shrimp are a highly prized seafood harvested from coastal tropical and warm-temperature waters throughout the world. Bottomfish of many species are harvested in large quantities in the shrimp trawling operation. The species composition of this by-catch varies with time and area, but it is generally dominated by sciaenids, pomadasyids, sparids, synodontids, serranids, and bothids. Determination of the nature and extent of ecological interactions between shrimp and bottomfish is important in managing the shrimp fisheries and in exploiting the bottomfish resources.

Direct interaction can be limited to predation, competition or scavenging. Scavenging may be of significance because of the quantity of the trawl catch that is discarded. A large proportion of the bottomfish catch is discarded due to the higher value of shrimp. Some shrimp also are discarded due to size requirements, and are lost during the bottomfish culling process. In this paper we investigate the likely magnitude of shrimp/bottomfish interactions by presenting 1) an extensive review of the biological information on shrimp and bottomfish assemblages, 2) an analytical model to assess the impacts of discards, and 3) an energy flow model of the shrimp/bottomfish ecosystem that evaluates the possible effect on shrimp of reducing discards by two alternative methods. We then make some qualitative conclusions about the importance of considering interactions between shrimp and bottomfish in the management of these resources.

Biological review of shrimp/bottomfish assemblages

Distribution of major shrimp grounds and associated fishes

Penaeid shrimps support commercially valuable fisheries in many areas of the world which lie between 35° north and south of the equator (Turner, 1977). At least 97 species in the family Penaeidae are of commercial interest (Holthuis, 1980). The Food and Agriculture Organization of the United Nations publishes yearly summaries of global fisheries statistics. The 1978 summary identifies 21 species of penaeid shrimps which contributed to overall shrimp landings of 1 474 176 metric tons (Table 1, FAO, 1979). For all shrimps, the major catch areas were Western Central Pacific, Western Indian Ocean, Western Central Atlantic, and Northwest Pacific, in order of decreasing catch. By region, the predominant shrimps were: Western Central Pacific—*Penaeus merguiensis*, *Metapenaeus spp.*, and *Penaeus* spp.; Western Indian Ocean—unspecified Natantia; Western Central Atlantic—*Penaeus aztecus*; Northwest Pacific—*P. japonicus*. Although penaeids are taken world-wide, they represent only a small portion of total fisheries production. In fact, for countries which reported penaeid shrimps specifi-cally (rather than as 'Natantia'), penaeids comprised less than 10% of the total fisheries catch in 18 of 21 countries (*Table 1*). Penaeids were a significant portion of the total fishery catch only in Honduras (35·7% by weight), Guatemala (51·4%), and El Salvador (72·2%). Countries with large overall fisheries generally showed proportionally low penaeid catches; Japan, 0·1%; U S A, 3·5%; Thailand, 1·5%; Indonesia, 2·9%; Spain, 0·2%.

Very little is known concerning the bottomfish assemblages associated with shrimping grounds outside of the Gulf of Mexico. Allsopp (M. S.) estimates a weight ratio of discarded fish to shrimp in Caribbean shrimp fisheries ranging between 3:1 and 20:1. Off the Amazon River delta of Brazil,

Table 1
RECENT CATCHES OF PENAEID SHRIMPS FROM 1978 WORLD CATCH RECORDS (FAO, 1979)

Total catch of all shrimps: 1 474 176 metric tons

Major species	Weight (mt)	Primary areas
Penaeus aztecus	63 624	U S A
Penaeus merguiensis	40 098	Indonesia (70%)
Penaeus duorarum	25 347	U S A (65%), Cuba (20%)
Penaeus monodon	17 599	Indonesia (90%)

Major species and countries (both in decreasing order of abundance)

Penaeus spp. (unspecified)	U S A, Philippines, Panama, Honduras, Cuba, Colombia, Costa Rica, Guatemala, Sierra Leone, El Salvador, Peru
P. aztecus	U S A
P. merguiensis	Indonesia, Thailand, Papua-New Guinea
Metapenaeus spp.	Indonesia, Thailand, Korea, Papua-New Guinea
P. duorarum	U S A, Cuba, Nigeria, Spain, Gambia
P. monodon	Indonesia, Thailand, Papua-New Guinea
Xiphopenaeus/Trachypenaeus	El Salvador, Panama, Colombia, Guatemala
P. kerathurus	Italy, Spain
P. japonicus	Japan, Korea, Papua-New Guinea
P. semisulcatus	Thailand
P. brevirostris	Panama, Costa Rica, Guatemala, El Salvador
Sicyonia brevirostris	U S A
Parapenaeus longirostris	Spain, U S S R
P. chinensis	Korea
Plesiopenaeus edwardianus	Spain
P. californiensis	Guatemala, El Salvador
Artemesia longinaris	Argentina

Major areas and species

Western Central Pacific	P. merguiensis, Metapenaeus spp., Penaeus spp. > P. monodon > P. japonicus, P. semisulcatus, P. latisulcatus
Western Central Atlantic	P. aztecus > P. duorarum > Xiphopenaeus, Trachypenaeus, Sicyonia
Northwest Pacific	P. japonicus > Metapenaeus spp., P. chinensis
Eastern Central Pacific	Xiphopenaeus > Penaeus spp. > P. brevirostris > P. californiensis
Eastern Indian Ocean	Metapenaeus spp., P. merguiensis > P. semisulcatus > P. monodon, P. latisulcatus
Southwest Atlantic	Artemesia
Eastern Central Atlantic	P. duorarum > >> P. kerathurus, Parapenaeus, Plesiopenaeus
Mediterranean/Black Sea	P. kerathurus > > > Parapenaeus, Plesiopenaeus
Southeast Atlantic	Parapenaeus > P. kerathurus

Table 1 (continued)

Penaeid catch related to total fishery catch (mt) by country

Area	Country	Species	Shrimp	Total	%
Africa	Gambia	P. duorarum	183	10 795	1·7
	Nigeria	P. duorarum	1 916	518 567	0·4
	Sierra Leone	Penaeus spp.	143	50 080	0·3
North Amer.	Costa Rica	P. brevirostris	420	14 491	7·4
		Penaeus spp.	189		
		Xipho/Trachy	461		
	Cuba	P. duorarum	5 300	213 170	3·6
		Penaeus spp.	2 300		
	El Salvador	Xipho/Trachy	3 849	5 487	72·2
		Penaeus spp.	115		
	Guatemala	Xipho/Trachy	998	3 074	51·4
		Penaeus spp.	583		
	Honduras	Penaeus spp.	2 288	6 405	35·7
	Panama	Xipho/Trachy	3 378	113 768	7·8
		Penaeus spp.	3 552		
		P. brevirostris	1 982		
	USA	P. aztecus	63 624	3 511 719	3·5
		Penaeus spp.	35 279		
		P. duorarum	16 910		
		X. kroyeri	3 771		
		Sicyonia	1 825		
South Amer.	Argentina	Artemesia	241	537 323	0·1
	Colombia	Xipho/Trachy	2 984	63 965	7·1
		Penaeus spp.	1 550		
Asia	Indonesia	P. merguiensis	27 856	1 655 000	3·9
		Metapenaeus spp.	19 318		
		P. monodon	16 967		
	Japan	P. japonicus	3 857	10 752 163	0·1
	Korea	P. japonicus	3 037	2 350 778	0·3
		Metapenaeus spp.	2 436		
		P. chinensis	1 124		
	Philippines	Penaeus spp.	23 197	1 558 383	1·5
	Thailand	Metapenaeus spp.	16 707	2 264 000	1·5
		P. merguiensis	11 685		
		P. semisulcatus	4 042		
		P. latisulcatus	1 620		
		P. monodon	514		
Europe	Italy	P. kerathurus	6 694	401 958	1·7
	Spain	P. duorarum	1 038	1 379 882	0·2
		Parapenaeus	942		
		Plesiopenaeus	509		
		P. kerathurus	17		
Oceania	Papua-New Guinea	P. merguiensis	557	74 186	1·2
		Metapenaeus	213		
		P. japonicus	34		
	Australia	'Natantia'	18 807	122 947	15·3

Nomura and Filho (1968) found fish to shrimp ratios of 2:1 to 1:3, and the bottomfishes were primarily elasmobranchs, ariids, triglids, lutjanids, and soleids. Recent surveys by U S National Marine Fisheries Service in the same area of the Amazon delta (OREGON II cruise 84, Nov.– Dec. 1977) found lutjanids, balistids, pomacanthids, and elasmobranchs predominating offshore over hard bottoms, and sciaenids, pomadasyids, ariids, and balistids abundant inshore over soft bottoms. The average fish to shrimp weight ratio was 9:1 at the time. Several surveys have been conducted around India. Pruter (1964) found dasyatids, sciaenids, synodontids, polynemids, lutjanids, ariids, and pomadasyids dominated the by-catch in areas where shrimp catches were small. In areas where shrimp catches were relatively large along the Indian coast, sciaenids, percids, elasmobranchs, engraulids, clupeids, and trichiurids are abundant (Muthu *et al*, 1975).

In the Gulf of Mexico, the three dominant commercial species, *Penaeus aztecus*, *P. setiferus*, and

P. duorarum, are generally found throughout the Gulf, but each has distinctive seasons and areas of maximum abundance (Osborne *et al*, 1969). *Penaeus aztecus* is caught primarily off the Texas and Louisiana coasts, *P. setiferus* off the Mississippi Delta area, and *P. duorarum* on the Tortugas–Sanibel grounds of Florida and the Campeche Bank area of Yucatan, Mexico. Several other genera are exploited to a lesser extent. Brusher *et al* (1972) discussed the distributions of 9 species in the general *Sicyonia*, *Trachypenaeus*, *Xiphopenaeus*, *Parapenaeus*, and *Solenocera* on the Texas–Louisiana continental shelf. Huff and Cobb (1979) described the general ecologies of 9 species in the general *Trachypenaeus*, *Solenocera*, *Sicyonia*, *Mesopenaeus*, and *Metapenaeopsis* on the western Florida shelf.

Although the species compositions of the bot-tomfishes found on the Gulf shrimp grounds vary somewhat with time and area, eight species (primarily sciaenids) characterize the Mississippi Delta grounds, seven species dominate in the Campeche pink shrimp grounds, and 13 species characterize the Texas and Campeche brown shrimp grounds (*Table 2*). A recent NMFS survey from Tampa Bay south to the Tortugas (OREGON II cruise 85, January 1978) found the six most abundant fish families to be pomadasyids, sparids, synodontids, seranids, bothids, and sciaenids.

The average ratio of fish to shrimp in shrimp catches on the Mississippi Delta grounds is approximately 14 to 1 (unpublished data, National Marine Fisheries Service, Pascagoula, Mississippi). Directed fisheries for the fish species occur in this area, most of which are harvested for pet food. Croaker (*Micropogonias undulatus*), the principal

Table 2

RELATIVE ABUNDANCES OF BOTTOMFISHES ON SHRIMP GROUNDS IN THE GULF OF MEXICO. SPECIES ARE RANKED BY RELATIVE ABUNDANCE: 1 = MOST ABUNDANT, R = RELATIVELY RARE, DASH = NOT FOUND

| Species | Brown shrimp (P. aztecus) grounds – Hildebrand, 1954 | | | | |
| | South Texas | | | | Campeche |
	Nov	Jan	May	July	July–Aug
Syacium gunteri	1	1	2	2	4
Cynoscion nothus	2	R	3	5	R
Cynoscion arenarius	2	8	5	10	10
Cyclopsetta chittendeni	3	2	4	4	9
Prionotus rubio	4	7	R	9	6
Synodus spp.	5	6	6	6	3
Serranus atrobranchus	7	3	R	R	1
Centropristis philadelphica	R	4	8	R	–
Stenotomus caprinus	10	5	R	R	R
Porichthys porosissimus	R	10	R	R	2
Peprilus burti	R	R	1	1	R
Micropogonias undulatus	R	R	R	3	R
Lepophidium graellsi	R	R	R	R	5

| Species | Pink shrimp (P. duorarum) grounds | | |
| | Campeche – Hildebrand, 1955 | | Tortugas |
	Feb	July	
Haemulon aurolineatum	1	5	?
Chloroscombrus chrysurus	2	2	see
Eucinostomus gula	3	3	text
Diplectrum formosum	4	6	
Prionotus scitulus	5	–	
Stenotomus caprinus	–	1	
Syacium gunteri	8	4	

| Species | Mississippi Delta-NMFS OREGON II Cruises 101, 10 | | | |
| | White shrimp grounds | | Brown shrimp grounds | |
	Fall, 1979	Spring, 1980	Fall, 1979	Spring, 1980
Micropogonias undulatus	1	4	1	1
Arius felis	2	1	R	–
Leistomus xanthurus	3	R	2	3
Cynoscion arenarius	4	3	4	4
Chloroscombrus chrysurus	5	R	R	–
Stenotomus caprinus	R	–	3	5
Peprilus burti	–	5	5	2
Cynoscion nothus	R	2	R	R

species caught incidentally in this area, are harvested for making surimi (fish paste), for a limited fresh fish market, and by recreational fisheries throughout the northern Gulf of Mexico. In the Mississippi Delta area, combined annual landings by these directed fisheries are approximately one tenth the size of the discards by the shrimp fleet.

Competitive interactions
Competition among species only occurs when a resource is in limited supply. Two basic types of competition have been recognized: 1) interference, in which one individual physically prevents the use of a resource by another individual either by aggression or mere physical presence, and 2) exploitation, in which one individual utilizes a limited resource before another individual arrives. Competition can be intraspecific or interspecific. Resources may be food, habitat, or time. Schoener (1974) has reviewed various kinds of resource partitioning, for example, species which overlap in habitat tend to have different foods, or species which have similar foods may feed at different times.

Among the penaeid shrimps and bottomfishes, sympatric species in each group often differ in their positions along such resource gradients as spawning season and depth, time or tide of migration, substrate type, temporal activity, and diet. Penaeids and bottomfishes may be in competition for certain resources when their life history stages overlap. At present, there have been no definitive explorations into potentially competitive interactions between fishes and shrimps beyond artificial laboratory studies.

Predatory interactions
All available evidence points to omnivory by penaeid shrimps, which consume varying proportions of sediment, detritus, algae, and benthic organisms. The bottomfishes, however, represent a diverse assemblage of trophic types ranging from herbivores through carnivores. The predatory interactions between penaeids and bottomfishes are basically limited to fish attacking shrimp. There are no published accounts of the reverse, but since shrimp are omnivores they are likely to feed on disabled, dying or dead fishes (*eg*, discarded bycatch). The impact of discarding will be discussed later in this paper. A question not yet satisfactorily answered is to what extent shrimp stocks are affected by bottomfish predation upon them.

Quantitative assessments of fish predation on penaeid shrimps are limited, and problems inherent in gut contents analyses are many. Yanez-Arancibia *et al* (1976) studied the feeding of *Galeichthys* (=*Aurius*) *caerulescens* in western Mexico estuaries and found this catfish preyed mainly upon fishes and crabs. Shrimp identified as *Penaeus* spp. averaged only 7% by volume of the stomach contents and were only found in certain seasons. Bell *et al* (1978) studied the foods of an Australian scorpaenid (*Centropogon australis*) in seagrass meadow and found only a 4·3% frequency occurrence of *Penaeus* in fish stomachs, yet penaeids in that region prefer seagrass habitats over bare substrates. In Japanese estuaries, the penaeids most often eaten by fish predators are the small, non-commercial species whereas the larger commercial shrimps do not figure to any great extent in fish diets (Kakuda and Matsumoto, 1978; Kosaka, 1977, 1978).

A great deal of qualitative and quantitative information has been gathered in Gulf of Mexico studies, but it is primarily derived from estuarine investigations and rarely are prey shrimps identified beyond the categories 'shrimps' or 'penaeids'. Qualitatively, of the 42 fish species listed in *Table 3*, only 11 species have a 40% or greater frequency of occurrence of 'penaeids' (not just *Penaeus*) in their diets. Unfortunately, frequency of occurrence does little to quantify the importance of any food item in any diet. When some form of quantitative assessment was made (*Table 4*), only 8 (and probably only 4) of the 26 species examined made 'penaeids' 40% or more (by volume or weight) of their diets. In one synoptic quantitative study wherein prey shrimps were identified beyond the category of 'penaeids' in offshore fishes (*Table 5*), only shrimps of the genera *Sicyonia*, *Solenocera*, *Parapenaeus*, and *Trachypenaeus* were found. The genus *Penaeus* was not detected in the diets of 26 abundant offshore fishes, even though *Penaeus* inhabited the same waters. Recent studies of trawl-susceptible fishes in offshore Gulf waters (Divita *et al*, 1983; Sheridan and Trimm, 1983; NMFS, Galveston, TX, unpubl. data) found migrating juvenile and subadult *Penaeus* in less than 1% of the fish stomachs examined.

Information concerning large potential predators of both penaeids and bottomfishes is quite limited. Tunas (*Tunnus*, *Euthunnus*, *Katsuwonus*) are mainly piscivorus (*Table 4*). Mackerels (*Scomberomorus*), cobia (*Rachycentron*), and bluefish (*Pomatomus*) feed to some extent on penaeids as well as fishes. Sharks may also be penaeid predators but seem to prefer bottomfishes (Bass *et al*, 1973). Of the 16 species of *Carcharhinus* examined

Table 3
QUALITATIVE ANALYSES OF FISH PREDATION ON PENAEID SHRIMPS BASED ON INSHORE (I) AND OFFSHORE (O) STUDIES

Fish predators	Percent Frequency of Occurrence				
	0	1–20	21–40	41–60	61–100
Anchoa hepsetus	I				
Anchoa mitchilli	I				
Harengula jaguana	I				
Caulolatilus chrysops	I				
Menticirrhus littoralis	I				
Katsuwonus pelamis	O				
Larimus fasciatus	I				
Thunnus thynnus	O				
Thrichiurus lepturus	I				
Raja spp.	O				
Paralichthys albigutta	I	I			
Synodus foetens	I	I			
Leistomus xanthurus	I	I			
Stellifer lanceolatus	I	I			
Thunnus albacares	O	O			
Caranx hippos		I			
Caulolatilus microps		I			
Euthynnus alletteratus		O			
Porichthys porosissimus		I			
Prionotus tribulus		I			
Stenotomus caprinus		O			
Thunnus alalunga		O			
Cynoscion arenarius		I	I		
Bairdiella chrysoura		I	I		
Urophycis floridanus		I	I		
Micropogonias undulatus	I	I, O	I		
Menticirrhus americanus	I	I	I		
Paralichthys lethostigma		I	I	I	
Lutjanus campechanus			I		
Prionotus scitulus			I		
Scomberomorus maculatus			I		
Carcharhinus sp.			I		
Rachycentron canadum				I	
Menticirrhus sp.				I	
Scomberomorus cavalla		I		I	
Arius felis	I	I		I	I
Bagre marinus		I		I	
Centropristis melana					I
Cynoscion nebulosus					I
Pomatomus saltatrix					I
Oligoplites saurus					I
Diplectrum formosum					I

See references section to specific literature

Table 4
QUANTITATIVE ANALYSES OF FISH PREDATION ON PENAEID SHRIMPS BASED ON INSHORE (I) AND OFFSHORE (O) STUDIES. ? = PENAEIDS NOT DIFFERENTIATED FROM OTHER SHRIMPS

Fish Predators	Percent by volume, weight or number				
	0	1–20	21–40	41–60	61–100
Anchoa mitchilli	I				
Harengula jaguana	I				
Opisthonema oglinum	I				
Leiostomus xanthurus	I				
Rhomboplites aurorubens	I				
Trachinotus carolinus	I				
Bellator militaris	O				
Prionotus salmonicolor	O				
Saurida brasiliensis	O				
Paralichthys lethostigma	I	I			
Cynoscion arenarius	I	I			
Micropogonias undulatus	I	I			
Bairediella chrysoura	I	I			
Arius felis	I	I			
Anchoa hepsetus	O	I?			
Synodus foetens		I?			
Prionotus roseus		O			
Prionotus scitulus		I, O			
Prionotus tribulus		O			
Prionotus alatus		O			
Lutjanus campechanus		I	I		
Ancyclopsetta quadrocellata	I	I		I	
Citharichthys spilopterus		I			I
Oligoplites saurus		I?			I?
Trachinotus falcatus		I?		I?	
Haemulon plumieri		I?		I?	
Orthopristis chrysoptera			I?	I?	I?
Prionotus ophryas					O
Caranx hippos					I

See references section for specific literature

bottomfish discards on shrimp stocks, assuming certain interactions between the two stocks exist.

Impacts of interactions

In this section we will investigate analytically the impact of discarding of both shrimp and bottomfish and other interactions on the production of these populations. The analysis is meant to be illustrative in that relative changes rather than absolute values are of concern; nevertheless, some care was taken to parameterize the hypothetical models so that differences between the populations would be meaningful.

Discard model
Let us assume that the shrimp and bottomfish can be characterized as two discrete stocks which have population biomasses of P_1 and P_2, respectively and which are exploited by a common fishery. Assume their dynamics are depicted by a simple population growth model which expresses the

off South Africa (which has 5 species of commercial penaeids, Joubert and Davies, 1966), only 5 species fed on *Penaeus* and only to a minor extent (average, 5% frequency of occurrence).

The biological review suggests that man may be the major predator of penaeid shrimps. The consequences of interactions of man and bottomfish with penaeid shrimp will be explored in the following sections by means of a population dynamics model and an energy flow model based upon the biological review.

Both models evaluate the potential influence of

Table 5

RESULTS OF A QUANTITATIVE STUDY OF FISH FEEDING ON THE GULF OF MEXICO CONTINENTAL SHELF (ROGERS, 1977). SIZE = SIZE OF FISHES. % VOL = PERCENTAGE OF VOLUME OF FISH STOMACH CONTENTS ATTRIBUTED TO SHRIMP. A TOTAL OF 4 550 STOMACHS WERE EXAMINED

Species	*Size (mm : SL)*	*% Vol*	*Dominant shrimps*
Anchoa hepsetus	26–125	0	
Saurida brasiliensis	51–125	0	
Halieutichthys aculeatus	26–75	0	
Ogcocephalus parvus	51–125	0	
Chloroscombrus chrysurus	101–150	0	
Stenotomus caprinus	26–125	0	
Micropogonias undulatus	51–125	0	
Bollmannia communis	26–75	0	
Peprilus burti	26–75	0	
Prionotus stearnsi	26–100	0	
Etropus crossotus	26–125	0	
Symphurus civittatus	51–150	0	
Symphurus plagiusa	101–125	0	
Synodus foetens	50–200	3	Sicyonia
Porichthys porosissimus	26–100	8	Sicyonia, Parapenaeus
Cynoscion arenarius	26–100	15	Trachypenaeus
Trichopsetta ventralis	76–125	18	Trachypenaeus, Parapenaeus
Cynoscion nothus	26–175	19	Sergestids, Trachypenaeus
Centropristis philadelphica	26–225	23	Sicyonia, Sergestids
Syacium gunteri	51–150	29	Trachypenaeus, Carideans
Prionotus rubio	26–175	29–50	Trachypenaeus, Sicyonia
Diplectrum bivittatum	26–125	48	Trachypenaeus, Solenocera
Serranus atrobranchus	26–125	49	Trachypenaeus, Sicyonia
Lepophidium graellsi	101–225	54	Carideans
Cynoscion nebulosus	26–75	57	Trachypenaeus, Sergestids
Citharichthys spilopterus	51–125	62	Trachypenaeus

change in P_i as the production minus harvest plus reassimilation if discards of P_i plus reassimilation of discards of P_j. In more mathematical terms this can be expressed as a modification to the logistic growth model

$$dP_i/dt = (a_iP_i - b_iP_i^2) - q_ifP_i$$
$$+ P_i(c_{ii}d_iq_ifP_i)(P_i(c_{ji}d_jq_jfP_j)) \quad (1)$$

where f = fishing effort; q_i = catchability coefficient for stock i; a_i, b_i = production parameters for stock i; d_i = proportion of catch of stock i which is discarded; and c_{ji} = rate of biological conversion of biomass of discards of stock j to biomass of stock i per unit biomass of stock i; (subscript 1 refers to shrimp; subscript 2 refers to bottomfish, ij). When the stocks are in equilibrium (P_i^*), then

$$P_i^* = \frac{(a_i - q_{if})(b_jdk_jC_{jj}q_jf) + (a_j - q_jf)(dc_{ji}q_if)}{(b_i - d_ic_{ii}q_if)(b_j - d_jd_{jj}q_jf) - d_id_jc_{ji}c_{ij}q_iq_jf^2} \quad (2)$$

The equilibrium biological production is (N_i^*)

$$N_i^* = P_i^*(a_i + d_jc_{ji}q_jfP_j^*) - (P_i^*)^2(b_i - d_ic_{ii}q_{if})$$

This model is characterized by the parabolic production function curve in which the maximum equilibrium yield, when there is no discarding, occurs at population sizes which are one-half of their respective carrying capacities. In reality we expect that the production curve for shrimp would be exactly parabolic. Indeed density dependence implied by this model may not be readily demonstrated; the parabolic curve may be extremely flat in shrimp. Thus, parameter values were chosen to mimic this condition. However, we seek to determine the relative effect of the discard conversion terms on the shape of the production curves, ie, changes in maximum sustainable yield (MSY), changes in the effort required to produce MSY, and changes in the population size at MSY. Note that (1) assumed that there is no direct competition or predation between P_1 and P_2. Interaction occurs only indirectly through the consumption of discards. As we discussed previously in this paper, evidence of competition and/or predation between shrimp and bottomfish is not clear cut. We will discuss the possible effects of these factors subsequently.

Parameterization

The production parameters assumed for equation (1) were chosen as follows: $a_1 = a_2 = b_1 =$

$q_1 = 1; b_2 = q_2 = 0.5$. This particular choice of parameters implies several relationships for the two stocks: when there is no discarding ($d_i = d_j = 0$); the biomass of bottomfish when there is no fishing is twenty times that of shrimp $P_{2\max} = 20.0$; $P_{1\max} = 1.0$); the maximum sustainable yield of bottomfish is twenty times that of the shrimp ($\text{MSY}_2 = 5.00; \text{MSY} = 0.25$); MSY for both are 25 percent of their respective carrying capacities. Also, bottomfish MSY occurs at population size of 10.0 and effort of 1.0 ($P_2^* = 10.0, f = 1.0$); whereas for shrimp it is $P_i^* = 0.5, f = 0.5$. The relative scale of these population parameters approximately corresponds to the ratio of shrimp and bottomfish catches observed.

Several alternatives were chosen for the discard and biological conversion rates d_i and c_{ij}). These alternatives span a range which we feel is ecologically meaningful. Note that these two rates act in tandem in the model to increase production. Thus, the effect on the population dynamics may be considered to be one parameter ($d_i c_{ij}$). The values of d_i which we used ranged from 0.0 to 1.0, ie discard rates of zero percent to 100 percent. The alternatives for c_{ij} which we tested in the model were $c_{ij} = 0.005$ or $c_{ij} = 0.01$. These imply that the assimilation of discard biomass *per unit biomass* of the stock is either 0.5 percent of 1.0 percent. For example, if the stock size of bottomfish is 15.0 and for shrimp it is 0.5 and all shrimp are discarded, then the ratio of additional biomass assimilated by bottomfish to the shrimp discards is approximately ten percent. This agrees with a trophic efficiency of ten to twenty percent that is often hypothesized in ecological literature (Ryther, 1969). However, the above assumes that there are no sources of forage

other than discards. This is, of course, untrue; therefore, we expect that these parameters are overestimated. We will discuss the ramifications of this bias later.

Results of the discard analysis

The parameter sets described were tested in the discard model and optimum effort (f_{opt}) calculations were generated numerically. In these results we report on six combinations of the parameters, which span the outcomes for the parameter ranges (*Tables 6* and 7). Note that yield is the difference between biomass caught and biomass discarded. Therefore, the maximum equilibrium production ($\text{MGP}_i = \max N_i^*$), will be larger than the maximum equilibrium yield (MSY_i) by a factor of $1/(1 - d_i)$.

The net effect of discarding of catch and then its reassimilation into the population is an increase in the maximum productivity of the populations and the shift of the maximum to higher population and effort levels than in the no discard case. If there are no discards of shrimp and a large discard rate of bottomfish (eg, *Table 7: Case II*), then the reassimilation can be directly translated into shrimp yield. However, in the above case in which reassimilation rates are probably higher than expected, the increase in yield is only eight percent. Other discarding practices may increase shrimp productivity, but yield is reduced by the amount discarded to levels below the no discard situation (see *Table 7: Cases III* and *VII*). Additionally, the shrimp population size which produces the maximum productivity for a given discard policy only increases by a maximum of three percent (*Case II*). Similarly, the maximum effort increase (at shrimp MSY) is only

Table 6

DESCRIPTION OF ALTERNATIVE PARAMETER SETS TESTED IN DISCARD MODEL

Production Parameters (constant throughout this analysis)
$a_1 = 1.0; b_1 = 1.0; a_2 = 1.0; b_2 = 0.05; q_1 = 1.0; q_2 = 0.5$

Discard Parameters

Case I: No discards $d_1 = d_2 = 0$

Case II: No discards of shrimp, high discard rate for bottomfish; high conversion by shrimp and bottomfish
$d_1 = 0.0; d_2 = 1.0; c_{11} = c_{21} = c_{12} = c_{22} = 0.10$

Case III: Low discard rate for shrimp; high discard rate for bottomfish; low conversion rates by shrimp; and by bottomfish
$d_1 = 0.1; d_2 = 1.0; c_{11} = c_{21} = 0.005; c_{12} = 0.005$

Case IV: High discard rate for shrimp; moderate discards for bottomfish low conversion by shrimp; and by bottomfish
$d_1 = 0.5; d_2 = 0.5; c_{11} = c_{21} = 0.005; c_{12} = c_{22} = 0.01$

Case V: High discard rate for shrimp; moderate discards of bottomfish; high conversion by shrimp; high conversion by bottomfish
$d_1 = 0.5; d_2 = 0.5; c_{11} = c_{21} = 0.01; c_{12} = c_{22} = 0.01$

Case VI: Low discard rate for shrimp; high discards of bottomfish; high conversion by shrimp; high conversion by bottomfish
$d_1 = 0.1; d_2 = 1.0; c_{11} = c_{21} = 0.01; c_{12} = c_{22} = 0.01$

Case VII: Low discard rate for shrimp; high discards of bottomfish; high conversion by shrimp; low conversion by bottomfish
$d_1 = 0.1; d_2 = 1.0; c_{11} = c_{21} = 0.01; c_{12} = c_{22} = 0.005$

Table 7

IMPACT OF DISCARDING PRACTICES ON PRODUCTION OF MODEL SHRIMP AND BOTTOMFISH POPULATIONS (EQUATION (1)). SEE TEXT FOR DETAILS OF MODEL AND DEFINITION OF NOTATION. PARAMETER ALTERNATIVES ARE DESCRIBED IN Table 6. NOTE SUBSCRIPT 1 REFERS TO SHRIMP AND 2 REFERS TO BOTTOMFISH. ALSO MGP_1 IS THE MAXIMUM EQUILIBRIUM BIOLOGICAL PRODUCTION FOR STOCK i ($MGP_1 = $ MAX N_1^*)

Parameter Alternatives	Shrimp Maximization							Bottomfish Maximization						
	MSY_1 at MGP_1	f given MPG_1	P_1 given MPG_1	MGP_1	P_2 given MGP_1	N_2^* given MGP_1	Y_2 given MGP_1	MSY_2 at MPG_2	P_2 given MGP_2	f given MGP_2	$MGP2$	P_1 given MGP_2	N_2^* given MGP_2	Y_1 given MGP_2
Case I: No discards $d_1 = d_2 = 0$	0·250	0·500	0·500	0·250	15·000	3·750	3·75	5·00	10·000	1·000	5·000	0	0	0
Case II: $d_1 = 0; d_2 = 1·0$	0·270	0·525	0·515	0·270	15·431	4·128	0	0	10·615	1·050	5·573	0·006	0·006	0·006
Case III: $d_1 = 0·1; d_2 = 1·0$	0·233	0·520	0·500	0·260	15·200	3·951	0	0	10·28	1·025	5·267	0	0	0
Case IV: $d_1 = d_5 = 0·5$	0·128	0·510	0·500	0·255	15·110	3·852	1·926	2·565	10·157	1·010	5·129	0·003	0·003	0·001
Case V: $d_1 = d_2 = 0·5$	0·131	0·520	0·501	0·261	15·220	3·958	1·979	2·643	10·327	1·020	5·267	0·006	0·006	0·003
Case VI: $d_1 = 0·1; d_2 = 1·0$	0·244	0·540	0·503	0·271	15·439	4·169	0	0	10·615	1·050	5·573	0·006	0·006	0·005
Case VII: $d_1 = 0·1; d_2 = 1·0$	0·234	0·520	0·501	0·260	15·008	3·902	0	0	10·157	1·010	5·129	0·016	0·016	0·014

eight percent (*Case VI*). The effect of discard policy on bottomfish MSY's are similar. However, note that harvesting bottomfish at levels approximating MSY with a gear common to shrimping would likely cause overexploitation of the shrimp.

The qualitative results in *Table 7* show the minimal effect of reassimilation of discards on shrimp and bottomfish productivity. The results are based, of course, on a rather arbitrary choice of the reassimilation rates and discard rates. Specifically, we have assumed that all of the discards were dead. If some were released alive, then effectively the reassimilation rate would be underestimated. However, it is unlikely that a significant number of discards are in fact released alive. We also assumed that the discards were converted (at a rate of 0·5 or 1·0 percent) only to shrimp or bottomfish biomass. We did not consider the consumption by populations other than shrimp and bottomfish. Our estimates represent an upper bound. If consumption by other species occurs, our estimates are too high, and the impact of reassimilation of discards is even less than presented in *Table 7*.

In general shrimp production is likely to be dominated by environmental influences, which cause large variations in year class strength. Thus it is unlikely that reassimilation of discards would be detectable using fisheries related data. Additionally, any benefits of altering management strategies to account for the reassimilation would also be undetected.

There are at least two aspects of shrimp and bottomfish interaction which were not considered in this analysis. The first is the effect of gear saturation. For example, culling large quantities of bottomfish from the shrimp catch may effectively reduce the catchability of the shrimp stocks. This could be incorporated into equation (1). The net effect of gear saturation caused by an increased biomass of another stock would reduce the maximum sustainable yield and the effort which produces it. Any regulation or gear development which might alter the stock size of bottomfish would thus affect the production of shrimp. At this point we make no suppositions as to the strength of this relationship.

Another important interaction is the effect of predation. Although we have noted in the previous section that predation rates of bottomfish on shrimp are probably small, the disparate magnitudes of their biomass may cause a significant effect. We will evaluate the possible effects of predation and other forms of interaction through an alternative model in the next section.

Energy-flow model of shrimp – bottomfish ecosystem

An energy-flow ecosystem model was developed to quantify the present role of bottomfish discards in the north-central Gulf of Mexico ecosystem and to evaluate the changes that might occur should discards be reduced. The fish discards of the shrimp fishery are generally viewed as a wasted resource (Technical Consultation on Shrimp By-catch Utilization, 1982). Regulations to reduce the quantity of discards have been implemented in some parts of the world (ie Guyana and Indonesia) and have been considered for the U S Gulf of Mexico (GMFMC, 1981). Because the volume of fish now captured in shrimp nets and thrown back to sea is so great in the north-central Gulf of Mexico, any reduction in the magnitude of what is captured or discarded may cause changes in other parts of the system and may even affect shrimp stocks and shrimp harvests.

Two methods of reducing discards have been considered by the Gulf of Mexico Fishery Management Council (GMFMC, 1981). The first is to utilize a greater portion of the by-catch. The second is to catch fewer fish by employing a specially-designed trawl with a lower catch-efficiency for bottomfish. The first would decrease the amount of dead fish returned to the system. The second would result in less dead fish and more living fish in the system. The ecosystem model was designed to test the theoretical effect of reducing discards by either of these methods.

The first step in the modeling effort was to draw a diagram depicting all the major components of the system and their relationships (Figure 1). The model system consists of 12 compartments, 11 of which represent trophic groups, in terms of biomass. The compartments are related by one-way flows of energy, primarily occurring as the feeding of one trophic group upon another, indicated by connecting lines. All possible trophic interactions between commercial penaeid shrimp, bottomfish, and other species of this system are covered. Fishery harvests of shrimp, bottomfish, menhaden (pelagic fish), and mackerels (migratory pelagics) are also included. The discard of bottomfish by shrimp vessels is indicated by the line leading from the bottomfish compartment to the compartment of high nitrogen organic material. High-nitrogen organic material, solely of animal origin, is distinguished from low-nitrogen organic material, consisting mainly of terrestrial plants that have been washed into the sea and phytoplankton that has sunk to the bottom. High-

nitrogen organic material, low-nitrogen organic material, and the benthic organisms that feed on them are the food sources of shrimp shown in the model. Bottomfish feed on these same foods, as well as on shrimp.

The compartments of the system are connected not only by trophic relationships but also by the cycling of nitrogen and other essential minerals. These nutrients, which stimulate primary production, are initially incorporated into living tissue by the phytoplankton and are then passed up the food chain, from which they are gradually remineralized and released, again becoming available to primary producers. The remineralization and release of nitrogen from animal groups and dead organic material is indicated in the diagram by dotted lines.

Several possible ways that shrimp standing stocks could be affected by reducing discards are obvious from the model diagram (Figure 1). If discards were reduced through either greater utilization of the by-catch or the use of trawls with a lower catch-efficiency for fish, the quantity of dead fish flesh available as food to shrimp would be lower. Additionally, the rate of remineralization and release from dead fish flesh of nitrogen and other nutrients essential to primary productivity would be lower. This might reduce the rate of production of phytoplankton, which, when it sinks to the bottom, becomes shrimp food.

The connecting lines representing trophic relationships indicate that bottomfish stocks could interfere with shrimp stocks through both predation and competition. Bottomfish not only prey on shrimp to at least some extent (although gut analyses suggest that the rate per fish is very low) but also feed on the same food eaten by shrimp. Changing the biomasses of either living or dead bottomfish in the system could have other indirect effects on shrimp that are not readily apparent from looking at the model diagram but may be observable from model simulations.

The model assumes that shrimp stocks and the other plant and animal stocks in the system are resource limited. Although there are no data directly relating shrimp production by natural systems to food supplies, the location of major fisheries in areas of the world's oceans where either primary productivity is high or the production of widespread areas is concentrated suggests that most, if not all, fisheries are resource limited.

The second step in model development was to quantify the compartments and flows of energy (as biomass) and nitrogen. The model was quantified for steady-state (inputs = outputs) conditions using an iterative top-down flow-balancing proce-

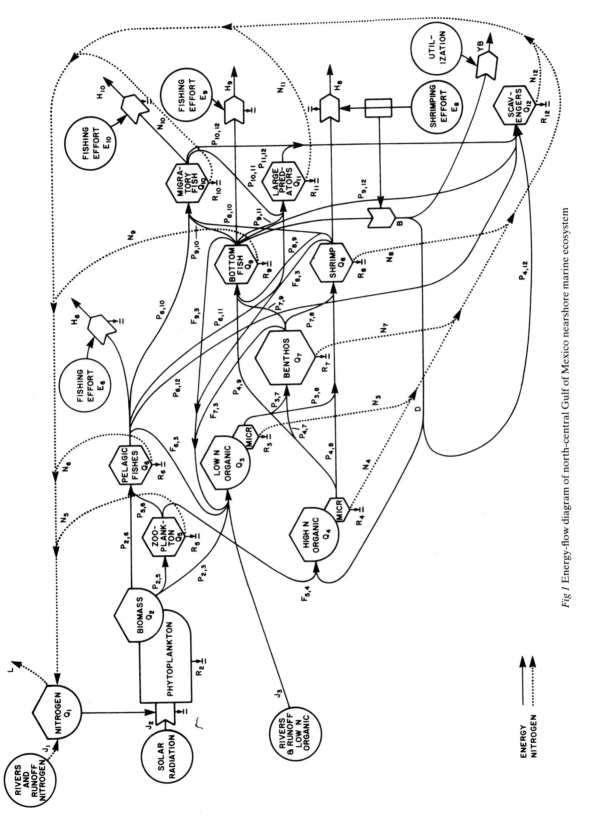

Fig 1 Energy-flow diagram of north-central Gulf of Mexico nearshore marine ecosystem

ENERGY
NITROGEN

245

dure in which feeding flows to each animal compartment were calculated on the basis of total outflows from that compartment, including respiration, harvests, and predation (Browder, 1983).

The procedure started with the highest trophic level in the system and worked backward from the direction of energy flow. Feeding flows to a predator from alternative food sources were assumed to be proportional to the relative biomasses of these sources, except in cases of selectivity. Selectivity was approximated by the differential 'weighting' of two or more feeding flows to the same predator.

Essential inputs for calculating feeding flows

and setting rate coefficients by this method were (1) biomass of each trophic group, (2) respiration rate coefficients of each animal group, (3) either the respiration rate coefficient or outside flows to each food-chain-base compartment (ie phytoplankton and low-nitrogen organic material), (4) assimilation coefficient for each type of food of each predator, and (5) selectivity weighting factors for alternative prey of each predator. Values entered into the model (not calculated by the flow-balancing procedure) are given with derivation in *Tables 8* and *9*.

Assigned weighting factors (*Table 9*) were on a scale of 0·001 to 1·0, with 0·001, when used with

Table 8

INITIAL DRY WEIGHTS (Q) (MG/M^2), EXOGENOUS INFLOW (J) (MG/M$^2 \cdot$ DAY), RESPIRATION RATE COEFFICIENTS (C_R) (MG/MG WEIGHT · DAY), AND HARVESTING RATE COEFFICIENTS (C_H), (MG/MG WEIGHT · DAY), BODY NITROGEN/DRY WEIGHT RATIOS (C_n), AND RATIOS OF NITROGEN RELEASED IN EXCREMENT TO ORGANIC MATTER BURNED IN METABOLISM (B)

		Q	J	C_R[a]	C_H[b]	C_N[c]	B
1	Nitrogen	1 260·0[d]	112[p]				
2	Phytoplankton	1 125·0[e]	2711[q]	0·768[s]			0·06[ii]
3	Low-nitrogen organic material	218 850·0[f]	3037[r]	0·01872[t]		0·05[gg]	
4	High-nitrogen organic material	33·4[g]		0·2071[u]		0·0756[hh]	
5	Zooplankton	145·9[h]		0·32[v]		0·0165[c]	0·073[jj]
6	Pelagic fish (menhaden)	2 966·0[i]		0·020[w]	0·01389[dd]	0·0453[c]	0·0924[kk]
7	Benthos	8 000·0[j]		0·081[x]		0·0353[c]	0·1202[ll]
8	Shrimp	79·4[k]		0·0414[y]	0·002778[dd]	0·0959[c]	0·1083[mm]
9	Bottomfish	2011·0[l]		0·018[z]	0·0004061[ee]	0·1159[c]	0·1297[nn]
10	Migratory pelagics (mackerel)	32·4[m]		0·006[aa]	$1·023 \times C_{H8}$[ff]	0·1023[c]	0·1250[oo]
11	Large predators (dolphin)	6·6[n]		1·38[bb]		0·1096[c]	0·1308[pp]
12	Large scavengers (sharks)	24·3[o]		0·0068[cc]		0·1098[c]	0·1445[qq]

[a] $Q \times C_R = R =$ respiration rate
[b] $Z \times C_H = H =$ harvesting rate
[c] $R \times C_N = N =$ nitrogen release rate. The nitrogen release rate was assumed to be proportional to respiration of animals (and of the microbes decomposing organic material). C_N values were estimated from body nitrogen/dry organic matter ratios by calculating steady state nitrogen flows corresponding to steady state organic matter flows. In the calculations, nitrogen not used in growth or predation was divided between excrement and feces in the proportions 0·8 and 0·2, respectively, in all animals except zooplankton. For zooplankton, the apportionment was 0·2 for excrement and 0·8 for feces to reflect the high-nitrogen concentration of zooplankton fecal pellets suggested by the literature. C_{N3} (the coefficient for low-nitrogen organic matter) was estimated rather than calculated in the above manner.
[d] from concentration of inorganic nitrogen (NO$_2$—N and NO$_3$—N) in waters offshore of the Mississippi Delta (126 mg/m^3) (Sklar, 1976) and estimated depth of photic zone (10 m).
[e] from concentration of chlorophyll a offshore Mississippi Delta (18 mg/m^2) (Sklar, 1976), ratio of carbon to chlorophyll (25) (Parsons *et al*, 1977), and ratio of dry organic matter to carbon (2·5) (Parsons *et al*, 1977).
[f] from concentration of dry organic material in shallow-water Mississippi Delta marine sediments (10 mg/g) (Hausknecht, 1980); depth of sediment in which more than 80% of mecifauna occurs (3 cm) (Tietjen, 1969); and weight of inorganic fraction per centimeter thick square meter area, calculated by method of Bennett and Lambert (1977), assuming specific gravity of 2·7 for Mississippi Delta sediments (Bennett, pers. comm., for clay).
[g] estimated to be equal to daily flow of zooplankton fecal pellets and discards.
[h] from number of individuals per unit volume (2 432/m^3) (Reitsema, 1980), dry weight per individual (6 × 10^{-3}mg) (Conover, 1959) and estimated depth of photic zone.
[i] Louisiana menhaden landings in 1975 (4·5 × 10^{14}mg), assumed ratio of all coastal herring to menhaden (2), dry weight/wet weight ratio (0·2) (Parsons *et al*, 1977) and assumed annual fishing mortality of 1·0; divided by Louisiana bottom area inside 93 meters (50 fathoms).
[j] from Parker *et al* (1980).

Table 8 (continued)
INITIAL DRY WEIGHTS (Q) (MG/M^2), EXOGENOUS INFLOW (J) (MG/M$^2 \cdot$ DAY), RESPIRATION RATE COEFFICIENTS (C_R) (MG/MG WEIGHT $\cdot$ DAY), AND HARVESTING RATE COEFFICIENTS (C_H), (MG/MG WEIGHT $\cdot$ DAY), BODY NITROGEN/DRY WEIGHT RATIOS (C_n), AND RATIOS OF NITROGEN RELEASED IN EXCREMENT TO ORGANIC MATTER BURNED IN METABOLISM (B)

[k] Louisiana shrimp landings in 1975 ($2 \cdot 41 \times 10^{13}$ mg), converted to dry weight ($0 \cdot 2$) and divided by Louisiana bottom area inside 93 meters ($6 \cdot 069 \times 10^{10}$m^2) (Patella, 1975).

[l] estimated bottomfish catch, including discards, in 1975, for area inside 93 meters from Pt. Au Fer, Louisiana, to Perdido Bay, Florida (430×10^{12}mg); converted to dry weight ($0 \cdot 2$); multiplied by an assumed annual fishing mortality of $0 \cdot 8$; and divided by area ($3 \cdot 42 \times 10^{10}$m^2) (Patella, 1975).

[m] from annual U S king and Spanish mackerel landings in 1975 ($2 \cdot 27 \times 10^{12}$ mg and $4 \cdot 5 \times 10^{12}$ mg) (NMFS, 1978); king and Spanish mackerel annual fishing mortality ($0 \cdot 41$ and $0 \cdot 2756$) (GMSAFMC, 1980); estimated fraction of total mackerel stocks that occur in the Gulf of Mexico ($0 \cdot 67$), ratio of total migratory biomass to mackerel biomass ($2 \cdot 0$), dry weight–wet weight conversion ($0 \cdot 2$) and estimated bottom area of Gulf of Mexico inside 93 meters (50 fathom) ($1 \cdot 81 \times 10^{11}$m^2) (Patella, 1975).

[n] from density of bottlenose dolphin at Louisiana coastal sites ($0 \cdot 44 \times 10^{-6}$ (Leatherwood *et al*, 1978), estimated average weight of bottlenose dolphin (150 kg wet weight), dry weight to wet weight conversion ($0 \cdot 2$), and factor to correct for better-than-average densities thought to have occurred in sampling area ($0 \cdot 5$).

[o] estimated average biomass of elasmobranchs in Gulf of Mexico ($24 \cdot 29$ mg/m^2) (L. Rivas, Nova University, Ft. Lauderdale, FL, pers. comm.), converted to dry weight ($0 \cdot 2$), and multiplied by an assumed near-shore concentration factor of 5.

[p] concentration of inorganic nitrogen in Mississippi River water (2 000 mg N/m^3) (Sackett, 1972) times rate of freshwater inflow from Mississippi and Atchafalaya Rivers ($1 \cdot 9 \times 10^9$ m^3/day) (Sackett, 1972), divided by estimated area of bottom under the immediate influence of Mississippi and Atchafalaya discharge ($3 \cdot 42 \times 10^{10}$m^2).

[q] rate of gross primary productivity, equal to net primary productivity per unit weight ($1 \cdot 642$) (Sklar, 1976) plus respiration rate coefficient ($0 \cdot 768$) (Ryther and Guillard, 1962), multiplied by phytoplankton biomass.

[r] estimated annual carbon export from Barataria Bay per square meter of inshore water (15×10^4 mg C) (Happ *et al*, 1977) times Louisiana inshore water area ($1 \cdot 367 \times 10^{10}$m^2) (Perret *et al*, 1971), divided by Louisiana offshore water area to 93 m; plus concentration of dissolved organic carbon in Mississippi River water (20 mg/liter) times annual volume of Mississippi and Atchafalaya discharge ($6 \cdot 9 \times 10^{14}$ liters/day), divided by area inside 93 m from Pt. Au Fer, LA, to Perdido Bay, FL ($3 \cdot 42 \times 10^{10}$m^2)—all multiplied by $2 \cdot 5$ to convert carbon to organic matter and divided by 360 to express in terms of days.

[s] Ryther and Guillard (1962) for 4, eurythermal diatom (*Cyclotella nana*, 3H).

[t] set in flow-balancing procedure to approximate value for zooplankton fecal pellet decomposition calculated from Johannes and Satomi (1966) ($0 \cdot 1834$).

[u] calculated in flow-balancing procedure.

[v] Conover (1959) for 6 g D.W. copepod.

[w] Hettler (1976) for 226 g W.W. menhaden.

[x] Pamatmat (1980) 1 g W.W. (assumed 20 × anaerobic rate).

[y] Bishop *et al* (1980) for $6 \cdot 7$ g brown shrimp (assumed 12 hrs activity).

[z] Hoss (1974) for 100 g pinfish

[aa] Brill (1979) for 1 kg skipjack tuna.

[bb] Irving *et al* (1941) for 150 kg bottlenose dolphin.

[cc] Brett and Blackburn (1978) for 900 g spiny dogfish.

[dd] based on landings (NMFS, 1978).

[ee] harvesting coefficient for directed fishery, based on landings (NMFS, 1980).

[ff] harvesting coefficient for bycatch, based on estimated discards (GMFMC, 1980) (by-catch is proportional to the harvesting rate coefficient for shrimp).

[gg] estimate for river detritus.

[hh] calculated for initial mix of zooplankton fecal pellets and fish flesh.

[ii] Strickland (1960) for mixed taxa.

[jj] Parsons *et al* (1977) for zooplankton.

[kk] Sidwell (1981) for menhaden.

[ll] Darnell and Wissing (1975).

[mm] Sidwell (1981) for commercial penaeid shrimp.

[nn] Darnell and Wissing (1975) for pinfish.

[oo] estimated from a value for mackerel from Sidwell (1981) for muscle tissue only.

[pp] Sidwell (1981) for bottlenose dolphin muscle.

[qq] Sidwell (1981) for *Sphyrna blochii* muscle.

NOTE: Constants used in the model but not shown on the table are given below with information sources.

K_{m1} (Michaelis-Menten coefficient for the effect of inorganic nitrogen concentration on phytoplankton production) = 120 (based on a half-saltwater constant of $1 \cdot 5$ moles-liter for *Asterionella japonica*, a coastal diatom, as determined by laboratory experiments of Eppley and Thomas (1969). K_{m1} occurs in the equation for calculating J_2.

K (coefficient for loss of nitrogen from the system in currents and by denitrification) = $0 \cdot 189$ (calculated for steady state conditions (inflows = outflows) on basis of inflows and other outflows. L, rate of loss of nitrogen from the system, is equal to $Q_1 \times K$. For initial conditions, $L = 238$.

Table 9

ASSIMILATION COEFFICIENTS (A) AND WEIGHTING FACTORS (W) FOR NORTH CENTRAL GULF OF MEXICO ECOSYSTEM MODEL (VALUES NOT FOOTNOTED WERE ASSUMED)

FROM	1		2		3		4		5		6		7		8		9		10		11		12	
TO	A	W	A	W	A	W	A	W	A	W	A	W	A	W	A	W	A	W	A	W	A	W	A	W
1	–	–	–	–	–	–	–	–	–	–	–	–	–	–	–	–	–	–	–	–	–	–	–	–
2	–	–	–	–	–	–	–	–	–	–	–	–	–	–	–	–	–	–	–	–	–	–	–	–
3	–	–	–	–	–	–	–	–	–	–	–	–	–	–	–	–	–	–	–	–	–	–	–	–
4	–	–	–	–	–	–	–	–	–	–	–	–	–	–	–	–	–	–	–	–	–	–	–	–
5	–	–	0.70[a]	1.0	–	–	–	–	–	–	–	–	–	–	–	–	–	–	–	–	–	–	–	–
6	–	–	0.42	1.0	–	–	0.6	1.0	–	–	–	–	–	–	–	–	–	–	–	–	–	–	–	–
7	–	–	–	–	0.5	0.008[e]	0.9	1.0	–	–	–	–	–	–	–	–	–	–	–	–	–	–	–	–
8	–	–	–	–	0.2[b]	0.001	0.4[c]	1.0	–	–	–	–	0.4[c]	1.0	–	–	–	–	–	–	–	–	–	–
9	–	–	–	–	–	–	0.86[d]	1.0	–	–	–	–	0.7	1.0	0.86[d]	1.0	–	–	–	–	–	–	–	–
10	–	–	–	–	–	–	–	–	–	–	0.8	1.0	–	–	0.7	1.0	1.0	–	–	–	–	–	–	–
11	–	–	–	–	–	–	–	–	–	–	0.8	1.0	–	–	–	–	1.0	1.0	0.7	1.0	–	–	–	–
12	–	–	–	–	–	–	–	–	–	–	0.8	0.1	–	–	–	–	1.0	0.1	1.0	0.1	1.0	0.001	–	–

[a] Parsons *et al* (1977) for copepods.
[b] Jones (1973) for brown shrimp, away from shore.
[c] Jones (1973) for brown shrimp, at edge of shore.
[d] Darnell and Wissing (1975) for polychaetes consumed by pinfish.
[e] Approximated in calculation to set the respiration rate coefficient for high-nitrogen organic material.

1.0 for alternative prey, indicating intense selection against a prey type. Absence of selectivity was indicated by assigning a weighting factor of 1.0 to all prey of the same predator. Assignment of weighting factors was based on qualitative information about a predator and its alternative prey. In the initial quantification of the model, weighting factors were set so that all organisms feeding on low-nitrogen organic material selected against this food in favor of alternatives and so that large scavengers selected dead fish over living prey. Otherwise, all weighting factors were set at 1.0. Although the low frequency of shrimp in the stomach contents of bottomfish—even those caught in trawls with shrimp—suggests that bottomfish may concentrate on other organisms, the model was initially quantified to indicate no selectivity against shrimp in order that the worst possible condition of predation by bottomfish on shrimp might be evaluated.

Initial values for all flows of energy and nitrogen in the system, most of which were calculated by the flow-balancing procedure, are given in *Table 10*. Flows to and from the inorganic nitrogen (Q_1) high-nitrogen organic material (Q_4), benthos (Q_7), shrimp (Q_8), and bottomfish (Q_9) compartments are directly relevant to evaluating the effect of reducing bottomfish discards on shrimp biomass and harvests. (Other flows may have indirect effects, but these can only be evaluated with simulations.)

The nitrogen compartment received inflows from river and runoff (J_1), from the microbial breakdown of low-nitrogen organic material (N_3), from the breakdown of high-nitrogen organic material (N_4), and in animal excrement (N_5 to N_{12}). Outflows were uptake by phytoplankton, in association with their growth ($P_{1,2}$), and losses to the system in currents (L). The major contribution was from the remineralization of low-nitrogen organic material, most of which entered the system in rivers and as coastal runoff. This contribution was more than two orders of magnitude higher than that from the breakdown of high-nitrogen organic material, approximately half of which was made up of discards. The combined contribution from animal excrement was also more than two orders of magnitude greater than that from high-nitrogen organic material. Recycling from animals alone was almost sufficient to satisfy the daily phytoplankton uptake rate, although it might not have been adequate to promote maximum photosynthesis. Nitrogen in river and runoff waters also was sufficient to replenish nitrogen taken up in daily growth of phytoplankton at the calculated rate. The quantity of nitrogen released from discards and other high-nitrogen sources was less than one one-thousandth the stock of inorganic nitrogen in the environment.

In the model, shrimp received energy from three sources: benthos (Q_7), low-nitrogen organic material (Q_3), and high-nitrogen organic material (Q_4). Despite the fact that high-nitrogen organic material was weighted 1 000 to 1 over low-nitrogen organic material, the latter was 50 times more important as a food source of shrimp [25 times

248

Table 10

MATRIX OF INITIAL STEADY-STATE FLOW FROM SOURCES TO COMPARTMENTS, FROM COMPARTMENTS TO SINKS, AND BETWEEN COMPARTMENTS.[a,b]
(FLOW RATES WERE CALCULATED IN THE ITERATIVE TOP-DOWN FLOW-BALANCING PROCEDURE, UNLESS OTHERWISE INDICATED)

To \ From	Source	Q1	Q2	Q3	Q4	Q5	Q6	Q7	Q8	Q9	Q10	Q11	Q12	Organic	Nitrogen
Sinks[c,d]	—	(238·6)[e]	864	4102	6·984	46·69	59·32	648·0	3·288	36·00	0·1944	9·108	0·1657	5776	(238·6)
Harvest[d]	(112·0)[e]	—	—	—	—	—	4·119	—	0·2206	0·8112	0·0209	—	—	—	(349·4)
Q1	(110·8)[d,e]	—	—	(205·1)	(0·5280)	(0·7694)	(2·687)	(22·83)	(0·3198)	(4·169)	(0·0199)	(0·9980)	(0·0182)	—	(110·8)
Q2	2711	—	—	—	—	—	—	—	—	—	—	—	—	5473	—
Q3	3037	—	1614	—	—	27·81	88·69	690·7	6·208	7·724	0·0395	1·226	0·0220	—	—
Q4	—	—	—	—	—	—	—	—	—	5·686[d,g]	—	—	—	5·686	—
Q5	—	—	92·70	—	—	—	—	—	—	—	—	—	—	92·70	—
Q6	—	—	140·4	—	—	18·20	—	—	—	—	—	—	—	158·6	—
Q7	—	—	—	1376	26·25	—	—	—	—	—	—	—	—	1402·3	—
Q8	—	—	—	0·2720	0·0415	—	—	9·942	—	—	—	—	—	10·26	—
Q9	—	—	—	—	0·2246	—	—	53·80	0·5341	—	—	—	—	54·56	—
Q10	—	—	—	—	—	—	0·1899	—	0·0051	0·1280	—	—	—	0·3230	—
Q11	—	—	—	—	—	—	6·132	—	—	4·135	0·0670	—	—	10·33	—
Q12	—	—	—	—	0·0021	—	0·1101	—	—	0·0743	0·0012	0·0000025	—	0·1877	—
Total organic matter	5748	—	2711	5478	33·50	92·70	158·5	1402	10·26	54·56	0·3230	10·33	0·1877	25700	—
Total nitrogen	(112·0)	(349·4)	—	(205·1)	(0·5280)	(0·7694)	(2·687)	(22·83)	(0·3198)	(4·169)	(0·0199)	(0·9980)	(0·0182)	—	(699)

Q1 = nitrogen
Q2 = phytoplankton
Q3 = low-nitrogen organic material
Q4 = high-nitrogen organic material
Q5 = zooplankton
Q6 = pelagic forage fish (menhaden)

Q7 = benthos
Q8 = commercial penaeid shrimp
Q9 = bottomfish
Q10 = coastal pelagics (mackerels)
Q11 = marine mammals (dolphin)
Q12 = large scavengers (sharks)

a Values are milligram dry weight per square meter per day, except those in parentheses, which are milligrams nitrogen per square meter per day.
b Respiration rates in milligrams dry organic matter equivalent per day.
c Values independent of flow-balancing procedure unless otherwise indicated.
d Loss of nitrogen from system in currents and denitrification, calculated by flow balancing.
e Uptake of nitrogen by phytoplankton in photosynthesis.
f Discards.

more important if the different assimilation rates (*Table 10*) are considered]. Benthos, which was weighted equally with high-nitrogen organic material as a food source, was 200 times more important, according to flow rates. Bottomfish fed on the same types of food eaten by shrimp. The flow of each to bottomfish was approximately five times greater than the flow to shrimp.

Energy flowed from the shrimp compartment (Q_8) in predation by bottomfish (Q_9), predation by migratory pelagics (Q_{10}), harvests of man, and respiration. Predation by bottomfish was three times the harvest rate and one-fifth the respiration rate of shrimp. The flow of energy from shrimp to bottomfish was two orders of magnitude lower than the flow of energy to bottomfish from the benthos (Q_7).

The rate of predation of bottomfish on shrimp in the model was highly dependent upon the weighting factors for bottomfish feeding on alternative prey. Variation in the calculated predation rate was directly proportional to the ratio of the weighting factor for shrimp to that of the alternative prey. Predation rates of bottomfish on shrimp were 0·0538, 0·00538, and 0·000538 for shrimp weighting factors of 0·1, 0·01, and 0·001 respectively (weighting factors for alternative prey were 1·0).

The third step in model development was write a computer program in which the mathematical relationships suggested by the model diagram were incorporated into a set of integral (Euler numerical integration) equations. In these equations, all flow rates to animal compartments were donor-recipient controlled and of the form:

$$P = c_{i,j}Q_iQ_j,$$

where c was the rate coefficient, Q was compartment, i indicated prey, and j indicated predator. Gross primary productivity was a function of the form:

$$J = S(Q/K_m + Q),$$

where S was maximum gross primary productivity, K_m was the nitrate–nitrogen content of the water at half maximum velocity, and Q was nitrogen. Nitrogen released in excrement or decomposition was a function of respiration rate and the nitrogen concentration in the source material. All other flows in the model were proportional to donor biomass or were simple algebraic relationships. A more detailed mathematical description of the model was given in Browder (1983).

The computerized model was used to simulate the biomasses of the trophic groups as they changed over time in response to a reduction in the rate of discarding. The effect of reducing discards by each of the two suggested methods was tested by resetting one coefficient to approximate the test condition, after initially quantifying the flow rates for present steady-state conditions. For each test, the model was run for a simulated 5-yr period, iterating 10 times per day. Under test conditions, the biomasses, simulated over time, moved from initial steady-state levels to a new steady state, whereas, when the model was run for present conditions (not resetting either coefficient), biomasses were constant throughout the 5-yr period. The direction and magnitude of the change in steady-state biomasses indicated the effect of the test condition. In quantifying the model, inputs such as inorganic nitrogen and detritus were held constant over time to enable all changes in biomass levels to be attributable to the conditions being tested.

Using the computerized model, simulations of shrimp biomass were produced for two test conditions: (1) one-half of the by-catch utilized and (2) fish catch-efficiency of shrimp trawls reduced by one-half. The latter test was run twice, once with no selectivity against shrimp (shrimp weighting factor = 1·0) and, the second time, with moderate selectivity (shrimp weighting factor = 0·01) by bottomfish against shrimp.

Under the assumption of no selection against shrimp by bottomfish, a decrease in shrimp biomass resulted from decreasing bottomfish discards by either method. A 25 percent decrease occurred with by-catch utilization, whereas, when the special trawls were used, the decrease was only 8 percent. When moderate selectivity was assumed, results of reducing discards by the two methods diverged even more. The use of trawls with a lower catch efficiency for fish did not have a detrimental effect on shrimp biomass over the long term. In this simulation, shrimp biomass declined briefly but rebounded to former levels before the end of the second year. Although the higher biomass of bottomfish caused by decreased fishing mortality increased the pressure on shrimp from both predation and competition, this negative effect was outweighed by concurrent responses to the change elsewhere in the system that were beneficial to shrimp. The main factor appears to have been a chain of events beginning with an increase in marine mammal biomass resulting from the increased supply of bottom-fish prey. Marine mammals, having increased, fed more heavily on menhaden and other pelagic fish as well as on bottomfish, leading to a decline in pelagic fish biomass. The resultant reduction in predation pressure on zooplankton by pelagic fish caused zooplankton to proliferate and produce more fecal pellets, increasing the supply of high-nitrogen food for shrimp.

The decrease in shrimp biomass that followed reducing discards by means of utilization of one-half the by-catch was due to a decrease in the supply of food for shrimp. (There had been no increase in either predation or competition, because fishing pressure on bottomfish had not been reduced.) The rate that shrimp fed on high-nitrogen organic material decreased by about one-third when utilization of one-half the by-catch was instituted. The rate that shrimp fed on low-nitrogen organic material also was slightly lower. Nitrogen remineralization decreased slightly when discarding was reduced through by-catch utilization and slightly depressed primary productivity in the model system, despite the large quantity of nitrogen entering the system in river water.

The nitrogen remineralization rate was greatest when shrimp trawls with reduced catch-efficiency for fish were used, suggesting that bottomfish promote a greater rate of remineralization when alive than dead. The increased remineralization associated with the use of gear with a lower efficiency for catching fish did not increase primary productivity – possibly because the system was already operating in close proximity to the nitrogen-saturation point. (Sensitivity tests of the effect of nitrogen-related variables on these results have not yet been performed.)

The compartments of this system are linked by many routes, and it was impossible to evaluate which pathways were causally important by merely looking at the diagram and by comparing the magnitudes of direct flow rates. Direct effects were, in some cases, outweighed by indirect effects that were not obvious from the diagram. The computer model kept track of flows throughout the system and allowed their net effects to be observed.

Model results, at this point, are highly theoretical. How well the model reflects the behavior of the real system remains to be determined. A weakness of the model is that it contains several parameters that have not been measured and could only be grossly estimated, or, in some cases, arbitrarily set. Further sensitivity testing is needed to determine the dependence of model results on these parameters, and field or laboratory studies are needed to quantify those parameters to which model results are sensitive. The demonstrated sensitivity of results to selectivity weighting factors points to the need to measure the rate of predation by bottomfish on shrimp and on alternative prey relative to the biomasses of shrimp and the alternative prey in the habitats where they occur together.

Conclusions

The nature of shrimp and bottomfish interactions was evaluated using three approaches: 1) a biological review to define possible ecological connections; 2) a classical population dynamics approach to determine effects of discards on shrimp harvests, assuming that discards are a source of food to shrimp; and 3) an energy-flow model to evaluate the effect on shrimp stocks of changing discard practices, given the multiple connections between shrimp, bottomfish, and the other components of the ecosystem.

The biological review revealed that only a few quantitative studies of the stomach contents of bottomfish species have been performed and most did not separate commercial penaeid shrimp from other penaeids. The limited information available indicated a low-frequency of occurrence of commercial penaeid shrimp in the stomachs of bottomfish species and suggested that man may be the major predator of these penaeids.

The population dynamics model suggested that even the discard practices most favorable to shrimp would increase shrimp harvests only 8% over the case with no discards. Since assimilation rates in the model were deliberately overestimated, the actual benefit of discards to shrimp production probably is less. Environmental 'noise' undoubtedly would prevent changes this small from being detected in fisheries data.

The hypothetical effect of interference with shrimping operations by an increased bottomfish stock was examined using a simple extension to the population dynamics model. The interference resulted in both a reduction in shrimp MSY and a reduction in the effort required to achieve MSY. No attempt was made to quantify these reductions, as this would require more extensive modeling of fishing operations.

The computerized energy-flow ecosystem model suggested that, theoretically, shrimp biomass would decline 25 percent if discards were reduced by utilization of one half the by-catch, but would decline by only 8 percent if discards were reduced through the use of new trawls one-half as efficient in catching fish. The 8 percent reduction occurred only if the 'worst case' of bottomfish predation on shrimp – no selectivity by bottomfish against shrimp – were assumed. The model predicted that, if moderate selection against shrimp by bottomfish were occurring, the introduction of special trawls

251

would result in no long term effect on shrimp stocks and shrimp harvests.

Initial model results indicated that determining the rate of predation by bottomfish on shrimp and on alternative prey relative to the biomasses of each in the environment may facilitate an understanding of shrimp and bottomfish interactions and the potential impact on shrimp of reducing bottomfish discards.

References

ALLSOPP, W H L M S. Fish by-catch from shrimp trawling. The
1980   main protein resource for Caribbean Atlantic countries: reality and potential. Unpubl. report to the Inter-American Development Bank. Round Table on Non-traditional Fishery products for Mass Human Consumption, Washington, D.C., 15–19 Sept. 1980. 32 p.

ARMITAGE, T M and ALEVIZON, W S. The diet of the Florida
1980   pompano (*Trachinotus carolinus*) along the east coast of central Florida. *Florida Sci.* 43:19–26.

BASS, A J, D'AUBREY, and KISTNASAMY, N. Sharks of the east
1973   coast of southern Africa. I. The genus *Carcharhinus* (Carcharhinidae). So. Afr. Assoc. Mar. Biol. Res., Oceanogr. Res. Inst. Investig. Rep. No. 33. 168 p.

BEARDEN, C M. A contribution to the biology of king whitings,
1963   genus *Menticirrhus*, of South Carolina. Contrib. Bears Bluff Lab. No. 38, 27 p.

BEAUMARIAGE, D C. Age, growth, and reproduction of king
1973   mackerel, *Scomberomorus cavalla*, in Florida. Fla. Mar. Res. Publ. 1. Dept. Nat. Resources, Tallahassee. 45 p.

BELL, J D, BURCHMORE, J J and POLLARD, D A. Feeding ecology
1978   of a scorpaenid fish, the fortescue *Centropogon australis*, from a *Posidonia* habitat in New South Wales. *Aust. J. Mar. Freshwat. Res.* 29: 175–186.

BENNETT, R H and LAMBERT, D N. Rapid and reliable technique
1971   for determining unit weight and porosity of deep-sea sediments. Mar. Geol. 11: 201–207.

BISHOP, J M, GOSSELINK, J C and STONE, J H. Oxygen consump-
1980   tion and hemolymph osmolality of brown shrimp, *Penaeus aztecus*. Fish Bull. 78: 741–757.

BRADLEY, E and BRYAN, C E. Life history and fishery of the red
1975   snapper (*Lutjanus campechanus*) in the northwestern Gulf of Mexico: 1970–1974. *Proc. Gulf Caribb. Fish. Inst.* 27: 77–106.

BRETT, J R and BLACKBURN, J M. Metabolic rate and energy
1978   expenditure in the spiny dogfish *Squalus acanthias*. *J. Fish Res. Bd. Canada* 35: 816–821.

BRILL, R W. The effect of body size on the standard metabolic
1979   rate of skipjack tuna, *Katsuwonus pelamis*. Fish. Bull. 77: 494–498.

BROWDER, J A. A simulation model of a near-shore marine
1983   ecosystem of the north-central Gulf of Mexico. *In:* Turgeon, K (Ed). Marine Ecosystem Modeling. Procedures of a Workshop held April 6–8, 1982, Frederick, Maryland, NOAA (National Environmental and Satellite Data and Information Service) Publication. Washington, D.C.

BRUSHER, H A, RENFLO, W C and NEAL, R A. Notes on the
1972   distribution, size, and ovarian development of some penaeid shrimps in the northwestern Gulf of Mexico, 1961–1962. *Contrib. Mar. Sci.* 16: 75–87.

CARR, W E S and ADAMS, C A. Food habits of juvenile marine
1973   fishes occupying seagrass beds in the estuarine zone near Crystal River Florida. *Trans. Amer. Fish. Soc.* 103: 511–540.

CHAO, L N and MUSICK, J A. Life history, feeding habits, and
1977   functional morphology of juvenile sciaenid fishes in the York River estuary, Virginia. *Fish. Bull.* 75: 657–702.

CONOVER, R J. Regional and seasonal variation in the respira-
1959   tory rate of marine copepods. *Limnol. Oceanogr.* 4: 259–269.

DARNELL, R M. Food habits of fishes and larger invertebrates of
1958   Lake Ponchartrain, Louisiana, and estuarine community. *Pub. Inst. Mar. Sco. Univ. Texas* 5: 353–416.

DARNELL, D M and WISSING, T W. Nitrogen turnover and food
1978   relationships of the pinfish *Lagodon rhomboides* in a North Carolina estuary. P 81–110. *In:* F. J. Vernberg (ed.). Physiological ecology of Estuarine Organisms. University of South Carolina Press, Columbia. 396 p.

DIVITA, R, CREEL, M and SHERIDAN, P F. Foods of coastal fishes
1983   during brown shrimp *Penaeus aztecus*, migration from Texas estuaries, *Fish. Bull.* 81 (in press).

DOOLEY, J K. Systematics and biology of the tilefishes (Per-
1978   ciformes: Branchiostegidae and Malacanthidae), with descriptions of two species. *NOAA Tech. Rept. NMFS CIRC-411.* 78 p.

DRAGOVICH, A. Review of studies of tuna food in the Atlantic
1969   Ocean. *U S Fish Wildl. Serv. Spec. Sci. Rept. Fish.* 593. 21 p.

DRAGOVICH, W. The food of bluefin tuna (*Tunnus thynnus*) in
1970a   the western North Atlantic Ocean. *Trans. Amer. Fish. Soc.* 99: 726–731.

DRAGOVICH, A. The food of skipjack and yellowfin tunas in the
1970b   Atlantic Ocean. *U S Fish and Wildl. Service Fishery Bull.* 68: 445–460.

DRAGOVICH, A and POTTHOFF, T. Comparative study of food of
1972   skipjack and yellowfin tunas off the coast of West Africa. *Fish. Bull.* 70: 1087–1110.

EPPLEY, R W and THOMAS, W H. Comparison of half-saturation
1969   constants for growth and nitrate uptake of marine phytoplankton. *J. Phycol.* 5: 375–379.

FONTENOT, F J, JR and ROGILLIO, H E. A study of estuarine
1970   sport fishes in the Biloxi marsh complex, Louisiana. *La. Wildl. Fish. Comm., F-8 Compl. Rep.* 172 p.

FOOD AND AGRICULTURE ORGANIZATION OF THE UNITED
1979   NATIONS. Catches and landings, 1978. *FAO Yearbook of Fishery Statistics, Vol. 46.* FAO, Rome, 372 p.

FOX, L S and WHITE, C J. Feeding habits of the southern
1969   flounder, *Paralichthys lethostigma*, in Barataria Bay, Louisiana. *Proc. La. Acad. Sci.* 31: 31–38.

GRIFFIN, W L and WARREN, J P. Costs and return data;
1978   groundfish trawlers of northern Gulf of Mexico. Unpubl. Report to Gulf of Mexico Fishery Management Council. Tampa, Florida.

GRIMES, C B. Diet and feeding ecology of the vermillion
1979   snapper, *Rhomboplites aurorubens* (Cuvier), from North Carolina and South Carolina waters. *Bull. Mar. Sci.* 29: 53–61.

GMFMC (Gulf of Mexico Fishery Management Council).
1980   Draft fishery management plan for groundfish in the Gulf of Mexico. Tampa, Fla.

GMSAFMC (Gulf of Mexico and South Atlantic Fishery
1980   Management Council). Fishery Management Plan, Final Environmental Impact Statement, and Regulatory Analysis for the Coastal Migratory Pelagic Resources.

GUNTER, G. Studies on marine fishes of Texas. *Publ. Inst. Mar.*
1945   *Sci. Univ. Texas* 1: 9–190.

HAPP, G, GOSSELINK, J G and DAY, J W, JR. The seasonal
1977   distribution of organic carbon in a Louisiana estuary. *Estuarine Coastal Mar. Sci.* 5: 695–705.

HARRIS, A H and ROSE, C D. Shrimp predation by the sea
1968   catfish, *Galeichthys felis*. Trans. Amer. Fish. Soc. 97: 503–504.

HAUSKNECHT, K A. Describe surficial sediments and suspended
1980   particulate matter. Vol. V *In:* W. B. Jackson and G. M. Faw (eds.), Biological/chemical survey of Texoma and Capline sector salt dome brine disposal sites off Louisiana, 1978–1979. *NOAA Tech. Mem. NMFS-SEFC-20,* 56 p.

HENWOOD, T, JOHNSON, P and HEARD, R. Feeding habits and
1978   food of the long-spined porgy, *Stenotomus caprinus* Bean. *Northeast Gulf Sci.* 2: 133–137.

HETTLER, W F. Influence of temperature and salinity on routine
1976   metabolic rate and growth of young Atlantic menhaden. *J. Fish. Biol.* 8: 55–65.

HILDEBRAND, H H. A study of the brown shrimp (*Penaeus*
1954   *aztecus* Ives) grounds in the western Gulf of Mexico. *Publ. Inst. Mar. Sci. Univ. Texas* 3: 233–366.

HILDEBRAND, H H. A study of the fauna of the pink shrimp
1955   (*Penaeus duorarum Burkenroad*) grounds in the Gulf of Campeche. *Publ. Inst. Mar. Sci. Univ. Texas* 4: 169–232.

HOLTHUIS, L B. FAO Species Catalog Vol. 1 – Shrimps and
1980   Prawns of the World. *FAO Fish. Synopsis* No. 125. 271 p.

HOSS, D E. Energy requirements of population of pinfish
1974   *Lagodon rhomboides* (Linnaeus). *Ecology* 55: 848–855.

HUFF, J A and COBB, S P. Penaeoid and sergestoid shrimps
1979   (Crustacea: Decapoda). Mem. Hourglass Cruises, Vol. V, Part IV. *Fla. Dept. Nat. Res., Mar. Res. Lab.,* St. Petersburg, Fla. 102 p.

IRVING, L, SCHOLANDER, P F and GRINNELL, S W. The respira-
1941   tion of the *Tursiops truncatus. J. Cell. Comp. Physio.* 17: 145–168.

JOHANNES, R E and SATOMI, M. Composition and nutritive
1966   value of fecal pellets of a marine crustacean. *Limnol oceanogr.* 11: 191–197.

JOUBERT, L S and DAVIES, D H. The penaeid prawns of the St.
1966   Lucia Lake system. *So. Afr. Assoc. Mar. Biol. Res., Oceanogr. Res. Inst. Investig. Rep.* no. 13. 40 p.

KAKUDA, S and MATSUMOTO, K. On the food habits of the white
1978   croaker *Argyrosomus argentatus. J. Fac. Fish. Anim. Husb. Hiroshima Univ.* 17: 133–142.

KINCH, J C. Trophic habits of the juvenile fishes within artificial
1979   waterways—Marco Island, Florida. *Contrib. Mar. Sci.* 22: 77–90.

KJELSON, M A, PETERS, D S, THAYER, G W and JOHNSON, G N.
1975   The general feeding ecology of postlarval fishes in the Newport River estuary. *Fish. Bull.* 73: 137–144.

KNAPP, F T. Menhaden utilization in relation to the conserva-
1949   tion of food and game fishes of the Texas Gulf coast. *Trans. Amer. Fish. Soc.* 79: 137–144.

KOSADA, M. On the ecology of the penaeid shrimp,
1977   *Metapenaeopsis dalei* (*Rathbun*), in Sendai Bay. *J. Fac. Mar. Sci. Technol. Tokai Univ.* 10: 129–136.

KOSAKA, M. Ecological notes on the penaeid shrimp,
1979   *Trachypenaeus curvirostris,* in Sendai Bay, Japan. *J. Fac. Mar. Sci. Technol. Tokai Univ.* 12: 167–172.

LANE, E D. A study of the Atlantic midshipman, *Porichthys*
1967   *porosissimus,* in the vicinity of Part Aransas, Texas. *Contrib. Mar. Sci.* 12: 1–53.

LEATHERWOOD, S, GILBERT, J R and CHAPMAN, D G. An
1978   evaluation of some techniques for aerial census of bottlenosed dolphins. *J. Wildl. Manage.* 423: 239–250.

LIST, R J. Smithsonian Meteorological Tbles. Sixth Revised
1971   Edition. Vol. 114. Smithsonian Collections. Smithsonian Institution Press. Washington, D. C. 527 p.

MOFFETT, A W, MCEACHRON, L W and KEY, J G. Observations
1979   on the biology of sand seatrout (*Cynoscion arenarius*) in Galveston and Trinity Bays, Texas. *Contrib. Mar. Sci.* 22: 163–172.

MUTHU, M S, NARASIMHAM, K A, RAO, S S, SASTRY, Y A and
1975   RAMALINGAM, P. On the commercial trawl fisheries off Kakinada during 1967–70. *Indian J. Fish.* 22: 171–186.

NMFS (National Marine Fisheries Service). Fisheries of the
1979   United States, 1978. Current Fishery Statistics No. 7800. *U S Dept. Comm. NOAA.* Washington, D.C. 120 p.

NOMURA, H and FILHO, J F. A shrimp exploratory survey in
1968   northeastern and northern Brazil, with some biological observations on *Penaeus aztecus. FAO Fish. Rep.* 57: 219–231.

OGURA, N. Decomposition of dissolved organic matter derived
1972   from dead phytoplankton. p. 508–515. *In:* A. Y. Takenouti (ed.), Biological Oceanography of the Northern Pacific Ocean. Publ. Idemitsu Shoten (Tokyo).

OSBOURNE, K W, MAGHAN, B W and DRUMMOND, S B. Gulf of
1969   Mexico shrimp atlas. U S Dept. Interior, *Bur. Commer. Fish. Circ.* No. 312. 20 p.

OVERSHEET, R M and HEARD, R W. Food of the red drum,
1978   *Schiaenops ocellata,* from Mississippi Sound. *Gulf Res. Repts.* 6: 131–135.

PAMATMAT, M M. Facultative anaerobeosis of benthos. p. 69–
1980   92. *In:* K. R. Tenore and B. C. Coull (eds.), Marine Benthic Dynamics. University of South Carolina Press, Columbia, South Carolina.

PARKER, R H, CROWE, A L and BOHME, L S. Describe living and
1980   dead benthic (macro- and meigo-) communities. Vol. I. *In:* W. B. Jackson and G. M. Faw (eds.), Biological/ chemical survey of Texoma and Capline sector salt dome brine disposal sites off Louisiana, 1978–1979, *NOAA Tech. Mem. NMFS-SEFC-25,* 103 p.

PARSONS, T R, TAKAHASHI, J and HARGRAVE, B. Biological
1977   Oceanographic Processes, Pergamon Press. New York. 103 p.

PATELLA, F. Water surface areas within statistical subareas used
1975   in reporting Gulf coast shrimp data. Marine Fisheries Review 37(12): 22–24.

PERRET, W S, BARRETT, B B, LATAPIE, W R, POLLARD, J F,
1971   MOCK, W R, ADKINS, G B, GAIDRY, W J and WHITE, C J. Cooperative Gulf of Mexico Estuarine Inventory and Study, Louisiana. Phase I, Area Discription. *Louisiana Dept. Wildl. Fish,* 175 p.

POWELL, A B and SCHWARTZ, F O. Food of *Paralichthys*
1979   *dentatus* and *P. lethostigma* (Pices: Bothidae) in North Carolina estuaries. *Estuaries* 2: 267–279.

PRUTER, A T. Trawling results of the R/V ANTON BRUUN in
1964   the Bay of Bengal and Arabian Sea. *Commer. Fish. Rev. 26* (11A, Suppl): 27–34.

REID, G K, JR. An ecological study of the Gulf of Mexico fishes
1954   in the vicinity of Cedar Key, Florida. *Bull. Mar. Sci. Gulf Caribb.* 4: 1–94.

REITSEMA, L A. Determine seasonal abundance, distribution,
1980   and community composition of zooplankton. Vol. II *In:* W. B. Jackson and G. W. Faw (eds.), Biological/ chemical survey of Texoma and Capline sector salt dome brind disposal sites off Louisiana, 1978–1979. *NOAA Tech. Mem. NMFS-SEFC-26,* 133 p.

RIVAS, L R. (Unpubl. report). Estimates of biomass for pelagic and coastal sharks in the Gulf of Mexico. Southeast Fisheries Center, National Marine Fisheries Service, Miami, Florida.

ROELOFS, E W. Food studies of young sciaenid fishes, *Micropo-*
1954   *gon* and *Leiostomus,* from North Carolina. *Copeia* 1954: 151–153.

ROGERS, R M, JR. Trophic interrelationships of selected fishes
1977   on the continental shelf of the northern Gulf of Mexico. Ph.D. Diss. Texas A and M Univ., College Station, Texas, 229 p.

ROSS, S T. Patterns of resource partitioning in searobins (Pices:
1977   Triglidae;. *Copeia* 1977: 561–571.

ROSS, S T. Trophic ontogrony of the leopard searobin,
1978   *Prionotus scitulus* (Pices: Triglidae). *Fishery Bull.* 76: 225–234.

RYTHER, J H. Photosynthesis and fish production in the sea.
1969   *Science* 166: 72–76.

RYTHER, J H and GUILLARD, R R L. Studies of marine plank-
1962   tonic diatoms. III. Some effects of temperature on respiration of five species. *Can. J. Microbiol.* 8: 447–453.

SACKETT, W M. *In:* S. Z. El-Sayed, W. M. Sackett, L. M.
1972   Jeffrey, A. O. Fredericks, R. P. Saunders, P. S. Conger, G. A. Fryxell, K. A. Steidinger, and S. A. Earle, Chemistry primary productivity, and benthic algae of the Gulf of Mexico. Folio 22, Serial Atlas of the Marine Environment. American Geographical Society, New York.

SCHOENER, T W. Resource partitioning in ecological com-
1974   munities. *Science* 185: 27–39.

SHANE, S H. Occurrence, movements, and distribution of
1980   bottlenose dolphin, *Tursiops truncatus,* in southern Texas. *Fish. Bull.* 78: 593–602.

SHERIDAN, P F. Food habits of the bay anchovy, *Anchoa*
1978 *mitchilli*, in Apalachicola Bay, Florida. *Northeast Gulf Sci.* 3: 1–15.

SHERIDAN, P F. Trophic resource utilization by three species of
1979 sciaenid fishes in a northwest Florida estuary. *Northeast Gulf Sci.* 3: 1–15.

SHERIDAN, P F and TRIMM, D L. Summer foods of Texas coastal
1983 fishes relative to age and habitat. *Fishery Bull.* 81 (In Press).

SIDWELL, V D. Chemical and nutritional composition of
1981 finfishes, whales, crustaceans, mollusks, and their products. *NOAA Tech. Mem. NMFS F/SEC-11.* U S Department of Commerce. Southeast Fisheries Center, Miami, Florida. 432 p.

SIKORA, W B, HEARD, R W and DAHLBERT, M D. The occur-
1972 rence and food habits of two species of hake, *Urophycis regius* and *U. floridanus*, in Georgia estuaries. *Trans. Amer. Fish. Soc.* 101: 513–525.

SKLAR, F H. Primary productivity in the Mississippi Delta bight
1976 near a shallow bay estuarine system in Louisiana. Louisiana State University, Ph.D. diss. Baton Rouge, Louisiana. 96 p.

SPRINGER, V G and WOODBURN, K D. An ecological study of the
1960 Tampa Bay area. *Fla. Dept. Nat. Res., Mar. Res. Lab. Prof. Papers Ser.* No 1, 104 p.

STICKNEY, R R, TAYLOR, G L, and HEARD, R W, III. Food habits
1974 of Georgia and estuaries fishes. I. Four species of flounders (Pleuronectiformes: Bothidae). *Fish. Bull.* 72: 515–525.

STOKES, G M. Life history studies of southern flounder
1977 (*Paralichthys lethostigma*) and Gulf flounder (*P. albigutta*) in the Aransas Bay area of Texas. Texas Parks Wildl. Dept., Tech. Ser. No. 25. 37 p.

STRICKLAND, J D H. Measuring the production of marine
1960 phytoplankton. *Fish. Res. Bd. Canada Bull.* 122: 172 p.

TABB, D C and MANNING, R B. A checklist of the flora and
1961 fauna of northern Florida Bay and adjacent brackish waters of the Florida mainland collected during the period July, 1957 through September, 1960. *Bull. Mar. Sci. Gulf Caribb.* 11: 552–649.

TECHNICAL CONSULTATION ON SHRIMP BY-CATCH UTILIZATION.
1982 Fish By-Catch . . . Bonus from the Sea. Report from a meeting in Georgetown, Guyana, October 27–30, 1981. 163 p.

TIETJEN, J H. The ecology of shallow-water meiofauna in two
1969 New England estuaries. *Oecologia* 2: 251–291.

TURNER, R E. Intertidal vegetation and commercial yields of
1977 penaeid shrimp. *Trans. Amer. Fish. Soc.* 106: 411–416.

TYLER, A V. Food resource division among northern, marine
1972 demersal fishes. *J. Fish. Res. Bd. Canada* 29: 997–1003.

WELSH, W W and BREDER, C M, JR. Contributions to life
1923 histories of Sciaenidae of the eastern United States coast. *Bull. U S Bur. Fish.* 39: 141–201.

YANEZ-ARANCIBIA, A, CURIEL-GOMEZ, J and DE YANEZ, V L.
1976 Prospeccion biologica y ecologica del bagre marine *Galeichthys caerulescens* (Gunter) en el sistema lagunar costero de Guerrero, Mexico (Pices: Ariidae), *an. Centro Cienc. Mar. Limnol. Univ. Nat. Auton. Mexico* 3 (1): 135–180.

Do discards affect the production of shrimps in the Gulf of Mexico?

D H Cushing

Introduction

In the shrimp fishery in the Gulf of Mexico the smallest shrimp are often discarded and so are much larger quantities of juvenile demersal fish. The total weight of material discarded is probably much greater than the quantity of shrimp landed. The shrimp stocks have been exploited without recruitment failure for two decades or more. The question has been raised whether the discards provide enough food to counter potential recruitment overfishing. It is a question which cannot be answered decisively unless it were shown that recruitment had failed under the stress of fishing. An attempt is made here to describe the relevant production mechanisms in relation to the quantities discarded.

The food of penaeids

The large penaeids of Australian and Indonesian waters feed on crustacea, vegetation, polychaetes, annelids and smaller molluscs; smaller crustacea are eaten whole (FAO synopses 3–8). *Penaeus setiferus* is a selective particulate feeder which searches the sand grains and passes the bits it likes forward to the mouth. The gut contents are macer-

ated and difficult to identify. But the white shrimp likes polychaetes and is cannibal in captivity (Lindner and Cook, 1970). Hood *et al* (1971) showed that the bacterial content of the gut was high and that the bacteria were chitinoclastic. The assimilation efficiency is low on a carbohydrate diet, also when the algal mat is the main source of food. But such efficiency was high with protein or lipid diet. The algal mat, however, produces a heavy detritus (Condrey *et al*, 1972). Qasim and Easterson (1974) fed *Metapenaeus monceros* (Fabricius) on detritus and found high assimilation efficiencies. Small or juvenile shrimps, 10–15 mm in length will feed on *Artemia nauplii*, but those of 20–30 mm in length will search in the detritus. Animals larger than 50 mm tend to feed on animal food. *Penaeus duorarum duorarum* is omnivorous. Early larval stages feed on microplankton. Juveniles feed on crustacea and polychaetes. Guts contain debris, larger algae, diatoms, dinoflagellates nematodes, polychaetes, ostracods, copepods, mysids, isopods, amphipods, carideans, molluscs and fish scales (Costello and Allan, 1970). Because the food is triturated and because the animals feed on sand or mud it is difficult to establish their food preferences. However, the mysis stages are planktonic feeders, the post larvae and juveniles feed on the detritus from the algal mat and larger animals eat any animal food from the mud and sand.

A brief life history

The adults spawn offshore and the planktonic larvae presumably live there as well. Post larvae are found inshore perhaps because their later mysis stages sink into the lower levels of the water column. Selective tidal transport (de Groot; Harden Jones) describes a migration by which animals stay on the bottom during one tidal cycle and move or are drifted in midwater on the other. With this mechanism, juvenile shrimp may find their way into the estuaries. Gunter and Edwards (1970) showed that catches of white shrimp in Texas were correlated with rainfall one or two years before the catch, but not with river discharge. Catches are predominantly composed of recruiting shrimp. Berry and Baxter (1970) were able to show that stock sizes could be forecast from indices of abundance of juveniles. Difference in catches depend upon differences in recruitment, itself determined before or during early juvenile stages. If recruitment were established then, it depends upon planktonic food for the larvae or upon detritus for the juveniles. Then discards of fish may augment the food of adult shrimps; those of small shrimps may do so also, but they probably augment the food of recruits, merely because the bits are smaller.

Distribution of shrimps and the discard problem

Phares (this meeting) tabulates landings of shrimps, brown, pink and white, by season, region and depth range for the year 1978 in hundreds of pounds, tailweight. They were converted to tons wet weight kms^{-2} for the areas 7–12 (West Florida), 13–17 (Louisiana) and 18–21 (Texas). From Figure (1) in Rothschild and Parrack (this meeting), areas of depth ranges 0–10 fms (including 'inshore' within the estuaries) and 11–20 fms were estimated in km^2. Tons wet weight kms^{-2} were converted to gCm^{-2} (Bougis, 1974, gives: dry weight = 0·13 wet weight and carbon = 0·40 dry weight, for crustacea, so carbon = 0·052 wet weight). Results are given in *Table 1*.

Phare's table was also used to establish the quantities of shrimp in areas where small shrimp are available (from the seasonal distributions of small shrimp in Table 2 in Rothschild and Parrack, this meeting) in 0–10 fms, in *Table 2*.

Small shrimp cannot be sold and so are protected by minimum landing size (MLS) and seasonal clo-

sures of certain areas; in Texas, the MLS is $39\,lb^{-1}$, whole weight, in Florida $47\,lb^{-1}$, whole weight and in other states, $68\,lb^{-1}$, whole weight. Consequently the small shrimps are discarded dead. Table 12 in Rothschild and Parrack (this meeting) give some information on quantities discarded during the middle sixties, 10% or 50% by weight. Maximal estimates of the discards of small shrimps are given by taking half the quantities given in *Tables 1* and *2*. They are estimated in tailweight and in *Table 3* the quantities are doubled to estimate whole weight.

Table 1
TAILWEIGHT OF SHRIMP LANDED IN $gCm^{-2}yr^{-1}$ OF ALL THREE SPECIES

Depth range \ Areas	7–12	13–17	18–21
0–10 fms (and inshore)	0·0011	0·054	0·0018
11–20 fms	0·0029	0·011	0·0293

Table 2
TAILWEIGHT OF THE SMALLEST SHRIMP LANDED IN $gCm^{-2}yr^{-1}$ OF EACH SPECIES

Species \ Areas	7–12	13–17	18–21
Pink	0·0007		
Brown	0·006	0·028	0·016
White	0·0023	0·018	0·0094

Table 3
WHOLEWEIGHT OF SHRIMPS DISCARDED IN $gCm^{-2}yr^{-1}$ (FROM *Table 2*)

Species \ Areas	7–12	13–17	18–21
Pink	0·0007		
Brown	0·006	0·028	0·016
White	0·0023	0·018	0·0094
(FROM *Table 1*)			
All shrimps	0·0010	0·054	0·0018

Some information on the quantity of bottom fish discarded is given in Sheridan *et al* (this meeting). I shall assume that within the depth range 11–20 fms, it is ten times the quantity of shrimp landed in $gCm^{-2}yr^{-1}$. I have assumed that most fish are discarded in the outer part of the zone because they have to have grown into the juvenile stages in order to account for the weight. The results are given in *Table 4*.

Table 4
QUANTITIES OF FISH DISCARDED AS TEN TIMES THE WHOLE WEIGHT
OF SHRIMP LANDED IN $11-20$ FMS IN $gCm^{-2}yr^{-1}$

Depth range \ Areas	7–12	13–17	18–21
11–20 fms	0·058	0·22	0·586

Such are perhaps maximal estimates of small shrimp discarded in the inner zone and of juvenile demersal fish discarded in the outer zone.

Discards as part of the benthic production in the Gulf of Mexico

Walsh *et al* (1981), in their study of the enrichment of the continental shelves as consequence of production processes, have made a study of the Gulf of Mexico and the influence of the rivers, predominantly the Mississippi. In most marine organisms the ratio of carbon to nitrogen, C/N < 6. Because nitrogen is recycled more quickly than carbon, in the detritus, C/N > 10. Sediment trap data show that higher C/N ratios are found immediately after the spring bloom (Smetacek *et al*, 1978). From the distribution of the C/N ratio in the surface sediments of the Gulf of Mexico, it may be concluded that the shrimp fishery is found in a region of high detritus.

Walsh (in press) has constructed budgets of benthic production on the shelves of Louisiana/Texas and of West Florida:

Production in the Gulf of Mexico in $gCm^{-2}yr^{-1}$

	Louisiana/Texas	West Florida
Primary production	100	c 30
Detritus pool	78	30
Meiobenthos	30	12
Macrobenthos	3	1·2
Shrimp	0·3	0·08

Let us assume that juvenile shrimp eat detritus and that adults eat meiobenthos, macrobenthos and fish discards. We then construct the following table of potential food in $gCm^{-2}yr^{-1}$:

	Louisiana/Texas		Export	West Florida		Export
	Shrimp			Shrimp		
	Juveniles	Adults		Juveniles	Adults	
Detritus	78		48	30		17.6
Meiobenthos		} 33			} 13.2	
Macrobenthos						
Discards	0.74	0.86		0.009	0.08	

Thus the discards comprise a small proportion of the total food available in this structure. The column labelled 'Export' includes the quantities of detritus exported to the continental slope. We also need to know what other animals compete for the apparent excess of food.

Walsh (in press) has analyzed the food web in the following way, in $gCm^{-2}y^{-1}$:

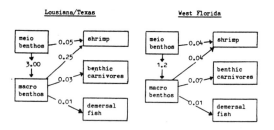

From Walsh's analysis the production of shrimp from the two benthic sources, meio and macro, is roughly equivalent to that of their competitors, benthic carnivores and demersal fish. Then the discard problem can be expressed in the following way in $gCm^{-2}yr^{-1}$:

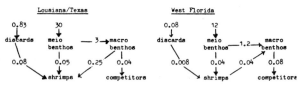

In this analysis shrimps in Louisiana/Texas depend primarily upon macrobenthos; only one sixth of the total food is in the meiobenthos. In West Florida, shrimps depend equally upon meio- and macrobenthos. In Louisiana/Texas, demersal competitors take only a small proportion of the macrobenthos as compared with the shrimps. But in West Florida the demersal competitors take twice as much from the macrobenthos as do the shrimps which is perhaps why the shrimps depend to a greater degree upon the meiobenthos there.

In West Florida, the proportion of discards in the shrimp food is low, <10%. In Louisiana/Texas, the proportion is higher, c23%. The quantity discarded shown above in both regions is predominantly of fish, estimated as (Tailweight) × 2 × 10. It is perhaps an overestimate, in which case the proportions as food are overestimated.

The effects of discards were assumed to augment food either to recruits or adults and not to reduce predation. Sheridan *et al* (this meeting) suggest that penaeids in the Gulf of Mexico are not eaten much by fish. Ursin (1973) showed that a predator is often about one hundred times heavier than its prey, or about four or five times longer. In fact the shrimps and fish are about the same size, both in the estuaries and in the shallow waters offshore.

256

The distributions in length of fish must overlap so that the largest fish eat the smallest shrimp and relaxed predation should augment recruitment, but probably only a little.

Discussion

The argument presented in this paper depends upon a separation of energy flow, as $gCm^{-2}yr^{-1}$, into recuit and into adult shrimp. The greatest proportion of discards ends probably in adult or near adult shrimps. If the percentage of food is really as high as 10% or 20%, as suggested above, it represents food added to the adult population. One of the great gaps in fisheries science is a description of the population dynamics of unexploited stocks, in particular the degree of density dependence in growth. As the data series lengthen into decades, evidence starts to appear that in exploited fish stocks growth is mildly density dependent (Houghton and Flaxman 1981). Perhaps in unexploited stocks, growth is sharply density dependent. But in stocks as heavily exploited as the shrimp in the Gulf of Mexico, the degree of density dependence must be low. Then there is a low need for any additional food represented by the discards taken by the adult shrimps. It is of course possible thatn an increment of food ends as an increment of fecundity. But if the processes that determine recruitment are density dependent, the added eggs and subsequent larvae would end as food for planktonic predators in offshore waters. But if the stocks are heavily exploited, the density dependence of the recruitment process may be relatively low, in which case added eggs may indeed lead to increased recruitment. However, the argument is speculative because the link between food and fecundity has not been established in fish or shrimp.

Discards provide a smaller proportion of the food of recruits, $\leqslant 1\%$. The quantity of detritus is high. It is not known whether recruitment is determined during the larvae stage offshore or during the juvenile stage in the estuaries or close inshore. The primary production is not very high, $100gCm^{-2}yr^{-1}$, about the same as that in the North Sea. It is of course spread across the euphotic zone, some metres inshore, and perhaps twenty metres offshore. The detritus pool, of $78gCm^{-2}yr^{-1}$, is on the surface (or near surface) of the seabed. Hence the amount of food on the seabed is very much greater than that in the euphotic layer by perhaps an order of magnitude. Thus an animal that uses the detritus has an advantage and for that reason, recruitment may be determined during the juvenile

stages, dependent on detritus. The meiobenthos production comprises 40% of the detritus pool and the juvenile population of shrimp is a very small proportion of that quantity. It is of course possible that the juvenile shrimp feed on part of the meiobenthos.

The question was raised whether the quantities discarded counter potential recruitment overfishing. The discards probably do not affect the production of recruits, although the arguments presented here are not decisive. It is remarkable that the stocks have withstood heavy fishing for so long, but the animals are very fecund and perhaps the heavy quantity of detritus provides the food needed to sustain recruitment. Predation may be relaxed but the potential augmentation of recruitment cannot be estimated. There is no conclusion to this paper save that the question in the title requires much uncollected information to provide a decisive answer.

References

BERRY, R J and BAXTER, K N. 'Predicting brown shrimp abun-
1970   dance in the north western Gulf of Mexico.' *FAO Fish. Rep.* 57. 3. 775–798.

BOUGIS, P. 'Écologie du placton marin II Le zooplacton.'
1974   Masson et Cie, 200 pp.

CONDREY, R E, GOSSELINK, J G and BENNETT, H J. 'Comparison
1972   of the assimilation of different diets by *Penaeus setiferus* and *P. aztecus*.' *Fish. Bull.* 70. 4. 1281–1292.

COSTELLO, T J and ALLAN, D M. 'Synopsis of biological data on
1970   the pink Shrimp *Penaeus duorarum duorarum* Burkenroad 1939.' *FAO Fish. Rep.* 57. 4. 1499–1538.

GUNTER, G and EDWARDS, J C. 'The relation of rainfall and
1970   fresh water drainage to the production of the penaeid shrimps *Penaeus fluviatilis* SAY and *Penaeus aztecus* IVES in Texas and Louisiana waters.' *FAO Fish. Rep.* 57. 3. 875–892.

HARDEN JONES, F R, GREER WALKER, M and SCHOLES, P.
1979   'Selective tidal stream transport and the immigration of plaice (*Pleuronectes platessa* L.) in the southern North Sea. *J. Cons. int. Explor. Mer.* 38(1) 331–7.

HOOD, M A, MEYERS, S P and COLMER, A R. 'Bacteria of the
1971   digestive tract of the white shrimp, *Penaeus setiferus*.' *Bact. Proc. G.* 147.

HOUGHTON, R G and FLAXMAN, S. 'The exploitation pattern,
1981   density dependent catchability and growth of cod (*Gadus morhua*) in the west-central North Sea.' *J. Cons. int. Explor. Mer.*

LINDNER, M J and COOK, H L. 'Synopsis of biological data on
1970   the white shrimp *Penaeus setiferus* (Linnaeus) 1767.' *FAO Fish. Rep.* 57. 4. 1439–1470.

PHARES, P (this meeting). 'The relationship between catch and time of fishing for the Pink, Brown and White Shrimp of the northern Gulf of Mexico.'

QASIM, S Z and EASTERSON, D C V. 'Energy conversion in the
1974   shrimp *Metapenaeus monoceros* (Fabricius).' *Ind. J. Mar. Sci.* 3. 131–4.

ROTHSCHILD, B J and PARRACK, M L (this meeting). 'The U S Gulf of Mexico shrimp fishery.'

SHERIDAN, P F, BROWDER, J A and POWERS. J E (this meeting). 'Ecological interactions between Penaeid shrimp and associated bottom fish assemblages.'

SMETACEK, V, VON BROCKEL, K, ZEITZSCHEL, B and ZENK, W.
1978   'Sedimentation of particulate matter during a phyto-

plankton spring bloom in relation to the hydrographical régime.' *Mar. Biol.* 47. 211–226.

URSIN, E. 'On the prey size preferences of cod and dab'. *Meddr.*
1973   *Danm. Fisk. og Havunders.* N.S. 7. 85–89.

WALSH, J J (in press). 'Death in the sea: enigmatic loss processes.' *Progress in Oceanography.*

WALSH, J J, PREMUZIC, E T and WHITLEDGE, T E. 'Fate of
1981   nutrient enrichment on continental shelves as indicated by the C/N content of bottom sediments.' *In:* Ecohydrodynamicised J. C. Nihoul, 13–50 Elsevier Amsterdam, Oxford, New York.

The *Nephrops* fishery in the northeast Atlantic

D H Cushing

Abstract

The *Nephrops* fishery in the Northeast Atlantic yields about 40 000 tons each year. The prawns live on patches of mud or sandy mud in depths of 75–150 m. They have usually been caught by cutters using small meshed trawls (40 mm meshes in the cod end and elsewhere). Mesh selection experiments assuming selection factors of 0·3 or 0·5 have shown that a long term gain can be sustained if the mesh sizes in the trawls were increased from 40 mm to 70 mm. This assessment is based on the assumption that $M = 0·2$; if $M = 0·4$ or $0·6$ no long term gain can be shown. However, Jones (1980) has shown with an independent estimate of stock in the Firth of Forth off the Scottish coast that $M = 0·1$ or $0·2$. He assumed a 50% survival of the small shrimp discarded. There is, of course, a short term loss estimated in any assessment of this nature but because the prawns grow quickly and increase in value with weight a reasonable short term loss can be tolerated.

In most *Nephrops* fisheries, immature white fish have been discarded in large quantities, whiting in the Irish Sea and hake in the Bay of Biscay. It was estimated that forty million immature whiting were discarded each year in the fishery off Northern Ireland. If these fish had been allowed to grow and die in the fishery for adults, a catch of 6 000 tons might have been added to the usual annual catch of 10 000 tons. Such increments could probably be obtained elsewhere in the Irish Sea but to a somewhat lesser degree because the whiting nursery tends to be restricted to the northeastern Irish Sea. In the Bay of Biscay, immature hake grow on La Grande Vasière, a prawn ground off the southern coast of Brittany. They are also found on muddy patches in the Celtic Sea where the prawns grow. Because there is a directed fishery for small hake off the northern coast of Spain, three quarters of the annual catch of hake is less than 20 cm in length. Part of this problem can be mitigated by increasing the mesh sizes in the prawn trawls. More generally such increases would yield bigger prawns for the *Nephrops* fishermen and allow better survival for immature whiting and hake to be exploited later in their lives by fishermen for white fish.

Reference

JONES, R. Estimate of natural mortality for the Firth of Forth
1980   stock of *Nephrops*. Appendix 3 to Report of the *Nephrops* Working Group. CM 1980 K2 Mimeo.

Environmental factors

Catch prediction of the banana prawn, *Penaeus merguiensis,* in the south-eastern Gulf of Carpentaria

D J Staples
W Dall
D J Vance

Abstract

The banana prawn, *Penaeus merguiensis* is the major species fished in the Gulf of Carpentaria, a tropical area characterised by a monsoonal wet season of about three months and a very low rainfall in the remainder of the year. A marked seasonal cycle of temperature also exists, especially in the south-eastern region, due to its proximity to the interior of a large continental land mass. Commercial catch statistics for the south-eastern Gulf show a ten-fold variation in annual catch over the last decade. Effort (boat days) has remained relatively constant over the period and because the exploitation rate is high (78–86%), fluctuations in total catch reflect real change in abundance. Adult banana prawns in this region spawn offshore at depths of 18–24 m with a spring peak around September–October and another peak around February–March. The extensive estuarine river systems provide nursery grounds for the postlarvae. Life history studies have shown that it is the October spawning which contributes mainly to the commercial catch. Postlarvae from this brood reach the estuaries in November and begin emigrating offshore with the onset of the wet season in December. Early juvenile prawns can tolerate almost freshwater, but late juveniles have a lower lethal salinity limit of about 7%. During periods of low rainfall only a small number of the larger juveniles emigrate, but emigration increases with rainfall until eventually all sizes emigrate. The magnitude and duration of emigration, therefore, depends on the amount and timing of rainfall. The fishing season commences in mid-March, immediately following the wet season and there is a good correlation between rainfall and commercial catch. Forward step-wise regression analysis was used to investigate this relationship further and to examine the added effects of river discharge, temperature, onshore and longshore wind components. The best model to explain catch variance was based on a simple regression model of summer plus autumn rainfall: $C = 2.626R - 950.5$ where C is catch in tonnes (t) and R is the sum of summer plus autumn rainfall in mm. This model explains 80% of the observed variance.

For predictive purposes, only those variables which can be collected before the fishing season are of use. Spring rainfall alone accounts for 37% of the observed variance and gives an approximate prediction three and a half months before the season. As data for each successive summer month's rainfall is added, by the end of summer the explained variance reaches 50%. Catch can be predicted with a standard error of ±19% by the equation: $C = 1.864R - 66.8$ where C is catch in t and R is sum of spring plus summer rainfall in mm. Accuracy of prediction decreases with lower and higher rainfall.

Introduction

Prediction of catches is an often-stated aim of the fisheries manager, but there are few fisheries where reliable predictions are made regularly. Predictions demand that the critical environmental and biological factors which cause annual fluctuations be identified, normally a laborious and time-consuming, long-term task. Management agencies are often unwilling or unable to allocate the necessary manpower and funds for this research and so prediction tends to be a neglected aspect of management.

Developing a prediction model for penaeid prawns may not be as formidable a task as it first appears. Because the commercially important Penaeidae are short-lived species, mainly events of the preceding 12–18 months need be considered. Given a geographically stable environment without major man-induced pollution, the very high fecundity of penaeid prawns tends to ensure a high and fairly reliable return of postlarvae to the nursery grounds. Thus it is often the survival after settlement of the postlarvae which determines the size of the coming season's catches. In some cases, estimates of the size of postlarval of juvenile populations have been considered adequate for prediction purposes (Berry and Baxter, 1969; Rao and Gopalakrishnayya, 1974; Roessler and Rehrer, 1971; Yokel et al, 1969). Such predictions assume that the juvenile populations are subject to uniform year-to-year environmental influences, which is rarely true. Certainly the diversity and numbers of predators may cause predation pressure to remain fairly constant from year to year, but other likely causes of variation in juvenile abundance are the effects of weather, the two most obvious being temperature and extreme salinity fluctuations, due to rainfall.

Since the Penaeidae are primarily tropical-subtropical animals whose range extends into warm temperate waters, it is not surprising that temperatures below 18–20°C are sub-optimal (eg Aldrich et al, 1968; Dall, 1958; Ford and St. Amant, 1971; Hunt et al, 1980; Wickins, 1976; Wiesepape et al, 1972; Zein-Eldin and Aldrich, 1965). Commercial fishing is rare in waters cooler than 18°C. Thus, in climates where temperatures are below 20°C for a large part of the year, temperature may be the dominant factor influencing juvenile prawn abundance (Barrett and Gillespie, 1973, 1975; Hunt et al, 1980; Williams, 1969). While low temperature may not be so critical in tropical and subtropical areas, proximity to large land masses may cause appreciable falls below 20°C in estuaries and other enclosed waters in the cooler months of the year (Price, 1979; Staples, 1980a), and possible effects in such areas should be considered. In regions of high rainfall, however, particularly when it tends to be seasonal, salinities in estuaries and embayments may fall to extremely low levels so that drop in salinity becomes the dominating environmental factor, which in turn may influence prawn abundance. A number of investigators have sought to correlate rainfall directly with prawn catches, the main effect of the rainfall being to enhance emigration from nursery grounds (Glaister, 1978; Gunter and Hildebrand, 1954; Gunter and Edwards, 1969; Hildebrand and Gunter, 1952; Ruello, 1973; Subramanyam, 1964).

Because of the periodicity of natural fluctuations in cases where time lags are introduced into the model, correlations are easy to obtain but, to be valid, such correlations need to be rigorously tested. This can only be achieved by having sound biological and ecological information for the species being investigated. Multivariate analyses should also be used to enable a more realistic model, including temperature and other likely variables, to be developed (Driver, 1976; Hunt et al, 1980).

The banana prawn (Penaeus merguiensis) fishery in the south-eastern Gulf of Carpentaria (Fig 1) was established in 1968 and now contributes approximately one-third to the Gulf of Carpentaria prawn fishery which is the largest in Australia. Because the prawns tend to school, the fishery rapidly became a high return, high effort and capital intensive industry. Normally the short season (in recent years less than 2 months) opens in mid-March and over the last ten years effort (boat days) has remained fairly constant. The considerable fluctuations in annual catch that have been recorded are therefore largely due to real fluctuations in fishable stocks. Hence prediction was one of the goals of the Tropical Prawn Research Project when it was started by CSIRO in mid-1974.

The Gulf of Carpentaria lies well within the tropics and is subjected to extremes of weather. From late March to December there is little or no rain, with strong prevailing east to south-east winds coming across the arid interior of the continent. In December–January a monsoonal trough develops across northern Australia and usually persists more or less continuously for two to three months. Winds are variable but tend to be north to northwest and the characteristic heavy rain squalls normally deposit in excess of 1 000 mm of rain during this period. Stream run-off is consequently heavy and extensive flooding of vast areas of flat Gulf country is usual. Run-off of the flood water is rapid and after the wet season the country soon returns to its typical dry savannah condition. Winter temperatures are relatively low. Thus it seemed likely that the wet season, which immediately precedes the banana prawn fishery, would be likely to have a major effect on abundance, with perhaps a secondary effect from relatively low winter temperatures in June–July. Indeed, the industry had arrived at the conclusion that a flood year presaged a good catch whereas a low rainfall season meant a poor harvest.

Biological studies on the banana prawn have been carried out as part of the CSIRO's Tropical Prawn Project since 1975. The Project has included studies on the reproductive biology of the prawn, the temporal and spatial distribution of larvae,

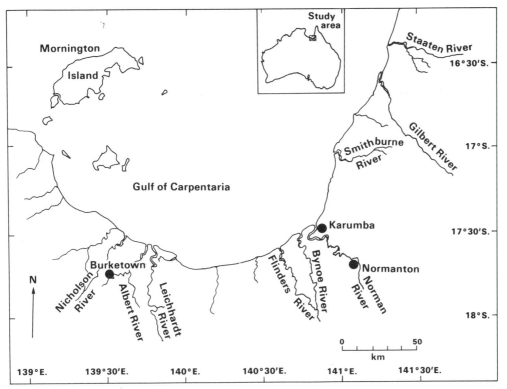

Fig 1 South-eastern Gulf of Carpentaria (after Staples, 1980a)

juvenile prawn ecology in the Norman River and adjacent rivers in the south-eastern Gulf, distribution and abundance of adult prawns with supporting studies on prawn physiology, feeding and behaviour. The aim of this paper is to show how these results have been used to develop a multivariate model aimed at explaining the maximum amount of variance observed in the commercial prawn catches. The predictive value of the multivariate model will then be assessed.

Methods

The centre of these studies was a field station at Karumba on the Norman River, Gulf of Carpentaria (*Fig 1*).

Basic biological information categories used to develop the multivariate model were:

(*1*) A detailed knowledge of temporal and spatial parameters of the life history of *P. merguiensis* and its ecology, including response to lowered salinity.

(*2*) Postlarval immigration rate. This gives a measure of the variation in annual spawning magnitude and larval survival. This was estimated from 1975 to 1979 by deploying an array of 0·5 m × 0·5 m, 1 mm mesh nets across the Norman

River throughout the incoming tide (Staples, 1980a).

(*3*) Emigration rate of juveniles. Rate of emigration from the estuary gives a measure of the interaction between fluctuations of the number of juvenile prawns and the proportion of that population emigrating. This was also measured by an array of nets, 1·0 m × 0·5 m, 2 mm mesh, on this occasion across the ebb tide (Staples, 1980b).

(*4*) Commercial catch statistics. These give the most comprehensive and reliable measure of adult prawn abundance and were collected by the CSIRO scientific logbook program, systematic commercial catch sampling and returns from processing plants (Lucas *et al*, 1979; Somers and Taylor, 1981).

Environmental data

A meteorological station, together with current meter and tide recorder were set up at the Karumba Field Station. River water samples were analysed regularly for temperature, salinity, dissolved oxygen and nutrients (Staples, 1980a). Rainfall, temperature and wind data for ten years (1970–79) from Normanton and other stations of the Bureau of Meteorology were collated and analysed (Vance, Staples and Kerr, in prep.).

Originally rainfall data were based on monthly rainfall totals collected from all Bureau of Meteorology recording stations in the south-eastern Gulf of Carpentaria region. Later analyses used data collected from Normanton only since there was a high correlation among stations. Water temperature data were available only from the Karumba Field Station, but air temperatures at Normanton gave a good record of seasonal and year-to-year temperature changes, correlating well with estuarine water temperatures ($r = 0.964$, $n = 20$) and offshore south-eastern Gulf water temperatures ($r = 0.892$, $n = 11$).

Multivariate analysis

A forward stepwise multivariate regression analysis was used to explain the maximum amount of variance observed in annual catch of banana prawns in the south-eastern Gulf. Variables were included in the analysis on their respective contribution to the explained variance of the commercial catch. The size of the data set was limited to the 10 years of reliable catch statistics, 1970–79. Because of the small number of data points involved, the percentage of explained variance increases rapidly to 100% as the number of independent variables included approaches the sample size, even when no true correlations are involved. Variables were therefore included in the analysis not only on their significance but also if they added more to the explained variance than that expected on the basis of the number of variables and observations (Morrison, 1976).

Data of the type used in these analyses may not follow normal distributions, and to satisfy the criteria of normality that are inherent in standard statistical procedures, both log and square root transformations were applied.

Results

Life history of P. merguiensis in the south-eastern Gulf of Carpentaria

The marked seasonality in rainfall, water temperature and salinity in the south-east Gulf of Carpentaria is shown in *Figure 2*. Approximately 80% of the annual rainfall fell during December, January and February. Average estuarine salinities ranged from near 0% at the mouth of the river just after

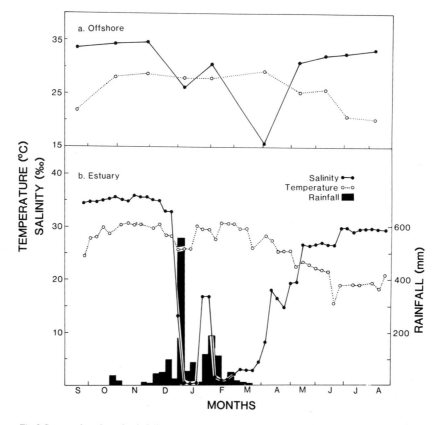

Fig 2 Seasonal cycles of rainfall, water temperature and salinity in the south-eastern Gulf of Carpentaria, 1975–76. (a) 55 km offshore; and (b) Norman River estuary. (After Staples, 1980a)

262

the onset of the wet season to 36% at the end of the dry season in November. Water temperature ranged from 15°C in June to 32·7°C in February. Offshore results showed similar but reduced annual cycles in both temperature and salinity. Through the influence of the summer wet season period water temperature does not increase much after November but remains around this level for a six-month period. Winter low temperatures are of a much shorter duration.

The banana prawn normally spawns in the shallow, open waters of the south-eastern Gulf in depths of 18–24 m (P. Crocos, unpublished). Although some spawning occurs through most of the year there are two main peaks, one around September–October and the other around February–March (Staples, 1979; Crocos, unpublished). Larval stages are passed through in 10–14 days depending on temperature (Rothlisberg, unpublished). Diurnal vertical migratory behaviour coupled with net transport of currents in the various water layers transports the postlarvae inshore (Rothlisberg, in prep.).

Postlarvae from the September–October spawning penetrate the Norman River up to 85 km from the mouth reaching a peak in November and two–three months later at a carapace length of 5–10 mm begin emigrating (Staples, 1980a, b) (*Figure 3*). Thus the bulk of the population is only 6–7 months of age at the opening of the fishing season in mid-March. Sexual maturity is reached at about 8 months of age. The autumn brood resulting from this spawning is much less consistent than the spring brood (indicated by dashed lines in *Figure 3*) and appeared as a large cohort in the river in only one of the four years studied. The bulk of this autumn brood overwinters in the estuary and apparently is subject to heavy mortality (Staples, 1979). As a result, some late juveniles and young adults are evident in the estuary from August onwards, but do not provide a recognisable cohort to the fishery (*Fig 3*), although they may contribute to the September–October spawning stock. Thus the fishery has the characteristics of a single annual cohort, even though there are two well-defined spawning periods.

Postlarval immigration patterns
Figure 4 shows the pattern of immigration for 1975–76 (Staples, 1980a). Postlarvae first appear and rapidly reach a peak in November, but appreciable immigration is continued until February when the effects of freshwater discharge prevent postlarvae from entering the river. Analyses of the effects of other factors on larval immigration (cycli-

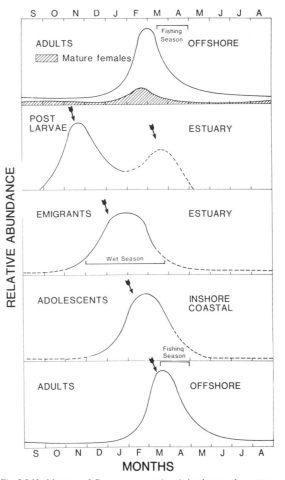

Fig 3 Life history of *Penaeus merguiensis* in the south-eastern Gulf of Carpentaria: Seasonal distribution of abundance of the main life history stages. ($\rightarrow$) indicates progression of the main cohort in spring(——) and the less consistent cohort in autumn (....)

cal events – tidal cycles and moon phase – and environmental variables) are in progress, but the net immigration rates (mean number entering/m^2) for the mid-October to mid-January period proved

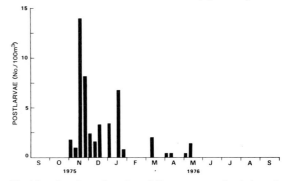

Fig 4 Immigration of postlarval *Penaeus merguiensis* into the Norman River 1975–76 (after Staples, 1980a)

to be remarkably similar for subsequent years. They were 2·652 ± 0·992 for 1975–76; 2·372 ± 0·740 for 1976–77; and 2·300 ± 0·980 for 1977–78 (Staples, unpublished).

Juvenile emigration patterns

The relationship between rainfall and juvenile prawn emigration is clearly seen for the 1975 to 1978 period (*Fig 5*). The 1975–76 season was characterised by three main periods of rain in December, January and February, each associated with a peak of prawn emigrations. In 1976–77, rainfall was more uniform and this was also reflected in the pattern of prawn emigration. The low rainfall of 1977–78 resulted in very few emigrants. During periods of low rainfall, only the largest juveniles emigrate, but as the rainfall increases the size range of emigrating prawns increases until eventually prawns of all sizes move out. Dall (1981) found that late juveniles and early adults had a lower lethal salinity limit of about 7%, whereas small juveniles could survive salinities of almost zero. These differences are shown in their respective osmoregulatory abilities (*Fig 6*). Thus the larger prawns would tend to respond immediately to lowered salinities by moving out of the rivers, whereas the smaller juveniles could remain until freshwater conditions prevailed. Hence the physiological evidence also supports the conclusion that banana prawn emigration is enhanced by high rainfall. Obviously a wave of emigration will be maximal if high juvenile population numbers coincides with a period of heavy rainfall and thus the timing and amount of rainfall can cause large variations in the number of prawns available to the offshore fishery. Unlike the immigration rates, mean emigration rate was found to be highly variable over a four year period, ranging from 0·533 ± 0·221/m²/min. in the high rainfall year of 1975–76, to 0·011 ± 0·004/m²/min. in the drought year 1977–78. Rainfall in the south-eastern Gulf has averaged 984 mm over the last 10 years with good rainfall in 1974, 1976 and 1979 (*Fig 7a*).

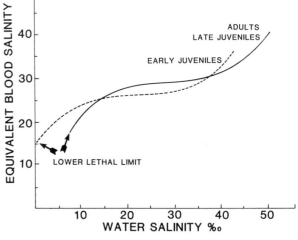

Fig 6 Osmoregulation of early juvenile *Penaeus merguiensis* compared with adults – late juveniles

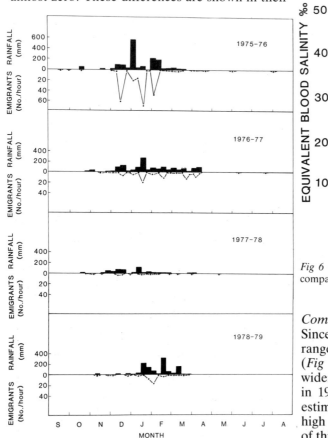

Fig 5 Relationship between weekly rainfall and emigration of juvenile *Penaeus merguiensis* from the Norman River 1975–79

Commercial catch patterns

Since 1971, total Gulf of Carpentaria catches have ranged from 1 800 in 1978 to 9 700 tonnes in 1974 (*Fig 7b*). The south-eastern Gulf showed even wider fluctuations ranging from 360 in 1978 to 3 850 in 1974, a ten-fold variation. Lucas *et al* (1979) estimated that the population is subject to the very high exploitation rate of 78–86%. The net effect of this heavy fishing pressure has been a decline in the average size of prawn taken, and a shortening of the fishing season over the 10 year period 1970–

1979. In this fishery, therefore, total catch, not catch per unit effort is the best estimate of prawn abundance. For the period 1970–79, there is a good correlation between annual rainfall in the south-eastern Gulf and catch of banana prawns in both the south-eastern region and the total Gulf (*cf* Fig 7a and 7b).

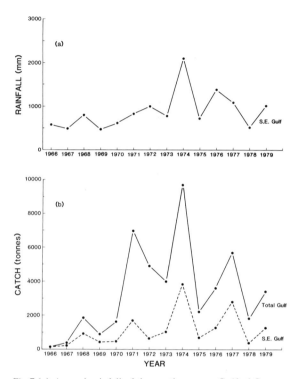

Fig 7 (a) Annual rainfall of the south-eastern Gulf of Carpentaria. (b) Annual yield of the south-eastern Gulf of Carpentaria

Multivariate analysis

Model development: the general model The above studies on the ecology of *P. merguiensis* indicated that the size of commercial catches was directly related to numbers of juveniles emigrating, which in turn appeared to be dependent on the timing and amount of rainfall during the wet season. Although rainfall and river discharge differences appeared to be the likely primary cause of variation in catch, temperature and wind components acting on earlier life history stages could also be important and were included in the analysis.

The model was developed by firstly calculating simple correlations between annual prawn catches for the south-eastern Gulf and the above variables singly for four seasons of the year. Secondly, a 16 × 16 matrix was computed among all variables to test for independence between them. Thirdly, the best model to explain prawn catch variance was

developed using forward stepwise multiple regression techniques.

Table 1 shows winter, spring, summer, autumn and total year catch correlations for rainfall, river discharge, temperature and two wind components. Correlations between spring, summer and autumn rainfall and annual catch are relatively high. River discharge depends on rainfall in the south-eastern Gulf and, as expected, the river discharge correlations show a similar pattern to that of rainfall. They are appreciably lower, however, and are thus much less use as predictive variables so are not considered in further analyses. Also in this analysis there are no significant correlations with either longshore or offshore winds, but they are considered further since they are independent variables. High correlation with annual catch also occurred with summer and winter temperatures. Log and square root transformations applied to this data altered correlation values slightly, but the overall patterns of significance in *Table 1* were not changed.

Table 1

SIMPLE CORRELATIONS OF ANNUAL *Penaeus merguiensis* CATCH WITH ANNUAL AND SEASONAL RAINFALL, RIVER DISCHARGE, MEAN TEMPERATURE AND WIND COMPONENTS

	Total Year	Winter	Spring	Summer	Autumn
Rainfall	0·84**	−0·36	0·60†	0·70*	0·66*
River discharge	0·56†	0·07	0·50†	0·56†	0·65*
Temperature	−0·33	0·54†	−0·18	−0·69*	−0·38*
Longshore Wind	—	0·40	−0·16	−0·40	−0·25
Onshore Wind	—	0·12	−0·24	0·32	0·13

**$P < 0.01$, *$P < 0.05$, †$0.05 < P < 0.15$

Several high correlations occurred in the 16 × 16 matrix, especially in the rainfall data and it was necessary to eliminate some variables and to combine others to create independent variables for analysis. For example, spring, summer and autumn rain were all highly correlated and single variables were created by summing firstly spring and summer rain and then summer and autumn rain. Both summer and winter temperatures were correlated with summer and autumn rainfall.

Forward stepwise multiple regression on all remaining variables indicates that the best model to explain catch variance is based on the sum of summer and autumn rainfall (*Table 2*). This combination alone accounts for over 80% of the catch variation. The next variable to be introduced into the model is the summer longshore wind component which explains a further 7·5% of the variation, but this effect is not significant. Summer and

Table 2

STEPWISE MULTIPLE REGRESSION FOR *Penaeus merguiensis* CATCHES, RAINFALL, AND WIND COMPONENTS. (R = MULTIPLE CORRELATION COEFFICENT), D.F = DEGREES OF FREEDOM

Step Number	Variable	Stepwise values				Multiple regression		
		F value to enter variable	Significance (each variable on entering)	R	R^2 change	Overall F	d.f.	Significance
1	Rain (summer + autumn)	28·912	0·001**	0·897	0·805	28·912	8	0·001**
2	Longshore wind (summer)	3·713	0·102ns	0·938	0·075	21·915	7	0·002**
3	Onshore wind (spring)	1·974	0·219ns	0·956	0·034	17·639	6	0·004**

** = $P < 0.01$, ns = not significant

winter temperatures which are simply correlated with catch (*Table 1*) are also correlated with rainfall and therefore add little to the explained variance after the rain variable has been included. It is possible that, with increasing numbers of observations, temperature effects, especially those of winter, may become more important. Thus the regression model becomes simply

$$C = 2.626R - 950.5$$

where C = catch and R = sum of summer and autumn rainfall in mm.

Predictive model

Some of the variables used in this analysis can be obtained only as the season progresses. For a model to be of use for predictive purposes, only those variables which can be collected before the autumn fishing season commences in mid-March can be considered. Although summer plus autumn rainfall provides the best explanation of catch variation, the autumn effect is based largely on March rainfall, and this variable is of no use as a predictive tool. However, because of high correlation among spring, summer and autumn rain, the correlations in *Table 1* suggest that spring rainfall might also be usable. Using this variable alone, a measure of

prediction is possible at the end of November, three and a half months before the fishing season begins (*Table 3*). As monthly rainfall values were added when they became available, the prediction would become more useful. By the end of summer the catch could be predicted with a standard error of ±19% in a year of average rainfall. With lower or higher rainfall values the accuracy would decrease. In this predictive model, as with the general model, a second independent variable does not add significantly to the explained variance and the best description is:

$$C = 1.864R - 66.8$$

where R = spring plus summer rainfall in mm and C = catch in t.

In conclusion, we must stress that this model is based on the data from only ten years. Obviously, further records over a number of years are necessary before the model can be tested adequately. Nevertheless, we feel that for prawn fishery prediction, this model is decidedly encouraging.

References

ALDRICH, D V, WOOD, C E and BAXTER, K N. An ecological
1968 interpretation of low temperature responses in *Penaeus*

Table 3

PREDICTION OF *Penaeus merguiensis* CATCHES FROM RAINFALL DATA COLLECTED AT MONTHLY INTERVALS BEFORE THE COMMERCIAL PRAWN SEASON (r = SIMPLE CORRELATION COEFFICENT)

Time of Prediction	Information needed	Regression Coefficient	Standard Error	Intercept	r	F value	d.f.	Significance
December	Spring rain (Sept., Oct. + Nov.)	7·969	3·731	830·09	0·603	4·562	9	0·065ns
January	Spring + Dec, rain	5·690	2·480	115·69	0·630	5·263	9	0·048*
February	Spring + Dec. + Jan. rain	1·981	0·773	398·06	0·672	6·569	9	0·038*
March	Spring + Summer rain	1·864	0·654	−66·83	0·710	8·132	9	0·024*

ns = not significant, * = $P = 0.05$

aztecus and *Penaeus setiferus* postlarvae. *Bull. Mar. Sci.*, 18:61–71.

BARRETT, B B and GILLESPIE, M C. Primary factors which
1973 influence commercial shrimp production in coastal Louisiana. *La. Wildl. Fish. Comm. Tech. Bull.* 9:1–28.

BARRETT, B B and GILLESPIE, M C. 1975 environmental condi-
1975 tions relative to shrimp production in coastal Louisiana. *La. Wildl. Fish. Comm. Tech. Bull.* 15:1–22.

BERRY, R J and BAXTER, K N. Predicting brown shrimp abun-
1969 dance in the northwestern Gulf of Mexico. *FAO Fish. Rep.* 57:775–798.

DALL, W. Observations on the biology of the greentail
1958 prawn, *Metapenaeus mastersii* (Haswell) (Crustacea Decapoda: Penaeidae). *Aust. J. Mar. Freshwater Res.* 9:111–134.

DALL, W. Osmoregulatory ability and juvenile habitat prefer-
1981 ence in some penaeid prawns. *J. Exp. Mar. Biol. Ecol.* 54:55–64.

DRIVER, P. Prediction of fluctuations in the landings of brown
1976 shrimp (*Crangon crangon*) in the Lancashire and West-ern Sea Fisheries District. *Estuarine and Coastal Marine Science* 4:567–573.

FORD, T B and ST. AMANT, L S. Management guidelines for
1971 predicting brown shrimp, *Penaeus aztecus*, production in Louisiana. *Gulf Carrib. Fish. Inst. 23rd Annu. Sess.*, Nov. 1970:149–161.

GLAISTER, J P. The impact of river discharge on distribution and
1978 production of the school prawn *Metapenaeus macleayi* (Haswell) (Crustacea: Penaeidae) in the Clarence River region, northern New South Wales. *Aust. J. Mar. Freshwater Res.* 29:311–323.

GUNTER, G and EDWARDS, J C. The relation of rainfall and
1969 fresh-water drainage to the production of the penaeid shrimps (*Penaeus fluviatilis* Say and *Penaeus aztecus* Ives) in Texas and Louisiana water. *FAO Fish. Rep.* 57:875–892.

GUNTER, G and HILDEBRAND, H H. The relation of total rainfall
1954 of the state and catch of the marine shrimp (*Penaeus setiferus*) in Texas waters. *Bull. Mar. Sci. Gulf Caribb.* 4:95–103.

HILDEBRAND, H H and GUNTER, G. Correlation of rainfall with
1952 Texas catch of white shrimp, *Penaeus setiferus* (Lin-naeus). *Trans. Am. Fish. Soc.* 82:151–155.

HUNT, J H, CARROLL, R J, CHINCHILLI, V and FRANKENBERG, D.
1980 Relationship between environmental factors and brown shrimp production in Pamlico-Sound, North Carolina. *N.C. Dep. Natural Resources Community Devel.*, *Div. Mar. Fisheries. Rep.* 33:1–29.

LUCAS, C G, KIRKWOOD, G and SOMERS, I. An assessment of the
1979 stocks of the banana prawn *Penaeus merguiensis* in the Gulf of Carpentaria. *Aust. J. Mar. Freshwater Res.* 30:639–652.

MORRISON, D F. 'Multivariate statistical methods'. McGraw-
1976 Hill, New York, 415 pp.

PRICE, A R G. Temporal variations in abundance of penaeid
1979 shrimp larvae and oceanographic conditions off Ras Tanura, Western Arabian Gulf. *Estuarine and Coastal Marine Science* 9:451–465.

RACEK, A A. Penaeid prawn fisheries of Australia with special
1957 reference to New South Wales. *Indo-Pac. Fish. Counc. Proc. 6th Annu. Sess.* 1955:347–359.

RACEK, A A. Prawn investigations in eastern Australia. *N.S.W.*
1959 *State Fish. Res. Bull.* 6:1–57.

RAO, K J and GOPALAKRISHNAYYA, C H. Penaeid prawn catches
1974 from Publicat Lake in relation to ingress of post-larvae and lake hydrography. *Indian J. Fish.* 21:445–453.

ROESSLER, M A and REHRER, R G. Relation of catches of
1971 postlarval pink shrimp in the Everglades National Park, Florida, to the commercial catches on the Tortugas grounds. *Bull. Mar. Sci.* 21:790–805.

ROTHLISBERG, P C. Vertical migration and its effects on disper-sal of penaeid shrimp larvae in the Gulf of Carpentaria, Australia. *Fish. Bull. U S* (in preparation).

RUELLO, N V. The influence of rainfall on the distribution and
1973 abundance of the school prawn *Metapenaeus macleayi* in the Hunter River Region (Australia). *Mar. Biol. (Berl.)* 23:221–228.

SOMERS, I F and TAYLOR, B R. Fisheries statistics (1968–1979)
1981 relating to the Declared Management Zone (DMZ) of the Australian northern prawn fishery. *Australian CSIRO Marine Laboratories Report.*

STAPLES, D J. Seasonal migration patterns of postlarval and
1979 juvenile banana prawns, *Penaeus merguiensis* de Man, in the major rivers of the Gulf of Carpentaria, Australia. *Aust. J. Mar. Freshwater Res.* 30:143–147.

STAPLES, D J. Ecology of juvenile and adolescent banana
1980a prawns *Penaeus merguiensis*, in a mangrove estuary and adjacent off-shore area of the Gulf of Carpentaria. I. Immigration and settlement of postlarvae. *Aust. J. Mar. Freshwater Res.* 31:635–652.

STAPLES, D J. Ecology of juvenile and adolescent banana
1980b prawns *Penaeus merguiensis*, in a mangrove estuary and adjacent off-shore area of the Gulf of Carpentaria. II. Emigration, population structure and growth of juveniles. *Aust. J. Mar. Freshwater Res.* 31:653–665.

STAPLES, D J and VANCE, D J. Effects of changes in catchability
1979 on sampling of juvenile and adolescent banana prawns, *Penaeus merguiensis* de Man. *Aust. J. Mar. Freshwater Res.* 30:511–519.

SUBRAMANYAN, M. Fluctuations in prawn landings in the
1964 Godavari estuarine system. *Indo-Pac. Fish. Counc. Proc.* 11:44–51.

VANCE, D J, STAPLES, D J and KERR, J D. Factors affecting year-to-year variation in the catch of banana prawns, *Penaeus merguiensis*, in the Gulf of Carpentaria. *Aust. J. Mar. Freshwater Res.* (in preparation).

WICKENS, J F. Prawn biology and culture. *Oceanogr. Mar. Biol.*
1976 *Annu. Rev.* 14:435–507.

WIESEPAPE, L M, ALDRICH, D V and STRAW, K. Effects of
1972 temperature and salinity on thermal death in postlarval brown shrimp *Penaeus aztecus*. *Physiol. Zool.* 45:22–33.

WILLIAMS, A B. Penaeid shrimp catch and heat summation, an
1969 apparent relationship. *FAO Fish. Rep.* 57:643–656.

YOKEL, B J, IVERSEN, E S and IDYLL, C P. Prediction of the
1969 success of commercial shrimp fishing on the Tortugas grounds based on enumeration of emigrants from the Everglades National Park estuary. *FAO Fish. Rep.* 57:1027–1039.

ZEIN-ELDIN, Z P and ALDRICH, D V. Growth and survival of
1965 postlarval *Penaeus aztecus* under controlled conditions of temperature and salinity *Biol. Bull.*, *Woods Hole* 129:199–216.

A note on environmental aspects of penaeid shrimp biology and dynamics

S Garcia

Abstract

Shrimps are short-lived animals living in highly variable inshore areas during the juvenile phase and are therefore subject to particularly strong environmentally driven variability in recruitment and stock size. This paper examines the likely consequences of this fact on the surplus yield production and stock-recruitment modelling underlining the high risk of generating artefactual models when the data series are short.

1 Introduction

Because of their very littoral distribution the coastal penaeid shrimps are exposed to important environmental factors of continental or marine origin. These factors may vary periodically with a high (diel, lunar, tidal cycles), medium (annual, seasonal cycles) or low frequency (climatic long term inter-annual cycles, sunspots, *etc*). They may also present trends when they are related to the effects of man's activity (pollution, land reclamation, *etc*).

The effects of these variations on the shrimps biology and dynamics are important and this paper is intended to underline their main consequences regarding the use and interpretation of production models and stock-recruitment relationships.

2 Environment and production models

Changes in environmental parameters are likely to affect recruitment and the extensive use of 'predictive models' for shrimps (see Garcia and Le Reste, 1981, for a review) indicates that the yearly abundance of these animals is, in fact, affected. When working with production models, these environmental effects appear as additional variance in the catch per unit effort/effort relationship used.

The fished shrimp stocks consisting essentially in one year class, the average abundance of any year is close to the equilibrium value and it can be assumed, if no important errors are involved in the measure of catch and effort, that the departures from the equilibrium curve are linked to environmental factors. An examination of the time series of the residuals will give some indication on the trend and amplitude of the variations from the expected equilibrium line and on the variability

pattern with time (see for example Garcia, 1983, or Garcia and Van Zalinge, 1982). An analysis of the correlation between the production model residuals and environmental parameters will show if a significant relationship exists between the two types of anomalies (Lhomme and Garcia, this meeting). When the residuals are important and well explained by environmental variables it would even be better to extract their effect before analysing the relationship between fishing effort and stock size.

When useful apparent correlations are identified, a multivariate regression between catch rates, efforts and environmental parameters is appropriate. The result will be a tridimensional (possibly multidimensional) 'production model' like the one proposed by Griffin, Lacewell and Nichols (1976) for *P. setiferus* in Louisiana

$$Y = 3365D^{-0.5064}(1.0 - 0.994746E)$$

where D = annual river discharge and E = fishing effort. This approach has been in fact generalized to other stocks of the Gulf of Mexico (Turner and Condrey, this meeting).

An important consequence of this type of situation is that a production model for shrimps, based on a limited set of data should always be considered with caution as it can lead to wrong estimates. *Fig 1* shows the apparent production curve for a stock when recruitment increases or decreases progressively because of 'environmental' reasons. It is clear that in the absence of knowledge of the actual curves the MSY is only one value among the possible ones, that f_{MSY} is biased and that it can be over or under estimated depending on the situation considered.

In terms of management, when the effect of an increase or decrease of effort is estimated from the apparent curves the actual results may be completely different from the expected ones, either lower or much higher, depending on the recruitment variability and on the trend. The apparent model does not anymore picture a reversible phenomenon.

3 Environment and stock-recruitment relationships

There are evidences that the level of recruitment is at least partly governed by environmental condi-

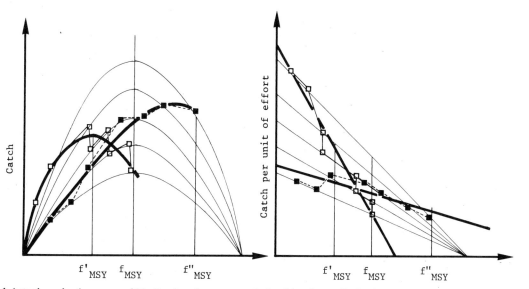

Fig 1 Actual production curves (thin lines) and apparent relationships (heavy line) when recruitment is affected by changes independent of fishing effort. Two situations are pictured, leading either to pessimistic f_{MSY} estimates (□) or to optimistic ones (■). All curves are drawn by eye

tions in the nursery areas (Ford and St Amant, 1971). There are also indications that recruitment is affected by the changes in the extension in nursery areas, (Doi *et al*, 1973; Browder and Moore, 1981; Ehrhardt *et al*, this meeting) by reclamation, damming, dredging of the lagoon passes, *etc*. In a strongly variable environment the shrimp annual recruitment should therefore be only loosely related to the parental stock size.

Very little was available before 1979 on the shape of the eventual relation. Neal (1975) and Le Reste and Marcille (1973) have speculatively suggested that it might be the case for shrimps. Hancock (1973) and Garcia and Le Reste (1981) underlined the lack of information on this subject.

More recently (Kirkwood, this meeting) has indicated that for a heavily exploited stock of *P. merguiensis* in Australia, the analysis of the residuals of the relation between rainfall and catches does not provide any evidence of trend in recruitment or recruitment overfishing despite the fact that the exploitation ratio has reached values of 0·75 to 0·85.

Bakun and Parrish (1981) indicate that, in general, when the nurseries have a limited and relatively constant biological capacity, the Beverton and Holt type of relationship is the most likely one.

The problem here is that we may have indeed a flat relationship of a Beverton and Holt type with an important environmental variability masking the existing relationship. In this last case the exact shape of the signal does not really matter because

it is the 'noise' which has to be studied and eventually understood for prediction purposes.

As in the case of production models, it is important to point out that when a trend in recruitment not related to fishing occurs, the results in terms of stock-recruitment relationship are strongly biased. It is obvious that with very short lived animals like shrimps which generation time is close to the basic time interval of one year, the spawners abundance is related to the abundance of the recruits which have generated it. There is a recruitment-stock relationship (see *Figure 2*, $R \rightarrow S$) which slope will depend on the level of fishing.

If the environment has an effect on the recruitment level, it means that instead of one stock-recruitment curve there exists in fact a family of them corresponding each to a different level of environmental constraints (*Fig 2*, actual $S \rightarrow R$).

If, in a limited time series, the recruitment starts decreasing (or increasing) it will be *immediately* (within 6 months) followed by a decrease (or increase) of the subsequent biomass and spawning stock. The result will be the existence of parallel (concurrent) decreasing (or increasing) trends in R and S and, of course, any attempt to relate directly the spawning stock in year n to the recruitment at the beginning of year $n + 1$ will show a relationship (see *Fig 2*, apparent $S \rightarrow R$) which is nothing more than one trajectory across a family of stock-recruitment curves. In fact, if the effort is stable and the shift from one $S \rightarrow R$ curve to the following is very progressive, the observed apparent $S \rightarrow R$

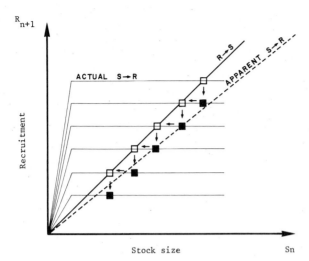

Fig 2 Actual stock-recruitment curves (thin lines) and apparent one when the recruitment drops progressively for non fishing related causes. Effort is kept constant for simplicity of the argument (only one recruitment-stock function). Progressively shifting $R \rightarrow S$ lines in relation to changes in the effort level will distort the apparent relationship and/or increase the variability without modifying the consequences. ■ = observed data points.

curve is in fact the $R \rightarrow S$ curve.

If effort and recruitment change at the same time, the apparent $S \rightarrow R$ given in *Figure 2* will be distorted and the apparent variability of the data increased.

The important consequence of this action of environmental changes on recruitment is that the observed trajectory may not be a stock-recruitment model but instead an environment-recruitment model for a given range of fishing mortalities. The 'model' observed *is not exactly reversible* if the level of fishing changes which means that there is no reason why in case the trends in effort or in recruitment become reversed, the reversion should follow back the same trajectory. In terms of management it means also that managing the stock (by reduction of fishing mortality, closed season, *etc*) in order to improve the spawning biomass may not give the expected results in terms of improved recruitment (because its decrease may not be due to fishing). The actual results may be worse or much better than expected. In case the recruitment trend is related to progressive destruction of the nurseries (mangrove deforestation, land reclamation) this type of management measure will simply fail because the model is not reversible at all and the current actual stock-recruitment curve may be among the lowest of the potentially possible ones.

With the above in mind, it is clear that the trends in recruitment (*Fig 3*) observed in Kuwait (Morgan

and Garcia, 1982), largely uncorrelated with fishing and possibly related to intensive land reclamation, or other wide scale environmental change may have produced an apparent and non reversible stock-recruitment relationship. The results obtained by Pauly (in Press) with data from 1963 to 1972 also show a downward nearly linear trend in recruitment from 1963 to 1968 and an upward one from 1969 to 1972. For the same reasons as above, the relationship given (*Fig 4*) may also be an artefact caused by trends in recruitment not related to fishing.

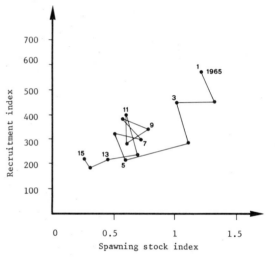

Fig 3 Apparent stock-recruitment relationship in Kuwait (from Morgan and Garcia, this meeting).

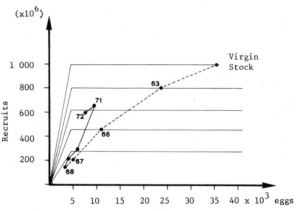

Fig 4 Stock-recruitment relationship in Gulf of Thailand penaeid shrimps (virgin stock is estimated by backward projection of the 1963–67 trend (from Pauly, in press). The dotted lines are entirely speculative and have been added by me to the original figure

270

Finally the stock-recruitment given by Ye (this meeting) for *Penaeus orientalis* raises the same problem because the data contain a positive trend in recruitment.[1]

The conclusion that can be drawn on this basis is that because of the fact that shrimp stocks are made by one single year class, there is a direct precise relation between the recruitment and the following stock. This relation will result with an apparent stock-recruitment relationship when a recruitment trend exists. This fact complicates seriously the stock-recruitment analysis where a limited number of years is available and leads to the necessity of using multivariate analysis involving environmental parameters.

---

[1]Other papers presented at the meeting on stock-recruitment of shrimps were not available to the author when preparing this note and will be analysed in a more detailed work on the subject (cf. Garcia, 1983)

References

BAKUN, A and PARRISH, R A. Environmental inputs to fishery
1981   population models for eastern boundary current regions. *In:* Report and supporting documentation on the workshop on the effects of environmental variations on the survival of larval pelagic fishes. *IOC Workshop Rep. Ser.*, (28):67–104.

BROWDER, J and MOORE, D. A new approach to determining the
1981   quantitative relationship between fishing production and the flow of freshwater to estuaries in Cross, R and Williams, D (Eds.) 1981. Proceedings of the National Symposium on Freshwater inflow to estuaries. Vol. I. USFWS. Office of Biological Services–FWS/OB5-81/04:403–430.

DOI, T, OKADA, K and ISIBASHI, K. Environmental assessment
1973   of survival prawn, *Penaeus japonicus*, in Tideland-I. Environmental conditions in Saizyo tideland and selection of essential characteristics. *Bull. Tokai Reg. Fish. Res. Lab.*, 76:37–52.

FORD, T P and ST AMANT, L S. Management guidelines for
1971   predicting brown shrimp, *Penaeus aztecus* production in Louisiana. *Proc. Gulf Caribb. Fish. Inst.*, 23:149–164.

GARCIA, S and LE RESTE, L. Life cycles, dynamics, exploitation
1981   and management of coastal penaeid shrimp stocks. *FAO Fish. Tech. Pap. No. 203:* 215 p.

GARCIA, S and VAN ZALINGE, N P. Shrimp fishing in Kuwait:
1982   methodology for a joint analysis of the artisanal and industrial fisheries. Paper presented at the workshop on the assessment of the shrimp stocks of the west coast of the Gulf between Iran and the Arabian peninsula, Kuwait, 17–22 October 1981. FI/DP/RAB/80/015:163p

GARCIA, S. The stock-recruitment relationship in penaeid
1983   shrimps: reality of artefacts and misinterpretations? *Oceanogr. trop.* 18(1): 25–48.

GRIFFIN, W L, NICHOLS, J P and LACEWELL, R D. Trends in
1973   catch/effort series: Gulf of Mexico shrimp fishery. *Tech. Rep. Texas Agric. Exp. Stn.*, 85 p.

KIRKWOOD, G P (in press). Modelling of the Gulf of Carpentaria prawn fisheries. Doc. presented at the Workshop on the scientific basis for the management of penaeid shrimps. Key West, 18–24 November 1981.

LHOMME, F and GARCIA, S (in press). Biologie et exploitation de la crevette penaeide *Penaeus notialis* (Perez-Farfante, 1967) au Sénégal. Doc. presented at the Workshop on the scientific basis for the management of penaeid shrimps. Key West, 18–24 November 1981.

LE RESTE, L and MARCILLE, J. Réflexions sur les possibilités
1973   d'aménagement de la pêche crevettière à Madagascar. *Bull. Madagascar*, (320):15 p.

MORGAN, G R and GARCIA, S. The relationship between stock
1982   and recruitment in the shrimp stock of Kuwait and Saudi Arabia. *Oceanogr. trop.* 17(2):133–137.

NEAL, R A. The Gulf of Mexico research and fishery on penaeid
1975   prawns. *In:* First Australian National Prawn Seminar, Maroochydore. Queensland, 22–27 November 1973, edited by Young, P C, Canberra, Australian Government Publishing Service, pp. 1–8.

PAULY, D (in press). A method to estimate the stock-recruitment relationship of shrimps. *Trans. Amer. Fish. Soc.*

YE, C C. The prawn (*Penaeus orientalis* Kishinouye) in Pohai
1984   Sea and their fishery. This volume.

Management

The limited entry prawn fisheries of Western Australia: research and management

B K Bowen
and D A Hancock

Abstract

Prawn fisheries are carried out in a number of discrete locations in the sounds and bays along the coast of Western Australia from Shark Bay north to Joseph Bonaparte Gulf. Biological and economic data have been collected since the beginning of the fishery in the 1960s and analysis by the standard production model methods show the stocks are heavily fished. Limits on the number of trawlers licensed to fish in the different areas have prevented the growth of gross over-capacity, and improved the economic performance of the fishery. Problems that remain include the growth of the effective effort per vessel, the difficulty of reducing effort if the optimum level is over-shot, the need to balance effort on different species in the same area, and the interaction with other fisheries.

Introduction

In 1973, available information on the prawn fisheries of Western Australia and the experience of ten years of their management by licence limitation was presented in detail in two contributions to the National Prawn Seminar in Queensland, Australia (Bowen, 1975; Hancock, 1974, 1975).

The present paper therefore attempts to provide an updating of the data presented then, together with additional analyses for stock assessment and more recent views and experiences of management regulations and of the limited entry system generally.

Procedurally, it has become the custom to review available fishery data as a basis for management decisions covering the ensuing three years (triennium), during which particular attention is given to the consequences of any additional licences permitted in previous years. There is a continuing research and monitoring programme which, together with the priority given to processing of data from the commercial fishery, allows early warning and, hopefully, correction of any untoward effects during the intervening period. Research, management, administration and inspection of limited entry fisheries in Western Australia are therefore dynamically involved on virtually a day-to-day basis.

The fisheries

The major fisheries are still to be found within the relatively enclosed bays and inlets of the central west coast of Western Australia (*Figure 1*). Shark Bay continues to be the major producer (*Tables 1* and *2*) with catches predominantly of western king prawns (*Penaeus latisulcatus*), smaller quantities of brown tiger prawns (*Penaeus esculentus*) and other minor species. Exmouth Gulf, an embayment some 320 kilometres farther north, supports a somewhat smaller fishery dominated by brown tiger prawns with smaller catches of western king prawns and endeavour prawns (*Metapenaeus endeavouri*) (*Table 1*). Penn (this volume) has recorded how the Exmouth fishery was initiated for banana prawns (*Penaeus merguiensis*) which are now of only minor importance there (*Table 3*).

There is, however, a small fishery for the latter species based on Nickol Bay, where smaller quantities of brown tiger and western king prawns are also taken (*Tables 1* and *4*). The most important species in the small coastal fishery off Onslow is the brown tiger prawn (*Table 1*) together with smaller quantities of other species and highly variable, but usually small, catches of banana prawns. During

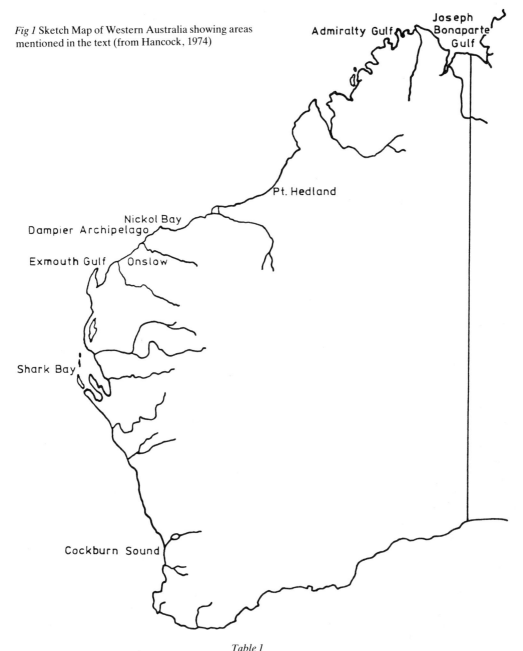

Fig 1 Sketch Map of Western Australia showing areas mentioned in the text (from Hancock, 1974)

Table 1
LANDINGS FROM WESTERN AUSTRALIAN PRAWN FISHERIES (KG × 10⁻³) – 1980

	Western king	Brown tiger	Endeavour	Banana	Total	Maximum No. of boats*
Shark Bay	1 398·3	252·7	16·9	—	1 667·8	35
Exmouth Gulf	215·9	646·5	191·1	S†	1 053·5	23
Onslow	5·2	72·5	5·1	0·1	82·9	9
Nickol Bay	21·4	35·4	1·2	118·0	176·0	13
Admiralty Gulf	—	—	—	S	S	
TOTAL	1 640·8	1 007·1	214·3	118·1	2 980·3	

* Maximum number of boats fishing during any month, but may be present for only part of the year.
† S – small quantities – no records available

Table 2
PRAWN LANDINGS – SHARK BAY, 1962–80
Catches are given in kilograms $\times 10^{-3}$, heads on. Effort is measured in hours trawled. The number of boats is the maximum number fishing during any month

Year	Western king Catch	Brown tiger Catch	Endeavour Catch	Total Catch	Total nominal effort	Boats fishing
1962	104·8	46·8	—	151·6	2 420	4
1963	359·4	243·7	—	602·1	9 898	22
1964	505·6	406·6	—	912·2	13 960	28
1965	442·8	396·8	—	839·6	17 861	28
1966	261·2	406·5	—	667·7	19 211	29
1967	227·8	673·3	—	901·1	31 644	30
1968	413·9	499·3	—	913·2	36 379	29
1969	798·3	460·3	—	1 258·6	37 210	27
1970	1 042·7	732·2	—	1 774·9	48 667	32*
1971	937·0	608·5	—	1 545·6	46 483	32*
1972	1 382·8	369·1	—	1 751·9	51 522	31
1973	1 185·6	636·1	—	1 821·7	51 474	33*
1974	1 432·6	667·7	—	2 100·3	51 814	32
1975	1 382·8	770·0	—	2 152·8	55 134	35
1976	1 510·8	771·4	—	2 282·3	61 340	35
1977	1 070·9	550·1	36·3	1 657·4	58 757	34
1978	1 370·6	729·1	12·7	2 112·4	57 244	35
1979	1 439·5	659·5	38·2	2 137·2	62 655	35
1980	1 398·3	252·7	16·9	1 667·8	57 786	35

NOTE: The 1980 data in this table may be subjected to minor revision following more detailed analysis
* Boats exceed number licensed through standby arrangements

Table 3
PRAWN LANDINGS – EXMOUTH GULF, 1963–80
Catches are given in kilograms $\times 10^{-3}$, heads on. Effort is measured in hours trawled. The number of boats is the maximum number fishing during any month

Year	Western king Catch	Brown tiger Catch	Banana Catch	Endeavour Catch	Total Catch	Total nominal effort	Boats fishing
1963	1·0	15·1	51·9	S	68·0	1 799*	12
1964	16·6	33·5	60·2	S	110·3	2·063*	6
1965	16·1	135·2	56·9	S	208·2	8 380*	13
1966	72·0	420·5	39·1	S	531·6	11 097*	15
1967	41·3	704·3	22·1	S	767·7	16 651*	17
1968	167·0	212·4	S	S	379·4	17 667	17
1969	76·7	472·9	S	104·9	654·5	26 245	17
1970	208·1	887·7	S	295·4	1 391·2	38 764	20
1971	135·1	233·7	S	150·0	518·8	29 706	20
1972	364·3	672·5	S	210·4	1 247·2	45 039	22
1973	278·1	596·1	0·4	277·2	1 151·8	47 296	22
1974	206·1	515·3	0·6	222·8	944·8	41 478	22
1975	312·1	1 239·4	2·3	450·4	2 004·2	45 066	22
1976	232·9	745·0	17·4	285·9	1 281·2	49 726	22
1977	340·5	639·2	1·2	236·5	1 217·4	51 035	22
1978	376·8	882·7	0·1	423·0	1 682·6	54 388	22
1979	271·5	572·4	S	328·4	1 172·3	51 097	23
1980	215·9	646·5	S	191·1	1 053·5	52 710	23

* estimates   S – small quantities only
NOTE: The 1980 data in this table may be subject to minor revision following more detailed analysis.

Table 4
PRAWN LANDINGS – NICKOL BAY AND ASSOCIATED AREAS 1966–80

Catches are measured in kilograms $\times 10^{-3}$, heads on. Effort is measured in hours trawled. The number of boats is the maximum number fishing during any month

Year	Banana	Brown tiger	Western king	Endeavour	Total	Nominal effort	Boats fishing
1966	10·6				10·6	344	1
1967	476·2				476·2	13 560	22
1968	53·8				53·8	3 219	9
1969	50·3				50·3	710	7
1970	120·9				120·9	2 446	10
1971	185·6				185·6	1 268	10
1972	49·2				49·2	401	5
1973	240·2				240·2	2 209	14
1974	180·6				180·6	1 164	15
1975	176·1				176·1	1 792	11
1976	452·2				452·2	1 655	15
1977	95·4				95·4	2 036	14
1978	92·6	19·6	5·1		117·3	4 700	13
1979	20·4	86·2	32·1	6·1	144·8	13 745	16
1980	118·0	35·4	21·4	1·2	176·0	8 827	13

NOTE: 1980 data may need minor revision following more detailed analysis.

the early 1970's negotiations were completed between the Government and a private company to promote a feasibility study of the commercial fishing of banana prawns in the Admiralty Gulf area (Hancock, 1974), but despite some encouraging catches there and in Joseph Bonaparte Gulf, these areas of the North West Shelf have been utilised mainly by trawlers proceeding between Western Australia and the Gulf of Carpentaria. Several other exploratory surveys off the west coast have failed to identify commercial quantities of prawns. A small stock of western king prawns occurs regularly in Cockburn Sound, where for various reasons it is protected from exploitation.

The fisheries of Shark Bay, Exmouth Gulf and Nickol Bay are managed by licence limitation.

Shark Bay

Table 2 gives details of the annual catches of prawns from Shark Bay since the commencement of the fishery in 1962. Minor variations from figures presented in earlier publications (Hancock, 1974; Slack-Smith, 1978) have resulted from improvement in analytical procedures.

A maximum catch of 2·28 m kg was recorded in 1976, with an average for the ten years to 1980 of 1·92 m kg. A lower than average catch of 1·66 m kg in 1977 was due mainly to smaller quantities of western king prawns, and of 1·67 m kg in 1980 to greatly reduced numbers of brown tiger prawns. The catch of brown tiger prawns was expected to show little improvement during the 1981 season (Penn, this volume).

There have been 35 vessels licensed to fish since

1975. Prior to this there were 25 (1963), 30 (1964–71) and 32 (1972–74). Between 1969 and 1972 most of the inefficient older vessels were replaced by new properly designed prawn trawlers. Between 1972 and 1977 replacement of the few remaining older vessels was achieved. Since 1977 most of the replacements have been larger (>21* m) vessels which qualify for a ship building subsidy (Penn and Hall, in prep.).

Exmouth Gulf

Although the catch exceeded 2 m kg in 1975 (*Table 3*) Exmouth Gulf catches are generally lower than Shark Bay (ten year average 1·23 cf 1·92 m kg) and are also more variable (ten year range 0·51 to 2·00 m kg compared with 1·55 to 2·28 m kg in Shark Bay). This variability is associated largely with catches of brown tiger prawns, the major species (ten year average 0·67 m kg). Western king prawns averaged 0·27 m kg and endeavour prawns 0·28 m kg. The latter will be an underestimate of the average annual catch, some of which was discarded when demand and prices were poor.

In Exmouth Gulf, 15 licences were issued in 1965, then increased to 17 in 1967, and again to 20 (+3 standby vessels) for 1970 and 1971. The standby arrangement was phased out in 1972 and the number of full licences increased to 22 for the period 1972 to 1978. For the current triennium, 1979–81, one provisional licence has been issued

*Loaded waterline length (equivalent to approximately 24 m overall length)

bringing the total to 23. This licence will be reviewed at the end of the 1981 season.

During the early years of the fishery, vessels were exclusively converted rock lobster boats, and it was not until the early 1970's that these were replaced by specifically designed prawn trawlers of approximately 19 m overall length (Penn and Hall, in prep.). Since 1979 four vessels have been replaced with larger vessels which qualify for the shipbuilding subsidy (Penn and Hall, in prep.).

Nickol Bay

Catch statistics for Nickol Bay are given in *Table 4*. Up to and including 1977, catches were based on banana prawns, but from 1978 brown tiger prawns, with smaller quantities of western king prawns, from more distant grounds have assumed importance. Catches have been very variable, with a ten year range of 0·02 to 0·45 m kg and an average of 0·16.

Boat licences in this fishery were limited in 1971 to 13. Since 1975 there have been 16 licensed vessels, though during 1980 only 13 boats were fishing at any one time. Fishing commenced in 1966 with one vessel followed by numbers ranging from 7 to 22 fishing in any one month during the years before limited entry.

Sources of data

Commercial fisheries production monthly return

It is mandatory for all licensed professional fishermen to provide monthly returns, which are collected by the Western Australian Department of Fisheries and Wildlife under the Fisheries Act 1905 and processed by the Australian Bureau of Statistics. Prawn fishermen are required to list, separately for each species of prawn trawled, the area fished (by 1° statistical blocks), the average number of shots per day, the average number of hours fished per day, the number of days searching (relevant only to banana prawns) and fishing, and total weight landed during the month, in the categories of 'whole' or 'headed'.

Prawn processors' monthly return It is mandatory for prawn processors to provide details of their suppliers (home anchorage, registered boat number, live weight supplied by them) and quantities sold.

Research logbooks Commencing in 1963, each prawn vessel skipper is expected to complete a daily research logsheet giving details of net type and specifications, fishing position at the start and finish of the day's fishing, together with, for each shot, the area listed (by 1/6° statistical blocks), named trawl ground, start time and duration of haul, depth at start and end, and weight of catch by species. An innovation in recent years has been to record the main fishing preference (intention to fish) for the day, by species of prawn, for scallops or with 'no preference'. These records proved to be less useful for analysis than those of habitat preference. The research logsheet scheme (Hancock, 1973) has continued to provide essential data for management. The ingredients for the success of this system (Hancock, 1974) have remained as:

(*a*) Sustained personal contact with the fishermen, and early verbal validation of anomalous records.

(*b*) Computer validation for spurious log book entries.

(*c*) Validation between log books and fishermen's and processors' returns available.

(*d*) Rapid availability of data for management.

(*e*) Prompt feedback of information to participants.

Rapid availability of data (*d*) is of considerable importance to management by limited entry and it will be noted that assessment can here be made to include fully analysed data for the last completed prawn season (1980), which would not, by necessity, be possible from available official statistics.

Factory measurements Commenced in 1963 to provide size frequency measurements of prawns landed from each fishing region in Shark Bay, which are required for weight to number conversions.

Monitoring the stocks From 1974 to 1976 annual surveys were made by research vessel, (Penn, pers. comm.) of prawn distribution and abundance in Shark Bay. In 1978, 1980 and 1981 these surveys were extended to include scallop fishing grounds. Similar surveys of the Exmouth fishery were carried out in 1975, 1976 and 1977. In addition, from 1978, the growth of juvenile prawns in Shark Bay has been measured aboard commercial and research vessels to monitor the effects of closure of an extended nursery area on the size of prawns entering the fishery and to determine an appropriate date for opening the extended nursery area to fishing.

Economic studies Conducted periodically and results published by the Commonwealth Department of Primary Industry (Anon, 1970, 1975; Meany, 1979; Owen, 1981).

Research programmes Continuing and past studies have included factors affecting the distribution of juveniles in nursery areas and of adult prawns, the distribution of spawning stock, transport of larvae, growth of juveniles, catchability and population dynamics for fishery assessment (see List of References). For the most part research on W.A. prawn fisheries has been undertaken by the State Department which has the constitutional authority for managing fisheries in State territorial waters. However, research on the banana prawn fisheries to the north of the State was, until 1974, undertaken by the Commonwealth Scientific and Industrial Research Organisation (CSIRO) (Hynd, unpub). That organisation has in recent years confined its prawn research to the fisheries of the Northern Territory and Queensland.

Information from the fishing industry In addition to the contributions made by the fishing industry to the data requirements listed above, there is a continuing verbal and written dialogue between the fishing industry and the W.A. Department of Fisheries and Wildlife and its Research and Inspection Branches.

Biology and life history

The biology and life history of species of importance to Western Australian prawn fisheries were described in a comprehensive review by Walker (1975). Full accounts have been given of the biology and life history of western king prawns in Cockburn Sound (Penn, 1975, 1976) of western king and brown tiger prawns in Shark Bay (Penn and Stalker, 1979) and of brown tiger prawns in Exmouth Gulf (White, 1975a and b). It is important to note that the western king prawn is generally a temperate water species extending north of Shark Bay and Exmouth Gulf. The brown tiger prawn is a tropical species which is at the southern end of its range in Shark Bay, while the southern limit of the distribution of banana prawns is at Exmouth Gulf.

The attached List of References contains titles of all publications concerning prawns in Western Australia, whether referred to in the text or not. Not listed are the unpublished works of the late J. S. Hynd (CSIRO) on banana prawns in W.A. and papers published by CSIRO on banana prawns in other parts of Australia.

Administration of the fishery and the research programmes

The prawn fisheries of Shark Bay and Exmouth Gulf are administered by the Western Australian Department of Fisheries and Wildlife under the Western Australian Fisheries Act. The purpose of the Act is 'for the regulation of the fishing industry and fish farming, and for the conservation and management of fisheries and aquatic animal and plant life, and for purposes connected therewith' and the regulating authority resides with the Minister for Fisheries and Wildlife.

The fishery was first developed in the early 1960's after a period of test fishing by a Departmental research vessel. Interest in the potential of Shark Bay as a prawn fishery was shown by at least one prawn fisherman from the east coast of Australia who was reported as having ideas of organising for up to 50 east coast prawn trawlers to travel to Shark Bay to take advantage of the resource potential. However, interest was also developing within Western Australia especially by the owners of a whaling company at Carnarvon situated at the northern end of Shark Bay. This station closed after the 1962 season because of a lack of whales due to extreme overfishing. The company was thus conscious of the overfishing problem in a general sense and urged the Government of Western Australia to take appropriate action to ensure that a cautious approach was adopted to the development of a prawn fishing fleet. This overture to Government was generated at least to some extent by the threat of large numbers of east coast prawn trawlers being brought to Shark Bay.

The response of Government was to take a decision to limit the number of prawn trawlers even though the fishery was very much in its infancy. Licence limitation as an aid to management was a relatively new concept, especially for a fishery which had not yet been developed, and it is interesting to speculate whether it would have been introduced if one of the parties interested in the prawn resource had not been forced to close down its whaling station because of overfishing and if there had been no report of a substantial fleet from the east coast moving to Shark Bay. A decision to limit entry to the Shark Bay fishery in 1963 was followed by a decision to limit entry also to the Exmouth Gulf fishery north of Shark Bay in 1965.

Nearly twenty years of experience has now been gained in relation to the administration of prawn limited entry fisheries, and its relative success has resulted in other fisheries being considered for limited entry. In retrospect, the developing Shark Bay prawn fishery was one which had many of the attributes required for the successful administration of a limited entry philosophy. Even so, its introduction has not guaranteed success in ensuring the maintenance of the prawn stocks at a consistently high level.

Some of the major attributes were:

(a) The whole of the life cycle, or at least that segment likely to be exploited, of the prawn species involved in the fishery was encompassed within the limited entry boundary.

(b) The fleet had not developed to the extent that there was already more catching capacity than required to successfully exploit the stocks.

(c) The stock had the potential to provide a resource which could be exploited by the authorized fleet for at least the major portion of the year.

(d) With the exception of a seasonal fishery on snapper, which was then being exploited only by limited entry rock lobster fishermen from farther south, there were no fisheries in the area to which the prawn trawlers might be attracted as a sideline activity and thus be a cause for concern by other fishermen.

Experience has shown that the administration of a limited entry fishery is much more difficult if any one of these four attributes do not apply. Experience from the Shark Bay prawn fishery and the rock lobster fishery has also demonstrated that there is merit in identifying maintenance of the stocks as the prime objective of a limited entry policy, whilst being aware that such a policy will have considerable economic ramifications, rather than set as a prime objective that of economic management.

Controlling the combination of size at first capture and fishing effort, and understanding the effect of fishing on the resource is difficult enough for a fisheries administrator, but more difficult still is the problem of coping with economic factors such as changing costs and market prices.

One of the advantages of a limited entry fishery is that there is a precise record of the operators who comprise the industry and they develop a responsible concern for the fishery of which they are a part. There is thus considerable scope for meaningful discussions between government and industry aimed at maintaining the fishery and gaining maximum benefit from the resource. Provided the research programmes can be shown to be competently carried out and directed towards the essential element of fisheries management, indus-try is receptive to changes in management strategy even though some changes will be of an experimental nature rather than as a result of a clearly demonstrated benefit.

The cost to the State management authority of administration, inspection and research specifically relating to the commercial prawn fisheries amounted to ca A\$183 000 in 1980–81, which represented 1·5% of the gross local value (A\$12·2 m) of the catch. Of this 8% was for administration, 29% for inspection and 63% for research, including data processing. It is of interest to compare these figures with the quite different figures for the W.A. rock lobster fishery in 1979–80 (Hancock, 1981) ie A\$1·35 m, representing 2·5% of the value of the catch, and comprising 18% for administration, 61% for inspection and 21% research. This reflects the lower total and percentage cost of administering and controlling the prawn fishery. Despite the higher percentage cost of prawn research it is less than half the absolute cost of rock lobster research. Total revenue to the Government from the prawn fishery amounted to \$113 000. This includes a small amount from personal fishing licences of \$14 and boat licence fees of \$30. The limited entry licence fees of \$1 800 per vessel in Shark Bay and Exmouth Gulf, and \$200 in Nickol Bay are paid into a trust fund for fisheries research and development.

Objectives of management

Various definitions of the management objectives for the Western Australian prawn fisheries have been given and referred to in the literature, eg Bowen (1975) listed the main considerations as economic viability of the fishing unit, economic viability of the processing establishment and the effect of exploitation on the prawn resources of the Bay. Hancock (1974) suggested that 'In terms of practical management the desired result will probably be a compromise between the attainment of the greatest benefit to the community in terms of maximising the yield from the fishery, and maintaining an adequate return on capital by the individual boats; ie somewhere between the maximum economic yield and the maximum sustainable'.

There may be merit in giving a more precise definition of the objectives of management for the limited entry prawn fisheries in the light of the added experience of management. This definition will need to take cognisance of the relatively recent view of the possibility of recruitment overfishing (Penn, this volume) as well as economic considerations.

278

In simple terms, then, the prime objective must be the maintenance of the resource at a level approaching the maximum sustainable yield, while giving proper attention to the economic viability of the fishing units with a view to maintaining a profitable industry.

In achieving this objective, a continuing flow of information is required on:

(*a*) the dynamics of the fishery, especially in relation to the possibility of growth and recruitment overfishing;

(*b*) the economics of the fishing units, including the value of the licence goodwill; and

(*c*) the economics of the processing and support facilities.

Management measures

Limited entry Limitation of the number of boat licences was introduced into W.A. as a management philosophy in 1963, when the prawn fishery of Shark Bay and the rock lobster fishery of Western Australia were made the first two limited entry fisheries in Australia, at a time when there were few examples of licence limitation in the world. Licence limitation was extended to the prawn fisheries of Exmouth Gulf in 1965, and to Nickol Bay in 1971. This form of management has been used for only two other fisheries in Western Australia, the Australian salmon (*Arripis trutta esper*) and abalone fisheries.

Closed season The length of the prawn season is not regulated, though summer (Nov–Feb) fishing is controlled by strong winds and low abundance of prawns.

Closed areas There is a permanently closed nursery area in Shark Bay, to which in recent years has been added the temporary closure of an extended area to allow juvenile prawns to achieve an acceptable market size before the closure is lifted in mid-April. In Exmouth Gulf, there is a permanently closed nursery area, and a temporary nursery area which is closed from 30 November to 28 February. This temporary nursery zone often remains closed to April by industry agreement.

Vessel size There are no regulations governing vessel size which is largely limited by economics. However, since 1977 there has been a tendency towards larger boats to take advantage of the Government boatbuilding subsidy for vessels over 21 m.

Number and size of nets In 1976, as a curb on future increases in vessel size, a limit of 14·63 metres (48 ft or 8 fathoms) was placed on the total headline length of trawls, and the number of trawls limited to two per vessel in Shark Bay and Exmouth Gulf. There are no mesh size restrictions for prawn nets, but mesh size is controlled for scallop trawling in Shark Bay.

Standardised and effective fishing effort

To be of any value in stock production models the unit of fishing effort needs to be apportioned to the individual species and to be standardised so that catch per unit of effort figures can be considered to be more representative of abundance as a time sequence.

Hancock (1974) summarised procedures for the standardisation of catch and fishing effort data leading to estimates of fishing effort, and Hall and Penn (1979) followed by Penn and Hall (in prep.) have modified these and presented them in greater detail, including the use of the data in stock production models.

In summary these procedures are as follows:

(*a*) Tabulate monthly catch by each boat, separately for each species.

(*b*) Tabulate monthly catch for each region (trawling area) of the fishery.

(*c*) Calculate relative fishing powers for each boat in the fishery using the method of Gulland (1956).

(*d*) For those boats which fished in any two successive years with the same skipper and engine, relate the relative fishing power in pairs of years to provide conversion factors to standardise fishing effort against 1973 effort levels.

(*e*) After standardising the fishing effort of each boat, obtain estimates of average catch rate for each species for each region of the fishery, using multiple regression to estimate catch rates in those regions which were not fished in some months.

(*f*) Convert catch rates and monthly catches to numbers using the average weight of the species for each region and month.

(*g*) Calculate the average catch rate (numbers/hour) for the whole area for each month using regional catch rates weighted by the area within each region. (As indicated by Hall and Penn (1979), further work is planned to apportion the effort directed towards each species, and subsequently weighting regional catch rates by area occupied by suitable habitat (sediment type within each region)).

(*h*) Estimate the average monthly catch rate for

the period from January to October.

(*i*) Divide this estimate into the total annual catch to give an estimate of annual effective effort for the species.

Penn and Hall (in prep) have defined this estimate of effective effort as 'The ratio of catch (in numbers) of the species to the average catch per standardised unit of effort, where the average catch rate is calculated over all areas for each month and then averaged over the months from January to October. This effort unit is corrected for the effects of concentration (spatial and temporal) of fishing effort on the stock.'

There remains to be borne in mind the effects of various unquantifiable influences. While vessel size, seasonal catch rates and, in some measure, area distribution of fishing effort, and any increases in net size up to the permitted maximum, have been given due weight in calculations, changing levels of operator efficiency and technology may be reflected in additional unrecorded changes in effective effort. It might be expected that the latter would now be of minor importance in the highly developed fisheries which have been operating for nearly twenty years in Shark Bay and Exmouth Gulf, but a recent tendency which is likely to increase is improved location of stock aggregations using sophisticated navigation aids.

Assessment of the fishery

(1) *Past history*

Hancock (1974) has described the introduction of the limited entry system of management into the prawn fisheries of Western Australia. The approach adopted in 1963 for Shark Bay and in 1965 for Exmouth Gulf was similar to that subsequently recommended by Gulland (1972) *ie*, to issue licences in some arbitrary number not expected to exceed the long term optimum, and by a process of iteration to gradually improve estimates of that optimum on the basis of data gathered, accordingly modifying (in this case increasing) the number of licences, in a stepwise fashion.

In Western Australia in recent years there has been a triennial review followed by any appropriate change in the number of licences. Hancock (1974) summarised the procedures for assessing the desirable number of licences for the 1975–1977 triennium, while Penn and Hall (in prep) have presented full details of a similar assessment to set the number of licences for 1978–1980. This included reference to economic data supplied by the Commonwealth Department of Primary Industry.

The recommendation was made that the number of Shark Bay licences for the 1975–1977 triennium should be 35, the additional 3 licences being issued as Provisional Licences with no obligation for renewal at the end of three years. This was to allow for a trial assessment of the practical and economic consequences of the increase before reaching a decision about the permanent status of the new licences, and recognized that a step-by-step approach need not always be forward, but should be flexible enough for a step backward should any unpredictable changes occur. In the event, from the assessment of data up to and including 1977, it was concluded that the indications of an expanding fishery were less obvious than from previous reviews and the option of further additional licences was not considered. The alternative of withdrawing the 3 provisional licences was rejected in favour of making them permanent. The current situation will be discussed below.

In Exmouth Gulf, a suggestion to increase the number of licences from 22 by 3 provisional licences for 1975–1977 was not adopted because of the past history of variable recruitment (*Figure 2*) and industry's consequent concern about economic security. However, one provisional licence was approved for the 1979–1981 triennium, bringing the total to 23.

(2) *Assessment based on raw data*

Figure 2 gives annual catch and total nominal (uncorrected) fishing effort data for Shark Bay and Exmouth Gulf, as a time series. *Figure 3* uses the same data to relate annual catch and nominal fishing effort, and provides a further eight years of data since Hancock (1975).

Several general conclusions can be drawn from these data:

Shark Bay

(*a*) While *Figure 3* gives a suggestion that total catch has continued to increase with increasing nominal fishing effort, *Figure 2* shows that catches have levelled off in recent years.

(*b*) Western king prawn catches appear to be increasing with increasing effort (*Figure 3*) but there has been little change in catch level over the past nine years.

(*c*) Catches of brown tiger prawns in Shark Bay, where it is the minor species, are quite variable and appear not to have increased for the past ten years, either annually (*Figure 2*) or as a function of increasing nominal effort (*Figure 3*). During this period there was an unusually low catch of brown tiger prawns in 1972, which was compen-

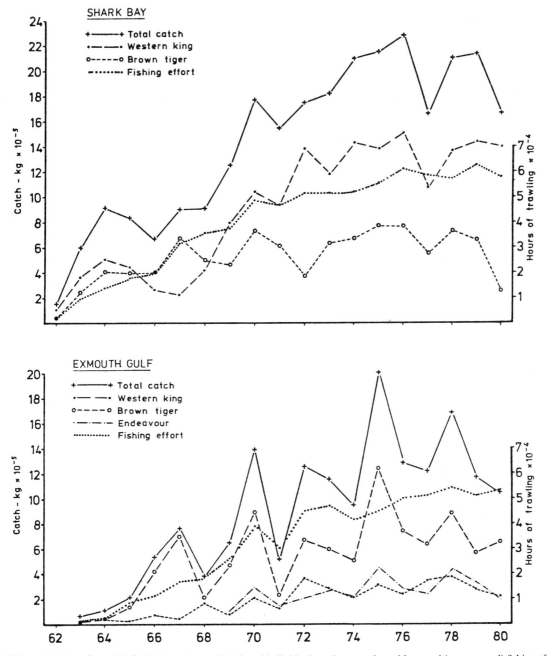

Fig 2 Annual catches (by weight) of all species combined, and individual species, together with annual (uncorrected) fishing effort (in trawling hours $\times 10^{-4}$) Shark Bay (1962–80) and Exmouth Gulf (1963–80)

sated for by good catches of western king prawns (*Figure 2*). Average catches followed until 1978 following which the catch of brown tiger prawns declined in 1980 to the lowest recorded since the commencement of this fishery. The 1981 catch of this species is also expected to be of a similar order.

(*d*) The low total catch in 1977 was obtained at an unusually low catch per unit of effort, and was a feature of catches of both species.

(*e*) Nominal fishing effort in Shark Bay has changed a relatively small amount since 35 vessels were permitted to fish in 1975. However it will be seen from sect. 3 that effective effort has shown a marked increase especially in recent years.

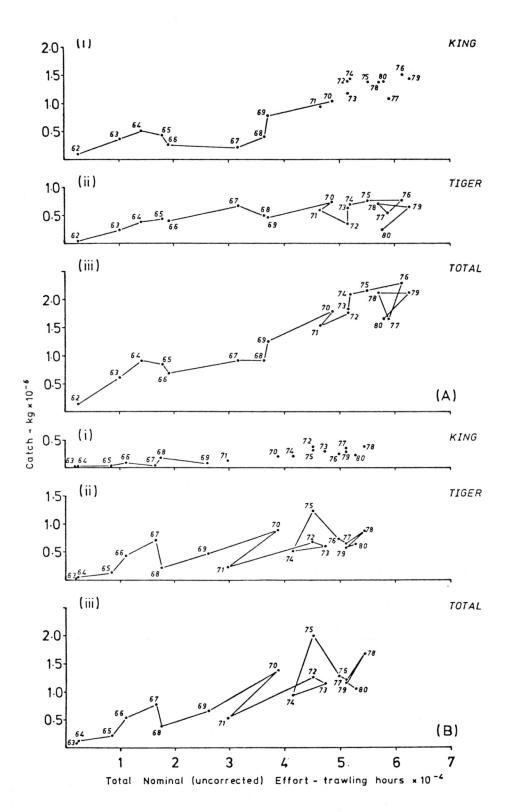

Fig 3 Annual catches by weight of (i) western king, (ii) brown tiger, (iii) total, prawns for (A) Shark Bay 1962–1980 and (B) Exmouth Gulf 1963–1980, related to annual nominal (uncorrected) fishing effort

Exmouth Gulf

(*a*) From *Figures 2* and *3* it can be seen that annual catches are much more variable than in Shark Bay. This variability comes mainly from the major species, brown tiger prawns.

(*b*) Total catches from Exmouth Gulf appear to have increased with increasing fishing effort (*Figure 3*), while *Figure 2* is more suggestive of a general decline since 1975. The catches of brown tiger prawns appear to have been mainly responsible for this decline, and when related to nominal effort the suggestion is that there has been no recent increase in catch with increasing effort.

(3) *Assessment based on effective fishing effort*
Figure 4 shows the estimates of effective effort for the two species calculated using Penn and Hall's (in prep) method described above, related to annual catches in numbers of each species. This analysis has so far been undertaken only for Shark Bay.

A range of functions was fitted to the data, based on the assumption of equilibrium for a single year-class stock. These were the generalised production model (GENPROD) (Pella and Tomlinson, 1969) the asymptotic yield model (*ie*, Genprod with $m = 0$), Fox's (1970) model (Genprod with $m = 1$) and Schaefer's (1957) well known stock production model (Genprod with $m = 2$). Residual sums of squares were calculated (*Table 5*), and approximate 95% confidence limits obtained using the method of Bard (1974, p. 204). Further details of methods are given in Penn and Hall (in prep).

The generalised production model appeared to given an acceptable representation for both species, and in the case of the western king prawn was virtually coincident with the Fox model (*Table 5*). For brown tiger prawns Genprod and asympto-

tic yield models gave similar results over the range of the data.

The fits of the generalised production model to the data, together with approximate 95% confidence limits have been given in *Figure 4*, together with fitted Schaefer curves. These will allow comparison with the results of Rothschild and Parrack (this meeting) for the Gulf of Mexico fishery.

Table 5 indicates a maximum sustainable yield (MSY) for western king prawns of around 50 million individuals for an optimum fishing effort of 90 000 effective hours trawling from the Genprod and Fox models, and 80 000 from Schaefer's. This compares with the current level of effective fishing effort of *ca.* 100 000 hours.

The MSY indicated for brown tiger prawns is around 21–22 million individuals (*Table 5*) for an optimum fishing effort of 62 000 hours (Schaefer) or 54 000 hours (Fox). (The Genprod fit had an *m* value close to that of the asymptotic yield function, and therefore produced a virtually infinite optimum effort figure). Effective effort exceeded 80 000 trawling hours in 1979 before falling to around 64 000 in 1980 at the time of the lowest catch of brown tiger prawns (*Figure 2*). It has been noted that low catches were experienced in 1972, but this was at a time of only modest fishing effort for both species. The 1980 result must therefore be viewed against the much higher level of fishing effort which preceded it, and the fact that 1981 is also expected to provide less than average brown tiger prawn catches.

Penn (this volume) has made the very important observation that the level of effort expended on the reduced available stock during 1980 could not have occurred without the availability of the second species on the same grounds which made the combined catchrate economically viable. In other

Table 5

RESULTS OF FITTING FOUR STOCK PRODUCTION MODELS TO CATCH (NUMBERS) AND EFFECTIVE EFFORT DATA FOR (i) WESTERN KING AND (ii) TIGER PRAWNS FROM SHARK BAY, 1962–80

	Genprod	Asymptotic Yield	Fox	Schaefer
(i) *Western king*				
m	1·058	0	1	2
Maximum sustainable yield (numbers $\times 10^{-6}$)	50·4	75·9	50·4	50·9
Optimum effort (hours trawled $\times 10^{-3}$)	89·6	∞	91·3	79·9
Residual sum of squares	6·993E14	7·366E14	6·994E14	7·253E14
(ii) *Brown tiger*				
m	0·0001*	0	1	2
Maximum sustainable yield ($\times 10^{-6}$)	22·5	22·6	20·9	21·4
Optimum effort (hours trawled $\times 10^{-3}$)	84·900	∞	53·7	62·2
Residual sum of squares	4·651E14	4·651E14	5·955E14	7·024E14

* Lower bound for m allowed in estimation run

words, a species may continue to be fished beyond the point here falling catchrates would otherwise have discouraged continued exploitation. High value species, such as the brown tiger prawn, would be specially vulnerable to this situation. More generally, Penn (pers. comm.) has noted that the fishing effort directed towards areas suitable for brown tiger prawns (the minor species) is sometimes relatively greater than for western king prawns. This may occur when western king prawn catch rates are poor as in 1966 and 1967 (*Figure 2*), or it may also be that the fishing effort directed towards brown tiger prawn habitat yields also western king prawns to give an eocnomically viable combined catch rate.

(4) *Economic assessment*

The definitions given under the section 'objectives of management' indicate clearly the need for an understanding of fishery economics. The prawning industry expects, and prefers, to make its own economic decisions. However, proper management of the resource must relate such economic decisions to the wellbeing of the stocks, which cannot be achieved from assessments based only on quantitites caught and fishing effort expended. Economic data and opinion provided by the Commonwealth Department of Primary Industry (Anon, 1970, 1975; Meany, 1979 and Owen, 1981) are therefore of the utmost importance to the triennial assessments of the fishery.

DPI's reports have given figures for percentage return on capital as 'market value, without concession', which are quoted as follows:

	1966–7	67–8	71–2	72–3
Shark Bay	29·2	25·8	21·6	25.2
Exmouth Gulf	37·2	32·7	loss	30·0

	73–4	74–5	75–6	76–7
Shark Bay	24·4	11·8	50·0	39·0
Exmouth Gulf	15·4	10·0	43·2	19·7

For Shark Bay, the latter three years surveyed, which averaged 33·5 percent on capital, ranged from 11·8 in 1974–75 to 50·0 percent in 1975–76. In 1974–75 the price of $1·83 per kg did not compensate for rising fuel and vessel costs, but by 1976–77 the price had doubled to $3·69 per kg.

The comparable percentage return on capital, when the resale value of the concession was included, for 1976–77 was 18·5 (cf 39·0 for market value without concession). A significant comment (Anon, 1975) for the 1971–72 to 1973–74 period was that within each financial year, rates of return were widely spread, with nearly 20 percent of trawlers achieving in excess of 50 percent return on capital each year, while a smaller proportion recorded losses on their operations.

Owen (1981) commented that the return on capital represents a higher rate than that available on 'gilt-edged' investments (Government bonds, *etc.*). However, noting the risk factor, *eg*, a series of low production years, low prices or damage or loss of a vessel, he believed the return must be considered 'reasonable'.

However, in the years since Owen's (1981) assessment *ie*, 1978–80. the additional operating costs related to fuel and related products, and declining prices (1980–81) for prawns resulting from exchange rate variations and competition in Japan between Chinese and Australian products, together have made the fisheries generally less profitable. These changes coupled with the over-capitalisation of concessions particularly through the building of larger vessels allowed by the highly profitable years of 1978–79, appear to have resulted in an overall decline in profitability which may be difficult to reverse.

The comparable figures for Exmouth Gulf (market value, without concession) for 1974–75 to 1976–77 were 10·0, 43·2 and 19·7 percent respectively, which are lower than for Shark Bay, but which to Owen (1981) 'demonstrate that the fleet has certainly achieved more than satisfactory returns over the survey period'. It should be noted (*Table 3*) that catches during the three survey years were at or above the 10 year average, and that annual catches from Exmouth Gulf are generally more variable than for Shark Bay. It can be seen from the table above that in 1971–72, when poor catches were experienced, there was a small average loss.

(5) *Current assessment*
Shark Bay

Several trends in the Shark Bay asssessment which need to be viewed with some concern are (1) the recent increase in effective fishing effort (*Figure 4*), by the same number of boats, resulting from increased net and vessel size, (2) the decline in brown tiger prawn catches, particularly since this has followed a decade of catches which have not increased with increasing fishing effort, (3) the current effective effort levels which, at least for western king prawns are in excess of the optimum required for MSY, and (4) the apparently less healthy economic situation of recent years which will reflect the consequences of (1) and (2). Short lived downturns in catches have occurred in previous years, *eg*, for western king prawns in 1965–68 and brown tiger prawns in 1967–69 and 1970–72,

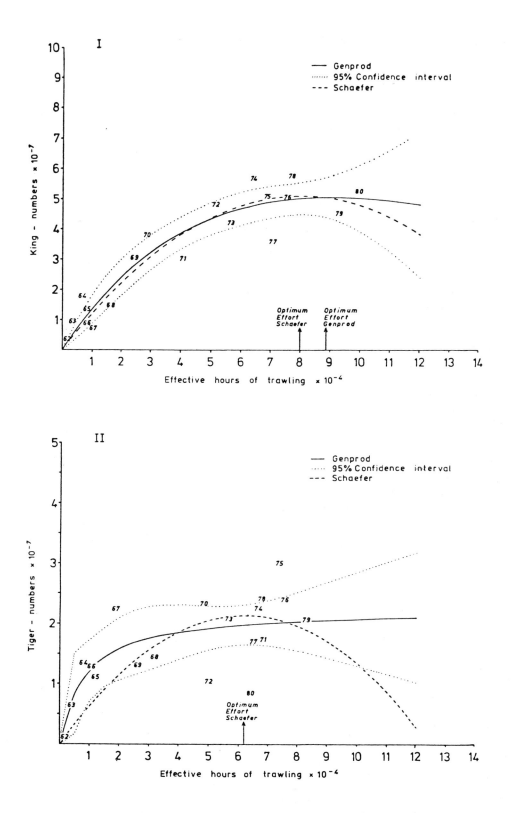

Fig 4 Relationship between annual catches in numbers of I western king and II brown tiger prawns, in Shark Bay, 1962–1980 and effective fishing effort. Curves fitted by Genprod, with appropriate 95% confidence limits, and by the Schaefer model are shown.

but these could usually be explained by increased effort for the alternative species, and in any case took place at lower effective effort levels.

In view of the above trends, which will be monitored and evaluated carefully, there is unlikely to be any justification for increasing the number of boat licences in the foreseeable future.

Exmouth Gulf
In Exmouth Gulf, where the fishery appears to be in a similar state of exploitation to Shark Bay, the situation is compounded by the greater variability of annual catches. Additional data analysis is required, but a preliminary view would be to discourage any further expansion of fishing effort.

Effectiveness of management measures

(*1*) The fishery has remained viable and annual catches from both Shark Bay and Exmouth Gulf have remained relatively stable in recent years despite increased effective fishing effort.

(*2*) The number of licences for Shark Bay was increased to 35 in 1975. During the following few years the level of effective effort remained less than the optimum effort required for maximum sustainable yield predicted by GENPROD for both western king and brown tiger prawns. However, despite an unchanged number of boats the effective effort rose during 1979, and, in the case of western king prawns, in 1980 exceeded the estimated optimum effort (*Figure 4*). Escalation of effective fishing effort within a limited entry fishery is also a feature of the Western Australian rock lobster fishery (Hancock, 1981). In the Shark Bay prawn fishery it will have resulted for the most part from the introduction of larger vessels with to some extent an increase in size of net (see below). However, at least up until 1977 the fisheries of both Shark Bay and Exmouth Gulf were considered to be economically viable (Owen, 1981).

(*3*) Temporarily and permanently closed nursery areas have been relatively successful in protecting the physical environment of nursery areas from trawling damage, but their value in limiting catches to a size range of commercial usefulness has been very much dictated by market demand *ie*, in years of demand for small prawns, operators concentrate on nursery borders whereas in other years they would have fished larger prawns more distant from nursery areas. Yield per recruit consequences of the different fishing strategies have not been computed.

(*4*) *Closed season* The need for a closed season for W. A. prawn fisheries has been reviewed from time to time but to date no obvious advantage has emerged. Hancock (1972) concluded 'there is no evidence to suggest that, under the conditions which have existed in the Shark Bay fishery in recent years, a close season from 1 September to the end of February (requested by fishermen) would have contributed significantly either to the survival of spawning females and subsequent success of recruitment, or to improved catches following lowered fishing effort on young prawns entering the fishery from nursery grounds.' This assessment still holds good, since recruitment to the trawl grounds does not usually occur before March (Penn, pers. comm.) and subsequently in number of spawners occurs before September (see also (*9*) below).

(*5*) *Number and size of nets* The maximum headline length and the restriction to two nets were introduced in 1976 as a measure to place an upper limit on the unit of effort (a similar restriction limiting pot size has been employed in the W.A. rock lobster fishery). The choice of headline length of 14·63 metres was however a compromise between the intention to control fishing effort from this cause and to accommodate existing gear. In the event there has been a tendency when replacing gear, despite a previous apparent preference for a smaller size (12·8 m), to upgrade nets towards the maximum permitted.

(*6*) *Vessel size and fishing power* The decision to restrict the number and size of nets was intended to discourage the building of large replacement vessels qualifying for the shipbuilding subsidy, which had previously (1971–73) proven uneconomic in both Shark Bay and Exmouth Gulf. However, the shipbuilding subsidy has still encouraged the building of larger vessels which have utilised the maximum net size permitted and caused significant increases in effective fishing effort especially in 1979–80 (*Figure 4*). The control on nets has therefore been only partially successful in containing fishing effort, but noting that there would undoubtedly otherwise have been a tendency towards more nets per vessel as in the Gulf of Carpentaria.

Research and management considerations

Research
(*1*) *Stock recruitment relationships* Boerema (MS) believed that there is little danger of a fishery for penaeid prawns developing to such a level that insufficient spawners remain. Owen (1981) concluded that 'Biologists generally agree that the penaeid species, because of their high fecundity

and short life span, withstand considerable fishing pressure without either lower recruitment or smaller size of prawns taken resulting in a reduction in total catch'. These widely held views, which put the emphasis on economic, rather than biological, management, are now being challenged (Penn, this volume; Rothschild and Parrack, 1984). Only future events will show whether the tiger prawn stocks in Shark Bay are showing recruitment overfishing as a consequence of excessive fishing effort.

(2) *Growth overfishing* Also becomes a possibility when excessive effort is concentrated into times (early season) and areas (close to nursery grounds) in which young, small prawns predominate.

(3) *Variability of data* Environmental variables *eg*, temperature, salinity, rainfall, cause variability in recruitment which can hinder interpretation of trends in catch and effort data as well as exaggerating any consequences of recruitment overfishing. Variability of catch data also seems to be a feature of higher effort levels.

(4) *Quantification of effective fishing effort* While seasonal and spatial catchability influences, vessel size/fishing power and net size trends can be allowed for in the estimate of effective effort, only qualitative recognition can be given to improved operator and technological efficiency. This could result in a failure to recognize insidious increases in effective fishing effort in stock production models.

(5) *Nominal and effective effort* It can be seen above that assessments based on nominal fishing effort can be quite misleading and mask changes in effective effort. Hancock (1981) similarly found for rock lobsters that yield analyses based on nominal effort gave much more optimistic results than for effective effort.

(6) *Stock production models* Traditionally accepted models have been fitted and, except where data are highly variable, *eg*, brown tiger prawns, in Exmouth Gulf, give reasonably consistent results. Their application to short lived species should give meaningful equilibrium representations. However, they must be considered to be *descriptive* rather than *predictive*.

(7) *Separation of fishing effort in mixed species fisheries* Hall and Penn (1979) have addressed this problem for Shark Bay, but use of the separated effort data requires an intimate knowledge of species requirements for, and distribution of, habitat within trawling grounds, which is not yet available for Exmouth Gulf. Knowledge of 'intention to fish' (see above) would then assume importance.

(8) *Nursery area closures* The experience of Shark Bay is that these are of value in protecting the physical environment but their practical effectiveness in adjusting commercial size has proved to be a function of current marketability. Yield per recruit predictions will remain essentially subjective pending useful natural mortality estimates, and will require an intimate knowledge of migration paths and seasonal fishing patterns.

(9) *Closed season issues* While the view has been that a closed season would serve little useful conservation purpose in Shark Bay, this will need to be reviewed carefully in the light of suggestions that penaeid recruitment may become affected by intense exploitation.

(10) *Collection of research data* The value of detailed information through a logbook system has been highlighted above, one of its greatest advantages being the timely availability of fully analysed data for management purposes. However, it is costly in labour and expense to maintain effectiveness. A recent attempt to reduce some of the detail required, *eg*, to use engine hours and landed weights rather than daily log records has proved inopportune (Penn, pers. comm.) because it unexpectedly coincided with a time of lowered brown tiger prawn catches which required investigation based on detailed data. On the other hand the annual cost of data collection of less than $70 000 from prawn fisheries valued at first sale at approaching $20 million seems well justified.

(11) *Additional licences* The quite arbitrary system of 3 year reviews and augmenting the number of licences by three, has been quite successful in providing fishing effort increments adequate for statistical interpretation, despite annual variability of stocks. However, the dangers in this practice of over-shooting optimum effort where effort per licence is not fully controlled are clearly to be seen, and not only is there a requirement for a statistically significant increment in fishing effort, but also for an adequate period of time for assessment (see also (2) below).

Management

(1) The problem of *escalating effective effort* within a fishery in which the number of boats is limited appears to be common (see also Morgan, 1980; Hancock, 1981; for W.A. rock lobsters). Clark (1980) reasons 'each licensed fisherman will be motivated to undertake any action that will increase his share of the catch.' This can occur contrary to economic wisdom, and in Western Australian prawn fisheries has manifested itself in increasing net sizes and larger, more powerful,

boats and to some extent, particularly during the early 1970's, prolonging the fishing season.

(2) Clark (1980) has commented that limited entry systems are prone to *overshooting MSY*, and experience from Shark Bay has shown that this will still happen, despite carefully modulated increases in the number of licences, unless the fishing unit is fully controlled. The importance of caution in increasing licences at a time of high fishing effort is underlined by the Shark Bay experience, where effective effort remained steady for four years when 3 licences were added in 1975 to give 35, but in the following two years (1979 and 1980) vessel replacements caused effective effort to increase greatly (*Figure 4*).

(3) Reduction of fishing effort within a limited entry fishery is not easy (see also for rock lobsters – Hancock, 1981) particularly where concessions have acquired a high transfer value. While the issue of provisional licences is a theoretical means of allowing a retreat from over-exploitation in a limited entry fishery, the withdrawal of such licences has yet to be put into practice.

In practical terms, once a provisional licence is issued it becomes politically difficult to withdraw that licence, except where the vessel involved has been taken from an economically successful fishery, *eg*, for rock lobster, to which it can be returned with no additional capital expenditure. In practice high capital cost is likely to be involved in providing specialist boat and gear. Such decisions therefore put a greater responsibility on data acquisition and analysis, and while acknowledging the need for full exploitation, the conclusion on issue of additional licences must be to defer decision making while there is any doubt about the outcome. This view is in accord with one of the principles drawn up at Airlie in 1975 (*cit.* Gulland, 1977) ie, 'Management decisions should include a safety factor to allow for the facts that knowledge is limited and institutions are imperfect'.

The alternatives to reducing the number of licences involve controls on individual efficiency, which must only be considered against the background of economic viability of the individual units, or reducing individual catching units to an economic size.

(4) The high market value of transferable licences has added substantially to the capitalisation of the industry. The result, particularly for new entrants to the fishery with large bank loans to service, will be for competition to achieve maximum catches with associated pressure on the stocks.

(5) The traditional view for uncontrolled fisheries is that at high levels of fishing effort, biological overexploitation is likely to be averted by boats leaving the fishery because of its economic unattractiveness, to fish alternative stocks. However, at levels which appear to be at least approaching overexploitation, especially for the minor species in Shark Bay, the industry was on the average, at least until the 1976–77 economic survey, apparently economically viable. More recent experience may differ from this view, and certainly the largest vessels have been economically less successful. It is also true that some years ago individual boats made losses within the limited entry fishery at a time when the average return on capital was good.

(6) Of concern is the knowledge that fishing may be maintained on low stock levels of one species by virtue of supporting stocks of another prawn species, and therefore contribute to overfishing in a situation which otherwise would be self regulating. Presumably any by-catch would have the same effect.

(7) Similarly a requirement for licensed vessels to continue to fish in the concessional area could be instrumental in maintaining fishing pressure on otherwise unattractive stocks. Vessels licensed to fish Shark Bay and Exmouth Gulf were obliged to fish in the concessional area from 1 May to 31 July but this regulation was rescinded in 1976.

(8) A factor of importance to understanding the implications of vessel size in W.A. prawn fisheries particularly, is that the duration of economic fishing season is inversely related to the size of the vessel (Penn and Hall, in prep). For example in Shark Bay outside of the peak catching season, catch rates fall below the economic break-even point of vessel operation. The larger vessels will therefore need to seek some alternative fishery to maintain economic viability. In earlier years the Gulf of Carpentaria provided this safety valve, but fishing for prawns there is now confined to vessels with a Gulf limited entry fishing concession and includes only a few Shark Bay boats. Apart from limited banana prawn stocks which have been located in coastal waters of Admiralty Gulf and farther north, exploratory fishing has so far failed to locate fishable stocks in Western Australian waters. Consequently there are now few alternatives but to lay up boats until the following season.

(9) Limitation of entry into selected fisheries, particularly if they are successful, seems to set in train two reactions by the fishing public (a) to demand more limited entry fisheries, often for quite inappropriate situations, and (b) to want licensed fishermen to be barred from alternative

288

open fisheries on the ground that they are already operating profitably and do not need further income. (a) is often a suggested way of achieving (b). So the scallop fishermen of Shark Bay would like to preclude prawn fishermen from fishing scallops, the snapper fishermen of Shark Bay believe licensed rock lobster fishermen should not be allowed to trap snapper, and so on.

The prime objective of a limited entry policy is the maintenance of the resource. Economic considerations follow, but the economic viability for individual operators is not guaranteed. Fishermen should therefore not be so constrained by an explosion of inappropriate limited entry fisheries to the extent that they lose the opportunity to operate on other resources. When the prawn limited entry fishery licences were first approved the scallop resource was not being exploited, and the snapper resource in the northern part of Shark Bay was being exploited by fishermen who already had a rock lobster limited entry authorization.

Subsequently, the Shark Bay snapper and scallop fishermen have made a decision to expend money to exploit the resources, being aware that they were already being exploited by rock lobster and prawn limited entry boats To now prevent the prawn and rock lobster boats from moving outside their prime fisheries would have an implicit overtone of Government guaranteeing their profitability from the limited entry fisheries. This would require a degree of economic management by Government to an impracticable degree.

(*10*) It is worth recording that the establishment of a 'limited entry fishery' at a time when fishing was profitable has brought with it a level of co-operation and mutual understanding between managers, research and the members of the limited entry fishery which is not always a feature with other sections of the industry. However, this brings with it a special responsibility for data collection, analysis, interpretation and implementation which has its response by the fishing industry in the high level of information provided and the exchange of ideas.

(*11*) The Inspection Branch has had its difficulties in attempting to police lines over water, *eg*, the boundaries of nursery areas, but these are common to all fisheries. However, in a limited entry fishery the potential for removal of a valuable licence should act as an additional deterrent.

(*12*) A limited entry system which freely permits transfer of concessions and associated capital value of a licence over and above the intrinsic value of the vessel has drawn criticism from some quarters (Meany, 1978). It is argued that excess profits

(economic rent) from a common property resource should somehow be better enjoyed by the community. Administrators of new limited entry fisheries will need to make a decision on conditions of licence transfer. However, the view in Western Australia is that fishermen who have enjoyed concessional rights since 1963, have, in the course of 18 years, by now earned them. New entrants are required to buy those rights, albeit at an inflated price depending on the current profitability of the fishery, and therefore gain nothing. However, the cost of the licence may make the cost of entry beyond that of potential boat owners, including experienced new fishermen wishing to buy boats of their own.

Conclusion

The comments of Bowen (1975) remain appropriate to the present discussion, *ie*,

'The use of licence limitation regulations as a fisheries management tool does provide a method of partially controlling the exploitation rate and capital input. For fisheries such as that on the western rock lobster and Shark Bay prawns it has been of benefit to the State and to the industry. However, its implementation requires close contact between industry, the scientist and the administrator.

If successfully operated, the measure provides economic security and high profitability, but it also impinges on personal liberties and freedom of choice of fishermen. The manager has to ensure that biological, economic and administrative understanding is at a level which ensures that the fisheries involved are fully developed, monopolies are not created, both Government and industry benefits financially, and that if new licences or authorisations are granted the system of selection is one which is accepted by those interested in obtaining entry into the limited fishery.'

Acknowledgements

The authors are indebted to Mr J Penn for critical help and advice with this manuscript, to Mr N Hall for data processing and statistical advice, to Mr P Rogers for data on administrative costs, to Mr M Cliff for editorial assistance and to Mrs M Isaacs for word processing.

References

ANON. An economic survey of the Shark Bay and Exmouth
1970   Gulf prawn fisheries 1966–67 and 1967–68. *Fish. Branch Dep. Primary Ind. Fish. Rep.* 6, 46 pp.

ANON. Western Australian prawn fisheries – an economic sur-
1975   vey. *Fish. Div. Aust. Dep. Agric. Fish. Rep.* 13, 53 pp.
BARD, Y. 'Nonlinear Parameter Estimation'. Academic Press,
1974   New York, 341 pp.
BOEREMA, L K. Provisional note on shrimp assessment and
MS   management. First draft, 6 June 1972, 13 pp.
BOWEN, B K. The economic and sociological consequences of
1975   licence limitation. National prawn seminar. Maroochy-
       dore, Queensland, 1973, 270–275.
CLARK, C W. Fishery management and fishing rights. Report of
1980   the ACMRR working party on the scientific basis of
       determining management measures. Hong Kong, 10–
       15 December 1979. Appendix 3, 101–113. *F.A.O. Fish.
       Rep.* 236, 149 pp.
FOX, W W, Jr. An exponential surplus yield model for optimis-
1970   ing exploited fish populations. *Trans. Am. Fish. Soc.*
       99, 80–88.
GULLAND, J. A. On the fishing effort in English demersal
1956   fisheries. *Fishery Invest. Lond. Ser.* 2: 20(5): 1–41.
GULLAND, J A. Some introductory guidelines to management
1972   of shrimp fisheries. IOFC/DEV/77/24, F.A.O.,
       IIORSDP, Rome, 12 pp.
GULLAND, J A. Goals and objectives of fishery management.
1977   *F.A.O. Fish. Tech. Pap.* 166, 14 pp.
HALL, N G and PENN, J W. Preliminary assessment of effective
1979   effort in a two species trawl fishery for prawns in Shark
       Bay, Western Australia. *Rapp. P-V Reun. Cons. perm.
       int. Explor. Mer* 175: 147–154.
HANCOCK, D A. Shark Bay prawn fishery – consideration of a
1972   closed season proposal. *Fishing Industry News Service,
       Western Australia* 5(2): 33–35.
HANCOCK, D. A. Why Log Books? *Fishing Industry News
1973   Service, Western Australia* 6(1): 3–9.
HANCOCK, D A. The basis for management of the Western
1974   Australian prawn fisheries. *Fish. Bull. West. Aust.* 14,
       23 pp.
HANCOCK, D A. The basis for the management of West Austra-
1975   lian prawn fisheries. National prawn seminar.
       Maroochydore, Queensland, 1973, 252–269.
HANCOCK, D. A. Research for management of the rock lobster
1981   fishery of Western Australia. *Proc. Gulf Caribb. Fish.
       Inst.* 33.
MEANY, T F. Should licences in a restricted fishery be saleable?
1978   *Aust. Fish.* 37(8): 16–21.
MEANY, T. F. Limited entry in the Western Australian rock
1979   lobsters and prawn fisheries – an economic calculation.
       J. Fish. Res. Bd Can. 36: 789–798.

MORGAN, G. R. Increases in fishing effort in a limited entry
1980   fishery – the western rock lobster fishery 1963–1976. *J.
       Cons. int. Explor. Mer* 39: 82–87.
OWEN, K E. Western Australian prawn fisheries. Shark Bay
1981   and Exmouth Gulf – a report of an economic survey.
       Fish. Div. Dep. Primary Ind. Fish. Rep. 36, 80 pp.
PELLA, J J and TOMLINSON, P K. A generalised stock production
1969   model. *Bull. inter-Am. trop. Tuna Comm.* 13, 421–496.
PENN, J W. Tagging experiments with the western king prawn,
1975   *Penaeus latisulcatus* Kishinouye. I. Survival, growth
       and reproduction of tagged prawns. *Aust. J. Mar.
       Freshwat. Res.* 26: 197–211.
PENN, J W. Tagging experiments with the western king prawn,
1976   *Penaeus latisulcatus* Kishinouye. II. Estimation of
       population parameters. *Aust. J. Mar. Freshwat. Res.*
       27: 239–250.
PENN, J W. The behaviour and catchability of some
1984   commercially exploited penaeids and their relationship
       to stock and recruitment. This volume.
PENN, J W and HALL, N G. Stock assessment of the Western
(in   Australian limited entry prawn fisheries with special
prep) reference to the 1975–77 triennium. *Fish. Bull. West.
       Aust.*
PENN, J W and STALKER, R W. The Shark Bay prawn fishery
1979   (1970–76). *West. Aust. Dep. Fish. Wildl. Rep.* 38: 1–38.
ROTHSCHILD, B J and PARRACK, M L. The U.S. Gulf
1984   of Mexico shrimp fishery. NOAA/FAO Workshop on
       the scientific basis for the management of Penaeid
       shrimp: Florida, November 1981.
SCHAEFER, M B. A study of the dynamics of the fishery for
1957   yellowfin tuna in the eastern tropical Pacific Ocean.
       Bull. inter-Am. trop. Tuna Commn 11: 247–285.
SLACK-SMITH, R J. Early history of the Shark Bay prawn fishery,
1978   Western Australia. *Fish. Bull. West. Aust.* 20: 1–44.
WALKER, R H. Australian prawn fisheries. National prawn
1975   seminar. Maroochydore, Queensland, 1973, Appendix
       1, 284–303.
WHITE, T F. Factors affecting the catchability of a penaeid
1975a shrimp, *Penaeus esculentus*. National prawn seminar.
       Maroochydore, Queensland, 1973, 115–137.
WHITE, T F. Population dynamics of the tiger prawn, *Penaeus
1975b esculentus*. Ph.D. Thesis, University of Western
       Australia.

Introductory guidelines to shrimp management: some further thoughts

J A Gulland

Abstract

The paper reviews and updates an earlier paper on the same subject. In the ten years since that was published there has been a greater realization of the variety of objectives that can be pursued, and of the need to identify and resolve possible conflicts between them. Basic assessment methods have not changed, but more can be done with length-structured versions of analytic models, and the influence of environment, especially man-made changes in coastal areas are better recognized. The needs for management procedures to be introduced early in the development of a fishery, and for some direct control of fishery effort, are re-emphasized. An outline procedure for management, which includes arrangements for continued review, is presented.

Introduction

Ten years ago, at the initiative of Jack Marr, then leader of the FAO/UNDP Indian Ocean Programme, I prepared a short paper on 'Some Introductory Guidelines to Management of Shrimp Fisheries' (Gulland, 1972). This attempted to provide in simple 'cook-book' form some assistance to the countries around the Indian Ocean, several of which were beginning to develop their shrimp

290

fisheries, on how they might proceed to obtain the greatest benefits from these resources.

The main points in that paper were that the dynamics of most shrimp fisheries could be adequately described by a simple Schaefer-type production model; that in the absence of controls fishing would rapidly expand until benefits were dissipated by excess costs, and that therefore controls should be introduced, if possible, before the fishery expanded beyond the optimum level. Possible methods of control were examined. Limits on the size of mesh in the net, or the size of shrimp landed, raised difficulties, and the aim of avoiding 'growth overfishing' might be better achieved by closed areas or closed seasons. Control of the total amount of fishing is more important in avoiding losses, and control in terms of effort (eg, numbers of boats licensed) was preferred to control in terms of catch (eg, setting a Total Allowable Catch). Finally a simple step by step procedure was suggested that would help a manager of a newly developing shrimp fishery to guide the development towards the optimum situation, with the amount of fishing in balance with the productivity of the stock.

While this IOP paper appears to have been useful, especially for some of the newly-developing fisheries round the Indian Ocean, it is increasingly clear that it does not provide a sufficient guide to tackling more than a few of the numerous problems of present day shrimp fisheries. The present paper has been written as a first step to a more comprehensive guide, examining first the new factors which were not included in the earlier study, and then considering to what extent the necessary management actions can be summarized in a series of fairly simple procedures.

New factors

Objectives

The original guidelines recognized that maximizing the gross weight of the catch was likely to be a less important objective than increasing the net economic return from the fishery. In fact in most shrimp fisheries the choice of objectives is more than just between MSY and MEY. Even in the simplest fishery many people – boat owners, fishermen, processors, retailers and consumers – as well as the public interest as a whole, are concerned, and only rarely will their interests coincide. In many shrimp fisheries there are different groups within each stage, eg, small-scale artisanal or sports fishermen catching shrimp in the inshore or lagoon areas, and industrial scale trawlers working the offshore grounds. The public interest may be measured in direct economic terms, eg, by the value added by the shrimp industry as a whole (ie, including all stages up to the retailer), but often the shrimp fishery is seen as one way of pursuing important national policy objectives, eg, earning foreign exchange, or providing employment in isolated and poor rural communities. A significant objective of higher public administration is that shrimp fishing is conducted in an orderly fashion, without the open conflicts between different interests (eg, between traditional and mechanized fishermen in India) or the need for the lengthy high level arguments on detail that have been occurring over the North Sea fisheries.

Many of these objectives are almost certain to conflict; maximizing employment in a fishery on a limited resource must mean keeping the earnings of the individual fishermen low; solving social problems by encouraging the small-scale inshore fishermen who tend to catch smaller lower-priced shrimp will probably lead to lower export earnings than might be obtained from encouraging offshore fishing; on the other hand rural fisheries will require considerably less fuel and investments (especially in foreign currencies) than industrial fisheries. This underlines the fact that the objective will seldom be simply to maximize some kind of output, but will try to optimize the balance between inputs and outputs. As much attention needs to be paid to the various inputs, and ways of reducing them, as to increasing outputs.

Such conflicts and inconsistencies must be accepted. It is unrealistic to expect that economic or social analysis will result in a magic formula that will replace MSY or MEY by some other single yard-stick against which the success of a management scheme can be measured. Managers, and those providing advice to managers, must accept that they are pursuing multiple objectives. The management policies must be chosen so as to achieve some acceptable and balanced degree of progress towards the objectives as a whole. This is a more complex task than just maximizing the total yield, but is a common task of management (in the broad sense). In the fisheries field particular attention to this problem has been given by the group at the University of British Columbia in relation to the very varied objectives in salmon management. The first step, and a most important step, in management is therefore to identify what are all the objectives to be pursued (and not merely the more obvious ones), so that they can be borne in mind in the subsequent stages of analysis, decision-making and implementation.

Resource assessment

The description of the essential dynamics of a shrimp stock by a simple yield-effort curve, and the assessment of such a stock by applying a simple production model is proving unsatisfactory for many purposes. While the simple models will undoubtedly continue to be used, and to give valuable insight into the nature of the problems, and of the actions that are needed especially when information is scarce, there are many situations when a more complicated model is needed.

In part this is associated with the complexity of objectives already noted. If the manager and his scientific adviser have to deal with conflicts between different groups of fishermen catching different sizes of shrimp, an age or size-structured model is essential, probably one that takes account of the pattern of movement between the grounds fished by the two groups of fishermen. Detailed models are also needed to look at some of the other objectives, *eg*, changes in value, which are affected by the size composition of the catch.

Age-structured models, including those that can incorporate variations in mortality from month to month, now exist, from which the yield-per-recruit, or other measures of output (catches by age- or size-class) can be obtained (but always as a function of recruitment) for any desired pattern of input, *ie*, fishing effort/mortality in different fisheries, seasonal closures, *etc*. In any one fishery the information on some of the parameters (*eg*, of growth or natural mortality) may be sparse, but there seems to be enough similarity between shrimp stocks for an extrapolation from one stock to another to give results that are usable, at least to a first approximation.

The need to modify the simple models also arises from the complexities of the natural system, in particular the fact that conditions are not constant, and that recruitment may be affected by fishing. These interact; the natural variation in the system does make it more difficult to detect whether the recruitment is falling because of a fishery-induced reduction in spawning stock.

Apart from obscuring relations of greater interest, natural variations may be of minor significance. In determining policies it will often be sufficient to look at the results of alternative management actions using the mean, or most probable, values of the parameters subject to variation, *eg*, annual recruitment. However stochastic modelling of populations show that the average value of the outputs (*eg*, annual catch) using an analysis that takes account of variation will not necessarily be the same as the output using constant values.

Further, the nature and extent of the variation to be expected is itself an output that can be of practical interest. An isolated community with little alternatives may be more concerned that their catches and incomes do not fall below a certain value, even in the worst year, than increasing the average catch over a long period. Problems of marketing and distribution may mean that catches that vary little from year to year are worth more than higher but more variable catches.

The most important variation is in recruitment, and it is more important to look at some of the causes of variation in recruitment than to incorporate an element of random recruitment variation in a stochastic model. The causes of change can be classed as: natural changes in the environment; man-made changes in environment (*eg*, mangrove cutting); and changes in the abundance of the spawners. Though this is probably the order of increasing practical importance to the administrator and policy-maker, it is also probably the order of decreasing attention from scientists, at least as reflected in the literature (for a good review of the literature on this and other points the reader is referred to Garcia and Le Reste, 1981). Within the first class changes in recruitment have most often been related to rainfall, river run-off or similar factors. The biological explanation is reasonably straightforward, and elucidation of this can enable better forecasts of catches to be made for the forthcoming season, as well as reducing the unexplained variation about, say, a stock-recruit regression.

Since the immediate coastal zones (estuary, lagoons, *etc*.) are the main nursery areas for shrimp, and also zones particularly liable to severe change from man's varied activities (reclamation for agriculture, or housing, pollution, *etc*), shrimp fisheries are likely to suffer from these varied non-fishery activities. Particular concern has been voiced in relation to the cutting of mangroves. In Southeast Asia this destruction – for timber, construction of fish ponds, *etc* – seems to have accounted for a high proportion of the original area of mangroves. Direct evidence of damage to fisheries is, however, sparse, though this is more an indication of the poor supply of data than of the lack of damage. Established correlations between mangrove area and shrimp production (Turner, 1977) would suggest that extensive damage is quite probable.

A possible relation between recruitment and the size of the spawning stock has received remarkably little attention. The implicit assumption has been that recruitment is independent of adult stock, *ie*, the average recruitment is the same at all sizes of spawning stock. The scientific justification for this is obscure, though a common reason is that by making it, analysis is made much simpler and there has so far been no obvious case of recruitment overfishing among shrimp stocks. In this the state of shrimp assessment practice bears some similarity with that of herring assessment in the North Atlantic in the 1960's. A real problem is that measurement of the spawning stock is not easy. The indices of abundance, *eg*, catches per unit effort, usually available refer to the average over a season. Since the shrimps that spawn are mostly those that survive through to the end of the season, such an index can overestimate the spawning stock to a serious extent, which will increase with the intensity of fishing. Even when not biased, the size of the spawning stock will not usually be estimated accurately. As Walter and Ludwig (1981) have pointed out, such variability can make it difficult to recognize any underlying stock-recruitment relation, or the existence of recruitment overfishing until it has become serious. Recently shrimp scientists have paid more attention to the possibility of average recruitment being affected by adult stock. In view of the serious practical implications of such a relation, if it exists, even more attention should be paid in the future. A further discussion of the stock-recruitment problems and the difficulties of establishing the true relation, is given by Morgan and Garcia (1982).

Despite the advantages of age-structured models over the simpler production models, the latter are being widely used, and will undoubtedly continue to be used. Partly this is due to the greater convenience, but partly because in many fisheries the available information is sufficient only for such simpler models. It is therefore worth considering how reliable these models are, and how they might give misleading results.

As generally applied they relate the catch (or catch per unit effort) in one season to the effort applied during that season. Given that the fishery during each season is based largely on a single brood of shrimp that recruit around the beginning of the season the observed yield/effort curve will, unless recruitment varies, be like the yield per recruit curve. Since, in the offshore fisheries on larger shrimps, for which most assessments have been made, the potential for further growth of individual shrimp is only moderate, the yield-per-recruit curve is flat-topped. It is therefore hardly surprising that the empirically derived yield-effort curve is often found to be flat-topped.

The problem is that, as with all empirical analysis, examination of the relation between catch and effort is good for explaining events over the range of values experienced, but can be misleading when extrapolated beyond that range. In fact extrapolation of the yield-per-recruit curve is reliable, and all will be well, if recruitment does not vary. However, if recruitment can be affected, extrapolation can be dangerous. As my colleague, Serge Garcia, has pointed out, in such cases a small increase in effort can cause a catastrophic fall in catch. Here I should draw attention to the experiences of fishery scientists in the North Sea and the contrast between their analyses of herring on the one hand, and cod and plaice on the other. Cod has a yield per recruit curve that is clearly peaked, so that before there was a threat from recruitment overfishing the scientists were able to point to clear evidence of growth overfishing, and warn about the dangers of too high effort. For herring, the yield-per-recruit curve is flat, growth overfishing cannot occur, and the scientists had no danger signals to point to (and indeed some were saying that herring could not be overfished) until shortly before recruitment failed and the fishery collapsed. I do not want to push the analogy too far, but it is at least disturbing and suggests that the results of production models should be used with great caution in predicting the effects of further increase in the amount of fishing.

Management actions
The original guidelines reviewed the expected usefulness of the traditional methods of management for controlling shrimp fisheries. Most of these have held true, though the evaluation that mesh regulation 'does not appear to be very useful in a shrimp fishery' needs some qualification and expansion. In the narrow sense of making minor adjustments to the effective size at first capture it is probably true. Difficulties of enforcement have tended to make mesh regulations in fin-fisheries ineffective except where the fishermen fully support the measures The selection of shrimp is so imprecise, *ie*, so many larger shrimp would be released by a mesh size large enough to release most 'undersized' shrimp that it would be difficult to persuade fishermen that they should use a larger mesh. On the other hand, shrimp fisheries catch many fin-fish, and may also use extremely small mesh sizes. There are several ways of tackling the 'by-catch' problem, but where large quantities of small juveniles of potentially valuable species are caught – either to be discarded dead, or to be brought ashore and sold as 'trash fish' – the use of a larger mesh size could be beneficial to the economy

as a whole irrespective of the effect on shrimp. The changes envisaged – from stretched meshes of perhaps 20–25 mm, up to 40–45 mm – should have immediately observable benefits – reduced drag, increased proportion of valuable species in the catch – so that they can be readily demonstrated to the fishermen, *eg*, by using two similar boats with different mesh sizes.

This question of the value of mesh regulation is just another example of the multispecies nature of most shrimp fisheries, and the need to assess the value of an action in the light of its impact on all species. There may be better ways of protecting juvenile shrimp (*eg*, closed areas/seasons), and it is the possible impact on fisheries of fin-fish that can make the use of mesh regulation attractive.

The conclusion that management must include control of the total fleet capacity has certainly been borne out by later practical experience, as well as by more fully developed theory. Without control of capacity, *ie*, some method of limiting entry the potential net benefits – which are large in many shrimp fisheries – will be dissipated in one way or another (*eg*, high fuel costs). At the same time it is clear, especially from a number of meetings held in the USA, Canada and Australia, that limiting entry is a complicated business, even on a simple fishery with not too wide a range of type and size of vessel engaged in it. It is therefore not just a matter of deciding to limit entry, but of deciding what form of limitation is appropriate to the conditions of the particular fishery. Since substantial benefits should be generated, attention should be given to how these benefits should be allocated (existing fishermen, consumers of shrimp, local communities, or the country as a whole). An explicit allocation may not always be made, but the form of limitation will always have consequences which will result in implicit allocation to one interest or another – unless the benefits are dissipated by excessive administrative and enforcement costs.

A serious omission in the guidelines was methods of influencing the development of a shrimp fishery other than the typical 'management' controls and regulations. There are many other actions of governments and similar agencies (including offices other than the fishery departments) – investment incentives, tax reliefs, subsidies, *etc* – which usually have much more influence on the level of fishing effort (and particularly of fishing capacity) than the management measures in the narrow sense. As has been pointed out several times, notably by a working party of ACMRR (FAO, 1979, 1980), it is important to take a wide view of the fishery management, and

for governments to use all the tools available to them to move the fishery in the desired direction. It makes no sense, in an over-exploited fishery, for one department to be applying controls (quotas, closed seasons, *etc*) to deal with the effects of excess capacity, when another is granting subsidies which, while helping with the short-term problems of fishermen, do nothing to discourage excess capacity, and continue the long-term problems.

The fact that subsidies, by reducing the incentives for excess capacity to leave the fishery, have the long-term effects of maintaining or worsening the problem they are meant to alleviate, is a good example of the need to look at the interactions, and long-term effects of all fishery decisions. Management measures are particularly likely to distort the pattern of fishing as fishermen seek to maximize their personal returns within the constraints set by these measures. These distortions will usually, though not necessarily always, reduce the efficiency of the fishery, and hence reduce the benefits from the regulation. Catch quotas, without added input controls, *eg*, allocations to individual fishermen or fishing companies, typically lead to shorter and shorter open seasons. Limits on the number of boats lead to more powerful vessels. Limits on the size (*eg*, length) of vessel can lead to peculiar designs (in the extreme almost round vessels) so as to maximize catching capacity within the specified rules. Some distortion of the pre-regulation pattern of fishing must be accepted, but enough analysis of the likely response of fishermen to proposed rules should be made to ensure that the distortion is not such as to nullify most of the expected benefits from the regulations. Distortion is likely to be reduced if licensing scheme recognized many categories of vessels (on the basis of size, horse-power, *etc*) and allows some movement between categories, provided the overall fishing capacity is not increased, *eg*, building a new large vessel has to involve scrapping three small vessels.

One response of many fishermen to regulations is to avoid complying with them. It is therefore essential, before introducing any regulation, to consider how it will be enforced. The most important action is to explain to the fishermen reasons for the rules – or better, to involve the fishermen in the discussions leading up to the formulation of the rules. If they do not believe in the rules, enforcement is likely to be nearly impossible, but even if they do, some control will be needed. Some rules, *eg*, on closed areas, can only be enforced by control at sea on the fishing grounds; others, *eg*, total catch quotas, require continuing close monitoring of all landings, while for others, *eg*, size limits, random checks at main landing places may be sufficient. The difference in costs between special patrol vessels and aircraft on the one hand, and a few port inspectors on the other, is very great, and the costs of enforcement (and the degree to which it may be possible to achieve an effective degree of enforcement at all) is an important, though generally neglected, aspect of choosing between alternative management measures.

Suggested procedures

Objectives

Since it is no longer correct to assume that the aim of management is simply to maximize the total catch, the first step in management should be to identify what is the aim. Since further the aim is likely to be complex, with varied and incompatible objectives, this identification must include at least some qualitative idea of how conflicting objectives should be weighed against each other. To what extent should the wish to improve the well-being of individual fishermen be favoured at the expense of providing employment for as many fishermen as possible? In the ideal world it might be possible to express all the objectives on some common scale, or at least to provide a quantitative weighting between different objectives, such that it would be possible to say that some particular policy would be the best, ie, maximize the 'benefits' obtained, as expressed in terms of the combined objectives.

This is to ask too much; in practice the choice has to be made by the policy maker, perhaps at a high government level, in the light of current national objectives, and political and social pressures. The best that the technical fishery adviser can do is to try and ensure that the policy maker is provided with the best information, ie, the costs and benefits to be expected from alternative policies. The essential first step then reduces to making the correct identification of what these costs (inputs) and benefits (outputs) are, to ensure that no important ones are left out; if possible to rank them in a rough order of importance; and to determine where they may conflict. The basic list is likely to be similar in all countries, but the ranking will vary very much from country to country. The inputs (costs) are likely to include capital (probably distinguishing local financing from that needing foreign exchanges, eg, for engines); labour (in which the opportunity costs, ie, the alternative jobs, would often need attention); running costs (especially the foreign exchange costs of imported fuels); and the administrative costs (including those of enforcement). Outputs (benefits) include the size of the catch (gross weight; gross value); economic indices (total value added; gross (or net) earnings of foreign exchange, prices of shrimp to the consumer, etc), and social measures (employment, especially in isolated communities; reduction of conflicts; contribution to national food supply).

Assessments

Given these inputs and outputs, the next task is that of expressing the outputs as a function of inputs. In the simple old-fashioned approach this implies no more than the biological assessment of the resource, and expressing the total catch (output) as a function of the amount of fishing, or fishing effort/fishing mortality (input). A biological assessment is still essential, but needs to be widened to take account of the greater variety of inputs (both biological, eg, in the sizes of shrimp caught, and socio-economic), and in the variety of outputs.

In the previous section the weaknesses of the production model were stressed. Here it should be stressed that in any shrimp fishery the biologist has to provide the manager with some advice about the resources, updated at regular intervals. If there is only enough data to use production models, then production models should be used, though the conclusions and advice should be set out in a way that make clear the possible limitations (this really should be done for any model, and the advice based on it). At the same time steps should be taken to improve the data base, eg, to collect size and growth data. In the extreme case there may not be even enough data to apply production models, eg, the fishery has only existed for a short time, during which the effort has changed very little. Even then it should be possible to make some sort of assessment which will provide the policy maker with guidance that is better than if he was given nothing scientific, and relied only on guesswork and optimism. For instance a rough idea of the potential yield, and the number of standard trawlers required to give full exploitation, can often be gained from looking at the size of the ground (or perhaps of the adjacent mangrove areas, cf, Turner, 1977) compared with other, and better known shrimp fisheries. Similarly the usefulness of an age-structured model, as applied to fisheries for which data are limited, can be increased by using, to a first approximation, parameters (eg, of growth or natural mortality) estimated by comparison with better studied stocks. Results from the application of age-structured models to data-rich fisheries can also be used to provide insight into the likely behaviour of data-poor fisheries, eg, on the likely form of the interaction between inshore and offshore fisheries, even when no formal assessment of the latter is made.

The two-dimensional (yield as a function of the amount of fishing) biological assessment has to be widened to take account of the extent to which fishing mortality may vary with the age, size or other characteristics. In principle this could involve an $n + 1$-dimensional analysis, if there are n distinguishable biological classes. In practice it will be sufficient to follow the conventional yield-per-recruit analysis, and add one dimension (age or length at first capture, or some surrogate for it, such as mesh size, or opening date of the season). Alternatively, if there are separate fisheries on each group, then the yield can be expressed as a function of the possible combinations of fishing effort, ie, an $x + 1$ dimensional representation, where x is the number of fisheries (cf, Figure 82 of Garcia and Le Reste, 1981).

The biological assessment needs also to present some information related to the variability of the system. The simplest is probably an estimate of the variance of the yield under any particular fishing regime as well as the mean yield. Other information may also be useful, eg, the probability that the yield in a particular year will fall below a given level.

The final complication to be included in the biological assessment is to take account of all the stocks, of both shrimp and fin-fish, that might be affected. To a first approximation this can probably be done by adding together each individual species (or stock) analysis. In doing this the effects on non-target species must be included. In particular the impact of the by-catch of fin-fish on the actual (or potential) fishery on these species needs to be taken into account. This first approximation may be unreliable to the extent that there are significant biological interactions between species. For example, it has been suggested that since some of the large fish are predators on shrimp, increased catches of these species, either in a directed fishery, or as part of the shrimp by-catch, could led to catches of shrimp greater than predicted on the basis of a single-species model. At the present time the data is seldom, if ever, good enough to make quantitative assessments of these interactions, but the possibilities should be kept in mind, and as appropriate incorporated in the comments and reservations included in the assessment advice.

More important for the design of immediate management schemes and considering the present state of art and practice, the inputs and outputs need also to be analysed in economic and social terms. A given fishing mortality can be exerted in many different ways; the use of a few powerful trawlers will require more capital, and more fuel, but less labour, and possibly less other costs to generate the same effective effort as a fleet of small inshore vessels. Careful analysis may be needed to establish the true cost of some of the inputs. For example, the labour costs of the small number of men on a sophisticated trawler based in a large port with high demand for semi-skilled labour may be much higher than that of the large number of small-scale fishermen with little other employment opportunities. Similarly the costs of fuel, engines, etc should be the real costs, and not the apparent costs after subsidies or taxes to encourage or discourage fishing have been applied.

Assessment of the outputs (benefits) needs equal care. In addition to the immediate measures (weight and value of total catch, amount of export earnings, etc), several derived measures – eg, the marginal returns of different kinds, the returns of different kinds (cash income, jobs created) per dollar of capital invested, etc – will probably be found useful.

Identification

Using the assessments, the next stage is to identify what fishing pattern is 'best', according to the prescribed objectives, and the directions in which the current fishery should be modified in order to move towards this optimum state. As pointed out earlier, deciding what is 'best' when there are conflicting objectives may require political decisions at a level higher than the Department of Fisheries. The fishery adviser in such situations can only point out the consequences of alternative actions. It is assumed here that he has received sufficient guidance on priorities to resolve the more serious conflicts in choosing what is best.

The simplest situation is where the sole objective is achieving MSY, and only one type of vessel operates, so that the pattern of fishing can be completely described by the number of these vessels. Then all that is involved at the identification stage is to determine the number of vessels required to achieve the MSY, and whether the present fleet needs to be increased or decreased.

With more complex inputs and outputs the identification is more difficult. For example it may be clear that the stock should be exploited close to the shoulder of the yield/effort curve, where the marginal yield becomes very small, but this can leave considerable choice on how that effort should be exerted – by a few powerful vessels, or by many small vessels. Normally the most economically efficient (least cost) mix of inputs will be chosen, though this does require that the costs are properly measured, *eg*, by looking at the opportunity costs of labour and capital. Only occasionally will economic efficiency in this sense mean high technology; sophisticated vessels are expensive, and need careful maintenance, and in most developing countries the necessary skilled labour is scarce, and needed elsewhere, while there is a surplus of the less skilled labour needed to man traditional vessels.

Because of the complexities of real fisheries, it may not be possible to identify a particular fishing pattern as producing uniquely the best result. This will usually not be serious. Even if a unique pattern could be identified as best, it would seldom be possible to adopt it immediately. The important matter is to identify the direction in which changes (*eg*, in the numbers of trawlers) should be made in order to move towards the optimum. The direction of movement will usually be clear, even if the position of the optimum is not, unless there are some wholly irreconcilable objectives.

A similar comment might be made about equilibrium and non-equilibrium conditions. The identification of the changes needed will often be made on the basis of comparison between a present situation (assumed to be in equilibrium) and some 'optimum' equilibrium position. It is highly unlikely that either the present fishing fleet, or the stocks is in equilibrium. It is more likely, but still improbable, that bearing in mind changes in the natural environment and human factors like market demand, catching and processing technology, *etc* the 'optimum' will remain constant over any long period of time. Identification of the direction of changes should therefore in principle recognize that the system is not in equilibrium, but generally an analysis of equilibrium conditions will give an accurate identification of the proper direction.

Actions
The actions usually thought of in relation to management are the direct controls to reduce the amount of fishing, in total or at certain times and places. Consideration also needs to be given to inducements to increase the amount of fishing (subsidies, *etc*) or indirect measures to reduce fishing (*eg*, promotion of alternative employment in fishing villages).

This paper is not the place to discuss development strategy in detail, but some points should be stressed. First, action taken to develop a fishery should not be taken without some idea of the size of the resource, and of the optimum size of the fishery, and some thought as to how the fishery will be managed (in the wide sense) so as to arrive, and stay, at the optimum state. Second, some actions that are thought of as essentially dealing with management (in the narrow sense) are useful tools also for development. For example companies will be more ready to invest in a new fishery if they are assured that the high catch rates (and profits) of the early years of a fishery will be maintained by a limitation of entry to the fishery. Third, most incentives to develop fisheries and to invest in new vessels, which are often vital in triggering off the growth of a new fishery (tax reliefs on profits ploughed back into more vessels, reduced duty on fuel or engines, *etc*) should be considered as temporary measures. They should probably be phased out as soon as the growth in a fishery becomes self-sustaining. In any case there is seldom justification for continuing measures to increase the amount of fishing once the stock becomes heavily exploited, and the problem is of too much fishing. Nevertheless, subsidies, in one form or another, are a common feature of fisheries on over-exploited stocks.

There are exceptions to this blanket criticism of subsidies. Temporary subsidies will often be needed while action is being taken to correct for past mis-management (for example, in buying out surplus fishing capacity (see following section)). Special assistance may be needed for some particular sector of the fishery (*eg*, small-scale fishermen in isolated villages), though this should be considered as a social measure, not principally concerned with the fishery as a whole, and its value should be judged against that of alternative measures to help the community concerned. Temporary assistance may also be needed, when there is a large natural variability, to help fishermen during a year of unusually low catches. The proper approach to this situation would, however, seem to be some form of insurance, funded by contributions during unusually good years.

The direct measures available to control or reduce the amount of fishing have been reviewed many times in relation to fisheries in general, and need not be reviewed again here. To a large extent the general considerations apply equally to shrimp. So far as control of the composition of the catch (especially of the effective size of first capture is concerned), mesh regulation is less effective for shrimp than for most fish, which experience a sharper selection. Against this the seasonal and geographical pattern of recruitment allows effective control to be achieved through closure of certain grounds (*eg*, inshore areas) or at certain times (around the season of peak recruitment).

In deciding on the action to control the total amount of fishing – number of boats, *etc* – an early step must be to determine how the benefits will be allocated. To some extent this should already have been considered at the previous stage, of identification of the optimum fishing strategy – should it be by a few large boats, or by many small-scale fishermen – but after the decision on the preferred size and composition of the fleet has been taken, there are still alternative methods of achieving the target size. These involve allocation of benefits. If under open access 100 trawlers would fish, but the optimum size of the fleet is only 75, who should be issued with the 75 licences? If benefits should go to the country as a whole, or to the national treasure, then high annual licences could be charged, so high that only about 75 applications are received. If licences are issued to existing fishermen (on the basis, say, of length of time in the fishery), and are transferable, then they will acquire a high value, and the benefits will go to the licence holder. Under any scheme, it may be necessary to provide support, *eg*, in terms of a buy-back scheme, to compensate those operating, and employed in, the 25 surplus trawlers.

Actions to encourage reduction of effort, other than direct controls, are aspects of fishery management that are too often neglected. They can be of two kinds – taxes or other measures to make fishing in the over-exploited sector less attractive and those that will make other activities, including fishing on other resources more attractive. While these measures are imprecise, so that it cannot be predicted, for example, exactly how many vessels will leave if there is a heavy tax, they do not require in general much enforcement – though if a tax is applied, its collection may be more difficult than in the case of land-based operations. If the difference between the current state of the fishery and its optimum is large, then the important matter is to start moving in the correct direction. For this, indirect measures are highly suitable. Fine tuning of the fishery, as it approaches the optimum, can be done through other, more precise, measures. Carefully applied, taxation, licence fees or similar levies, can be used to promote a selective change in the pattern of fishing. For example, licence fees for individual vessels, which to be unselective should probably be set in proportion to the average quantity caught by each class of vessel, could preferentially discourage one or other type of vessel by being higher for those vessels. Similarly a licence, or tax based on the weight landed could be used to discourage fishing on certain sizes or species of shrimp by being set higher for those categories.

Action to pull excess effort out of the fishery can be effective only in cases where the alternatives are so few that intervention initiated by or on behalf of the fishery authorities can significantly change their attractiveness. These are mainly fisheries out of isolated villages. For these an examination can be made of the relative effectiveness, for the local people, of trying to maintain employment in the shrimp fishery (*eg*, by limits being set on the types of gear that can be used) or of developing other activities.

Review and recycling
Management is a continuous process. No sooner is a biological assessment completed than information begins to arrive which makes it out of date. National objectives may change, and outside events like the increase in fuel costs may alter dramatically the economic performance of the industry, and what should be considered the optimum pattern of fishing. The procedures of review and recycling through the earlier stages are essential parts of a good management policy.

In the first instance, this recycling should be done immediately, before implementing whatever action might appear necessary from the first analysis. Indeed, immediate recycling is desirable after the earlier stages of identification, and possibly assessment. The initial listing of objectives is likely to be expressed in rather general terms. There may be little appreciation of the extent to which objectives may conflict, and therefore little attention paid to establishing priorities and relative weightings. After the influence of these objectives on the choice of actions – and especially the degree to which the balance between incompatible objectives affects the action – has become apparent, it is possible to re-examine the objectives, define them with greater precision, and establish more clearly the priorities between them. Again, after the choice of actions has been examined, and the practical limitations of some actions (*eg*, the costs of enforcement of catch quotas) have become clearer, it may be necessary to go back to the identification stage, and revise the ideas of what constitutes the optimum fishing pattern.

Once a management procedure is in place, then there should be a regular system for reviewing each stage. Probably arrangements should be made to carry out this review in different depths with different frequencies. Assessments will normally be reviewed each year, as the data for each season becomes available. After a few years this easily settles into a routine, adding one more point to the graph. Every four or five years, therefore, a special effort should be made to examine the assumptions and models used.

Similarly the detailed tactical actions, *eg*, the dates for opening the season, will be revised each year, but unless specific arrangements are made to review the basic strategy, it will soon become taken for granted, and can easily become inappropriate under changing conditions. The indepth review of the identification phase might best be done every four or five years at the same time as the assessments are carefully re-examined. A four or five year period is also appropriate for an indepth review of fishery objectives, and of the relative priorities that should be given to conflicting objectives.

References

FAO/ACMRR Working Party on the Scientific Basis of Determining Management Measures. Interim report of the ACMRR Working Party on the Scientific Basis of Determining Management Measures. Rome, 6–13 December 1978. *FAO Fish. Circ.* (718): 112 pp.

FAO/ACMRR Working Party on the Scientific Basis of Determining Management Measures. Report of the ACMRR Working Party on the Scientific Basis of Determining Management Measures. Hong Kong, 10–15 December 1979. *FAO Fish. Rep.* (236): 149 pp.

GARCIA, S and LE RESTE, L. Life cycles, dynamics, exploitation and management of coastal penaeid shrimp stocks. *FAO Fish. Tech. Pap.*, 203, 215 pp.

GULLAND, J A. Some introductory guidelines to management of shrimp fisheries. Rome, Indian Ocean Programme, IOFC/DEV/72/24.

MORGAN, G and GARCIA, S. The relationship between stock and recruitment in the shrimp stocks of Kuwait and Saudi Arabia. *Oceanogr. trop.* 17(2): 133–137.

TURNER, E. Inter-tidal vegetation and commercial yields of penaeid shrimp. *Trans. Am. Fish. Soc.* 106(5): 411–16.

WALTERS, C J and LUDWIG, D. Effects of measurement errors on the assessment of stock-recruitment relationship. *Can. J. Fish. Aquat. Sci.*, 38: 704–710.

An economic perspective of problems in the management of penaeid shrimp fisheries

John R Poffenberger

Abstract

The implications of recruitment and economic overfishing are discussed for penaeid shrimp fisheries in the context of a flat-topped sustained yield curve. Graphical analyses are used to show the differences in the economic characteristics of fisheries depicted by Schaeffer production models and flat-topped curves. The general conclusions are that beyond the apex of the Schaeffer curve more effort could be expected in flat-topped curve fisheries as they approach bionomic equilibrium. Also, vessel participation is a viable means of monitoring or correcting economic overfishing for flat-topped yield curve fisheries. The economic implications of fishery management issues, other than overfishing issues, are also discussed, *ie*, by-catch discards, transboundary migrations and uncertainty.

Introduction

Shrimp or prawns, as they are often referred to, provide important opportunities for commercial exploitation throughout the world. According to statistics prepared by the Food and Agricultural Organization (FAO) of the United Nations, 1·14 million metric tons of shrimp (live weight) were caught in 1978. These reported catches are of tropical marine species which are similar in biological characteristics to species of the genus *Penaeus*. In comparison, 2·3 million metric tons of crustaceans which include all shrimp species were caught in 1978. Thus, penaeid shrimp represented nearly 50 percent of the total reported catch of crustaceans during 1978. Fifty-nine countries reported shrimp catches of 100 metric tons or more. The reported shrimp catches by country and by statistical zone are presented in *Tables A.1* and *A.2* of the Appendix along with several other tables and figures which provide additional statistics and descriptions indicating the magnitude of penaeid fisheries throughout the world.

The magnitude and location of penaeid fisheries are not the discussion topics for this paper. The paper discusses issues and problems associated with the management of these important commercial fisheries and the potential economic consequences resulting from the failure to correct such management related problems. Not all of the problems presented herein are entirely economic ones. Biological problems such as recruitment and growth overfishing, shrimp-fish discards, *etc* are also discussed. However, since the main focus of this paper is an economic discussion of penaeid fisheries, the problem of economic overfishing is discussed at considerable length. A formal mathematical model of economic overfishing is not developed, however, suggestions are made regarding the management of penaeid shrimp fisheries based on a generally accepted sustainable yield curve. These suggestions are especially relevant to mature penaeid fisheries in which access is uncontrolled and a relatively competitive market environment exists.

The discussion is made more explicit by using examples from the literature on penaeid fisheries around the world. These examples are not presented as indications of the respective management strategy's success or failure, but as indications that such problems do exist in penaeid fisheries and that management agencies should consider solutions to the problems consistent with their respective management objectives.

2 Background

The general biological behavior of most tropical shrimp species around the world are similar (Gulland (1972) and Gross (1973)). Penaeids are short-lived species with their life cycle usually ranging from 12 to 18 months. The adults spawn offshore and the postlarvae move inshore to brackish water (lagoons and estuarine areas) for several months during their early development. Juvenile shrimp leave the inshore areas and are recruited to the offshore, commercial fisheries as they migrate into deeper water.

In most situations estimating the statistical relationship between historical fishing effort and catch has yielded, at best, inconclusive results. Using the generalized production model (Fox, 1975) has thus far failed to indicate the 'appropriate' shape of the sustainable yield curve. The parabolic or dome-shaped Schaefer model in which $m = 2 \cdot 0$ is completely inappropriate. A sustainable yield curve which tends to level off or flatten out at the top of the curve instead of reaching an apex and then decreasing to zero has been suggested.[1] The estimate of m for this model is zero. The difficulty with this theoretical model is that effort approaches infinity while the sustainable yield does not decrease. This situation is theoretically impossible and at some point the right-hand side of the yield curve must begin descending. Unfortunately data thus far acquired from most penaeid shrimp fisheries do not allow estimation of when or how rapidly the right side of the yield curve descends. Therefore, the right-hand or descending limb of the yield curve is not considered in this paper. The theoretical and analytical discussion for this paper are conducted in the horizontal apex of the sustain-able yield curve. That is, the fisheries are assumed to be operating in the flattened part of the yield curve.

Economic theory underlying the use of resources which are available for use by anyone (ie, no private ownership) indicates a general behavioral pattern which is applicable to penaeid fisheries.[2] Theory suggests that fishermen continue to enter a fishery as long as it is or perceived to be a profitable investment. Thus, increased participation would be anticipated in good years when shrimp are abundant, which would lead to the dispersion of profits in subsequent years. Furthermore, in years when the penaeid resources are not plentiful there is a considerable potential for financial losses. These general biological and economic behavioral characteristics lead to the types of issues which are discussed in the following sections.

3 Issues

Management related issues are discussed in two broad categories, overfishing problems and non-overfishing problems. The classical overfishing problems are recruitment, growth and economic. The non-overfishing problems which are considered crucial to various penaeid fisheries around the world are discards, transboundary migrations and uncertainty.

Overfishing

Recruitment overfishing is when a year-class recruitment is well below recruitment which normally would occur when the adult stock is high (Cushing, 1977). That is, the level of exploitation results in a smaller stock size than the amount needed to produce the maximum biological recruitment of young individuals to the exploited population on a sustained basis. Based on historical data the penaeid shrimp stocks in the Gulf of Mexico appear to be generally resistant to this type of overfishing as suggested by the leveling off of the yield curve (per the discussion in Section 2). Once effort has increased to a level which results in yields in the 'flat' portion of the curve then relatively large increases in effort would result in only modest increases in yield. The important point is that the sustainable yield would not decrease with increases in effort. Therefore, this model assumes that within the flat-topped part of the curve penaeid shrimp stocks are capable of maintaining

---

[1] This type of yield curve is suggested in Boerema (1980); although no statistical evidence is presented to support this representation.

[2] Even though many countries have declared areas of extended jurisdiction and closed these areas to foreign fishermen, they have not restricted the participation of their citizens in the fishery (the USA is an example). Thus, the theory is still applicable.

sustainable yields for a wide range of fishing effort. It should be clearly pointed out that this hypothesis is not suggesting that penaeid shrimp stocks are immune to recruitment overfishing or decreases in yield. The hypothesis of a flat-topped yield curve is proposing that sustained yields would increase very slowly with substantial increases in fishing effort.

Assuming that the yield curve representative of mature penaeid shrimp fisheries levels off, then within definable ranges management agencies would not have to be concerned with biological overfishing. Consequently there would not be any economic effects of biological overfishing if it is not expected to occur; however, this hypothesized yield curve does have some economic implications with respect to penaeid shrimp fisheries which are discussed in the economic overfishing section.

Growth overfishing is when the stocks are harvested at less than the potential maximum yield per recruit. This type of overfishing results in a disproportionate loss in total biomass relative to the potential loss in natural mortality if the cohorts had been permitted to grow. This situation probably occurs in a relatively large number of the penaeid fisheries throughout the world. It is caused by a combination of the offshore migration characteristic of growing shrimp and the 'catch-them-before-someone-else-does' situation of open access fisheries.

A situation of growth overfishing has been identified by the Gulf of Mexico Fishery Management Council in the recently enacted Fishery Management Plan (Gulf of Mexico Fishery Management Council, 1981b) for the shrimp fisheries in the Gulf of Mexico. The Council recommended that during the initial part of the shrimp season certain fishing areas should be closed in order to permit the juvenile shrimp leaving the estuary an opportunity to grow to a more marketable size.[3] Sufficient data were not available on the growth and natural mortality rates for the species in the closed areas to calculate the optimal size which would provide the maximum yield per recruit. Ex post facto evaluations of the Texas closure regulation place the estimated effects of the closure regulation at an increase of 3·9 million pounds valued at 9·4 million dollars during May 1981 through April 1982 (Poffenberger, 1982).

There are some important considerations which the fishery managers should include in their decision-making regarding this problem. First, this issue from an economic perspective is one of determining the most desirable (based on the management objective) trade-off between capturing smaller, less valuable shrimp at possibly lower operating costs versus the harvesting of larger, more valuable shrimp at higher operating costs.[4] Secondly, the demand for shrimp should be considered since a change in the quantity landed could have an effect on the price of shrimp. For example, if the quantity of large shrimp landed should increase, the price could decrease to such an extent that the total revenue to the fishery is less than the situation with growth overfishing. Finally, the managers may want to consider the social welfare benefits occurring to members of an artisanal fishery utilizing an easily accessible, shallow-water resource.

It is quite possible that penaeid fisheries (or fisheries in general) could be regulated or managed to provide social welfare benefits to specific sectors of a nation's population. For example, increased employment in coastal or depressed areas could result from increased fishing operations subsidized in those areas. A good example of this is the country of Panama and their realization of the importance of the shrimp fishery to the economy and specifically employment in Panama City. According to Gross (1973) the country offers financial support as well as tax incentives to assist in stimulating their fishery. Also, in many developing countries artisanal fisheries for shrimp exist in estuarine or near-shore areas where the young shrimp are available via fishing techniques requiring little or no capital investment.

For the most part, the economic consequences of this type of social objective would be measured in terms of the redistribution of income. For example, in the situation where artisanal fisheries coexist with established commercial fisheries, the income is being shifted away from the offshore commercial fishermen to the artisanal fishermen. This shift in income may not only be in terms of revenue to the artisanal fishermen, but their catch may also be a part of their diet which would otherwise have to be purchased. Thus, such a management strategy involves decisions based on the trade-offs between the socio-economic benefits

---

[3] The Council specified that two areas be closed. The waters off the coast of Texas are to be closed only during the initial part of the season, whereas the nursery areas in the Tortugas area off the west coast of Florida are closed all season. An additional reason for these closed seasons is the reduction or elimination of discarding small undersized shrimp which are not marketable. This problem is discussed in a subsequent section.

[4] Actually the analytical problem is considerably more complex than a simple either large or small solution. Since shrimp of all sizes are demanded and since different size shrimp have different market values, a multi-equation mathematical programming solution would be required.

accruing to one sector of the population and the reduced income to other sectors. There is also the potential for a reduction in the overall economic efficiency resulting from these trade-offs which managers should include in their decision making.

Economic overfishing is usually described as a departure from the maximum economic yield (MEY) of a fishery. MEY represents the maximum difference between the total cost curve (which may include estimates of opportunity costs) and the total revenue curve. Thus, economic overfishing usually occurs in an unregulated fishery in which the amount of fishing effort has increased beyond the point (on the effort axis) where the tangent of the total revenue curve is parallel to the total cost curve.[5] The result of this over-expansion of effort which in an open access fishery tends toward the bionomic equilibrium (*ie*, the point of intersection of the total revenue and total cost curves), is an inefficient use of the capital and labor inputs beyond the point of MEY. In other words, the economic profit or rate of return accruing to the fishery at MEY would be dispersed to zero at the point of bionomic equilibrium.

The fact that zero economic profits are anticipated in an equilibrium condition is not the problem associated with economic overfishing. Economic theory states that zero economic profits are to be expected at equilibrium in a perfectly competitive market. The problem arises as a result of the open access nature of uncontrolled fisheries in which capital and labor inputs enter the fishery, and beyond a point (MEY), the marginal return on these input factors begins to decrease. Thus, the problem of economic overfishing is the relatively inefficient use of capital and/or labor.

The above discussion describes the classical fishery's economic model which assumes a constant price for the output. Anderson (1973) presents a geometrical derivation using a dome-shaped yield curve which relaxes the constant-price assumption. He provides a six quadrant geometrical model which derives a backward bending supply curve (also see Clark, 1976, p. 153–157) and further derives a curve showing the relation between total revenue and fishing effort (see Figures 1 and 3 in Anderson). He labels this curve DMRC (double maxima revenue curve) because of the two points which represent the two levels of effort corresponding to the quantity of output where total

revenue is a maximum. He explains that the double maxima is a result of the dome-shaped nature of the Schaefer curve for which two levels of effort provide the same output.

Anderson's geometric derivations have been redrawn in *Figures 1* and *2* using a yield curve which levels off and does not decrease rather than the Schaefer model. As would be anticipated since these curves are assumed not to have a downward sloping symmetrical side, the supply curve is not backward bending and the total revenue-effort curve is not double humped. This geometric analysis does not, of course, alter Anderson's conclusions; but it does indicate that his results do not necessarily apply to the non-parabolic curves which are assumed to be representative of penaeid fisheries.

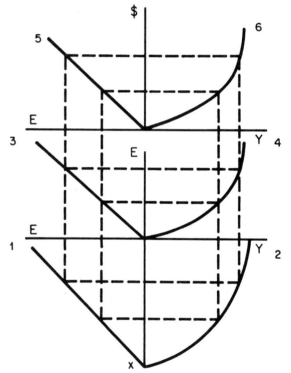

Fig 1 A sustained yield curve is derived (quadrant 4) from the constant relation between effort (E) and population size (Y) (quadrant 1) and the relation between growth rate (X) and population size (quadrant 2). This sustained yield and the relation between effort and total cost ($) (quadrant 5) are used to derive the relation between total cost and yield (quadrant 6).

However, a fishery which is represented by a non-parabolic yield curve does have a tendency towards greater levels of effort at bionomic equilibrium than the classical dome-shaped yield curve. This can be shown by assuming two hypothetical fisheries which are represented graphically by the

---

[5] This assumes a constant price for the shrimp no matter how much is landed. That is, the demand for shrimp is perfectly elastic. The assumption of constant prices is subsequently relaxed.

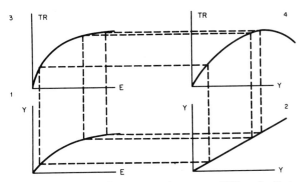

Fig 2 The relation between total revenue (*TR*) and effort (quadrant 3) is derived from the flat-topped yield curve (quadrant 1) and the total revenue-yield curve (quadrant 4).

two total revenue curves in *Figure 3*. This figure indicates that the bionomic equilibrium at the points of intersection of the same cost function are different. For the case shown in *Figure 3* where the equilibrium is on the downward sloping part of the dome-shaped yield curve, the level of effort for the other curve, E_2, is greater at equilibrium than the dome-shaped curve, E_1. Thus, it can be argued that for an uncontrolled penaeid fishery there is an inherent tendency towards more fishery participation (effort) than for fisheries targeting longer-lived species.

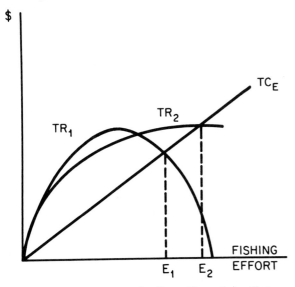

Fig 3 The potential difference in effort at bionomic equilibrium if the fishery is represented by a dome-shaped total revenue curve (*TR$_1$*) or a flat-topped revenue curve (*TR$_2$*) is shown by comparing E_1 to E_2. This assumes a constant output price.

The number of vessels can be used to monitor a potential over-expansion of participation into a shrimp fishery for two reasons. First, fishing effort is closely related to the number of vessels actively

fishing the resource. This is especially true for species like penaeids which are available for fishing only during certain parts of the day and certain seasons of the year (*ie*, limited temporal and areal availability of the species). Most likely vessels would already be fishing these times totally which would not permit an expansion of fishing time if the number of vessels remained the same. An indication of this can be demonstrated by dividing the total days fished by the number of vessels and comparing the variation of that ratio over time. For the US fishery in the Gulf of Mexico, the average number of days fished per vessel from 1962 to 1974 was 38 and the dispersion about that mean (*ie*, the standard deviation) was 3 days.[6] Although this does not provide valid statistical justification for using the number of vessels as an absolute proxy for fishing effort, it does indicate the logically close correlation between the two.

The second reason that vessels can be used to monitor expansion of fishing effort is the relatively small changes in yield which would result from increases in effort if the fishery is in the 'flattened' portion of the curve. This is important because fishery managers would not have to be concerned about the effects of larger more efficient vessels significantly increasing the amount of fishing effort. Therefore, even though an increase in fishing effort would occur, serious conservational problems (*ie*, recruitment overfishing) in the fishery would not be anticipated.

The previous discussion presented a brief description of the theory underlying economic overfishing. The important suggestions from this discussion are that in an unregulated situation penaeid fisheries probably have a tendency towards greater fishing participation and are less susceptible to biological overfishing than longer-lived species (*ie*, those which can adequately be represented by dome-shaped sustainable yield curves). Consequently, economic related management objectives may be the most relevant objectives for many of the penaeid fisheries throughout the world. Moreover, if the management decision is made to reduce economic overfishing, previous discussions in this section suggested that controlling the number of vessels participating in the fishery should prove adequate for management purposes. The difficult questions in this situation are to determine the appropriate number of vessels

[6] A fishing day is defined as 24 hours of actual fishing (*ie*, the amount of time the trawls are in the water). The data for days fished per vessel are from Christmas and Etzold (1977) – Table 8, p. 28.

and, if necessary, which fishermen are to be eliminated. For the most part, the determination of the number of vessels is an analytical problem which can be solved given reasonably good cost and revenue functions and a level of profit or rate of return considered appropriate by the fishery managers. This latter decision criteria from the managers is a crucial need in the solution of this analytical problem.

The question of who is to be eliminated is entirely a subjective or political determination. However, it certainly should be handled in a fair and equitable way. Perhaps one general suggestion for mature, well-established fisheries is that this management objective be a long-term one. That is, by establishing a long range goal the needed reduction could be an evolutionary process associated with a restriction on entry and the 'normal' retirement of vessels from the fishery. For developing penaeid fisheries, Gulland (1972) makes the useful suggestion that the amount of vessel participation be considered early in the managed development of the fishery because it is easier to restrict the number of vessels entering the fishery than reduce the number after they are already participating. An example of Gulland's suggestion is the Western Australian prawn fisheries in which entry has been limited since the beginning of these fisheries. Meany (1979) describes two results of this controlled development of these fisheries. First, there was no need for additional restrictions on fishing effort (p. 794). Secondly, he feels that this development has also helped control the characteristic overcapitalization of many open access fisheries (p. 797).

In summary, this section has suggested that penaeid fisheries may have a greater tendency toward economic overfishing than other fisheries. Furthermore, the correction of this problem may be handled appropriately by controlling or monitoring the vessel participation therefore allowing the individual fishermen or firms the freedom to establish the respective amounts of fishing effort, vessel size, *etc*, which provide the best rate of return individually. Problems other than overfishing are discussed in the next section.

Non-overfishing issues
Discards of unmarketable by-catch are essentially the result of the type of gear, *ie*, trawls, which are most often used in penaeid fisheries. Since a trawl collects nearly everything including small shrimp and lower-valued fish, an efficient solution historically has been to discard the unwanted animals and only land the high-valued shrimp. The economic consequences of destroying these resources are best evaluated in terms of the estimated opportunity cost of the destruction. For example, with respect to the fish discards, the magnitude of these costs depends on existing alternative uses of the fish resources. If there are established markets and commercial fisheries supplying these markets, then there are definite and measurable costs associated with these discards. On the other hand, if there are no commercial fisheries utilizing these stocks, then the costs of the discarded fish can be evaluated in terms of the effects on the ecological system. It could also be argued that opportunity costs are generated by discarding these non-utilized fish. These costs would be the lost opportunity for a viable fishery to develop which could exploit and market these otherwise wasted resources.

A prominent example of a shrimp-fish discard problem is the area in the north-central Gulf of Mexico. According to the Gulf of Mexico Groundfish Fishery Management Plan (Gulf of Mexico Fishery Management Council, 1981a), the incidental catch ratio of fish weight to total shrimp weight has ranged from 1.0:1.0 to 20.3:1.0. The estimated discard of fish by weight was 320 536 metric tons in 1975 (Gulf of Mexico Fishery Management Council, 1981a). Although this represents a large volume of resource, the opportunity cost associated with these discards should be measured in terms of the loss in existing or potential markets. The Groundfish FMP has a discussion of the existing markets for these resources and reference is made to it for further discussion.

The discards of under-size shrimp (*ie*, growth overfishing) is a problem because of the co-existence of marketable and non-marketable size shrimp in the same general fishing areas. The management question should be concerned with the amount of shrimp which would be available if the under-size shrimp were permitted to reach a marketable size (*ie*, the trade-off between growth and natural mortality). The economic consequence of this situation depends on the trade-off between the reduction in the number of commercial size shrimp available for capture and the potential revenue gained by the fishermen from the larger, more valuable (on a per pound basis) size of the remaining shrimp. In making a decision on this type of issue, fishery managers should also include enforcement and administrative costs. As an example of this type of analysis, reference is made to the analysis of the pink shrimp discards in Pamlico Sound, North Carolina, U S A by Waters, Danielson and Easley (1979).

Transboundary migration is a problem when

different countries with different management philosophies share a common stock which moves from one country's jurisdiction to another's. This type of situation in penaeid fisheries occurs when juveniles are present in the estuarine and inshore areas of one country and migrate to spawn in waters under another country's jurisdiction. The potential problem in this situation would be one of growth overfishing by one country before the under-size shrimp migrate to different territorial waters. From the national perspective of the country losing the resource, there is no strong incentive to prohibit the capture and sale of these sub-optimal size shrimp since they would be lost to the fishery if permitted to migrate. Consequently, the potential economic effects of this type of situation are difficult to prescribe in a general context. Different situations could be anticipated to have significantly different consequences depending on the migration patterns of the shrimp species and the existing relationships between the countries.

Uncertainty in the availability of a fish stock is present in nearly all exploited fishery resources. However, due to the high estuarine and environmental dependence and short life cycles of penaeids, these stocks tend to fluctuate more and are less predictable on an annual basis (based on catch statistics from previous years) than long-lived species. These fluctuations create an uncertain atmosphere both biologically and financially for penaeid fisheries. This is especially true for a fishery in which entry is unrestricted, and in environmentally good years when relatively large profits are realized there is incentive for increased entry into the fishery. Consequently, in years when the environmental conditions are less favorable and the total biomass is lower, the financial situation could be potentially disastrous.

There are two general ways management strategies could be used to approach this problem. First, fishing effort could be limited so that even in 'bad' years reasonable profits could be made. Or stated differently, the number of vessels in the fishery could be limited and permitted to accrue the excess profits during 'good' years as insurance against the returns in poorer years. Secondly, management could, to the extent possible, protect the estuarine areas from human pollution and alteration and therefore minimize some of the adverse effects on young shrimp.

An extreme example of the fluctuations which can exist in a penaeid fishery is the Nickol Bay fishery in Western Australia. Meany (1979) presents the annual catch statistics and explains that the number of licenses was established at thirteen.

However, there has not been one year since the number of licenses was established in 1971 that all of the licensed boats operated (p. 796).

4 Conclusion

The purpose of this paper was to present a discussion of the potential economic consequences of issues and problems associated with penaeid shrimp fisheries. A sustainable yield curve which levels off and does not have a descending right side was employed to describe the consequences of biological and economic overfishing. It was suggested that penaeid shrimp fisheries generally have not been affected by biological overfishing within the limits of historical levels of fishing effort. This is an important consideration because it essentially establishes the management criteria of these fisheries as almost entirely economic. Therefore, the problem of economic overfishing should be of paramount concern to management agencies.

The important indications discussed in the section on economic overfishing (and as a result of using a non-parabolic yield curve) are the following. First, the double humped, ambiguous revenue curve may not occur in penaeid shrimp fisheries when variable prices are included in the fishery's model. Second, more fishing participation (effort) can be expected in unregulated penaeid shrimp fisheries than in fisheries represented by a dome-shaped yield curve assuming identical cost functions. Finally, the number of vessels participating in a penaeid fishery was suggested as an adequate method of monitoring (or regulating) penaeid shrimp fisheries; adequate in the sense that fishery managers may not have to be concerned with long-term conservational problems (*ie*, biological overfishing) because of an increase in the fishing power of individual vessels.

The economic consequences of growth overfishing, shrimp – fish discards, transboundary migrations and uncertainty were also discussed. These problems are primarily of a policy or analytical nature. That is, once a management objective is established, analytical results can suggest appropriate regulations. A recent FAO report discusses possible regulations for the problems of growth overfishing and indirectly transboundary migrations.[7] This report also discusses the problem of over-expansion of fishing effort and suggests that it can be regulated in the following ways: a limitation on the catch, closed seasons and a limitation of the

---

[7] The reference for this FAO report is Boerema (1980). In this report he refers to growth overfishing as the protection of young shrimp (p. 147–149).

number of vessels. The paper did not consider the problem of too many vessels separately, but as a part of the economic overfishing problem. It is important to note that either of the first two methods suggested in the FAO report (*ie*, catch limitations and closed seasons) could reduce fishing effort, but only by making the existing fleet less efficient. A reduction in efficiency is essentially an additional cost and would probably decrease vessel profits. Therefore, from an economic criterion, these two regulatory options should be considered carefully.

In conclusion, the importance of this paper is the discussion of the economic consequences of potential management problems. The fundamental issue with respect to economic criteria in fishery management is that economic efficiency should be *considered* in the decision-making process. It may not always be appropriate to establish maximum economic yield or efficiency as the purpose for regulating shrimp fisheries. However, it is also inappropriate to cause fishermen to be less efficient and economically worse off if regulations are imposed which have not been well thought out or poorly analyzed with regards to their effects on the economic efficiency and, therefore, the costs of the fishing operation.

References

ANDERSON, L G. Optimum economic yield of a fishery given a
1973   variable price of output. *J. Fish. Res. Board Can.* 30(4): p. 509–518.

BOEREMA, L K. Expected effects of possible regulatory meas-
1980   ures in the shrimp fishery with special references to fisheries of the Guianas and Northern Brazil. *In* Jones and Villegas (ed.) Proceedings of the working group on shrimp fisheries of the Northeastern South America. *FAO/WECAF Rep.* (27): p. 144–151.

CHRISTMAS, J Y and ETZOLD, D J (eds). The shrimp fishery of
1977   the Gulf of Mexico United States: a regional management plan. Gulf Coast Res. Lab. Tech. Rep. 2. 128 p.

CLARK, C W. Mathematical bioeconomics of the optimal man-
1976   agement of renewable resources. John Wiley & Sons, Inc. Toronto, Canada. 352 pp.

CUSHING, D H. The problems of stock and recruitment. *In*
1977   Gulland (ed.) Fish population dynamics. FAO. Rome, Italy. p. 116–133.

FAO. Yearbook of fishery statistics. Vol. 46. Catch and land-
1979   ings, 1978. FAO. Rome, Italy. 375 pp.

FAO. Yearbook of fishery statistics. Vol. 47. Fishery com-
1979   modities, 1978. FAO. Rome, Italy. 279 pp.

FOX, W. W. Jr. Fitting the generalized stock production model
1975   by least-squares and equilibrium approximation. *Fish. Bull.* 73: p. 23–36.

GROSS, G B. Shrimp industry of Central America, Caribbean
1973   Sea, and Northern South America. *Marine Fisheries Review.* 35(3–4): p. 36–55.

GULF OF MEXICO FISHERY MANAGEMENT COUNCIL. Draft Fishery
1981A   Management Plan, Environmental Impact Statement and Regulatory Analysis for Groundfish in the Gulf of Mexico. Tampa, Florida. 39 pp.

GULF OF MEXICO FISHERY MANAGEMENT COUNCIL. Final
1981B   Environmental Impact Statement for the Fishery Management Plan for the Shrimp Fishery of the Gulf of Mexico. Tampa, Florida. 29 pp.

GULLAND, J A. Some introductory guidelines to management
1972   of shrimp fisheries. *FAO IOFC/DEV.*, (24): 12 pp.

MEANY, T F. Limited entry in the Western Australian rock
1979   lobster and prawn fisheries: an economic evaluation. *J. Fish. Res. Board Can.* 36: p. 789–798.

POFFENBERGER, J R. Estimated impacts of Texas closure regula-
1982   tion on ex-vessel prices and value, 1981 and 1982. NOAA Technical memorandum, NMFS-SEFC-111. Southeast Fisheries Center, Miami, Florida. 34 pp.

WATERS, J R, DANIELSON, L E and EASLEY, J E, Jr. An
1979   economic analysis of the shrimp discard problem in the Pamilco Sound. Economics Research Report No. 40. North Carolina State Univ., Raleigh, North Carolina. 47 pp.

Appendix

Table A.1
SELECTED SHRIMP CATCHES BY COUNTRY* (1978)

Country	Catch (MT)	Stat'l Zones	Country	Catch (MT)	Stat'l Zones
Angola	253	47	Kuwait	385	51
Australia	18 807	57; 71; 81	Liberia	1 563	34
Bahrain	2 000	51	Madagascar	5 560	51
Brazil	46 768	41	Malaysia	81 768	71
Brunei	483	71	Mexico	67 335	31; 77
Cameroon	1 400	34	Morocco	982	34
Colombia	4 252	77	Mozambique	4 800	51
Costa Rica	693	77	Nicaragua	6 600	31; 77
Cuba	8 816	31; 47	Nigeria	1 916	34
Ecuador	8 600	77	Other	66 601	57
Egypt	1 476	37; 51	Pakistan	19 177	51
El Salvador	115	77	Panama	6 118	77
Ethiopia	400	51	Papua N Guinea	922	71
France	174	34	Philippines	23 197	71
Gambia	184	34	Qatar	933	51
Ghana	423	34	Saudia Arabia	1 600	51
Greece	2 647	34; 37	Senegal	6 489	34
Guatemala	583	77	Sierra Leone	143	34
Guyana	3 175	34	Singapore	1 187	71
Honduras	2 418	31; 77	South Africa	882	51
Hong Kong	10 329	57	Spain	3 968	34; 37; 51
India	200 523	51; 57	Suriname	4 105	31
Indonesia	120 822	57; 71	Tanzania	680	51
Iran	3 400	51	Thailand	121 627	57; 71
Italy	9 460	27; 34; 37; 47	Trinidad	267	31
Ivory Coast	1 100	34	Tunisia	1 147	37
Japan	57 736	31; 57; 61	Turkey	464	37
Kampuchea Dm	400	71	USA	115 813	31
Korea Rep	20 187	31; 34; 41; 57; 61	Venezuela	3 820	31
			Viet Nam	62 000	71
			Total	1 139 673	—

Source: FAO 1979. Yb. Fish. Stat. Vol. 46. Catches and landings, 1978. Table B-45, p. 145–149.
* Catches are measured live weight. The species included in compilation of this table are listed in *Table A.1* of the Appendix.
MT = metric tons.

Table A.2
CATCH BY STATISTICAL ZONES* (FOR 1978 REPORTED IN METRIC TONS)

Zone 27	Zone 31	Zone 34	Zone 37
Atlantic Northeast 27	Atlantic Western Central 168 940	Atlantic Eastern Central 23 812	Mediterranean Black Sea 11 000

Zone 41	Zone 47	Zone 51	Zone 57
Atlantic Southwest 46 896	Atlantic Southeast 1 542	Indian Ocean Western 215 420	Indian Ocean Eastern 197 366

Zone 61	Zone 71	Zone 77	Zone 81
Pacific Northwest 10 454	Pacific Western Central 396 002	Pacific Eastern Central 63 784	Pacific Southwest 2 430

Source: See *Table 1*.
* For the geographical location of these statistical zones see *Figure A-1* in the Appendix.

Table A.3
WORLD IMPORTS AND EXPORTS OF CRUSTACEANS AND MOLLUSCS (1976–1978)

Continent	Unit*	Imports			Exports		
		1976	1977	1978	1976	1977	1978
Africa	MT	2 786	2 614	2 220	53 448	51 603	56 209
	$1 000	6 136	7 360	6 294	124 485	134 636	147 604
North America	MT	159 426	160 409	148 410	120 085	138 270	184 215
	$1 000	837 179	884 144	853 304	558 538	596 301	848 223
South America	MT	1 010	2 600	2 556	31 624	30 458	30 521
	$1 000	1 396	3 401	3 061	134 871	139 744	157 026
Asia	MT	431 410	431 538	530 545	346 245	343 986	360 852
	$1 000	1 200 064	1 297 968	1 790 062	898 113	962 550	1 083 167
Europe	MT	334 839	274 882	353 086	278 589	234 757	268 874
	$1 000	452 737	450 569	690 388	317 722	335 615	451 713
Ocenia	MT	3 620	3 484	2 878	14 874	17 635	20 598
	$100	13 591	13 869	12 170	117 829	156 334	178 750
World Total	MT	933 090	875 527	1 039 695	844 865	816 709	921 269
	$1 000	2 511 103	2 657 311	3 355 279	2 151 558	2 325 180	2 866 483

Source: FAO. 1979. Yb. Fish. Stat., Vol. 47. Fishery commodities, 1978. Table D2-1, p. 122–135.
* MT = metric tons; $1 000 U S dollars.

Table A.4
MAJOR IMPORTING AND EXPORTING COUNTRIES OF CRUSTACEANS AND MOLLUSCS

Country	Imports			Country	Exports		
	1976	1977	1978		1976	1977	1978
Belgium				Australia			
KMT	26·4	30·5	30·6	KMT	11·7	14·2	15·8
$1M USD	37·1	42·9	52·1	$1M USD	92·4	130·6	149·0
Canada				Hong Kong			
KMT	14·6	16·6	19·4	KMT	16·5	16·9	18·3
$1M USD	80·7	88·7	92·3	$1M USD	89·8	97·3	112·0
France				India			
KMT	73·4	77·6	98·8	KMT	49·7	54·9[a]	51·5[a]
$1M USD	130·9	139·6	229·4	$1M USD	185·9	198·6[a]	205·0[a]
Hong Kong				Indonesia			
KMT	28·7	26·6	28·9	KMT	35·2	34·9	34·9[a]
$1M USD	85·1	91·5	109·5	$1M USD	120·1	142·7	142·7[a]
Italy				Japan			
KMT	46·1	41·6	59·8	KMT	29·9	17·6	13·7
$1M USD	53·3	47·7	93·2	$1M USD	42·0	31·8	41·6
Japan				Korea			
KMT	337·1	330·7	418·9	KMT	61·6	74·0	58·0
$1M USD	1 084·7	1 173·3	1 635·3	$1M USD	89·6	127·7	123·2
Malaysia				Malaysia			
KMT	24·3	22·3	22·3[a]	KMT	50·4	34·6	62·4
$1M USD	6·4	5·6	6·0[a]	$1M USD	71·9	40·6	97·7[a]
Singapore				Mexico			
KMT	19·8	20·5	20·8	KMT	32·6	32·5	29·8
$1M USD	13·1	15·2	17·5	$1M USD	182·6	168·8	224·3
Spain				Thailand			
KMT	84·9	49·0	85·9	KMT	41·0	46·8	57·7
$1M USD	84·8	68·9	109·3	$1M USD	109·1	109·4	149·2
USA				USA			
KMT	142·5	141·8	126·7	KMT	24·4	33·9	55·0
$1M USD	750·6	789.5	755·2	$1M USD	101·9	142·8	277·4

Source: See Table 3.
KMT = 1 000 metric tons; $1M USD = one million U S dollars.
[a] Estimated by FAO.

Other books published by
Fishing News Books Ltd

Free catalogue available on request

Advances in aquaculture
Advances in fish science and
  technology
Aquaculture practices in Taiwan
Atlantic salmon: its future
Better angling with simple science
British freshwater fishes
Commercial fishing methods
Control of fish quality
Culture of bivalve molluscs
Echo sounding and sonar for fishing
The edible crab and its fishery in
  British waters
Eel capture, culture, processing and
  marketing
Eel culture
Engineering, economics and fisheries
  management
European inland water fish: a
  multilingual catalogue
FAO catalogue of fishing gear designs
FAO catalogue of small scale fishing
  gear
FAO investigates ferro-cement fishing
  craft
Farming the edge of the sea
Fibre ropes for fishing gear
Fish and shellfish farming in coastal
  waters
Fish catching methods of the world
Fisheries of Australia
Fisheries oceanography and ecology
Fisheries sonar
Fishermen's handbook
Fishery products
Fishing boats and their equipment
Fishing boats of the world 1
Fishing boats of the world 2
Fishing boats of the world 3
The fishing cadet's handbook
Fishing ports and markets
Fishing with electricity
Fishing with light
Freezing and irradiation of fish
Freshwater fisheries management
Glossary of UK fishing gear terms

Handbook of trout and salmon
  diseases
Handy medical guide for seafarers
How to make and set nets
Introduction to fishery by-products
The lemon sole
A living from lobsters
Making and managing a trout lake
Marine fisheries ecosystem
Marine pollution and sea life
Marketing in fisheries and aquaculture
Mending of fishing nets
Modern deep sea trawling gear
Modern fishing gear of the world 1
Modern fishing gear of the world 2
Modern fishing gear of the world 3
More Scottish fishing craft and their
  work
Multilingual dictionary of fish and fish
  products
Navigation primer for fishermen
Netting materials for fishing gear
Pair trawling and pair seining
Pelagic and semi-pelagic trawling gear
Planning of aquaculture development
Power transmission and automation
  for ships and submersibles
Refrigeration on fishing vessels
Salmon and trout farming in Norway
Salmon fisheries of Scotland
Scallop and queen fisheries in the
  British Isles
Scallops and the diver fisherman
Seafood fishing for amateur and
  professional
Seine fishing
Squid jigging from small boats
Stability and trim of fishing vessels
The stern trawler
Study of the sea
Textbook of fish culture
Training fishermen at sea
Trends in fish utilization
Trout farming manual
Tuna distribution and migration
Tuna fishing with pole and line